Complex Geometry

Springer-Verlag Berlin Heidelberg GmbH

Hans Grauert (photo by FotoStube Heiko Hornig, Göttingen)

Ingrid Bauer · Fabrizio Catanese
Yujiro Kawamata · Thomas Peternell
Yum-Tong Siu
Editors

Complex Geometry

Collection of Papers
Dedicated to Hans Grauert

Springer

Editors

Ingrid Bauer
Fabrizio Catanese
Thomas Peternell

Faculty for Mathematics
and Physics
University of Bayreuth
Universitätsstraße 30
95447 Bayreuth, Germany
e-mail:
ingrid.bauer@uni-bayreuth.de
fabrizio.catanese@uni-bayreuth.de
thomas.peternell@uni-bayreuth.de

Yujiro Kawamata
Department of Mathematical Sciences
University of Tokyo
Komaba, Meguro
Tokyo 153, Japan
e-mail: kawamata@ms.u-tokyo.ac.jp

Yum-Tong Siu
Department of Mathematics
Harvard University
1 Oxford Street
Cambridge, MA 02138, USA
e-mail: siu@math.harvard.edu

Printed with the kind support of the *Akademie der Wissenschaften zu Göttingen*

Library of Congress Cataloging-in-Publication Data applied for

Die Deutsche Bibliothek - CIP-Einheitsaufnahme
Complex geometry : collection of papers dedicated to Hans Grauert /
Ingrid Bauer ... ed. - Berlin ; Heidelberg ; New York ; Barcelona ; Hong Kong ;
London ; Milan ; Paris ; Tokyo : Springer, 2002
ISBN 978-3-642-62790-3   ISBN 978-3-642-56202-0 (eBook)
DOI 10.1007/978-3-642-56202-0

Mathematics Subject Classification (2000):
13C40, 13H10, 14C20, 14C25, 14D05, 14D15, 14D20, 14D22, 14E05, 14E08, 14E30, 14F17, 14J10,
14J15, 14J17, 14J28, 14J29, 14J40, 14J45, 14J60, 14K05, 14M07, 14M15, 14M20, 14Q20, 32C35,
32G05, 32G13, 32H50, 32J05, 32J15, 32J25, 32M05, 32M15, 32Q15, 32T05, 32W05, 37A25

ISBN 978-3-642-62790-3

This work is subject to copyright. All rights are reserved, whether the whole or part of the material
is concerned, specifically the rights of translation, reprinting, reuse of illustrations, recitation, broad-
casting, reproduction on microfilm or in any other way, and storage in data banks. Duplication of
this publication or parts thereof is permitted only under the provisions of the German Copyright Law
of September 9, 1965, in its current version, and permission for use must always be obtained from
Springer-Verlag. Violations are liable for prosecution under the German Copyright Law.

http://www.springer.de
© Springer-Verlag Berlin Heidelberg 2002
Originally published by Springer-Verlag Berlin Heidelberg New York in 2002
Softcover reprint of the hardcover 1st edition 2002

The use of general descriptive names, registered names, trademarks, etc. in this publication does not
imply, even in the absence of a specific statement, that such names are exempt from the relevant pro-
tective laws and regulations and therefore free for general use.

Typeset by the authors using a Springer TEX macro package
Cover design: *design & production* GmbH, Heidelberg

SPIN: 10868303   44/3142LK - 5 4 3 2 1 0 – Printed on acid-free paper

Preface

This volume contains a collection of research papers dedicated to Hans Grauert on the occasion of his seventieth birthday.

Hans Grauert is a pioneer in modern complex analysis, continuing the illustrious German tradition in function theory of several complex variables of Weierstrass, Behnke, Thullen, Stein, Siegel, and many others. When Grauert came on the scene in the early 1950's, function theory was going through a revolutionary period with the geometric theory of complex spaces still in its embryonic stage. A rich theory evolved with the joint efforts of many great mathematicians including Oka, Kodaira, Cartan, and Serre. The Cartan Seminar in Paris and the Kodaira Seminar provided important venues for its development. Grauert, together with Andreotti and Remmert, took an active part in the latter.

In his career he has nurtured a great number of his own doctoral students as well as other young mathematicians in his field from all over the world. For a couple of decades his work blazed the trail and set the research agenda in several complex variables worldwide.

Among his many fundamentally important contributions, which are too numerous to completely enumerate here, are:

1. The complete clarification of various notions of complex spaces.
2. The solution of the general Levi problem and his work on pseudoconvexity for general manifolds.
3. The theory of exceptional analytic sets.
4. The Oka principle for holomorphic bundles.
5. The proof of the Mordell conjecture for function fields.
6. The direct image theorem for coherent sheaves.
7. The theory of q-pseudoconvexity and finiteness theorems.
8. The deformation theory for singular complex spaces.

Many of the contributors to this volume participated in the International Conference on "Analytic and Algebraic Methods in Complex Geometry" in honor of Hans Grauert, held in Göttingen from April 3 to April 8, 2000. The Conference was sponsored by the Deutsche Forschungsgemeinschaft and the Akademie der Wissenschaften zu Göttingen, of which Hans Grauert has been a member since 1963 and served as President from 1992 to 1996.

VI Preface

The contributors consider themselves descendants of the mathematical heritage of Hans Grauert and feel deeply indebted to him for what they learnt directly from him or indirectly from his papers. It is their hope that this volume serves, in some small measure, as a sampler of the numerous active fields of current research in complex geometry stemming from the work of Hans Grauert.

To highlight the success of Hans Grauert as an educator and mentor as well as a researcher, in addition to a listing of the research publications of Hans Grauert we give below a complete list of his doctoral students. By itself the lengthy list of doctoral students does not render sufficient justice to him as a truly inspiring teacher with his caring, warm, albeit reserved, personality.

We, the editors of this volume, would like to express our gratitude to Roberto Pignatelli, who took care of many administrative tasks for the production of this volume. Without his tireless, generous efforts the project for this volume would not have come to fruition.

We would like also to give our thanks to Sarah Jaffe whose expertise in TeX and hard work helped integrate seamlessly the author-supplied TeX files with diverse styles into the present beautiful book format.

January, 2002   The Editors

1 List of Research Publications of Hans Grauert

1. Métrique kaehlérienne et domaines d'holomorphie. C. R. Acad. Sci. Paris 238 (1954), 2048–2050.
2. (with Reinhold Remmert) Zur Theorie der Modifikationen. I. Stetige und eigentliche Modifikationen komplexer Räume. Math. Ann. 129 (1955), 274–296.
3. Charakterisierung der holomorph vollständigen komplexen Räume. Math. Ann. 129 (1955), 233–259.
4. (with Reinhold Remmert) Fonctions plurisousharmoniques dans des espaces analytiques. Généralisation d'une théorème d'Oka. C. R. Acad. Sci. Paris 241 (1955), 1371–1373.
5. (with Reinhold Remmert) Plurisubharmonische Funktionen in komplexen Räumen. Math. Z. 65 (1956), 175–194.
6. Charakterisierung der Holomorphiegebiete durch die vollständige Kählersche Metrik. Math. Ann. 131 (1956), 38–75.
7. (with Reinhold Remmert) Konvexität in der komplexen Analysis. Nichtholomorph-konvexe Holomorphiegebiete und Anwendungen auf die Abbildungstheorie. Comment. Math. Helv. 31 (1956), 152–160, 161–183.
8. Généralisation d'un théorème de Runge et application à la théorie des espaces fibrés analytiques. C. R. Acad. Sci. Paris 242 (1956), 603–605.
9. Holomorphe Funktionen mit Werten in komplexen Lieschen Gruppen. Math. Ann. 133 (1957), 450–472.
10. Approximationssätze für holomorphe Funktionen mit Werten in komplexen Rämen. Math. Ann. 133 (1957), 139–159.
11. (with Reinhold Remmert) Sur les revêtements analytiques des variétés analytiques. C. R. Acad. Sci. Paris 245 (1957), 918–921.
12. (with Reinhold Remmert) Espaces analytiquement complets. C. R. Acad. Sci. Paris 245 (1957), 882–885.
13. (with Reinhold Remmert) Faisceaux analytiques cohérents sur le produit d'un espace analytique et d'un espace projectif. C. R. Acad. Sci. Paris 245 (1957), 819–822.
14. (with Reinhold Remmert) Singularitäten komplexer Mannigfaltigkeiten und Riemannsche Gebiete. Math. Z. 67 (1957), 103–128.
15. (with Reinhold Remmert) Komplexe Räume. Math. Ann. 136 (1958), 245–318.
16. (with Reinhold Remmert) Bilder und Urbilder analytischer Garben. Ann. of Math. (2) 68 (1958), 393–443.
17. On Levi's problem and the imbedding of real-analytic manifolds. Ann. of Math. (2) 68 (1958), 460–472.
18. Analytische Faserungen über holomorph-vollständigen Räumen. Math. Ann. 135 (1958), 263–273.
19. Die Riemannschen Flächen der Funktionentheorie mehrerer Veränderlichen. 1960 Proc. Internat. Congress Math., pp. 362–375, Cambridge Univ. Press, New York

VIII   Preface

20. On point modifications. 1960 Contributions to function theory (Internat. Colloq. Function Theory, Bombay, pp. 139–141, Tata Institute of Fundamental Research, Bombay

21. On the number of moduli of complex structures. 1960 Contributions to function theory (Internat. Colloq. Function Theory, Bombay, 1960), pp. 63–78, Tata Institute of Fundamental Research, Bombay

22. (with Heinrich Behnke) Analysis in non-compact complex spaces. 1960 Analytic functions, pp. 11–44 Princeton Univ. Press, Princeton, N.J.

23. Ein Theorem der analytischen Garbentheorie und die Modulräume komplexer Strukturen. Inst. Hautes Études Sci. Publ. Math. No. 5 (1960), 64 pp.

24. (with Ferdinand Docquier) Levisches Problem und Rungescher Satz für Teilgebiete Steinscher Mannigfaltigkeiten. Math. Ann. 140 (1960), 94–123.

25. (with Aldo Andreotti) Algebraische Körper von automorphen Funktionen. Nachr. Akad. Wiss. Göttingen Math.-Phys. Kl. II (1961), 39–48.

26. (with Aldo Andreotti) Théorème de finitude pour la cohomologie des espaces complexes. Bull. Soc. Math. France 90 (1962), 193–259.

27. Die Bedeutung des Levischen Problems für die analytische und algebraische Geometrie. 1963 Proc. Internat. Congr. Mathematicians (Stockholm, 1962), pp. 86–101, Inst. Mittag-Leffler, Djursholm

28. (with Reinhold Remmert) Über kompakte homogene komplexe Mannigfaltigkeiten. Arch. Math. 13 (1962) 498–507.

29. Berichtigung zu der Arbeit "Ein Theorem der analytischen Garbentheorie und die Modulräume komplexer Strukturen". Inst. Hautes Études Sci. Publ. Math. No. 16 (1963), 35–36.

30. Bemerkenswerte pseudokonvexe Mannigfaltigkeiten. Math. Z. 81 (1963), 377–391.

31. (with Hans Kerner) Approximation von holomorphen Schnittflächen in Faserbündeln mit homogener Faser. Arch. Math. 14 (1963) 328–333.

32. (with Hans Kerner) Deformationen von Singularitäten komplexer Räume. Math. Ann. 153 (1964), 236–260.

33. Über Modifikationen und exzeptionelle analytische Mengen. Math. Ann. 146 (1962), 331–368.

34. (with Helmut Reckziegel) Hermitesche Metriken und normale Familien holomorpher Abbildungen. Math. Z. 89 (1965), 108–125.

35. (with Wolfgang Fischer) Lokal-triviale Familien kompakter komplexer Mannigfaltigkeiten. Nachr. Akad. Wiss. Göttingen Math.-Phys. Kl. II (1965), 89–94.

36. Mordells Vermutung über rationale Punkte auf algebraischen Kurven und Funktionenkörper. Inst. Hautes Études Sci. Publ. Math. No. 25 (1965), 131–149.

37. (with Reinhold Remmert) Nichtarchimedische Funktionentheorie. 1966 Festschr. Gedächtnisfeier K. Weierstrass, pp. 393–476, Westdeutscher Verlag, Cologne

List of Research Publications of Hans Grauert

38. (with Reinhold Remmert) Über die Methode der diskret bewerteten Ringe in der nicht-archimedischen Analysis. Invent. Math. 2 (1966), 87–133.
39. The coherence of direct images. Enseignement Math. (2) 14 (1968), 99–119.
40. Affinoide Überdeckungen eindimensionaler affinoider Räume. Inst. Hautes Études Sci. Publ. Math. No. 34 (1968) 5–36.
41. (with Lothar Gerritzen) Die Azyklizität der affinoiden Überdeckungen. 1969 Global Analysis (Papers in Honor of K. Kodaira), pp. 159–184, Univ. Tokyo Press, Tokyo
42. (with Oswald Riemenschneider) Kählersche Mannigfaltigkeiten mit hyper-q-konvexem Rand. Problems in analysis (Lectures Sympos in honor of Salomon Bochner, Princeton Univ., Princeton, N.J., 1969), pp. 61–79, Princeton Univ. Press, Princeton, N.J., 1970.
43. (with Oswald Riemenschneider) Verschwindungssätze für analytische Kohomologiegruppen auf komplexen Räumen. Invent. Math. 11 (1970), 263–292.
44. (with Oswald Riemenschneider) Verschwindungssätze für analytische Kohomologiegruppen auf komplexen Räumen. 1970 Several Complex Variables, I (Proc. Conf., Univ. of Maryland, College Park, Md., 1970), pp. 97–109 Springer, Berlin
45. (with Ingo Lieb) Das Ramirezsche Integral und die Lˆsung der Gleichung $\bar{\partial} f = \alpha$ im Bereich der beschränkten Formen. Rice Univ. Studies 56 (1970) no. 2, 29–50 (1971).
46. Über die Deformation isolierter Singularitäten analytischer Mengen. Invent. Math. 15 (1972), 171–198.
47. Deformation kompakter komplexer Räume. Classification of algebraic varieties and compact complex manifolds, pp. 70–74. Lecture Notes in Math., Vol. 412, Springer, Berlin, 1974.
48. Der Satz von Kuranishi für kompakte komplexe Räume. Invent. Math. 25 (1974), 107–142.
49. (with Reinhold Remmert) Zur Spaltung lokal-freier Garben über Riemannschen Flächen. Math. Z. 144 (1975), no. 1, 35–43.
50. (with Gerhard Mülich) Vektorbündel vom Rang 2 über dem n-dimensionalen komplex-projektiven Raum. Manuscripta Math. 16 (1975), no. 1, 75–100.
51. (with Gerhard Mülich) Ergänzung zu der Arbeit: "Vektorbündel vom Rang 2 über dem n-dimensionalen komplex-projektiven Raum" (Manuscripta Math. 16 (1975), fasc. 1, 75–100). Manuscripta Math. 18 (1976), no. 2, 213–214.
52. Statistische Geometrie. Ein Versuch zur geometrischen Deutung physikalischer Felder. Nachr. Akad. Wiss. Göttingen Math.-Phys. Kl. II 1976, no. 2, 13–32.
53. (with Michael Schneider) Komplexe Unterräume und holomorphe Vektorraumbündel vom Rang zwei. Math. Ann. 230 (1977), no. 1, 75–90.

X   Preface

54. Über die Deformation von Pseudogruppenstrukturen. Colloque "Analyse et Topologie" en l'Honneur de Henri Cartan (Orsay, 1974), pp. 141–150, Asterisque, No. 32–33, Soc. Math. France, Paris, 1976.

55. Deformation komplexer Strukturen. Komplexe Analysis und ihre Anwendung auf partielle Differentialgleichungen (Tagung, Martin-Luther-Universität, Halle, 1976). Wiss. Beitr. Martin-Luther-Univ. Halle-Witteberg M 27 (1977), 24–27.

56. Kantenkohomologie. Compositio Math. 44 (1981), no. 1-3, 79–101.

57. Complex Morse singularities. Conference on Complex Analysis, Nancy 80 (Nancy, 1980), pp. 87–92, Inst. Élie Cartan, 3, Univ. Nancy, Nancy, 1981.

58. (with S. Leykum) Die pseudoeuklidische Geometrie als statistisches Gleichgewicht. Math. Ann. 255 (1981), no. 2, 273–285.

59. (with Michael Commichau) Das formale Prinzip für kompakte komplexe Untermannigfaltigkeiten mit 1-positivem Normalenbündel. Ann. of Math. Stud., 100, Princeton Univ. Press, Princeton, N.J., 1981, pp. 101–126.

60. Set theoretic complex equivalence relations. Math. Ann. 265 (1983), no. 2, 137–148.

61. Kontinuitätssatz und Hüllen bei speziellen Hartogsschen Körpern. Abh. Math. Sem. Univ. Hamburg 52 (1982), 179–186.

62. (with Ulrike Peternell) Hyperbolicity of the complement of plane curves. Manuscripta Math. 50 (1985), 429–441.

63. On meromorphic equivalence relations. Contributions to several complex variables, 115–147, Aspects Math., E9, Vieweg, Braunschweig, 1986.

64. Meromorphe Äquivalenzrelationen: Anwendungen, Beispiele, Ergänzungen. Math. Ann. 278 (1987), no. 1-4, 175–183.

65. Jetmetriken und hyperbolische Geometrie. Math. Z. 200 (1989), no. 2, 149–168.

66. (with Gerd Dethloff) Deformation of compact Riemann surfaces Y of genus p with distinguished points $P_1, \cdots, P_m \in Y$. Complex geometry and analysis (Pisa, 1988), 37–44, Lecture Notes in Math., 1422, Springer, Berlin, 1990.

67. Die Unendlichkeit in der Mathematik. Math. Semesterber. 37 (1990), no. 2, 153–156.

68. (with Gerd Dethloff) On the infinitesimal deformation of simply connected domains in one complex variable. International Symposium in Memory of Hua Loo Keng, Vol. II (Beijing, 1988), 57–88, Springer, Berlin, 1991.

69. Set theoretical real analytic spaces. Modern methods in complex analysis (Princeton, NJ, 1992), 183–189, Ann. of Math. Stud., 137, Princeton Univ. Press, Princeton, NJ, 1995.

70. Analytische und meromorphe Zerlegungen und der reelle Fall. Jahresber. Deutsch. Math.-Verein. 95 (1993), no. 4, 181–189.

71. Discrete geometry. Nachr. Akad. Wiss. Göttingen Math.-Phys. Kl. II 1996, no. 6, 20 pp.

72. Die Axiome der Elementargeometrie und die griechischen Theoreme. Math. Semesterber. 44 (1997), no. 1, 19–36.
73. Towards a discretization of quantum theory. Dedicated to Ennio De Giorgi. Ann. Scuola Norm. Sup. Pisa Cl. Sci. (4) 25 (1997), no. 3-4, 487–502 (1998).

XII   Preface

2   List of Doctoral Students of Hans Grauert

1. Fadlalla, Adib: *Die Caratheodorysche Metrik.* Göttingen, 1962.
2. Spilker, Jürgen: *Algebraische Körper von automorphen Funktionen.* Göttingen, 1963.
3. Fischer, Wolfgang: *Zur Deformationstheorie komplex-analytischer Faserbündel.* Göttingen, 1964.
4. Lieb, Ingo: *Komplexe Räume und komplexe Spektren.* Göttingen, 1965.
5. Trautmann, Günter: *Ein Kontinuitätssatz für die Fortsetzung kohärenter analytischer Garben.* Göttingen, 1965.
6. Richberg, Rolf: *Stetige streng pseudokonvexe Funktionen.* Göttingen, 1966.
7. Kalmbach, Gudrun: *Über niederdimensionale CW-Komplexe in nichtkompakten Mannigfaltigkeiten.* Göttingen, 1966.
8. Riemenschneider, Oswald: *Über den Flächeninhalt analytischer Mangen und die Erzeugung k-pseudokonvexer Gebiete.* Göttingen, 1966.
9. Finckenstein, Karl Graf Fink von: *Über zweidimensionale normale Singularitäten.* Göttingen, 1966.
10. Reckziegel, Helmut: *Hyperbolische Räume und normale Familien holomorpher Abbildungen.* Göttingen, 1967.
11. Diederich, Klas: *Das Randverhalten der Bergmanschen Kernfunktion und Matrik in streng pseudo-konvexen Gebieten.* Göttingen, 1967.
12. Denneberg, Dieter: *Divisoren in eindimensionalen affinoiden und affinen Räumen.* Göttingen, 1967.
13. Hund, Erwin: *Eigenschaften analytisch vollständiger Räume.* Göttingen, 1967.
14. Bartenwerfer, Wolfgang: *Über den Kontinuitätssatz in der nichtarchimedischen Funktionentheorie.* Göttingen, 1968.
15. Ramirez de Arellano, Enrique: *Ein Divisionsproblem in der komplexen Analysis mit einer Anwendung auf Randintegraldarstellungen.* Göttingen, 1968.
16. Kowitz, Siegfried: *Über die Existenz Steinscher Mannigfaltigkeiten zu vorgegebenen Homologiegruppen.* Göttingen, 1970.
17. Narasimhan, Ramabhadran: *Eine geometrische Eigenschaft von Rungeschen Gebieten.* Göttingen, 1971.
18. Pflug, Peter: *Eigenschaften der Forsetzungenvon in speziellen Gebieten holomorphen polynomialen Funktionen in die Holomorphiehülle.* Göttingen, 1972.
19. Braun, Robert-Burckhard: *Cauchy-Fantappie-Formeln und Dualität in der Funktionentheorie.* Göttingen, 1972.
20. Vormoor, Robert: *Topologische Fortsetzung bihilomorpher Funktionen auf dem Rande bei beschränkten streng-pseudokonvexen Gebieten im $\mathbb{C}^n$ mit C^∞-Rand.* Göttingen, 1973.
21. Schnitker, Jörg: *Algebraische Singularitäten analytischer Räume.* Göttingen, 1973.

List of Doctoral Students of Hans Grauert   XIII

22. Mülich, Gerhard: *Familien holomorpher Vektorraumbündel über der projektiven Geraden und unzerlegbare holomorphe 2-Bündel über der Projektiven Ebene.* Göttingen, 1974.

23. Skirde, Alfons: *Über ein Divisionsproblem bei differenzierbaren Funktionen komplexer Variablen.* Göttingen, 1974.

24. Commichau, Michael: *Deformation kompakter komplexer Mannigfaltigkeiten.* Göttingen, 1974.

25. Fritzsche, Klaus: *Ein Kriterium für die q-Konvexität von Restmengen in kompakten komplexen Mannigfaltigkeiten.* Göttingen, 1975.

26. Lichte, Magdalene: *Zur Berechnung der Dimension der ersten Bildgarbe $R^1\pi_*\mathcal{O}_X$ der Strukturgarbe einer Auflösung einer zweidimensionalen normalen Singularität.* Göttingen, 1976.

27. Spindler, Heinz: *Holomorphe Vektorraumbündel auf $\mathbf{P}^n$ mit trivialem Spaltungstyp.* Göttingen, 1982.

28. Leykum, Stefan: *Ein Satz der statistischen Geometrie.* Göttingen, 1980.

29. Riebesehl, Dieter: *Hyperbolische komplexe Räume und die Vermutung von Mordell.* Göttingen, 1980.

30. Hoppe, Hans Jürgen: *Stabile Vektorraumbündel auf rationalen Regelflächen.* Göttingen, 1980.

31. Peternell, Thomas: *Vektorraumbündel in der Nähe von kompakten komplexen Untermannigfaltigkeiten.* Göttingen, 1981.

32. Peternell, Mathias: *Ein Lefschetz-Satz für Schnitte in projektiv-algebraischen Mannigfaltigkeiten.* Göttingen, 1983.

33. Steinbiß, Volker: *Das formale Prinzip für reduzierte komplexe Räume mit einer schwachen Positivitätseigenschaft.* Göttingen, 1985.

34. Steinbiß, Ingrid: *Über Familien streng pseudokonvexer Umgebungen exzeptioneller analytischer Mengen.* Göttingen, 1988.

35. Dethloff, Gerd: *Wesentliche Singularitäten meromorpher Abbildungen.* Göttingen, 1988.

36. Siebert, Bernd: *Faserräume, geometrische Plattifikation und meromorphe Äquivalenzrelationen.* Göttingen, 1992.

37. Tada, Minuro: *Discrete analytic equivalence relations on complex spaces.* Göttingen, 1992.

38. Herbers, Guido: *Ein konstruktive Bewertung für die Spaltung komplexanalytischer Vektorbündel auf dem $\mathbf{P}^1$.* Göttingen, 1995.

39. Tüxen, Michael: *Reelle Räume.* Göttingen, 1996.

40. Armbrüster, Ulrich: *Fortsetzungsaussagen für Kohomologieklassen.* Göttingen, 1996.

XIV   Preface

3 Program of Göttingen Conference for the 70th Birthday of Hans Grauert

Mathematisches Institut, Georg-August-Universität Göttingen, Göttingen, April 4 – 8, 2000

Program

Tuesday, April 4, 2000

09:30 - 10:00   OPENING

10:00 - 10:50 Y.-T. Siu   "Hyperbolicity of generic high-degree hyper-surfaces in complex projective spaces"

11:20 - 12:10 H. Flenner   "A general semiregularity map and applications"

15:00 - 15:50 K. Diederich "Pluripotential Green functions with applications"

16:20 - 17:10 K. Oguiso   "Families of hyperkähler manifolds"

17:20 - 18:10 M. Kreck   "From Witten spaces to Calabi-Yau spaces via topological varieties"

Wednesday, April 5, 2000

09:10 - 10:00 J.-P. Demailly   "Subadditivity of multiplier ideals and asymptotic Zariski decomposition"

10:30 - 11:20 E. Viehweg   "On the Shafarevich conjectures for higher dimensional varieties over function fields"

11:40 - 12:30 F. Campana   Complex contact Fano manifolds

15:30 - 18:30   Shorter Communications:

        A. Gorodentzev: "Abelian Lagrangian algebraic geometry"

        K. Spallek: "Product-Decomposition of spacegerms"

        P. Katsylo: "The regularity of the moduli space of mathematical instantons MI(5)"

        P.-M. Wong: "Hyperbolic manifolds via branched covers"

        A. Bondal: "Operations with t-structures and perverse coherent sheaves"

        D. Mathieu: "Meromorphic equivalence relations and analytic families of cycles"

Program of 70th Birthday Conference of Hans Grauert   XV

M. Leenson: "On the Brill-Noether theory
           on K3 surfaces"
H. Hamm: "On the topology of fibres of algebraic maps"
B. Karpov: "Problems related to testing
          S-duality and exceptional vector bundles"
O. Debarre: "Linear spaces in real complete
           intersections"

Thursday, April 6, 2000

09:10 - 10:00 E. Looijenga   "A hyperbolic ball quotient
                          of dimension eight"

10:30 - 11:20 K. Hulek   Singular Prym varieties

11:40 - 12:30 T. Ohsawa   "An intrinsic relation between
                          extension and division problems"

15:30 - 16:20 G. Trautmann   "Milestones in Complex Analytic Geometry"

17:00 - 17:50 A. Huckleberry   "'Stützkurven' on one side or the other"

19:30   BANQUET AT HOTEL CLARION

Friday, April 7, 2000

09:10 - 10:00 J. E. Fornaess "Holomorphic dynamics in higher dimension"

10:30 - 11:20 D. Barlet   "Holomorphic functions on cycle's
                    space: the intersection method"

11:40 - 12:30 W. Barth   "Lines on projective surfaces"

15:30 - 16:20 B. Siebert   "Degenerations of Kodaira surfaces"

17:00 - 17:50 Y. Kawamata "On a relative version of Fujita's
                    freeness conjecture"

Saturday, April 8, 2000

09.30 - 10.20 Z. Jelonek   "Geometry of polynomial Mappings of C^n"

10:20 - 10:10 C. Ciliberto "Prym varieties and the canonical
                  map of surfaces of general type"

11:30 - 12:20 C. Okonek   "Towards a Poincare formula for
                  projective surfaces"

Financial support:

Deutsche Forschungsgemeinschaft, Bonn
Akademie der Wissenschaften, Göttingen
Gutingia Lebensversicherung, Göttingen

XVI   Preface

4 List of Participants of Göttingen Conference

1. Arghanoun, Ghazaleh (Göttingen)
2. Artmann, Benno (Göttingen)
3. Barlet, Daniel (Nancy)
4. Barth, Wolf (Erlangen)
5. Bauer, Ingrid (Göttingen)
6. Bauer, Stefan (Bielefeld)
7. Beyerstedt, Bernd (Göttingen)
8. Bondal, Alexei (Moscow)
9. Bothmer von, Hans-Christian (Regensburg)
10. Brückmann, Klaus-Peter (Halle)
11. Bunke, Ulrich (Göttingen)
12. Campana, Frederic (Nancy)
13. Canonaco, Alberto (Pisa)
14. Catanese, Fabrizio (Göttingen)
15. Chen, Meng (Göttingen)
16. Ciliberto, Ciro (Rome)
17. Colson, Pierre (Göttingen)
18. Dais, Dimitrios I. (Ioannina)
19. Debarre, Olivier (Strasbourg)
20. Demailly, Jean-Pierre (Grenoble)
21. Dethloff, Gerd (Brest)
22. Diederich, Klas (Wuppertal)
23. Fischer, Wolfgang (Bremen)
24. Flenner, Hubert (Bochum)
25. Fornaess, John-Erik (Michigan)
26. Franciosi, Marco (Pisa)
27. Frediani, Paola (Pisa)
28. Fritzsche, Klaus (Wuppertal)
29. Gorodentzev, A. (Moscow)
30. Grauert, Hans (Göttingen)
31. Greuel, Gert-Martin (Kaiserslautern)
32. Hamm, Helmut A. (Münster)
33. Heinz, Erhard (Göttingen)
34. Heinze, Joachim (Springer-Verlag Heidelberg)
35. Herbort, Gregor (Wuppertal)
36. Huckleberry, Alan (Bochum)
37. Hulek, Klaus (Hannover)
38. Jahnel, Jörg (Göttingen)
39. Jelonek, Zbigniew (Cracow)
40. Karpov, Boris (Moscow)
41. Katsylo, Pavel (Moscow)
42. Kawamata Yujiro (Tokyo)
43. Kneser, Martin (Goettingen)

List of Participants of Göttingen Conference   XVII

44. Knorr, Knut (Regensburg)
45. Königsberger, Konrad (München)
46. Kreck, Mathias (Heidelberg)
47. Krengel, Ulrich (Göttingen)
48. Kriete, Hartje (Göttingen)
49. Langer, Adrian (Warsaw)
50. Leenson, Maxim (Moscow)
51. Lehmkuhl, Thomas (Göttingen)
52. Lieb, Ingo (Northeim)
53. Liedtke, Christian (Göttingen)
54. Looijenga, Eduard (Utrecht)
55. Ludwig, Beate Ursula (Leipzig)
56. Mathieu, David (Bonn)
57. Meckbach, Sabine (Kassel)
58. Müller-Stach, Stefan (Essen)
59. Neumann, Frank (Göttingen)
60. Oeljeklaus, Eberhard (Bremen)
61. Oguiso, Keiji (Essen)
62. Ohsawa, Takeo (Kyoto)
63. Okonek, Christian (Zürich)
64. Olbrich, Martin (Göttingen)
65. Oliverio, Paolo A. (Cosenza)
66. Peternell, Thomas (Bayreuth)
67. Pidstrigach, Victor (Göttingen)
68. Pignatelli, Roberto (Göttingen)
69. Popp, Herbert (Mannheim)
70. Portelli, Dario (Trieste)
71. Porten, Egmont (Berlin)
72. Pragacz, Piotr (Torun)
73. Remmert, Reinhold (Münster)
74. Riemenschneider, Oswald (Hamburg)
75. Sankaran, Gregory K. (Bath)
76. Sarveniazi, Alireza (Göttingen)
77. Schmitt, Eberhard (Jena)
78. Schumacher, Georg (Marburg)
79. Schuster, Hans-Werner (München)
80. Seiler, Wolfgang K. (Mannheim)
81. Siebert, Bernd (Bochum)
82. Siu, Yum-Tong (Harvard)
83. Spallek, Karlheinz (Bochum)
84. Spindler, Heinz (Osnabrück)
85. Storch, Uwe (Bochum)
86. Stuhler, Ulrich (Göttingen)
87. Trapani, Stefano (Rome)
88. Trautmann, Günther (Kaiserslautern)

XVIII Preface

89. Trifogli, Cecilia (Oxford)
90. Tsushima, Ryuji (Heidelberg)
91. Viehweg, Eckart (Essen)
92. Vormoor, Norbert (Göttingen)
93. Weiss-Pidstrigach, Ysette (Göttingen)
94. Wiedmann, Stefan (Göttingen)
95. Wisniewski, Jaroslav (Warsaw)
96. Wong, Pit-Mann (Notre Dame)

Contents

Preface V
1 List of Research Publications of Hans Grauert VII
2 List of Doctoral Students of Hans Grauert XII
3 Program of Göttingen Conference for the 70th
   Birthday of Hans Grauert XIV
4 List of Participants of Göttingen Conference................... XVI

Even Sets of Eight Rational Curves on a K3-surface 1
Wolf Barth
0 Introduction... 1
1 Double Sextics with Eight Nodes............................... 3
2 Double Sextics with Eight Tritangents 4
3 Quartic Surfaces with Eight Nodes 12
4 Quartic Surfaces with Eight Lines 14
5 Double Quadrics with Eight Nodes 21
6 Double Quadrics with Eight Double Tangents 23
7 Comments 24
References 24

A Reduction Map for Nef Line Bundles 27
Thomas Bauer, Frédéric Campana, Thomas Eckl, Stefan Kebekus,
Thomas Peternell, Sławomir Rams, Tomasz Szemberg, Lorenz
Wotzlaw
1 Introduction.. 27
2 A Reduction Map for Nef Line Bundles 28
3 A Counterexample 34
References 36

**Canonical Rings of Surfaces Whose Canonical System has
Base Points** 37
Ingrid C. Bauer, Fabrizio Catanese, Roberto Pignatelli
0 Introduction.. 37
1 Canonical Systems with Base Points........................... 42
2 The Canonical Ring of Surfaces with $K^2 = 7$, $p_g = 4$
   Birational to a Sextic: From Algebra to Geometry.............. 49
3 The Canonical Ring of Surfaces with $K^2 = 7$, $p_g = 4$
   Birational to a Sextic: Explicit Computations................. 58
4 An Explicit Family 63
References 67
Appendix 1 69

XX   Contents

Appendix 2 70

Attractors 73
Araceli M. Bonifant, John Erik Fornæss
1   Introduction... 73
2   Endomorphisms 74
3   Hyperbolic Diffeomorphisms.................................... 77
4   Holomorphic Endomorphisms of $\mathbb{P}^k$ 79
References 83

A Bound on the Irregularity of Abelian Scrolls in Projective Space.. 85
Ciro Ciliberto, Klaus Hulek
0   Introduction... 85
1   Non-Existence of Scrolls 86
2   Existence of Scrolls.. 88
References 92

On the Frobenius Integrability of Certain Holomorphic p-Forms 93
Jean-Pierre Demailly
1   Main Results 93
2   Proof of the Main Theorem 95
References 97

Analytic Moduli Spaces of Simple (Co)Framed Sheaves 99
Hubert Flenner, Martin Lübke
1   Introduction... 99
2   Preparations 101
3   Simple $\mathcal{F}$-Coframed Sheaves 104
4   Proof of Theorem 1.1... 105
References 108

Cycle Spaces of Real Forms of $\mathrm{SL}_n(\mathbb{C})$ 111
Alan T. Huckleberry, Joseph A. Wolf
1   Background 111
2   Schubert Slices 114
3   Cycle Spaces of Open Orbits of $\mathrm{SL}_n(\mathbb{R})$ and $\mathrm{SL}_n(\mathbb{H})$ 121
References 132

On a Relative Version of Fujita's Freeness Conjecture 135
Yujiro Kawamata
1   Introduction.. 135
2   Review on the Hodge Bundles 137
3   Parabolic Structure in Several Variables 138
4   Base Change and a Relative Vanishing Theorem 141
5   Proof of Theorem 1.7... 144

References 146

Characterizing the Projective Space after Cho, Miyaoka and Shepherd-Barron 147
Stefan Kebekus
1 Introduction 147
2 Setup 148
3 Proof of the Characterization Theorem 150
References 154

Manifolds With Nef Rank 1 Subsheaves in Ω_X^1 157
Stefan Kebekus, Thomas Peternell, Andrew J. Sommese
1 Introduction 157
2 Generalities 158
3 The Case Where $\kappa(X) = 1$ 158
4 The Case Where $\kappa(X) = 0$ 159
References 163

The Simple Group of Order 168 and K3 Surfaces 165
Keiji Oguiso, De-Qi Zhang
0 Introduction 165
1 The Niemeier Lattices 168
2 Proof of the Main Theorem 170
References 180

A Precise L^2 Division Theorem 185
Takeo Ohsawa
0 Introduction 185
1 L^2 Extension Theorem on Complex Manifolds 186
2 Extension and Division 188
3 Proof of Theorem 189
References 191

Irreducible Degenerations of Primary Kodaira Surfaces 193
Stefan Schröer, Bernd Siebert
0 Introduction 193
1 Smooth Kodaira Surfaces 195
2 D-semistable Surfaces with Trivial Canonical Class 197
3 Hopf Surfaces 199
4 Ruled Surfaces over Elliptic Curves 207
5 Rational Surfaces and Honeycomb Degenerations 211
6 The Completed Moduli Space and its Boundary 217
References 221

XXII   Contents

Extension of Twisted Pluricanonical Sections with Plurisubharmonic Weight and Invariance of Semipositively Twisted Plurigenera for Manifolds Not Necessarily of General Type ... 223
Yum-Tong Siu

0   Introduction... 224
1   Review of Existing Argument for Invariance of Plurigenera 228
2   Global Generation of Multiplier Ideal Sheaves
    with Estimates 234
3   Extension Theorems of Ohsawa-Takegoshi Type from Usual Basic
    Estimates with Two Weight Functions 241
4   Induction Argument with Estimates 248
5   Effective Version of the Process of Taking Powers and Roots of
    Sections 256
6   Remarks on the Approach of Generalized Bergman Kernels........ 264
References 276

Base Spaces of Non-Isotrivial Families of Smooth Minimal Models 279
Eckart Viehweg, Kang Zuo

1   Differential Forms on Moduli Stacks 283
2   Mild Morphisms 287
3   Positivity and Ampleness 294
4   Higgs Bundles and the Proof of 1.4 301
5   Base Spaces of Families of Smooth Minimal Models 314
6   Subschemes of Moduli Stacks of Canonically Polarized Manifolds... 316
7   A Vanishing Theorem for Sections of Symmetric
    Powers of Logarithmic One Forms............................ 321
References 327

Uniform Vector Bundles on Fano Manifolds and an Algebraic Proof of Hwang-Mok Characterization of Grassmannians 329
Jarosław A. Wiśniewski

0   Introduction... 329
1   $\mathcal{M}$-Uniform Manifolds 331
2   Atiyah Extension and Twisted Trivial Bundles.................. 333
3   Characterization of Grassmann Manifolds 336
4   Characterization of Isotropic Grassmann Manifolds.............. 337
References 339

Even Sets of Eight Rational Curves on a K3-surface

Wolf Barth

Mathematisches Institut der Universität
Bismarckstr. 1 1/2, D 91054 Erlangen
E-mail address: `barth@mi.uni-erlangen.de`

Abstract The aim of this note is to characterize in geometric terms the property that eight disjoint, smooth rational curves L_i on a K3-surface Y form an even set, i.e. that their class $\sum_1^8 [L_i] \in NS(X)$ is divisible by two. This is done for algebraic K3-surfaces arising in one of the following six (most elementary) ways:

1. double plane branched over a sextic curve with eight nodes;
2. double plane branched over a sextic curve with eight triple tangents;
3. quartic surface with eight nodes;
4. quartic surface carrying eight skew lines;
5. double quadric branched over a type (4,4) curve with eight nodes;
6. double quadric branched over a type (4,4) curve with eight double tangents.

Table of Contents

0 Introduction 1
1 Double Sextics with Eight Nodes 3
2 Double Sextics with Eight Tritangents 4
3 Quartic Surfaces with Eight Nodes 12
4 Quartic Surfaces with Eight Lines 14
5 Double Quadrics with Eight Nodes 21
6 Double Quadrics with Eight Double Tangents 23
7 Comments 24
References 24

0 Introduction

A set $L_1, \ldots, L_k$ of smooth, disjoint rational curves on a K3-surface Y is called *even*, if the class of the divisor $L_1 + \cdots + L_k$ in $NS(Y)$ is divisible by 2. Nikulin [N] showed that in this case the number k of lines is either eight or 16. And 16 disjoint rational curves are always even, while for eight such curves this need not be the case. The aim of this note is to characterize even sets of eight disjoint rational curves on a K3-surface in geometric terms. This is done for the six most elementary geometric representations of K3-surfaces:

2000 *Mathematics Subject Classification*: 14J28.

1. Let Y be the minimal desingularisation of a double cover $Y' \to \mathbb{P}_2$, branched over a sextic curve S with eight nodes, and let $E_1, \ldots, E_8$ be the rational curves in Y resolving the nodes of Y'. Then $E_1, \ldots, E_8$ is even, if and only if the curve S decomposes into a smooth conic and a smooth quartic, the nodes on S being the intersections of these irreducible components.

2. Let $Y \to \mathbb{P}_2$ be a double cover branched over a smooth sextic curve S with eight triple tangents $T_1, \ldots, T_8$. The inverse image of each triple tangent T_i in Y decomposes into two smooth rational curves L_i and L_i'. Here it is possible to group these rational curves into two sets $L_1, \ldots, L_8$ and $L_1', \ldots, L_8'$, mutually disjoint and even, if and only if the 24 points of contact on S of the eight tritangents are cut out on S by a quartic curve and if the residual conic appearing naturally in this situation (see Sect. 2) is non-degenerate.

3. Let $Y' \subset \mathbb{P}_3$ be a quartic surface with eight nodes and no other singularities. Let $Y \to Y'$ be its minimal desingularization with $E_1, \ldots, E_8 \subset Y$ the rational curves resolving the nodes of Y'. Then the set $E_1, \ldots, E_8$ is even if and only if the eight nodes of Y' are associated, i.e., if they form a complete intersection of three quadric surfaces.

4. Let $Y \subset \mathbb{P}_3$ be a smooth quartic surface carrying eight skew lines $L_1, \ldots, L_8$. This set of lines on Y is even if and only if for one (and then in fact for each) division $\{L_1, \ldots, L_4\} \cup \{L_5, \ldots, L_8\}$ of these lines into two groups of four, either there are two skew lines $Z_1, Z_2 \subset Y$ with Z_1 meeting each line $L_1, \ldots, L_4$, missing each line $L_5, \ldots, L_8$, and Z_2 missing each line $L_1, \ldots, L_4$ and meeting $L_5, \ldots, L_8$, or there is a smooth quartic elliptic curve $Z \subset Y$ with $Z.L_1 = \cdots = Z.L_4 = 0$, $Z.L_5 = \cdots = Z.L_8 = 2$.

5. Let $Y' \to \mathbb{P}_1 \times \mathbb{P}_1$ be a double cover branched over a curve $S \subset \mathbb{P}_1 \times \mathbb{P}_1$ of type $(4,4)$ having eight nodes and otherwise smooth. Let $Y \to Y'$ be the minimal desingularization of the eight nodes on Y' with $E_1, \ldots, E_8$ the curves resolving these nodes. Then $E_1, \ldots, E_8$ is even if and only if S decomposes into two smooth curves of type $(2,2)$, the nodes of S being just the intersections of these two curves.

6. Let $Y \to \mathbb{P}_1 \times \mathbb{P}_1$ be a double cover branched over a smooth curve $S \subset \mathbb{P}_1 \times \mathbb{P}_1$ of type $(4,4)$. Assume S has eight double tangents $T_1, \ldots, T_8$, belonging to the lines ruling the surface $\mathbb{P}_1 \times \mathbb{P}_1$. Let $L_i, L_i' \subset Y$ be the rational curves over T_i. If it is possible to group these 16 rational curves in two even sets of eight disjoint ones, then four of the double tangents belong to one ruling, four to the other one, and the 16 points of contact on S of the eight double tangents are cut out by a curve of type $(2,2)$.

It is remarkable that the characterizations 1),3),5) of even sets of nodes are essentially trivial, while 2) is hard, constructing examples there is easy, and 4) again is trivial but constructing examples is hard. I give some examples

Even Sets of Eight Rational Curves on a K3-surface   3

for 4) relying on a Poncelet-2-property [BB, (1.1)] of quadrics in $\mathbb{P}_3$. The hard part of 2) is to show that any sextic curve meeting the conditions there leads to a K3-surface with eight disjoint even rational curves. In spite of many efforts I did not manage to do this directly. So I use the following, not very illuminating method: I show that the family of K3-surfaces in question is irreducible and I give one example where the rational curves indeed form an even set. Since this property is a topological one, it holds for all surfaces in the family. I do not prove here the reverse of 6), because I do not see any other possibility than the analog of the proof in 2), and I don't think this is worth while the effort.

1   Double Sextics with Eight Nodes

A *double sextic* is a double cover $\sigma' : Y' \to \mathbb{P}_2$ branched over a sextic curve $S \subset \mathbb{P}_2$. If S has at most A-D-E-singularities, then the minimal desingularization Y of Y' is a K3-surface. Here we consider the situation where S has precisely eight ordinary double points $s_1, \ldots, s_8$. The eight points $e_1, \ldots, e_8 \in Y'$ lying over $s_1, \ldots, s_8$ are ordinary double points A_1 (nodes) on Y'. Let $E_1, \ldots, E_8 \subset Y$ be the (-2)-curves resolving these nodes. Denote by $\sigma : Y \to \mathbb{P}_2$ the induced map.

Proposition 1.1. *The divisor $E_1 + \cdots + E_8 \in NS(Y)$ is divisible by 2 if and only if the sextic curve S splits as the union $C + Q$ of a smooth conic C and a smooth quartic Q meeting transversally in $s_1, \ldots, s_8$.*

Proof. Assume first that the divisor $E_1 + \cdots + E_8$ is even. Denote by $H \in NS(Y)$ the class $\sigma^* \mathcal{O}_{\mathbb{P}_2}(1)$. Then

$$\left(H - \frac{1}{2} \sum_1^8 E_i \right)^2 = 2 - 4 = -2 \, .$$

By Riemann-Roch on Y we have $\chi(H - \sum E_i/2) = 1$. Since $H.(\sum E_i/2 - H) = -2$, the divisor class $H - \sum E_i/2$ on Y is effective. Let $D \subset Y$ be an effective divisor representing this class. Since $D.E_i = 1$, it meets all the eight curves E_i. From the projection formula

$$\sigma(D).\mathcal{O}_{\mathbb{P}_2}(1) = D.H = 2$$

we conclude that $C := \sigma(D) \subset \mathbb{P}_2$ has degree two, while containing all the eight singularities $e_i \in S$.

Now C cannot be a double line, because eight singularities of a reduced sextic curve S cannot lie on a single line. If $C = L_1 + L_2$ splits into two distinct lines $L_i \subset \mathbb{P}_2$, then on one of these lines, say on L_1, there are at least four singularities of S. By $S.L_1 \geq 8$ the curve S splits off L_1, say $S = L_1 + M$ with a quintic curve M. But then M will have three singularities on L_2. So M

4   W. Barth

splits off L_2 and $S = L_1 + L_2 + Q$ with a quartic curve $Q \subset \mathbb{P}_2$. But then the number of double points of S will be at least 9, contradiction.

If, finally, C is irreducible, $S.C \geq 16$ imples that C is a component of S.

Assume now conversely, that $S = C + Q$ with C a smooth conic and Q a smooth quartic, both curves meeting transversally in eight points. Then $\sigma^* C = 2D + \sum n_i E_i$ with a smooth rational curve $D \subset Y$. From

$$0 = \sigma^* C.E_i = 2D.E_i - 2n_i \quad \text{and} \quad D.E_i = 1$$

we conclude $n_i = 1$ for $i = 1, \ldots, 8$. But then

$$\sum_1^8 E_i \sim 2(H - D)$$

is divisible by two in $NS(Y)$.   □

2   Double Sextics with Eight Tritangents

Let $\sigma : Y \to \mathbb{P}_2$ be a double sextic branched over a smooth sextic curve S. We denote by $S_Y \subset Y$ the copy of S lying in Y and mapped isomorphically onto S under σ.

By a *non-degenerate tritangent* to S we mean a line $T \subset \mathbb{P}_2$ touching S at three distinct points. The pre-image curve $\sigma^{-1}(T) \subset Y$ then splits into two smooth rational curves $L, L' \subset Y$. They meet at the three points lying over the three points of contact on T. Both curves L and L' are mapped isomorphically onto T.

By *eight non-degenerate tritangents* we mean eight distinct lines $T_1, \ldots, T_8$, all of which are non-degenerate tritangents to S. We say, *their points of contact are cut out by a quartic*, if there is a quartic curve meeting S in all the 24 points of contact of the eight tritangents. Denote the three points of contact on T_i by $p_{3(i-1)+1}, p_{3(i-1)+2}, p_{3(i-1)+3}$ and the divisor $\sum_1^{24} p_i$ on S by Λ. By the tangency property

$$\mathcal{O}_S(2\Lambda) = \mathcal{O}_S(8) \,,$$

and the points of contact are cut out by a quartic if and only if

$$\mathcal{O}_S(\Lambda) = \mathcal{O}_S(4) \,.$$

It seems non-trivial to construct examples of sextic curves with eight tritangents. It is fairly easy, however, to construct examples, where the points of contact are cut out by a quartic. Indeed, let $t_i = 0$ be the equations of the tangents T_i and $v = 0$ the quartic cutting out the points of contact. Both polynomials $t = t_1 \cdot \ldots \cdot t_8$ and v^2, when restricted to S, vanish precisely on the divisor 2Λ. So $t|S$ and $v^2|S$ differ by a multiplicative constant, which may be squeezed into t, say, to obtain $t|S = v^2|S$. Then

$$t - v^2 = 0$$

Even Sets of Eight Rational Curves on a K3-surface   5

is the equation of an octic curve splitting off S. So this equation vanishes on $S + C$ with some conic C. We call C *the residual conic.* Notice that V is determined uniquely by S and the tritangents, so this residual conic is uniquely determined too.

Proposition 2.1. *For eight nondegenerate tritangents to a smooth sextic with points of contact cut out by a quartic the following two properties are equivalent:*

i) *It is possible to label the rational curves lying over T_i as L_i and L_i' such that the eight curves $L_i \subset Y$ are disjoint.*
ii) *The residual conic is non-degenerate.*

Proof. "i)$\Rightarrow$ii)": Since the eight rational curves $L_1, \ldots, L_8$ are disjoint, no three tangents T_i will be concurrent. Consider the two possibilities for the residual conic to degenerate:

a) Let $C = M + M'$ with two distinct lines $M \neq M'$: None of these two lines can be a tritangent. In fact, by $t|C = v^2|C$, the quartic V would split off this tangent. But then V could not cut out the 21 other points of ontact, not on this fixed tritangent.

At most two tangents can meet at the vertex $M \cap M'$. So there must be some tangent, say T_1, intersecting one of the two lines, say M, transversally in a point different from the vertex. As $t|M = v^2|M$, the quartic V will pass through the intersection of T_1 and M. And $v^2|M$ will vanish there to the second order. So there must be another tangent, say L_2, passing through this point. Then, however, $t - v^2$ will vanish at least to the second order in this point, and the point would lie on S. Two distinct tritangents would meet on S, impossible.

b) Let $C = 2M$ with some line M. Again, this cannot be one of the tritangents T_i. Arguing as in a) we find that the tangents T_i would meet in pairs on M. Therefore it is possible to choose three lines, say T_1, T_2, T_3 such that any two among them meet outside M. The triangle T_1, T_2, T_3 is met by the quartic V in three points on M and in the three points of contact for the three tangents. But then these nine points of contact of this triangle with the branch curve S would be cut out by a cubic curve. This implies that the covering σ over the triangle is trivial, when the curves L_i and L_i', $i = 1, 2, 3$, are separated at their branch points. But then, the three curves $L_1, L_2, L_3 \subset Y$ could not be disjoint.

"ii)$\Rightarrow$i)": As in part a) above, one can exclude the possibility that two tangents T_i would meet in a point on C. The property $t|C = v^2|C$ therefore implies that the eight tangents T_i touch C in eight distinct points. No three among these points then are collinear. For each triangle T_i, T_j, T_k this implies that the nine points of contact of this triangle with S will not be cut out by a cubic curve. The covering σ will be non-trivial over this triangle (when the curves L_m and L_m' are separated over S). Therefore it is possible, to label the curves over T_i, T_j and T_k such that L_i, L_j and L_k are disjoint.

6   W. Barth

Arguing as in [BM, (4.4)] one finds that this is possible over all the eight tritangents. □

Fortunately, the property that the points of contact are cut out by a quartic is just what we need. This is shown in the next proposition.

Proposition 2.2. *Let $S \subset \mathbb{P}_2$ be a smooth sextic curve and $T_1, \ldots, T_8$ be eight non-degenerate tritangents to S. Then the following two properties are equivalent:*

i) *It is possible to label the 16 curves $L_i, L_i' \subset Y$ in such a way that the eight curves $L_1, \ldots, L_8$ are disjoint, and these curves form an even set.*
ii) *The divisor $\Lambda = p_1 + \cdots + p_{24}$ on S is cut out by a quartic curve and the residual conic is non-degenerate.*

Proof. Denote by $p_{3(i-1)+1}, p_{3(i-1)+2}$ and $p_{3(i-1)+3}$ the three points of contact on T_i and by Λ the divisor $p_1 + \cdots + p_{24}$ on S.

"i)$\Rightarrow$ii)": Because of

$$\mathcal{O}_S(2\Lambda) = \mathcal{O}_S(8) .$$

the class $\mathcal{O}_S(\Lambda)$ differs from $\mathcal{O}_S(4)$ by some 2-torsion class $\lambda \in \mathrm{Pic}(S)$. The assertion is equivalent to $\lambda = 0$.

By assumption there is a divisor class L on Y satisfying

$$\mathcal{O}_Y(2L) = \mathcal{O}_Y(L_1 + \cdots + L_8).$$

Denote by $\iota : Y \to Y$ the covering involution. Then

$$\mathcal{O}_Y(2\iota^* L) = \mathcal{O}_Y(L_1' + \cdots + L_8') .$$

The class $L + \iota^* L$ on Y is invariant under the involution and comes from $\mathbb{P}_2$ via σ, say $\mathcal{O}_Y(L + \iota^* L) = \sigma^* \mathcal{O}_{P_2}(k)$. Since

$$\mathcal{O}_Y(2L + \iota^* 2L) = \mathcal{O}_Y(L_1 + \cdots + L_8 + L_1' + \cdots + L_8') = \sigma^*(\mathcal{O}_{P_2}(8))$$

we have $k = 4$ and

$$\mathcal{O}_Y(L + \iota^* L) = \sigma^*(\mathcal{O}_{\mathbb{P}_2}(4)) .$$

Since ι acts on S_Y as the identity,

$$\mathcal{O}_{S_Y}(L) = \mathcal{O}_{S_Y}(\iota^* L), \quad \mathcal{O}_{S_Y}(2L) = \mathcal{O}_{S_Y}(4) .$$

This implies that $\mathcal{O}_{S_Y}(L)$ differs from $\sigma^*(\mathcal{O}_S(2))$ by some 2-torsion class $\mu \in \mathrm{Pic}(S_Y)$. Let now Λ_Y be the class $\sigma^* \Lambda$ on S_Y. It is cut out by the divisor $L_1 + \cdots + L_8$. But

$$\mathcal{O}_{S_Y}(L_1 + \cdots + L_8) = \mathcal{O}_{S_Y}(2L) = (\sigma^* \mathcal{O}_S(2) \otimes \mu)^{\otimes 2} = \sigma^* \mathcal{O}_S(4) .$$

This proves $\mathcal{O}_{S_Y}(\Lambda_Y) = \sigma^* \mathcal{O}_S(4)$ and $\mathcal{O}_S(\Lambda) = \mathcal{O}_S(4)$. The 2-torsion class $\lambda \in \mathrm{Pic}(S)$ indeed is trivial.

"ii)$\Rightarrow$i)": Now by assumption the divisor $\Lambda = p_1 + \cdots + p_{24}$ is cut out on S by some quartic curve $V \subset \mathbb{P}_2$ the residual conic C being smooth. By Proposition 2.1 it is possible to label the rational curves L_i, L_i' in such a way that they fall into two groups of eight disjoint lines.

Let as above t be the octic polynomial vanishing on $T_1 + \cdots + T_8$ and v the polynomial defining V, normalized such that $t - v^2 = 0$ on S. The tritangents $T_1, \ldots, T_8$ touch C, while V cuts out on C their eight points of contact. We show that the variety of all 9-tuplets

$$\Sigma := \{(S, \{T_1, \ldots, T_8\}) \in \mathbb{P}(H^0(\mathcal{O}_{\mathbb{P}_2}(6)) \times \mathrm{Sym}^8(\mathbb{P}_2^*)\}$$

appearing in such a situation is irreducible. By sending $(S, \{T_1, \ldots, T_8\})$ to the residual conic C, the variety Σ is $PGL(2, \mathbb{C})$-equivariantly fibered over the Zariski-open set in $\mathbb{P}_5$ of non-degenerate conics. So it suffices to show that $\Sigma_C \subset \Sigma$, the subset belonging to the same residual conic C, is irreducible.

We parametrize Σ_C as follows: Consider the Zariski-dense subset $U_1 \subset \mathbb{P}(H^0(\mathcal{O}_{\mathbb{P}_2}(4))$ of quartic curves not splitting off C as a factor. For such a $V \in U$ let $\{p_1, \ldots, p_8\}$ be its intersection with C. Denote by T_i the tangent to C at p_i and by $t_i \in H^0(\mathcal{O}_{\mathbb{P}_2}(1))$ a linear polynomial vanishing on T_i. Then $t := t_1 \cdot \ldots \cdot t_8 = 0$ cuts out the same divisor on C as the octic $2V$. So there is an equation $v = 0$ for V with $(t - v^2)|C \equiv 0$. If $c = 0$ is some equation for C, put $s := (t - v^2)/c$ and $S : s = 0$. Clearly $S, \{T_1, \ldots, T_8\}$ depend on V only and not on its equation. The map

$$V \mapsto (S, \{T_1, \ldots, T_8\}), \quad U_1 \to \mathbb{P}H^0(\mathcal{O}_{\mathbb{P}_2}(6)) \times \mathrm{Sym}^8(\mathbb{P}_2^*)$$

has an image containing Σ_C. To obtain as image the set Σ_C itself, we have to remove from U_1 the following Zariski-closed subsets leading to degenerate situations: quartic curves V for which

- not all the points $p, \ldots, p_8$ are distinct;
- V meets a tangent T_i in less than four points;
- the sextic curve S is singular.

After removing these Zariski-closed subsets from U_1 we obtain an irreducible variety U_2 mapping surjectively onto Σ_C. This proves that Σ_C is irreducible (but not that it is non-empty).

Since the evenness property of the lines $L_1, \ldots, L_8$ is a topological fact, it is invariant under deformation in an irreducible family. So it suffices to find one single example $(S, \{T_1, \ldots, T_8\}) \in \Sigma_C$ for which it holds. This example is given below, and so the assertion is proved. $\square$

Example. We fix

$$C : c = 0, \quad c = x_1^2 + x_2^2 - x_0^2,$$

as residual conic (the unit circle in affine coordinates $x_1/x_0, x_2/x_0$). On C we pick the following eight points p_i with their tangents $T_i : t_i = 0$ to C:

8   W. Barth

$$p_1 = (2 : \sqrt{3} : 1) , \qquad t_1 = \sqrt{3}x_1 + x_2 - 2x_0 ;$$
$$p_2 = (2 : 1 : \sqrt{3}) , \qquad t_2 = x_1 + \sqrt{3}x_2 - 2x_0 ;$$
$$p_3 = (2 : -1 : \sqrt{3}) , \qquad t_3 = -x_1 + \sqrt{3}x_2 - 2x_0 ;$$
$$p_4 = (2 : -\sqrt{3} : 1) , \qquad t_4 = -\sqrt{3}x_1 + x_2 - 2x_0 ;$$
$$p_5 = (2 : -\sqrt{3} : -1) , \qquad t_5 = -\sqrt{3}x_1 - x_2 - 2x_0 ;$$
$$p_6 = (2 : -1 : -\sqrt{3}) , \qquad t_6 = -x_1 - \sqrt{3}x_2 - 2x_0 ;$$
$$p_7 = (2 : 1 : -\sqrt{3}) , \qquad t_7 = x_1 - \sqrt{3}x_2 - 2x_0 ;$$
$$p_8 = (2 : \sqrt{3} : -1) , \qquad t_8 = \sqrt{3}x_1 - x_2 - 2x_0 .$$

The eight points p_i are cut out on C by the following quartic V, union of four lines,

$$V : v = 0 , \quad v = (2x_1 - x_0)(2x_1 + x_0)(2x_2 - x_0)(2x_2 + x_0) .$$

For $t := t_1 \cdot \ldots \cdot t_8$ we find

$$t - v^2 = 3 \cdot c \cdot s$$

with

$$s = 3 \left(x_1^6 + x_2^6\right) - 23 \left(x_1^4 x_2^2 + x_1^2 x_2^4\right) - 29x_0^2 \left(x_1^4 + x_2^4\right)$$
$$+ 38x_0^2 x_1^2 x_2^2 + 83x_0^4 \left(x_1^2 + x_2^2\right) - 85x_0^6 .$$

This situation is invariant under the symmetries

$$x_1 \leftrightarrow x_2, \quad x_i \mapsto -x_i.$$

We have to check that the situation is non-degenerate. First we check the sextic $S : s = 0$ for singularities. So we differentiate

$$\partial_0 s = 2x_0 \cdot \left(-255x_0^4 + 166x_0^2(x_1^2 + x_2^2) + 38x_1^2 x_2^2 - 29(x_1^4 + x_2^4)\right) ,$$
$$\partial_1 s = 2x_1 \cdot \left(83x_0^4 - 58x_0^2 x_1^2 + 38x_0^2 x_2^2 + 9x_1^4 - 46x_1^2 x_2^2 - 23x_2^4\right) ,$$
$$\partial_2 s = 2x_2 \cdot \left(83x_0^4 + 38x_0^2 x_1^2 - 58x_0^2 x_2^2 - 23x_1^4 - 46x_1^2 x_2^2 + 9x_2^4\right) ,$$

and evaluate the condition

$$\partial_0 s = \partial_1 s = \partial_2 s = 0 .$$

Here $x_1 = x_2 = 0$ implies $x_0 = 0$, as well as $x_0 = x_1 = 0$ implies $x_2 = 0$. So no two coordinates can vanish simultaneously. If $x_0 = 0$, $x_1 x_2 \neq 0$, then

$$9x_1^4 - 46x_1^2 x_2^2 - 23x_2^4 = 9x_2^4 - 46x_1^2 x_2^2 - 23x_1^4 = 0$$

leads to $x_1^2 = x_2^2 = 0$. And $x_1 = 0$, $x_0 x_2 \neq 0$ implies

$$-255x_0^4 + 166x_0^2 x_2^2 - 29x_2^4 = 83x_0^4 - 58x_0^2 x_2^2 + 9x_2^4 = 0 .$$

Adding three times the second equation to the first one we find

$$-2\left(3x_0^4 + 4x_0^2 x_2^2 + x_2^4\right) = 0 .$$

Inserting $x_2^4 = -3x_0^4 - 4x_0^2 x_2^2$ into both equations leads to

$$-168x_0^4 + 282x_0^2 x_2^2 = 56x_0^4 - 94x_0^2 x_2^2 = 0, \quad \text{i.e.} \quad 28x_0^2 = 47x_2^2 .$$

Using this to eliminate x_0^2 from both equations shows $x_2 = 0$. So we found: In a singularity of S all coordinates x_i are non-zero. At a singularity of S therefore

$$-255x_0^4 + 166x_0^2\left(x_1^2 + x_2^2\right) + 38x_1^2 x_2^2 - 29\left(x_1^4 + x_2^4\right) = 0 ,$$
$$83x_0^4 - 58x_0^2 x_1^2 + 38x_0^2 x_2^2 + 9x_1^4 - 46x_1^2 x_2^2 - 23x_2^4 = 0 ,$$
$$83x_0^4 + 38x_0^2 x_1^2 - 58x_0^2 x_2^2 - 23x_1^4 - 46x_1^2 x_2^2 + 9x_2^4 = 0 .$$

The difference between the second and the third equation is

$$32\left(x_1^4 - x_2^4\right) - 96x_0^2\left(x_1^2 - x_2^2\right) = 32\left(x_1^2 - x_2^2\right)\left(x_1^2 + x_2^2 - 3x_0^2\right) .$$

Inserting $x_2^2 = x_1^2$ or $x_0^2 = (x_1^2 + x_2^2)/3$ into the first equation leads to the vanishing of some coordinate x_i. This proves that the sextic S is nonsingular.

Next we have to check that the sextic S touches each line T_i at three distinct points. But for the point p_i, these points are cut out on T_i by the quartic V. So we compute the intersections of T_i with V. Because of the the symmetries, it suffices to consider T_1. On T_1 we have $x_2 = 2x_0 - \sqrt{3}x_1$. So the restriction of v to T_1 decomposes as

$$\left(2x_1 - x_0\right)\left(2x_1 + x_0\right)\left(3x_0 - 2\sqrt{3}x_1\right)\left(5x_0 - 2\sqrt{3}x_1\right) .$$

Obviously there are four distinct points of intersection on L_1. We have shown that the situation is non-degenerate. $\square$

By Proposition 2.1 the curves L_i and L_i' over T_i in the double sextic Y with branch locus S from the example above decompose into two groups of eight disjoint ones, say $L_1, \ldots, L_8$ and $L_1', \ldots, L_8'$. To show that $L_1, \ldots, L_8$ form an even set, and therefore to complete the proof of Proposition 2.2, we need

Proposition 2.3 (Recognition principle). *Let $Y \to \mathbb{P}_2$ be a double plane branched over some smooth sextic $S \subset \mathbb{P}_2$. Let $L_1, \ldots, L_8$ be disjoint rational curves mapping isomorphically onto tritangents T_i of S. If there is a smooth rational curve $D_Y \subset Y$ mapping isomorphically onto a non-degenerate conic $D \subset \mathbb{P}_2$, such that*

$$D_Y.L_1 = \cdots = D_Y.L_4 = 2, \quad D_Y.L_5 = \cdots = D_Y.L_8 = 0 ,$$

then the set $L_1, \ldots, L_8$ is even.

10   W. Barth

Proof. Denote by H the divisor class $\sigma^* \mathcal{O}_{\mathbb{P}_2}(1)$ on Y. The lattice generated by $H, L_1, \ldots, L_8$ is non-degenerate: In fact, if $M := \eta H + \lambda_1 L_1 + \cdots + \lambda_8 L_8 \perp H, L_1, \ldots, L_8$, we find

$$0 = M.L_i = \eta - 2\lambda_i \quad \Rightarrow \quad \lambda_1 = \cdots = \lambda_8 = \eta/2 \ ,$$

$$0 = M.H = 2\eta + \lambda_1 + \cdots + \lambda_8 = 6\eta \quad \Rightarrow \quad \eta + \lambda_1 = \cdots = \lambda_8 = 0 \ .$$

So

$$NS(Y) \otimes_{\mathbb{Z}} \mathbb{Q} = ((\mathbb{Z} \cdot H + \mathbb{Z} \cdot L_1 + \cdots + \mathbb{Z} \cdot L_8) \otimes_{\mathbb{Z}} \mathbb{Q}) \oplus ((H, L_1, \ldots, L_8)^\perp \otimes_{\mathbb{Z}} \mathbb{Q}) \ .$$

Therefore we may write in $NS(Y) \otimes_{\mathbb{Z}} \mathbb{Q}$

$$[D_Y] = \eta H + \sum_1^8 \lambda_i [L_i] + [D'] $$

with $\eta, \lambda_i \in \mathbb{Q}$ and D' a $\mathbb{Q}$-divisor on Y satisfying

$$D'.H = D'.L_1 = \cdots = D'.L_8 = 0 \ .$$

By assumption

$$D_Y.H = 2, \ D_Y.L_1 = \cdots = D_Y.L_4 = 2 \ , \ D_Y.L_5 = \cdots = D_Y.L_8 = 0 \ .$$

This implies

$$2\eta + \sum_1^8 \lambda_i = 2, \quad \eta - 2\lambda_i = 2 \, (i = 1, \ldots, 4) \ , \quad \eta - 2\lambda_i = 0 \, (i = 5, \ldots, 8) \ .$$

In particular

$$\lambda_i = \frac{\eta}{2} - 1 \text{ for } i = 1, \ldots, 4, \quad \lambda_i = \frac{\eta}{2} \text{ for } i = 5, \ldots, 8 \ .$$

Hence

$$6\eta - 4 = 2, \ \eta = 1, \quad \lambda_1 = \cdots = \lambda_4 = -\frac{1}{2} \ , \quad \lambda_5 = \cdots = \lambda_8 = \frac{1}{2} \ .$$

Now put

$$B := H + \frac{1}{2}(-L_1 - \cdots - L_4 + L_5 + \cdots + L_8) \ .$$

Then

$$B^2 = -2 \ , \ B.H = 2, \ B.D_Y = -2 \ , \ B.D' = 0 \ .$$

By assumption

$$D_Y^2 = -2, \ D_Y.H = 2 \ .$$

Consider the divisors

$$B_1 := B + H \text{ and } D_1 := D_Y + H .$$

For them we compute

$$B_1^2 = B_1.D_1 = D_1^2 = 4 .$$

By the Hodge index theorem the classes of B_1 and D_1 are linearly dependent (over $\mathbb{Q}$). From $B_1.H = 4 = D_1.H$ we conclude $B_1 = D_1$ and $B = D_Y$. In particular $[D'] = 0$ and

$$[L_1 + \cdots + L_4 - L_5 - \cdots - L_8] = 2 \cdot (H - [D_Y])$$

is divisible by 2. $\qquad\qquad\square$

We have to characterize the situation in the recognition principle in terms of curves in the plane $\mathbb{P}_2$ below. This is done by the following proposition. Its proof is obvious.

Proposition 2.4. *Let* $Y \to \mathbb{P}_2$, *and* $S, T_1, \ldots, T_8 \subset \mathbb{P}_2$ *be as in Proposition 2.3. Let* $D \subset \mathbb{P}_2$ *be a smooth conic touching* S *in six distinct points, meeting each triple-tangent* T_i *in two distinct points. Let* $L_i, L_i' \subset Y$ *be the two rational curves over* T_i *and* $D_Y, D_Y' \subset Y$ *the two rational curves over* D.

a) *The following properties are equivalent:*
   a1) D_Y *meets either* L_i *or* L_i' *in two points missing the other curve;*
   a2) *the covering restricted to* $D + T_i$ *is trivial (when separated at the nine branch points);*
   a3) *the nine points of contact of* S *with* $D + T_i$ *are cut out by a cubic curve.*
b) *Assume 16 the curves* $L_i, L_i' \subset Y$ *can be grouped such that* $L_1, \ldots, L_8$ *are mutually disjoint. Then for* $i \neq j$ *the following properties are equivalent:*
   b1) *Both* L_i *and* L_j *meet or miss* D_Y;
   b2) *the covering restricted to* $D + T_i + T_j$ *is trivial (when separated at the twelve branch points);*
   b3) *the twelve points of contact of* S *with* $D + T_i + T_j$ *are cut out by a cubic curve.*

Example (continued). Let $D \subset \mathbb{P}_2$ be the non-degenerate conic of equation

$$d = 0 , \quad d := 3 \left(x_1^2 + x_2^2 \right) - 17 x_0^2 .$$

This conic meets the sextic S in the two circle points $(0 : 1 : \pm i)$ and in the four points

$$\left(\pm\sqrt{3} : \sqrt{17} : 0 \right) , \quad \left(\pm\sqrt{3} : 0 : \sqrt{17} \right) .$$

One readily checks that both curves touch in these six points.

12   W. Barth

Next we verify property a) from proposition 2.4 for all triple tangents T_i. As D too is invariant under the group generated by the symmetries specified above, which acts transitively on the eight triple tangents, it suffices to do this for the triple tangent T_1. The three points, where T_1 touches S are

$$p_1 = \left(2 : 1 : 4 - \sqrt{3}\right), \, p_2 = \left(2 : -1 : 4 + \sqrt{3}\right), \, p_3 = \left(2\sqrt{3} : 5 : -\sqrt{3}\right).$$

One readily checks that the cubic F of equation

$$17\sqrt{3}x_0^3 - 34x_0^2x_1 - 3\sqrt{3}x_0x_1^2 - 16x_0x_1x_2 - 3\sqrt{3}x_0x_2^2 + 6x_1^3 + 6x_1x_2^2 = 0$$

cuts out the nine points of contact on $D + T_1$.

Let us now fix one of the two rational curves in Y lying over D and call it D_Y. The property just verified shows that all curves $L_i, L_i' \subset Y$ over the tritangents T_i either meet D_Y in two points or miss it.

Finally we check Property b) from Proposition 2.4. By the symmetries it suffices to consider $T_i = T_1$. Both cubics, $D + T_1$ and F pass through the nine points of contact on $D + T_1$. We compute (using MAPLE) the values of their defining polynomials on the three points of contact for each tritangent $T_2, \ldots, T_8$. The two triplets of values so obtained turn out linearly independent for T_4, T_5, T_8, while they are linearly dependent for T_2, T_3, T_6, T_7. So the twelve points of contact on $D + T_1 + T_j$ are not cut out by a cubic for $j = 4, 5, 8$, whereas they are cut out for $j = 2, 3, 6, 7$. By the symmetries the covering is non-trivial over $D + T_i + T_j$ whenever T_j is obtained from T_i by one of the three reflections $x_1 \mapsto -x_1, x_2 \mapsto -x_2$ or $x_1, x_2 \mapsto -x_1, -x_2$ while it is trival for the other tritangents T_j.

Now we relabel the curves $L_i, L_i' \subset Y$. Let L_1 be the curve meeting D_Y. The curves $L_2, \ldots, L_8$ then are determined by the property $L_1.L_j = 0$, $j = 2, \ldots, 8$, and then they are all disjoint (cf. Proposition 2.2). For $j = 4, 5, 8$ the curve L_j, being disjoint from L_1, meets D_Y (the covering on $D + T_1 + T_j$ is non-trivial). For $j = 2, 3, 5, 6$ the curve L_j disjoint from L_1 misses D_Y (the covering on $D + T_1 + T_j$ is trivial now). So indeed, four of the curves L_i meet D_Y in two points and the other four ones miss D_Y. The recognition principle works and the set $L_1, \ldots, L_8$ is even.

3   Quartic Surfaces with Eight Nodes

Let $Y' \subset \mathbb{P}_2$ be a quartic surface with eight nodes $e_i \in Y'$ and no other singularities. Let $\sigma : Y \to Y'$ be the minimal desingularization and $E_i \subset Y$ the (-2)-curve over e_i.

Proposition 3.1. *The divisor $\sum_1^8 E_i$ in $NS(Y)$ is even if and only if the eight points $e_1, \ldots, e_8 \in Y'$ are associated, i.e. if they form the complete intersection of three quadrics in $\mathbb{P}_3$.*

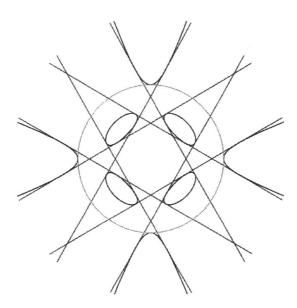

Proof. Assume first that $\sum_1^8 E_i$ in $NS(Y)$ is divisible by two. Put
$$L := \sigma^* \mathcal{O}_{\mathbb{P}_3}(1) - \frac{1}{2}\sum_1^8 [E_i] \ .$$
Then
$$L^2 = 4 - 4 = 0 \ .$$
Since $-L$ cannot be effective, by Riemann-Roch we have $h^0(L) \geq 2$. Let $s_1, s_2 \in H^0(L)$ be two linearly independent sections. Then
$$q_1 := s_1^2, \quad q_2 := s_1 s_2, \quad q_3 := s_2^2 \in H^0(2L)$$
are linearly independent too. For $i = 1, 2, 3$ let
$$C_i := \{q_i = 0\} + \sum_1^8 E_i \in |\sigma^* \mathcal{O}_{Y'}(2)| \ .$$
Since
$$\sigma^* : H^0(\mathcal{O}_{Y'}(2)) \to H^0(\sigma^* \mathcal{O}_{Y'}(2))$$
is bijective, the three curves C_i descend to curves $C_i' \subset Y'$ cut out on Y' by three quadrics $Q_i \subset \mathbb{P}_3$. Let the equations of these quadrics be $q_i' = 0$. By construction $q_1' q_3' - (q_2')^2$ vanishes on Y'. If $q_1' q_3' - (q_2')^2$ would vanish identically on $\mathbb{P}_3$, then the quadrics Q_1 and Q_3 would coincide with Q_2, in conflict with the linear independence of q_1, q_2, q_3. So $q_1' q_3' - (q_2')^2 = 0$ is an equation for the quartic surface Y'. Any point in $Q_1 \cap Q_2 \cap Q_3$ must

14   W. Barth

be a singularity of Y'. This shows that the eight points $e_1, \ldots, e_8$ form the complete intersection $Q_1 \cap Q_2 \cap Q_3$.

Asume now that the eight points e_i are associated, say $e_1, \ldots, e_8 = Q_1 \cap Q_2 \cap Q_3$ with three quadric surfaces Q_1, Q_2, Q_3. The divisor class

$$M := \sigma^* \mathcal{O}_{\mathbb{P}_3}(2) - \sum_1^8 [E_i]$$

has self-intersection $M^2 = 0$. $H^0(M)$ contains three sections q_i vanishing on $\sigma^{-1}(Q_i \cap Y')$. Since the three quadrics Q_1, Q_2, Q_3 vanish simultaneously only in the eight points e_i, their equations will be linearly independent. Then also q_1, q_2, q_3 will be linearly independent and $|M|$ cannot be a pencil. Since $|M|$ has no base curve, necessarily $M = 2L$ with a divisor class $L \in NS(Y)$. Hence

$$\sum_1^8 [E_i] \sim 2(L - \sigma^* \mathcal{O}_{P_3}(1))$$

is even. $\qquad\qquad\qquad\qquad\qquad\qquad\qquad\qquad\qquad\qquad\qquad\qquad\qquad\square$

4   Quartic Surfaces with Eight Lines

Let now $Y \subset \mathbb{P}_3$ be a smooth quartic surface with $L_1, \ldots, L_8 \subset Y$ eight skew lines. We say that four lines $M_1, \ldots, M_4$ among $L_1, \ldots, L_8$, with $N_1, \ldots, N_4$ the four other lines, satisfy the *evenness condition* if

E1) either there are two skew lines Z_1 and $Z_2 \subset Y$ with

$$Z_1.M_i = 1, \quad Z_1.N_i = 0, \qquad Z_2.M_i = 0, \quad Z_2.N_i = 1;$$

E2) or there is a smooth quartic elliptic curve $Z \subset Y$ with

$$Z.M_i = 0, \quad Z.N_i = 2;.$$

Proposition 4.1. *The following three properties are equivalent:*

1) *The divisor $\sum_1^8 L_i$ is divisible by two in $NS(Y)$.*
2) *There are four lines $M_1, \ldots, M_4$ among $L_1, \ldots, L_8$ satisfying the evenness condition.*
3) *Any four lines $M_1, \ldots, M_4$ among $L_1, \ldots, L_8$ satisfy the evenness condition.*

Proof. "1) $\Rightarrow$ 3)": By assumption the divisor $\sum M_i - \sum N_i$ is divisible by two in $NS(Y)$. Consider the class

$$E := \mathcal{O}_Y(1) + \frac{1}{2} \left[\sum M_i - \sum N_i \right].$$

It has degree 4, and $E^2 = 0$. Since $-E$ has negative degree, the class $-E$ cannot be effective. This implies $h^0(E) \geq 2$. There are two possibilities:

Even Sets of Eight Rational Curves on a K3-surface   15

1) Either the linear system $|E|$ has a fixed component, necessarily a line Z_2. And $E = [Z_2] + E'$ with $E' \sim \mathcal{O}(1) - [Z_1]$ being the class of a pencil of plane cubics on Y residual to a line $Z_1 \subset Y$. Then

$$\frac{1}{2}\left(\sum M_i - \sum N_i\right) \sim Z_2 - Z_1 \ .$$

In particular

$$(Z_2 - Z_1)^2 = \frac{1}{4}\left(\sum M_i^2 + \sum N_i^2\right) = -4 \ .$$

So the two lines Z_1 and Z_2 don't meet. And

$$Z_1.\left(\sum M_i - \sum N_i\right) = 2Z_1.(Z_2 - Z_1)$$
$$= \quad 4 \Rightarrow Z_1.M_i = 1 \ , \ Z_1.N_i = 0 \text{ for } i = 1,\dots,4 \ ,$$
$$Z_2.\left(\sum M_i - \sum N_i\right) = 2Z_2.(Z_2 - Z_1)$$
$$= -4 \Rightarrow Z_1.M_i = 0 \ , \ Z_1.N_i = 1 \text{ for } i = 1,\dots,4 \ .$$

Hence the lines M_i satisfy condition E1).

2) Or the linear system $|E|$ has as its general member a smooth quartic elliptic curve Z. Now

$$Z.M_i = \mathcal{O}(1).M_i + \frac{1}{2}M_i^2 = 0 \text{ and } Z.N_i = \mathcal{O}(1).N_i - \frac{1}{2}N_i^2 = 2 \ .$$

Hence the lines M_i satisfy E2).

"2) $\Rightarrow$ 1)": Assume first E1). There are two skew lines Z_1, Z_2 with $Z_1.M_i = Z_2.N_i = 1$, $Z_2.M_i = Z_1.N_i = 0$. Put $E_1 := 2Z_1 + \sum M_i$, $E_2 := 2Z_2 + \sum N_i$. Then $E_i^2 = 0$ and $E_1.E_2 = 0$. The linear system $|E_i|$ has no base curve: Such a curve would consist of components of E_i. Removing them would leave us with a curve E_i' such that $(E_i')^2 < 0$ and $dim|E_i'| = 0$. The divisor E_1 therefore is an elliptic fibre. Since it is simply-connected, it cannot be a multiple fibre [BPV, III (8.3)]. It follows that $|E_1|$ is an elliptic fibration on Y of degree six. The divisor E_2 is another fibre in the same fibration. Hence

$$E_1 \sim E_2, \quad \sum M_i - \sum N_i \sim 2(Z_2 - Z_1) \ ,$$

and $\sum M_i - \sum N_i$ is divisible by two in $NS(Y)$. But then also $\sum M_i + \sum N_i$ is divisible by two.

Assume next E2). There is a smooth elliptic quartic curve Z with $Z.M_i = 0$, $Z.N_i = 2$. The sublattice $L \subset NS(Y)$ generated by the classes of $\mathcal{O}(1)$, M_i, N_i is none-degenerate: In fact, if

$$C \sim \mathcal{O}(\eta) + \sum \mu_i M_i + \sum \nu_i N_i \perp \mathcal{O}(1), M_i, N_i \ ,$$

we find

$$0 = C.M_i = \eta - 2\mu_i,\ 0 = C.N_i = \eta - 2\nu_i \quad \Rightarrow \quad \mu_i = \nu_i = \eta/2\ ,$$

$$0 = C.\mathcal{O}(1) = 8\eta \quad \Rightarrow \quad \eta = \mu_i = \nu_i = 0\ .$$

This implies

$$NS(Y) \otimes_{\mathbb{Z}} \mathbb{Q} = (L \otimes_{\mathbb{Z}} \mathbb{Q}) \oplus \left(L^{\perp} \otimes_{\mathbb{Z}} \mathbb{Q}\right)\ .$$

Hence

$$Z \sim k\mathcal{O}(1) + \sum m_i M_i + \sum n_i N_i + D$$

with $k, m_i, n_i \in \mathbb{Q}$ and with a $\mathbb{Q}$-divisor $D \perp L$. From

$$0 = Z.M_i = k - 2m_i\ ,\quad 2 = Z.N_i = k - 2n_i$$

we conclude

$$m_i = \frac{k}{2}\ ,\quad n_i = \frac{k}{2} - 1\ .$$

Furthermore

$$4 = Z.\mathcal{O}(1) = 4k + \sum m_i + \sum n_i = 8k - 4$$

shows $k = 1$, $m_i = 1/2$, $n_i = -1/2$. Let now

$$Z' := Z - D \sim \mathcal{O}(1) + \frac{1}{2}\left(\sum M_i - \sum N_i\right).$$

Then

$$Z^2 = (Z')^2 = Z.D = D^2 = 0\ .$$

Let $D' = d \cdot D$, $d \in \mathbb{N}$, be a $\mathbb{Z}$-divisor. From $(D')^2 = 0$ and Riemann-Roch follows that either D' or $-D'$ is effective. Furthermore, $Z.D' = 0$ shows that either D' or $-D'$ consists of fibres in the elliptical fibration $|Z|$. But then $D \sim 0$, because $D \perp L$. So finally

$$\sum M_i - \sum N_i \sim 2(Z - \mathcal{O}(1))$$

is divisible by two. $\qquad\square$

It is one thing to analyze the fact that eight skew lines on a quartic surface are even. It is another thing to construct such sets of lines. Of course, there are the quartics from [BN, G, T] containing 16 skew lines. As Nikulin [N] observed, 16 such lines can be parametrized by the points of some affine space A of dimension four over the field $\mathbb{F}_2$. And then each zero-set of an affine-linear form $f : A \to \mathbb{F}_2$ parametrizes an even set of eight lines. There are 16 linear forms $\mathbb{F}_2^4 \to \mathbb{F}_2$, hence 32 affine linear forms $f : A \to \mathbb{F}_2$. Excluding $f \equiv 0$ (belonging to the even set of all 16 lines) and $f \equiv 1$ (characterizing the empty even set) we find 30 even sets of eight skew lines among these 16 lines.

It seems of independent interest to construct examples of quartic surfaces carrying eight even skew lines, not related to the quartics just mentioned. So let us analyze the situation, where the four lines $L_1, \ldots, L_4$ lie on a smooth quadric Q_1 while the four other lines $L_5, \ldots, L_8$ lie on a smooth quadric Q_2.

Proposition 4. Let the eight skew lines $L_1, \ldots, L_8 \subset Y$ on the smooth quartic surface Y form an even set, such that both sets $L_1, \ldots, L_4$ and $L_5, \ldots, L_8$ satisfy the evenness condition E2). Assume that the four lines $L_1, \ldots, L_4$ lie on a smooth quadric Q_1 while the four lines $L_5, \ldots, L_8$ lie on another smooth quadric Q_2. Assume further that the residual intersection of Y with Q_i, $i = 1, 2$, consists of four distinct lines, say

$$Y.Q_1 = L_1 + \cdots + L_4 + L'_1 + \cdots + L'_4 ,$$

$$Y.Q_2 = L_5 + \cdots + L_8 + L'_5 + \cdots + L'_8 .$$

Then these four quadruplets of lines can be reordered such that their intersection pattern is like this:

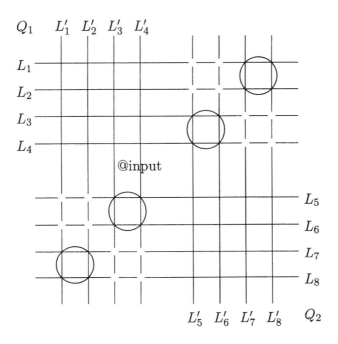

(Here the four circles symbolize the intersection curve $I := Q_1 \cap Q_2$.)

Proof. By assumption

$$\mathcal{O}_Y(2) \sim \sum_1^4 L_i + \sum_1^4 L'_i \sim \sum_5^8 L_i + \sum_5^8 L'_i.$$

The class

$$E \sim \mathcal{O}(1) + \frac{1}{2}\left(\sum_1^4 L_i - \sum_5^8 L_i\right)$$

18   W. Barth

of one of the elliptic fibrations from the evennes condition satisfies

$$2E \sim \mathcal{O}(2) + \sum_1^4 L_i - \sum_5^8 L_i \sim \sum_1^4 L_i + \sum_5^8 L'_i \,.$$

So the eight lines $L_1, \ldots, L_4, L'_5, \ldots, L'_8$ are the union of two elliptic fibres of degree four in $|E|$. This is possible only if they form two disjoint space quadrangles. And we can renumber these lines such that L_1, L_2, L'_7, L'_8 is one of these quadrangles, the other one being L_3, L_4, L'_5, L'_6.

In an analogous way one renumbers the lines $L_5, \ldots, L_8, L'_1, \ldots, L'_4$. $\square$

To proceed, we assume the elliptic curve $I := Q_1 \cap Q_2$ to be nonsingular. Then the pattern of a space quadrangle, formed by two lines in a ruling of Q_1 and two other lines in a ruling of Q_2, with its four vertices on I is a quite familiar variant of Poncelet's theorem [BB, (1.1)]: Let $r_1 \in \mathrm{Pic}(I)$ be the divisor class defined by the ruling $L_1, \ldots$ on Q_1 and $r'_2 \in \mathrm{Pic}(I)$ be the class defined by the ruling $L'_5, \ldots$ on Q_2. Then such space quadrangles with vertices on I exist if and only if

$$r_1 - r'_2 \in \mathrm{Pic}(I)$$

is a 2-torsion class. And then for the other rulings r'_1 on Q_1 and r_2 on Q_2 we have

$$\begin{aligned}
r_1 + r'_1 &= \mathcal{O}_I(1) \,, \\
r_2 + r'_2 &= \mathcal{O}_I(1) \,, \\
r_2 - r'_1 &= (\mathcal{O}_I(1) - r'_2) - (\mathcal{O}_I(1) - r_1) \\
&= r_1 - r'_2 \,.
\end{aligned}$$

So these two rulings satisfy the same 2-torsion condition and lead to space quadrangles with vertices on I in exactly the same way.

Based on this Poncelet property we now construct a smooth quartic Y and an even set of eight skew lines on Y in the following way:

Construction. Let Q_1 be a smooth quadric with R_1 and R_2 its two rulings. Fix a second quadric Q meeting Q_1 in a smooth elliptic quartic space curve I. The ruling R_1 defines on I four branch points $P_1, \ldots, P_4$ by the property $2P_i \sim R_1|I$. Fix a secant L of I joining two of these points, say $L \cap I = P_1 + P_2$. Then there is a unique quadric $Q_2 = \lambda Q_1 + \mu Q$ through I containing this secant. The ruling R'_2 of Q_2 defined by L has the property

$$R_1\,|I - R'_2|\,I \sim 2P_1 - (P_1 + P_2) \sim P_1 - P_2 \,.$$

So the two rulings R_1 and R'_2 satisfy the Poncelet 2-torsion condition.

At this point it is not clear that Q_2 is smooth. It may be one of the four cones through I. But if Q_2 were such a cone, with R_2' its uniqe ruling, then we would have

$$2\left(R_2'|I\right) \sim \mathcal{O}_I(1) .$$

And this would imply

$$2\left(R_1|I\right) \sim 2\left(R_2'|I\right) \sim \mathcal{O}_I(1) ,$$

and Q_1 would be a cone too. Since we assumed Q_1 smooth, this cannot happen, and Q_2 in fact will be smooth.

Let now $P_5, \ldots, P_8 \in I$ be the four branch points defined by the ruling R_2' on Q_2. Let $L_1 \subset Q_1$ be a line in the ruling R_1 not containing any one of the eight points $P_1, \ldots, P_8$. It meets the elliptic curve I in two distinct points. Let $L_7', L_8' \subset Q_2$ be the two lines in the ruling R_2' through these two points of intersection. By the choice of L_1, neither L_7' nor L_8' will be a tangent of I. And by the Poncelet property there will be a line L_2 in the ruling R_1 joining the two other points of intersection in $L_7' \cap I$ and $L_8' \cap I$.

We produced a space quadrangle L_1, L_2, L_7', L_8' with its four vertices on I. Denote these four vertices by $P_9, \ldots, P_{12}$. Then repeat the same construction starting with a line $L_3 \subset Q_1$ in the ruling R_1 not containing any one of the twelve points $P_1, \ldots, P_{12}$. This gives a second space quadrangle L_3, L_4, L_5', L_6' of lines $L_1, L_2 \in R_1$ and $L_5', L_6' \in R_2'$ with its vertices on I. Denote these vertices by $P_{13}, \ldots, P_{16}$.

Denote by R_1' the ruling on Q_1 complementary to R_1 and by R_2 the ruling on Q_2 complementary to R_2'. Let $P_{17}, \ldots, P_{24}$ be the eight branch points on I defined by these two rulings. Then let $L_5 \in R_2$ be a line not through any one of the 16 points $P_9, \ldots, P_{24}$. The Poncelet construction will give us lines $L_5, L_6 \in R_2$ and L_3', L_4' forming a space quadrangle with its vertices on I.

Let $P_{25}, \ldots, P_{28} \in I$ be the vertices of this last quadrangle. Let $L_7 \in R_2$ be a line not through one of the points $P_9, \ldots, P_{28}$. The Poncelet construction will produce a fourth space quadrangle consisting of lines $L_7, L_8 \in R_2$ and $L_1', L_2' \in R_1'$ with its vertices on I.

Altogether we got 16 distinct lines $L_1, \ldots, L_4, L_1', \ldots, L_4' \subset Q_1$ and $L_5, \ldots, L_8, L_5', \ldots, L_8' \subset Q_2$ forming four space quadrangles with vertices on I. We show next that their intersection pattern is exactly the one of the picture above. By construction all the lines which should meet by this picture do meet indeed. We have to show that there are no other incidences. This is clear for two lines on the same quadric. And a line L on one quadric meets the other quadric in its two points of intersection with I. From the eight lines on the other quadric there pass through these two points only the two lines belonging to the space quadrangle defined by L. So L does not meet any other lines.

We have so far 16 lines on the two quadrics Q_1, Q_2, but no quartic surface Y. We claim however: There is a linear system of (linear) dimension ≥ 2 of quartic surfaces containing all the 16 lines $L_1, \ldots, L_8'$. We prove this by counting constants: Let Y_0 be a quartic surface cutting out on Q_1 the eight lines $L_1, \ldots, L_4, L_1', \ldots, L_4'$. Then the linear system of quartics through these eight lines is

$$Q_1 \cdot Q + c \cdot Y_0$$

with Q an arbitrary quadric and $c \in \mathbb{C}$. This linear system has (linear) dimension eleven. Now we count the conditions that a quartic in this linear system contains the eight lines $L_5, \ldots, L_8, L_5', \ldots, L_8'$ too.

Consider the three lines L_5, L_6 and L_7. All three of them meet Q_1 in two points on the base locus of the linear system. So containing these three lines imposes at most $3 \times 3 = 9$ conditions on the quartics in the linear system. Each quartic in the linear system containing L_5, L_6 and L_7 will meet the lines $L_5', \ldots, L_8'$ in ≥ 5 points: Two on Q_1 in the base locus of the linear system and three more points on L_5, L_6, L_7. Hence this quartic will contain the four lines $L_5', \ldots, L_8'$. But then it meets L_8 in ≥ 6 points and it will contain this line too.

We proved: the 16 lines lie in the intersection of two quartics $Q_1 \cdot Q_2$ and $Y \neq Q_1 \cdot Q_2$. Here Y cannot split off any of the two quadrics Q_i. Indeed, if $Y = Q_1 \cdot Q$ with another quadric Q, then Q intersecting Q_2 in the eight lines $L_5, \ldots, L_8'$ would coincide with Q_2 and Y would equal $Q_1 \cdot Q_2$. This implies that the 16 lines $L_1, \ldots, L_8'$ are the complete intersection $Q_1 \cdot Q_2 \cap Y$ of two quartics. The linear system of quartics on these 16 lines then must have (linear) dimension two.

We still have to show that the linear system contains a smooth quartic. By Bertini the general quartic in the system will be smooth outside of the base locus $L_1, \ldots, L_8'$. Any quartic $Y \neq Q_1 \cdot Q_2$ in the linear system cuts out on the quadric Q_1 the eight lines $L_1, \ldots, L_4'$. Therefore it must be smooth on these lines except perhaps in the points $L_i \cap L_j'$, $i, j = 1, \ldots, 4$. But through each of these points there passes the quartic $Q_1 \cdot Q_2$, which is smooth there. This implies that there is precisely one quartic in the linear system singular at one of these 16 points. Similarly, for each of the 16 points $L_i \cap L_j'$, $i, j = 5, \ldots, 8$ there is precisely one quartic singular there. Avoiding the 32 quartics singular at these points, the general quartic Y in the linear system will be smooth.

And finally we have to check the evenness condition for the lines $L_1, \ldots, L_8 \subset Y$:

$$E \sim L_5 + L_6 + L_3' + L_4'$$

defines an elliptic fibration on Y with

$$E.L_i = 2 \, (i = 1, \ldots, 4), \quad E.L_i = 0 \, (i = 5, \ldots, 8).$$

Hence the eight lines $L_1, \ldots, L_8$ form an even set by Proposition 3.

Observe that the pair $Q_1 + Q_2$ of quadrics is not the only such pair in the linear system: Each one of the four lines L_1, L_2, L_7, L_8 meets each line L_1', L_2', L_7', L_8'. So there is a quadric P_1 containing these eight lines. Similarly the eight lines $L_3, L_4, L_5, L_6, L'_3, L'_4, L'_5, L'_6$ lie on a quadric P_2. The pair of quadrics $P_1 + P_2$ is another quartic containing the 16 lines. There are four such pairs of quadrics in the linear system, together with $Q_1 + Q_2$ five such pairs.

Proposition 4.2. *Let Q be a quadric containing four lines from the set $L_1, \ldots, L_8$. Then it is one of the quadrics, an irreducible component of one of the five quartics just specified.*

Proof. Let L_i, L_j, L_k, L_l be the four lines contained in Q. No three of them can belong to $\{L_1, \ldots, L_4\}$ unless $Q = Q_1$. So assume L_i, L_j lie on Q_1 while L_k, L_l lie on Q_2. Let $|M|$ be the ruling of Q containing the lines $L_i, \ldots, L_l$ and $|M'|$ the other ruling. Since Q meets Q_1 in the lines L_i, L_j, the intersection $Q \cap Q_1$ will also contain some line M'_1 belonging to the rulings $|M'|$ on Q and R'_1 on Q_1. This line M'_1 meets L_k and $L_l \subset Q_2$ in points on I. Through any point in I there is just one line from R'_1. So M'_1 is one of the four lines $L'_1, \ldots, L'_4$. And another one of these four lines is a second transversal for $L_i, \ldots, L_l$. This shows $\{L_k, L_l\} = \{L_5, L_6\}$ or $\{L_7, L_8\}$. Similarly one finds $\{L_i, L_j\} = \{L_1, L_2\}$ or $\{L_3, L_4\}$. $\qquad\Box$

Up to now we have not yet excluded the possibility thet all the smooth quartics in the linear system are Kummer surfaces of the type [BN, G, T] containing 16 skew lines. This is done by the following proposition.

Proposition 4.3. *Only finitely many quartics Y in the linear system contain a line L which is skew with $L_1, \ldots, L_8$.*

Proof. Such a line L will meet the quadrics Q_1 and Q_2 in points on the lines L'_i but not on $L_1, \ldots, L_8$. There Y and the quadrics meet transversally with L'_i being the intersection of their tangent spaces. This shows that L will meet both quadrics Q_1 and Q_2 in two distinct points. Therefore L meets two distinct lines from $L'_1, \ldots, L'_4$ and two lines from $L'_5, \ldots, L'_8$. Either these four lines from $L'_1, \ldots, L'_8$ do not lie on a quadric. Then L is one of their (one or two) transversals. There are finitely many such transversals, and finitely many quartics in the linear system containing one of these transversals. Or these four lines from $L'_1, \ldots, L'_8$ lie on a quadric Q. This quadric is one of the four quadrics specified by prop. 4.2 (with the roles of the L_i and L'_i reversed) containing two quadrangles like L_1, L_2, L'_7, L'_8 and L_7, L_8, L'_1, L'_2. The quartic Y containing the additional line $L \subset Q$ then will split off Q. Hence Y will be one of the four pairs of quadrics in the linear system different from $Q_1 + Q_2$. $\qquad\Box$

5 Double Quadrics with Eight Nodes

A double quadric is a K3-surface Y, birational to a double cover $\sigma' : Y' \to \mathbb{P}_1 \times \mathbb{P}_1$ of a quadric. This is branched over a curve $S \subset \mathbb{P}_1 \times \mathbb{P}_1$ of type

22   W. Barth

(4,4). Here we assume that S is smooth but for eight nodes $s_1, \ldots, s_8 \in S$. They correspond to eight nodes $e_1, \ldots, e_8 \in Y'$. Let $Y \to Y'$ be the minimal resolution of these nodes and $\sigma : Y \to \mathbb{P}_1 \times \mathbb{P}_1$ the induced map. Denote by $E_i \subset Y$ the rational curve over e_i.

Proposition 5.1. *The set $E_1 + \cdots + E_8$ is even if and only if the branch curve S splits as the union $S_1 + S_2$ of two smooth curves $S_i \subset \mathbb{P}_1 \times \mathbb{P}_1$ of type (2,2).*

Proof. Denote by H the divisor class $\sigma^*(\mathcal{O}_{\mathbb{P}_1 \times \mathbb{P}_1}(1,1))$. Then $H^2 = 4$ and $H.E_i = 0$.

Assume first that the set $E_1 + \cdots + E_8$ is even, say $[E_1 + \cdots + E_8] = 2L$ in NS(Y). Consider the class $D := H - L$ on Y. From $D^2 = 0$ one deduces $\chi(\mathcal{O}_Y(D)) = 2$. Since $D.H = 4$, the class D and not $-D$ is effective. If the pencil D would have a fixed curve D_0, being invariant under the covering involution, $D_0 = \sigma^{-1}C_0$ would come from below. The moving part of $|D|$ then would have a class

$$\sigma^* \mathcal{O}_{\mathbb{P}_1 \times P_1}(1,0) - L, \quad \sigma^* \mathcal{O}_{\mathbb{P}_1 \times P_1}(0,1) - L, \quad \text{or} \quad - L.$$

All three classes have negative self-intersection and cannot move. So there is no fixed curve. By $D.E_i = 1$ the class D is primitive and therefore defines an elliptic fibration on Y.

Let $\bar{S} \subset Y$ be the proper transform of S. Its class is $2(H - L)$, so $\bar{S}$ consists of two fibres of $D_1 \neq D_2 \in |D|$. Let $C_i := \sigma(D_i)$, $i = 1, 2$. Since

$$D.\mathcal{O}_{P_1 \times \mathbb{P}_1}(1,0) = D.\mathcal{O}_{P_1 \times \mathbb{P}_1}(0,1) = 2,$$

both these curves have type $(2,2)$. Hence $S = C_1 + C_2$ is a union of two curves of type $(2,2)$. Both, D_1 and D_2 meet each E_i, therefore both, C_1 and C_2 pass through all the eight points s_i. This shows $C_1 \cap C_2 = \{s_1, \ldots, s_8\}$. As S is smooth outside these points, the curves C_1 and C_2 are smooth, the nodes on S being their intersections.

Assume now that $S = C_1 + C_2$ with two smooth curves C_i of type $(2,2)$. Then

$$\sigma^*(C_1) = 2D_1 - \sum_{1}^{8} n_i E_i$$

with a smooth rational curve $D_1 \subset Y$. As in section 1, from

$$0 = \sigma^* C_1.E_i = 2D_1.E_i - 2n_i \quad \text{and} \quad D_1.E_i = 1$$

we conclude $n_i = 1$ for $i = 1, \ldots, 8$. So

$$[E_1 + \cdots + E_8] = 2 \cdot ([D_1] - \sigma^* \mathcal{O}_{\mathbb{P}_1 \times \mathbb{P}_1}(1,1))$$

is even. $\qquad\square$

6 Double Quadrics with Eight Double Tangents

Let now $\sigma : Y \to \mathbb{P}_1 \times \mathbb{P}_1$ be a double quadric branched over a smooth curve $S \subset \mathbb{P}_1 \times \mathbb{P}_1$ of type (4,4). By a *non-degenrate double tangent* to S we mean a curve T on $\mathbb{P}_1 \times \mathbb{P}_1$ in one of the rulings, i.e. of type (1,0) or (0,1), touching S in two distinct points. Its inverse image in Y splits into two smooth rational curves L and L'.

As usual, denote by $S_Y \subset Y$ the curve lying over S.

Assume that S has eight distinct non-degenerate double tangents $T_1, \ldots, T_8$. Here we consider the question, whether the rational curves L_i, $L_i' \subset Y$ lying over them can be grouped such that the eight curves L_i are disjoint and form an even set.

Let $T_1, \ldots, T_k$ be curves of type (1,0), while $T_{k+1}, \ldots, T_8$ have type (0,1). W.l.o.g. we may assume $k \geq 1$.

Proposition 6.1. *If the set $\{L_1, \ldots, L_8\}$ is disjoint and even, then $k = 4$ and the divisor of contact points is cut out on S by a curve of bi-degree (2,2).*

Proof. Let Λ on S be the divisor formed by the 16 points of contact on the eight tangents, i.e. 2Λ is cut out by $T_1 + \cdots + T_8$,

$$2[\Lambda] = \mathcal{O}_S(k, 8 - k) .$$

By assumption there is a class L on Y with

$$2L = [L_1 + \cdots + L_8].$$

Let $\iota : Y \to Y$ be the covering involution and put $L' := \iota^* L$. Then $L + L'$ comes from $\mathbb{P}_1 \times \mathbb{P}_1$. Since

$$2(L + L') = [L_1 + \cdots + L_8 + L_1' + \cdots + L_8'] = \sigma^* \mathcal{O}_{\mathbb{P}_1 \times \mathbb{P}_1}(k, 8 - k) ,$$

$k = 2l$ is even, and $L + L' = \sigma^* \mathcal{O}_{\mathbb{P}_1 \times \mathbb{P}_1}(l, 4 - l)$. Since $L_1.(L + L') = -2 + 8 - k$ we have $6 - 2l = 4 - l$, i.e. $l = 2$ and $k = 4$. Now $L|S_Y = L'|S_Y$, so

$$L|S_Y = \sigma^* \mathcal{O}_{S_Y}(1, 1) \otimes \mu$$

with some 2-torsion element $\mu \in \mathrm{Pic}(S_Y)$. This implies

$$[\sigma^* \Lambda] = 2L = (\sigma^* \mathcal{O}_Y(1, 1) \otimes \mu)^{\otimes 2} = \sigma^* \mathcal{O}_Y(2, 2)$$

and Λ is cut out on S by a curve of type $(2, 2)$. $\qquad\square$

Denote by V the curve of bidegree (2.2) from Prop. 6.1. As usual, there are equations $t_i = 0$ for T_i and $v = 0$ for V such that with $t = t_1 \cdot \ldots \cdot t_8$ the polynomial $t - v^2$ of bidegree (4,4) vanishes on S. Hence

$$t - v^2 = 0$$

is an equation for S. In this way, curves S with eight bitangents, the points of contact cut out by a curve of type (2,2), are easily constructed. I do not want to discuss here the question, whether the covering defined by S indeed contains an even set of eight rational curves lying over the bitangents.

24   W. Barth

7 Comments

1) In the 20-dimensional universe of K3-surfaces Y, it is one condition, that a class in $H^2(Y, \mathbb{Z})$ is represented by an algebraic curve. The surfaces, carrying a polarization H of fixed degree and in addition eight disjoint rational curves, are parametrized by some moduli space of dimension

$$20 - 9 = 11 \;,$$

because $H, L_1, \ldots, L_8$ span a lattice of rank 9. This is so independently from the fact, whether the eight rational curves form an even set or not.

Counting constants in Cases 1–4 (subtracting the dimension of the group acting) one indeed finds this number of eleven parameters. Although the situation *seems* more special than when the evenness-condition is not imposed, this turns out not to be the case.

In Cases 5 and 6 his number of parameters is ten, due to the appearance of a tenth independent algebraic class.

2) Of course, the proof of Proposition 2.2 is quite unsatisfactory. The reason is that the evenness condition is formulated in terms of divisors on the branch curve S. I did not find a way to pass from $S = S_Y$ to the surface Y and translate the condition into a condition for divisors on this surface, but for the proof by example of Proposition 2.2. A direct, preferable proof seems to me a major problem.

3) The next simple cases of K3-surfaces appearing geometrically, carrying a polarization of degree six (complete intersections of type (2,3) in $\mathbb{P}_4$) or eight (complete intersections of type (2,2,2) in $\mathbb{P}_5$) with eight nodes or carrying eight skew lines could also be studied. It would also be interesting to find evenness conditions in mixed cases, say k nodes and $8 - k$ lines.

4) I believe the results of this note to be new. At least I don't know any place where to find one of them. I thought it worth-while analyzing these situations, because evenness conditions or analogs thereof seem to be a key to questions like the maximal number of nodes on a surface of given degree in $\mathbb{P}_3$, or the maximal number of lines on such a surface. Of course, these questions are settled for K3-surfaces and start getting interesting in degree five (lines) or seven (nodes). Such statements will certainly be much harder to prove.

References

[BB] Barth, W., Bauer, Th., Poncelet theorems, *Expo. Math.* **14** (1996), 125–144.
[BM] Barth, W., Moore, R., On rational plane sextics with six tritangents, Volume in honour of M. Nagata, 45–58, 1987.
[BN] Barth, W., Nieto, I., Abelian surfaces of type (1,3) and quartic surfaces with 16 skew lines, *JAG* **3** (1994), 173–222.

[BPV] Barth, W., Peters, C., Van de Ven, A., Compact complex surfaces, *Ergeb. Math.* **4** (3) (1984), Springer.

[G] Godeaux, L., Sur la surface du quatrieme ordre contenant trente-deux droites, *Bull. Acad. Roy. Belg. V.s.* **25** (1939), 539–552.

[N] Nikulin, V. V., On Kummer surfaces, *Math. USSR-Izv.* **9**(2) (1975), 261–275.

[T] Traynard, M., Sur les fonctions theta de deux variables et les surfaces hyperelliptiques, *Ann. Sci. ENS III* **24** (1907), 77–177.

A Reduction Map for Nef Line Bundles

Thomas Bauer[1], Frédéric Campana[2], Thomas Eckl[1], Stefan Kebekus[1], Thomas Peternell[1], Sławomir Rams[3], Tomasz Szemberg[4], and Lorenz Wotzlaw[5]

[1] Institut für Mathematik, Universität Bayreuth, 95440 Bayreuth, Germany
E-mail addresses: `thomas.bauer@uni-bayreuth.de`
`thomas.eckl@uni-bayreuth.de`
`stefan.kebekus@uni-bayreuth.de`
`thomas.peternell@uni-bayreuth.de`
[2] Département de Mathématiques, Université Nancy 1, BP 239, 54507 Vandoeuvre-les-Nance Cédex, France
E-mail address: `campana@iecn.u-nancy.fr`
[3] Mathematisches Institut der Universität, Bismarckstrasse $1\frac{1}{2}$, 91054 Erlangen, Germany
E-mail address: `rams@mi.uni-erlangen.de`
[4] Universität GH Essen, Fachbereich 6 Mathematik, 45117 Essen, Germany
E-mail address: `mat905@uni-essen.de`
[5] Mathematisches Institut, Humboldt-Universität Berlin, 10099 Berlin, Germany
E-mail address: `wotzlaw@mathematik.hu-berlin.de`

Table of Contents

1. Introduction 27
2. A Reduction Map for Nef Line Bundles 28
   2.1 Construction of the Reduction Map 28
   2.2 Nef Cohomology Classes 31
   2.3 The Nef Dimension 32
   2.4 The Structure of the Reduction Map 32
3. A Counterexample 34
   References 36

1 Introduction

In [Ts00], H. Tsuji stated several very interesting assertions on the structure of pseudo-effective line bundles L on a projective manifold X. In particular he postulated the existence of a meromorphic "reduction map", which essentially says that through the general point of X there is a maximal irreducible L-flat subvariety. Moreover the reduction map should be almost holomorphic, i.e. has compact fibers which do not meet the indeterminacy locus of the reduction map. The proofs of [Ts00], however, are extremely difficult to follow.

2000 *Mathematics Subject Classification*: 14E05, 14J40, 14J60.

28   Th. Bauer et al.

The purpose of this note is to establish the existence of a reduction map in the case where L is nef and to prove that it is almost holomorphic —this was also stated explicitly in [Ts00]. Our proof is completely algebraic while [Ts00] works with deep analytic methods. Finally, we show by a basic example that in the case where L is only pseudo-effective, the postulated reduction map cannot be almost holomorphic —in contrast to a claim in [Ts00].

These notes grew out from a small workshop held in Bayreuth in January 2001, supported by the Schwerpunktprogramm "Global methods in complex geometry" of the Deutsche Forschungsgemeinschaft.

2   A Reduction Map for Nef Line Bundles

In this section we want to prove the following structure theorem for nef line bundles on a projective variety.

Theorem 2.1. *Let L be a nef line bundle on a normal projective variety X. Then there exists an almost holomorphic, dominant rational map $f : X \dashrightarrow Y$ with connected fibers, called a "reduction map" such that*

1. *L is numerically trivial on all compact fibers F of f with $\dim F = \dim X - \dim Y$*
2. *for every general point $x \in X$ and every irreducible curve C passing through x with $\dim f(C) > 0$, we have $L \cdot C > 0$.*

The map f is unique up to birational equivalence of Y.

This theorem was stated without complete proof in Tsuji's paper [Ts00]. Relevant definitions are given now.

Definition 2.2. Let X be an irreducible reduced projective complex space (projective variety, for short). A line bundle L on X is numerically trivial, if $L \cdot C = 0$ for all irreducible curves $C \subset X$. The line bundle L is nef if $L \cdot C \geq 0$ for all curves C.

Let $f : Y \to X$ be a surjective map from a projective variety Y. Then clearly L is numerically trivial (nef) if and only $f^*(L)$ is.

Definition 2.3. Let X and Y be normal projective varieties and $f : X \dashrightarrow Y$ a rational map and let $X^0 \subset X$ be the maximal open subset where f is holomorphic. The map f is said to be almost holomorphic if some fibers of the restriction $f|_{X^0}$ are compact.

2.1   Construction of the Reduction Map

2.1.1 A Criterion for Numerical Triviality In order to prove Theorem 2.1 and construct the reduction map, we will employ the following criterion for a line bundle to be numerically trivial.

A Reduction Map for Nef Line Bundles   29

Theorem 2.4. *Let X be an irreducible projective variety which is not necessarily normal. Let L be a nef line bundle on X. Then L is numerically trivial if and only if any two points in X can be joined by a (connected) chain C of curves such that $L \cdot C = 0$.*

In the remaining part of the present section we will prove Theorem 2.4. The proof will be performed by a reduction to the surface case. The argumentation is then based on the following statement, which, in the smooth case, is a simple corollary to the Hodge Index Theorem.

Proposition 2.5. *Let S be an irreducible, projective surface which is not necessarily normal, and let $q : S \to T$ be a morphism with connected fibers onto a curve. Assume that there exists a nef line bundle $L \in \mathrm{Pic}(S)$ and a curve $C \subset S$ such that $q(C) = T$ and such that*

$$L \cdot F = L \cdot C = 0$$

holds, where F is a general q-fiber. Then L is numerically trivial.

Proof. If S is smooth, set $D = C + nF$, where n is a large positive integer. Then we have $D^2 > 0$. By the Hodge Index Theorem it follows that

$$(L \cdot D)^2 \geq L^2 \cdot D^2 \ ,$$

hence $L^2 = 0$, since by our assumptions $L \cdot D = 0$. So equality holds in the Index Theorem and therefore L and D are proportional: $L \equiv kD$ for some rational number k. Since $0 = L^2 = k^2 D^2$ and $D^2 > 0$, we conclude that $k = 0$. That ends the proof in the smooth case.

If S is singular, let $\delta : \tilde{S} \to S$ be a desingularisation of S and let $\tilde{C} \subset \tilde{S}$ be a component of $\delta^{-1}(C)$ which maps surjectively onto C. Note that the fiber of $q \circ \delta$ need no longer be connected and consider the Stein factorisation

$$
\begin{array}{ccc}
\tilde{S} & \xrightarrow{\ \ \delta\ \ } & S \\
& \text{desing.} & \\
\tilde{q} \downarrow & & \downarrow q \\
\tilde{T} & \xrightarrow[\text{finite}]{} & T \ .
\end{array}
$$

It follows immediately from the construction that $\tilde{q}(\tilde{C}) = \tilde{T}$, that $\delta^*(L)$ has degree 0 on $\tilde{C}$ and on the general fiber of $\tilde{q}$. The argumentation above therefore yields that $\delta^*(L)$ is trivial on $\tilde{S}$. The claim follows. $\qquad\square$

2.1.2 Proof of Theorem 2.4 If L is numerically trivial, the assertion of theorem 2.4 is clear. We will therefore assume that any two points can be connected by a curve which intersects L with multiplicity 0, and we will show that L is numerically trivial. To this end, choose an arbitrary, irreducible curve $B \subset X$. We are finished if we show that $L \cdot B = 0$.

30   Th. Bauer et al.

Let $a \in X$ be an arbitrary point which is not contained in B. For any $b \in B$ we can find by assumption a connected, not necessarily irreducible, curve Z_b containing a and b such that $L \cdot Z_b = 0$. Since the Chow variety has compact components and only a countable number of components, we find a family $(Z_t)_{t \in T}$ of curves, parametrised by a compact irreducible curve $T \subset \mathrm{Chow}(X)$ such that for every point $b \in B$, there exists a point $t \in T$ such that the curve Z_t contains both a and b. We consider the universal family $S \subset X \times T$ over T together with the projection morphisms

$$S \xrightarrow{\;p\;} X$$
$$q \downarrow $$
$$T $$

Claim 1. *There exists an irreducible component $S_0 \subset S$ such that $p^*(L)$ is numerically trivial on S_0.*

Proof of Claim 1. As all curves Z_t contain the point a, the surface S contains the curve $\{a\} \times T$. Let $S_0 \subset S$ be a component which contains $\{a\} \times T$. Since $\{a\} \times T$ intersects all fibers of the natural projection morphism q, and since $p^*(L)$ is trivial on $\{a\} \times T$, an application of Proposition 2.5 yields the claim. $\qquad\square$

Claim 2. *The bundle $p^*(L)$ is numerically trivial on S.*

Proof of Claim 2. We argue by contradiction and assume that there are components $S_j \subset S$ where $p^*(L)$ is not numerically trivial. We can therefore subdivide S into two subvarieties as follows:

$$S_{\mathrm{triv}} := \{\text{union of the irreducible components } S_i \subset S$$
$$\text{where } p^*(L)|_{S_i} \text{ is numerically trivial}\}$$
$$S_{\mathrm{ntriv}} := \{\text{union of the irreducible components } S_i \subset S$$
$$\text{where } p^*(L)|_{S_i} \text{ is not numerically trivial}\}.$$

By assumption, and by Claim 1, both varieties are not empty. Since S is the universal family over a curve in $\mathrm{Chow}(X)$, the morphism q is equidimensional. In particular, since all components $S_i \subset S$ are two-dimensional, every irreducible component S_i maps surjectively onto T. Thus, if $t \in T$ is a general point, the connected fiber $q^{-1}(t)$ intersects both S_{triv} and S_{ntriv}. Thus, over a general point $t \in T$, there exists a point in $S_{\mathrm{triv}} \cap S_{\mathrm{ntriv}}$. It is therefore possible to find a curve $D \subset S_{\mathrm{triv}} \cap S_{\mathrm{ntriv}}$ which dominates T.

That, however, contradicts Proposition 2.5: on one hand, since $D \subset S_{\mathrm{triv}}$, the degree of $p^*(L)|_D$ is 0. On the other hand, we can find an irreducible component $S_j \subset S_{\mathrm{ntriv}} \subset S$ which contains D. But because $p^*(L)$ has degree 0 on the fibers of $p^*(L)|_{S_j}$, Proposition 2.5 asserts that $p^*(L)$ is numerically trivial on S_j, contrary to our assumption. This ends the proof of Claim 2. $\square$

We apply Claim 2 as follows: if $B' \subset S$ is any component of the preimage $p^{-1}(B)$, then $p^*(L).B' = 0$. That shows that $L.B = 0$, and the proof of theorem 2.4 is done. $\qquad\square$

2.1.3 Proof of Theorem 2.1 In order to derive Theorem 2.1 from Theorem 2.4, we introduce an equivalence relation on X with setting $x \sim y$ if x and y can be joined by a connected curve C such that $L \cdot C = 0$. Then by [Ca81] or [Ca94, Appendix], there exists an almost holomorphic map $f : X \dashrightarrow Y$ with connected fibers to a normal projective variety Y such that, two general points x and y satisfy $x \sim y$ if and only if $f(x) = f(y)$. This map f gives the fibration we are looking for.

If F is a general fiber, then $L|F \equiv 0$ by Theorem 2.4.

We still need to verify that $L \cdot C = 0$ for all curves C contained in an *arbitrary* compact fiber F_0 of dimension $\dim F_0 = \dim X - \dim Y$. To do that, let H be an ample line bundle on X and pick

$$D_1, \ldots, D_k \in |mH|$$

for m large such that

$$D_1 \cdot \ldots \cdot D_k \cdot F_0 = C + C'$$

with an effective curve C'. Then

$$L \cdot (C + C') = L \cdot D_1 \cdot \ldots \cdot D_k \cdot F$$

with a general fiber F of f, hence $L \cdot (C + C') = 0$. Since L is nef, we conclude $L \cdot C = 0$. $\qquad\square$

2.2 Nef Cohomolgy Classes

In Theorems 2.1 and 2.4 we never really used the fact that L is a line bundle; only the property that $c_1(L)$ is a nef class is important and even rationality of the class does not play any role. Hence our results directly generalize to nef cohomology classes of type $(1,1)$. To be precise, we fix a projective manifold (we stick to the smooth case for sakes of simplicity) and we say that a class $\alpha \in H^{1,1}(X, \mathbb{R})$ is nef, if it is in the closure of the cone generated by the Kähler classes. Moreover α is numerically trivial, if $\alpha \cdot C = 0$ for all curves $C \subset X$.

If $Z \subset X$ is a possibly singular subspace, then we say that α is numerically trivial on Z, if for some (and hence for all, see [Pa98]) desingularisation $\hat{Z} \to Z$, the induced form $f^*(\alpha)$ is numerically trivial on $\hat{Z}$, i.e. $f^*(\alpha) \cdot C = 0$ for all curves $C \subset \hat{Z}$. Here $f : \hat{Z} \to X$ denotes the canonical map. Similarly we define α to be nef on Z. If Z is smooth, this is the same as to say that $\alpha|Z$ is a nef cohomology class in the sense that $\alpha|Z$ is in the closure of the Kähler cone of Z.

32   Th. Bauer et al.

Theorem 2.6. *Let α be a nef cohomology class on a smooth projective variety X. Then there exists an almost holomorphic, dominant rational map $f : X \dashrightarrow Y$ with connected fibers, such that*

1. *α is numerically trivial on all compact fibers F of f with $\dim F = \dim X - \dim Y$*
2. *for every general point $x \in X$ and every irreducible curve C passing through x with $\dim f(C) > 0$, we have $\alpha \cdot C > 0$.*

The map f is unique up to birational equivalence of Y.

   In particular, if two general points of X can be joined by a chain C of curves such that $\alpha \cdot C = 0$, then $\alpha \equiv 0$.

2.3   The Nef Dimension

Since Y is unique up to a birational map, its dimension $\dim Y$ is an invariant of L which we compare to the other known invariants.

Definition 2.7. The dimension $\dim Y$ is called the nef dimension of L. We write $n(L) := \dim Y$.

As usual we let $\nu(L)$ be the numerical Kodaira dimension of L, i.e. the maximal number k such that $L^k \not\equiv 0$. Alternatively, if H is a fixed ample line bundle, then $\nu(L)$ is the maximal number k such that $L^k \cdot H^{n-k} \neq 0$.

Proposition 2.8. *The nef dimension is never smaller than the numerical Kodaira dimension:*
$$\nu(L) \leq n(L) \,.$$

Proof. Fix a very ample line bundle $H \in \mathrm{Pic}(X)$ and set $\nu := \nu(L)$. Let Z be a general member cut out by $n - \nu$ elements of $|H|$. The dimension of Z will thus be $\dim Z = \nu$, and since $L^\nu \cdot H^{n-\nu} > 0$, the restriction $L|_Z$ is big (and nef). Consequence: $\dim f(Z) = \nu$, since otherwise Z would be covered by curves C which are contained in general fibers of f, so that $L \cdot C = 0$, contradicting the bigness of $L|_Z$. In particular, we have $\dim Y \geq \dim f(Z) = \nu$ and our claim is shown. $\square$

Corollary 2.9. *The nef dimension is never smaller than the Kodaira dimension:*
$$\kappa(L) \leq n(L) \,.$$

2.4   The Structure of the Reduction Map

Again let L be a nef line bundle on a projective manifold and $f : X \to Y$ a reduction map. In this section we will shed light on the structure of the map in a few simple cases.

   At the present time, however, we cannot say much about f in good generality. The following natural question is, to our knowledge, open.

Question 2.10. Are there any reasonable circumstances under which the reduction map can be taken holomorphic? Is it possible to construct an example where the reduction map cannot be chosen to be holomorphic?

Of course, the abundance conjecture implies that the f can be chosen holomorphic in the case where $L = K_X$. Likewise, we expect a holomorphic reduction map when $L = -K_X$.

2.4.1 The Case Where L is Big If L is big and nef, then $n(L) = \dim X$. The converse however is false: there are examples of surfaces X carrying a line bundle L such that $L \cdot C > 0$ for all curves C but $L^2 = 0$. So in that case $n(L) = 2$, but $\nu(L) = 1$. The first explicit example is due to Mumford, see [Ha70]. This also shows that equality can fail in Proposition 2.8.

2.4.2 The Case Where $n(L) = \dim X = 2$ In this case, we have $L \cdot C > 0$ for all but an at most countable number of irreducible curves $C \subset X$. These curves necessarily have semi-negative self-intersection $C^2 \leq 0$.

2.4.3 The Case Where $n(X) = 1$ and $\dim X = 2$ Here we can choose Y to be a smooth curve. The almost-holomorphic reduction $f : X \to Y$ will thus be holomorphic. In this situation, we claim that the numerical class of a suitable multiple of L comes from downstairs:

Proposition 2.11. *If $n(X) = 1$, then there exists a $\mathbb{Q}$-divisor A on Y such that*
$$L \equiv f^*(A).$$

Proof. Since L is nef, it carries a singular metric with positive curvature current T, in particular $c_1(L) = [T]$. Since $L \cdot F = 0$, and since every $(1,1)-$form φ on Y is closed, we conclude that
$$f_*(T)(\varphi) = T(f^*(\varphi)) = [T] \cdot [f^*(\varphi)] = 0,$$
since $[f^*(\varphi)] = \lambda[F]$ for a fiber F of f.

Hence $f_*(T) = 0$. Let Y_0 be the maximal open subset of Y over which f is a submersion and let $X_0 = f^{-1}(Y_0)$. Then by [HL83, (18)]
$$T|X_0 = \sum \mu_i F_i,$$
where F_i are fibers of $f|_{X_0}$ and $\mu_i > 0$. Hence $\mathrm{supp}(T - \sum \mu_i F_i) \subset X \setminus X_0$, an analytic set of dimension 1. Hence by a classical theorem (see e.g. [HL83]),
$$T - \sum \mu_i F_i = \sum \lambda_j G_j$$

34   Th. Bauer et al.

where G_j are irreducible components of $X \setminus X_0$, i.e. of fibers, of dimension 1 and where $\lambda_j > 0$. Since $(\sum \lambda_j G_j)^2 = 0$, Zariski's lemma shows that $\sum \lambda_j G_j$ is a multiple of fibers, whence $L \equiv f^*(A)$ for a suitable $\mathbb{Q}$-divisor A on B.

Here is an algebraic proof, which however does not easily extend to higher dimensions as the previous (see 2.4.4).
Fix an ample line bundle A on X and choose a positive rational number m such that $L \cdot A = mF \cdot A$. Let $D = mF$. Then $L^2 = L \cdot D = D^2$ and $L \cdot A = D \cdot A$. Introducing $r = L^2$ and $d = L \cdot A$, the ample line bundles $L' = L + A$ and $D' = D + A$ fulfill the equation

$$(L' \cdot D')^2 = (r + 2d + A^2)^2 = (L')^2 (D')^2,$$

hence L and D are numerically proportional, and so do L and D. Consequently $L \equiv f(A)$ as before.   □

2.4.4 The Case When $n(L) = 1$ and $\dim X$ is Arbitrary The considerations of Sect. 2.4.3 easily generalize to higher dimensions: if X is a projective manifold and L a nef line bundle on X with $n(L) = 1$, then the reduction map $f : X \to Y$ is holomorphic and there exists a $\mathbb{Q}$-divisor on Y such that $L \equiv f^*(A)$.

2.4.5 The Case Where $n(L) = 0$ We have $n(L) = 0$ if and only if $L \equiv 0$.

3   A Counterexample

In [Ts00], H. Tsuji claims the following:

Claim 3.1 *Let (L, h) be a line bundle with a singular hermitian metric on a smooth projective variety X such that the curvature current Θ_h is non-negative. Then there exists a (up to birational equivalence) unique rational fibration*

$$f : X -- \to Y$$

such that

(a) *f is regular over the generic point of Y,*
(b) *$(L, h)_{|F}$ is well defined and numerically trivial for every very general fiber F, and*
(c) *$\dim Y$ is minimal among such fibrations.*

Tsuji calls a pair (L, h) numerically trivial if for every irreducible curve C on X, which is not contained in the singular locus of h,

$$(L, h).C = 0$$

holds. The intersection number has the following

A Reduction Map for Nef Line Bundles 35

Definition 3.2. Let (L, h) be a line bundle with a singular hermitian metric on a smooth projective variety X such that the curvature current Θ_h is non-negative. Let C be an irreducible curve on X such that the natural morphism $\mathcal{I}(h^m) \otimes \mathcal{O}_C \to \mathcal{O}_C$ is an isomorphism at the generic point of C for every $m \geq 0$. Then

$$(L, h).C := \limsup_{m \to \infty} \frac{\dim H^0(C, \mathcal{O}_C(mL) \otimes \mathcal{I}(h^m)/\text{tor})}{m} \, ,$$

where tor denotes the torsion part of $\mathcal{O}_C(mL) \otimes \mathcal{I}(h^m)$.

Of course, Claim 3.1 is trivial as stated —except for the uniqueness assertion. What is meant is that every curve C with $(L, h).C = 0$ is contracted by f. This is also clear from Tsuji's construction.

In order to make sense of Claim 3.1, statement (c) must be given the following meaning: if $x \in X$ is very general, then $(L, h) \cdot C > 0$ for all curves C through x.

There is an easy counterexample to the Claim 3.1: Take $X = \mathbb{P}^2$ with homogeneous coordinates $(z_0 : z_1 : z_2)$, $L = \mathcal{O}(1)$ and let h be induced by the incomplete linear system of lines passing through $(1 : 0 : 0)$. Then the corresponding plurisubharmonic function ϕ is given by $\frac{1}{2} \log(|z_1|^2 + |z_2|^2)$.

By [De00, 5.9], it is possible to reduce the calculation of $\mathcal{I}(h^m)$ to an algebraic problem: Since the ideal sheaf $\mathcal{J}_p$ generated by z_1, z_2 describes the reduced point $p = (1 : 0 : 0)$, it is integrally closed. Let $\mu : \mathbb{F}_1 \to \mathbb{P}^2$ be the blow up of $\mathbb{P}^2$ in p. Then $\mu^* \mathcal{J}_p$ is the invertible sheaf $\mathcal{O}(-E)$ associated with the exceptional divisor E. Now, one has $K_{\mathbb{F}_1} = \mu^* K_{\mathbb{P}^2} + E$. By the direct image formula in [De00, 5.8] it follows

$$\mathcal{I}(m\phi) = \mu_*(\mathcal{O}(K_{\mathbb{F}_1} - \mu^* K_{\mathbb{P}^2}) \otimes \mathcal{I}(m\phi \circ \mu)) \, .$$

Now, $(z_i \circ \mu)$ are generators of the ideal $\mathcal{O}(-E)$, hence

$$m\phi \circ \mu \sim m \log g \, ,$$

where g is a local generator of $\mathcal{O}(-E)$. But $\mathcal{I}(m\phi \circ \mu)) = \mathcal{O}(-mE)$, hence

$$\mathcal{I}(m\phi) = \mu_* \mathcal{O}_{\mathbb{F}_1}((1 - m)E) = \mathcal{J}_p^{m-1} \, .$$

Let $C \subset \mathbb{P}^2$ be a smooth curve of degree d and genus g. If $p \notin C$ then $\mathcal{O}_C(mL) \otimes \mathcal{J}_p^{m-1} = \mathcal{O}_C(mL)$, and $(L, h).C = L.C = d$. On the other hand, if $p \in C$, then

$$\mathcal{O}_C(mL) \otimes \mathcal{J}_p^{m-1} = \mathcal{O}_C(mL - (m - 1)p) \, .$$

The degree of this invertible sheaf is $md - m + 1 = m(d - 1) + 1$, hence by Riemann-Roch

$$\limsup_{m \to \infty} \frac{\dim H^0(C, \mathcal{O}_C(mL) \otimes \mathcal{I}(m\phi))}{m} = \limsup_{m \to \infty} \frac{m(d - 1) + 1 + 1 - g}{m}$$
$$= d - 1 \, .$$

36   Th. Bauer et al.

Consequently, Tsuji's fibration could only be the trivial map $\mathbb{P}^2 \to \mathbb{P}^2$: It can't be the map to a point, because there are curves with intersection number ≥ 1 on $\mathbb{P}^2$, and it can't be an almost holomorphic rational map to a curve, because $\mathbb{P}_2$ does not contain curves with vanishing self-intersection. This contradicts condition (c) in the claim.

It remains an open question if claim 3.1 is true without condition (a), and if in the case of singular hermitian metrics induced by linear systems, the fibration may be taken as the induced rational map.

References

[Ca81] F. Campana, Coréduction algébrique d'un espace analytique faiblement kählérien compact, *Inv. Math.* **63** (1981), 187–223.

[Ca94] F. Campana, Remarques sur le revetement universel des variétiés kähleriennes compactes, *Bull. SMF* **122** (1994), 255–284.

[De00] J.P. Demailly, Vanishing theorems and effective results in algebraic geometry, Trieste lecture notes, 2000.

[Ha70] R. Hartshorne, Ample subvarieties of algebraic varieties, *Lecture Notes in Mathematics* **156** (1970), Springer.

[HL83] R. Harvey, H.B. Lawson, An intrinsic characterization of Kähler manifolds, *Invent. Math.* **74** (1983), 169–198.

[Pa98] M. Paun, Sur l'effectivité numérique des images inverses de fibrés en droites, *Math. Ann.* **310** (1998), 411–421.

[Ts00] H. Tsuji, Numerically trivial fibrations, LANL-preprint math.AG/0001023, October 2000.

Canonical Rings of Surfaces Whose Canonical System has Base Points

Ingrid C. Bauer, Fabrizio Catanese, and Roberto Pignatelli *

Mathematisches Institut der Universität Bayreuth, Lehrstuhl Mathematik VIII,
Universitätstr. 30, 95447 Bayreuth
E-mail addresses: Ingrid.Bauer@uni-bayreuth.de
       Fabrizio.Catanese@uni-bayreuth.de
       Roberto.Pignatelli@uni-bayreuth.de

Table of Contents

0 Introduction... 37
1 Canonical Systems with Base Points......................... 42
2 The Canonical Ring of Surfaces with $K^2 = 7$, $p_g = 4$ Birational
  to a Sextic: From Algebra to Geometry...................... 49
3 The Canonical Ring of Surfaces with $K^2 = 7$, $p_g = 4$ Birational
  to a Sextic: Explicit Computations.......................... 58
  3.1 Rolling Factors Format.................................. 58
  3.2 The $AM(^tA)$-format.................................... 59
  3.3 The Antisymmetric and Extrasymmetric Format......... 62
4 An Explicit Family... 63
  References... 67
  Appendix 1... 69
  Appendix 2... 70

0 Introduction

In Enriques' book on algebraic surfaces ([Enr]), culminating a research's lifespan of over 50 years, much emphasis was set on the effective construction of surfaces, for instance of surfaces with $p_g = 4$ and whose canonical map is a birational map onto a singular surface Σ in $\mathbb{P}^3$.

The problem of the effective construction of such surfaces for the first open case $K^2 = 7$ has attracted the attention of several mathematicians, and special constructions have been obtained by Enriques [Enr], Franchetta [Fran], Maxwell [Max], Kodaira [Kod]. Until Ciliberto [Cil1] was able to construct

---

* The present cooperation took place in the realm of the DGF-Forschungsschwerpunkt "Globale Methoden in der komplexen Geometrie" and of the EAGER projectd. The third author would like to thank the University of Warwick for generous support and hospitality.
2000 *Mathematics Subject Classification*: 14J29, 14J10, 13C40, 13H10.

an irreducible Zariski open set of the moduli space of (minimal algebraic) surfaces with $K^2 = 7$, $p_g = 4$, and constituted by surfaces with a birational canonical morphism whose image Σ has ordinary singularities.

Later on, through work of the first two named authors and of Zucconi ([Ba], [Cat1], [Zuc]), the complete classification of surfaces with $K^2 = 7$, $p_g = 4$ was achieved, and it was shown in [Ba] that the moduli space consists of three irreducible components (two of them consist of surfaces with non birational canonical map). But, as in the previous work of Horikawa [HorI-V], [HorQ] who classified surfaces with $K^2 = 5, 6$, $p_g = 4$, a complete picture of the moduli space is missing (for instance, it is still an open question whether the moduli space for $K^2 = 7$, $p_g = 4$ has one or two connected components).

Usually, classifying surfaces with given invariants K^2, p_g, is achieved by writing a finite number of families such that every such surface occurs in precisely one of those families. Each family yields a locally closed stratum of the moduli space, and the basic question is how are these strata patched together.

Abstract deformation theory is very useful since Kuranishi's theorem [Kur] gives a lower bound for the local dimension of the moduli space, thus it helps to decide which strata are dominating a component of the moduli space.

In principle, the local structure of the moduli space [Pal] is completely described by a sequence of Massey products on the tangent cohomology of the surface, and Horikawa clarified the structure of the moduli space in the "easy" case of numerical 5-ics ($K^2 = 5, p_g = 4$) by using the Lie bracket $H^1(S, \Theta_S) \times H^1(S, \Theta_S) \to H^2(S, \Theta_S)$.

However, the analytic approach does not make us see concretely how do surfaces belonging to one family deform to surfaces in another family, and therefore Miles Reid, in the Montreal Conference of 1980 proposed to look at the deformations of the canonical rings for numerical 5-ics (cf. [Rei0]).

His program was carried out by E. Griffin [Gri] in this case, later on D. Dicks found an interesting approach to the question and applied it to the case of surfaces with $K^2 = 4, p_g = 3$ [Dic1], [Dic2]. His method was clearly exposed in an article by Miles Reid [Rei2], where he set as a challenge the problem to apply these methods to the hitherto still partially unexplored case of surfaces with $K^2 = 7$, $p_g = 4$.

In [Cat1] was given a method (of the so called quasi generic canonical projections) allowing in principle to describe the canonical rings of surfaces of general type. The method works under the assumption that the surface admits a morphism to a 3-dimensional projective space which is a projection of the canonical map, and is birational to its image Σ.

What happens when the canonical system has base points, in particular in our case of surfaces with $K^2 = 7$, $p_g = 4$?

Thus the first aim of this paper is to introduce a general method to calculate the canonical ring of minimal surfaces of general type whose canonical system has base points but yields a birational canonical map.

We will then apply this method in the case of the surfaces S with $K^2 = 7$, $p_g = 4$. We will compute the canonical ring of those minimal smooth algebraic surfaces S with $K^2 = 7$, $p_g = 4$, whose canonical system has just one simple base point and gives a birational map from S onto a sextic in $\mathbb{P}^3$: this is the only case, for these values of K^2, p_g, where the canonical system has base points, but yields a birational map.

What does it mean to compute a ring? As a matter of fact, using the computer algebra program Macaulay II, we will give three different descriptions of the above canonical rings. These presentations will allow us to deform explicitly the canonical ring of such a minimal surface (with $K^2 = 7$, $p_g = 4$ and with birational canonical map onto a sextic in $\mathbb{P}^3$) to the canonical ring of a surface with the same invariants but with base point free canonical system.

That these deformations should exist was already seen in [Ba], since it was proven there that the surfaces with $K^2 = 7$, $p_g = 4$ such that the canonical system gives a birational map from S onto a sextic in $\mathbb{P}^3$ form an irreducible family of dimension 35 in the moduli space $\mathfrak{M}_{K^2=7,p_g=4}$ and therefore they cannot dominate an irreducible component of the moduli space (by Kuranishi's theorem the dimension of $\mathfrak{M}_{K^2=7,p_g=4}$ in any point has to be at least $10\chi - 2K^2 = 36$).

Therefore it was clear from the classification given in [Ba] that this family has to be contained in the irreducible component of the moduli space whose general point corresponds to a surface with base point free canonical system (obviously then with birational canonical morphism).

Enriques [Enr] proposed to obtain this deformation starting by the surface of degree seven, union of the sextic surface (the canonical image) together with the plane containing the double curve: in our case, however, we see that the canonical images of degree seven do indeed degenerate to the union of the sextic canonical surface together with another plane, namely the tacnodal plane (cf. §3).

Now, although our method applies in a much more general setting, the complexity of the computations which are needed in every specific case grows incredibly fast.

We consider therefore a real challenge for our present days computer algebra programs to make it possible to treat surfaces with higher values of the invariant K^2.

We would however like to remark, that all our explicit computations are more "computer assisted computations" than computer algebra programs. That is: it would be almost impossible to do them without a computer algebra program, but on the other hand there are always several steps which have to be done by hand, because looking carefully with a mathematical eye we can see tricks that the computer alone cannot detect.

Our paper is organized as follows. In the first chapter we introduce under quite general conditions a naturally defined graded subring $\tilde{\mathcal{R}}$ of the canonical

40   I. C. Bauer et al.

ring $\mathcal{R}$, such that there is an exact sequence

$$0 \longrightarrow \tilde{\mathcal{R}}_m \longrightarrow \mathcal{R}_m \longrightarrow H^0(\tilde{S}, \mathcal{O}_{m\mathcal{E}}) \longrightarrow \mathbb{C}^r \longrightarrow 0 \,,$$

where $\mathcal{E}$ is an exceptional divisor on the surface $\tilde{S}$ obtained from S blowing up the base points of the canonical system. Then we introduce the "dual" module $M = \operatorname{Ext}^1(\tilde{\mathcal{R}}, \Gamma_*(\omega_{\mathbb{P}^3}))$.

The rough idea is now to calculate the subring $\tilde{\mathcal{R}}$ (and the dual module M) using the geometry of the canonical image of S. We proceed in each degree "enlarging" $\tilde{\mathcal{R}}$ to $\mathcal{R}$: we will see how the module M provides automatically a certain number of the "missing" generators and relations; the few remaining generators and relations have to be computed "by hand" by the above exact sequence.

In Chap. 2 we will run this program in the special case of surfaces with $K^2 = 7$, $p_g = 4$, whose canonical map has exactly one base point and is birational. The main result of this section leads to the following

Theorem 0.1 *The canonical ring of a surface with $p_g = 4$, $K^2 = 7$, such that $|K|$ has one simple base point and φ_K is birational, is of the form $\mathcal{R} := \mathbb{C}[y_0, y_1, y_2, y_3, w_0, w_1, u]/I$ where the respective degrees of the generators of $\mathcal{R}$ are $(1, 1, 1, 1, 2, 2, 3)$.*

There exist a quadratic polynomial Q and a polynomial B of degree 4 such that I is generated by the 2×2 minors of the matrix

$$A := \begin{pmatrix} y_1 & y_2 & y_3 & w_1 \\ w_0 & w_1 & Q(y_i) & u \end{pmatrix},$$

and by three more polynomials (of respective degrees $4, 5, 6$), the first one of the form $-w_1^2 + B(y_i) + \sum \mu_{ijk} y_i y_j w_k$, and the other 2 obtained from the first via the method of rolling (cf. section 2, in particular they have the form $-w_1 u + \ldots, -u^2 + \ldots$).

In Chap. 3 we will describe our canonical ring $\mathcal{R}$ in three different formats introduced by D. Dicks and M. Reid. It turns out that, in order to deform the surfaces with $K^2 = 7$, $p_g = 4$ whose canonical system has one base point and is birational, the third format (= antisymmetric and extrasymmetric format) is the most suitable one.

The result for this case is

Theorem 0.2 *Let S be a minimal (smooth, connected) surface with $K^2 = 7$ and $p_g(S) = 4$, whose canonical system $|K_S|$ has one (simple) base point $x \in S$ and yields a birational canonical map. Then the canonical ring of S can be presented as $\mathcal{R} = \mathbb{C}[y_0, y_1, y_2, y_3, w_0, w_1, u]/\mathcal{I}$, where $\deg(y_i, w_j, u) = (1, 2, 3)$ and the ideal of relations $\mathcal{I}$ of $\mathcal{R}$ is generated by the 4×4-pfaffians* *of the following antisymmetric and extrasymmetric matrix*

$$P = \begin{pmatrix} 0 & 0 & w_0 & Q(y_0, y_1, y_2) & w_1 & u \\ & 0 & y_1 & y_3 & y_2 & w_1 \\ & & 0 & -u + \overline{C} & y_3\overline{Q}_3(w_0, y_0, y_1, y_3) & Q\overline{Q}_3 \\ & & & 0 & y_1\overline{Q}_1(w_0, w_1, y_0, y_1, y_3) & w_0\overline{Q}_1 \\ & & & & 0 & 0 \\ & -\text{sym} & & & & 0 \end{pmatrix},$$

where $Q, \overline{Q}_1, \overline{Q}_3$ are quadratic forms of a subset of the given variables as indicated, and $\overline{C}$ is a cubic form. Moreover $\overline{C}$ does not depend on u and on the $w_i y_j$'s for $j \leq 1$, $\overline{Q}_1$ does not depend on y_3^2.

In the fourth chapter we will finally show how the above presentation of the canonical ring allows a deformation to the canonical ring S of a surface with $K^2 = 7$, $p_g = 4$ and free canonical system.

These last surfaces and their canonical rings are described in [Cat1], and it follows also from [B-E] that in this presentation the relations can be given by the 4×4-Pfaffians of a skew-symmetric 5×5 matrix. Our final result is

Theorem 0.3 *Let P be an antisymmetric and extrasymmetric matrix as in the previous theorem. Consider the 1-parameter family of rings $S_t = \mathbb{C}[y_0, y_1, y_2, y_3, w_0.w_1, u]/\mathcal{I}_t$ where the ideal $\mathcal{I}_t$ is given by the 4×4 pfaffians of the antisymmetric and extrasymmetric matrix*

$$P_t = P + \begin{pmatrix} 0 & t & 0 & 0 & 0 & 0 \\ & 0 & 0 & 0 & 0 & 0 \\ & & 0 & 0 & 0 & 0 \\ & & & 0 & 0 & 0 \\ & & & & 0 & t\overline{Q}_1\overline{Q}_3 \\ & -\text{sym} & & & & 0 \end{pmatrix}.$$

This is a flat family and describes a flat deformation of the surface corresponding to the matrix P to surfaces with $p_g = 4$, $K^2 = 7$ and with $|K|$ base point free. For $t \neq 0$, S_t is isomorphic to $\mathbb{C}[y_0, y_1, y_2, y_3, w_0.w_1]/J_t$ where J_t is the ideal generated by the 4×4 pfaffians of the matrix

$$\begin{pmatrix} 0 & y_1 & -y_3 & -t^2 w_0 & -tQ \\ & 0 & y_2 & t^3\overline{Q}_3 & tw_1 \\ & & 0 & t^2 w_1 & t^2\overline{Q}_1 \\ & & & 0 & -t^3 c - t^2 y_1 Q + t^2 y_3 w_0 \\ & -\text{sym} & & & 0 \end{pmatrix}.$$

42   I. C. Bauer et al.

Then we compare the above family with the one predicted by Enriques showing (as mentioned above) that it is a completely different type of degeneration.

Finally we have two appendices. In fact, although theoretically all the computation can be done by hands, it is better to use a computer program (as we did with Macaulay 2) to shorten the time needed and be sure that no computation's mistakes occured. We have put in the appendix the two Macaulay 2 scripts (without output) that we needed. The second appendix is interesting, because it shows how the computer suggested to us the 5×5 matrix appearing in Theorem 0.3.

1   Canonical Systems with Base Points

Let S be a minimal surface of general type defined over the complex numbers and let $|K_S|$ be its canonical system. If $H^0(S, \mathcal{O}_S(K_S)) \neq 0$, then $|K_S|$ defines a rational map

$$\varphi_{|K|} : S - - \to \mathbb{P}^{p_g - 1},$$

where $p_g = p_g(S) := \dim H^0(S, \mathcal{O}_S(K_S))$ is the geometric genus of S.

Throughout this paper we make the following

Assumption. $|K_S|$ has no fixed part.

Let $\pi : \tilde{S} \longrightarrow S$ be a (minimal) sequence of blowups such that the movable part $|H|$ of $|\pi^* K_S|$ has no base points. Then we have a commutative diagram:

$$\tilde{S} \xrightarrow{\varphi_{|H|}} \mathbb{P}^{p_g - 1}$$
$$\searrow{\pi} \quad \varphi_{|K_S|} \nearrow$$
$$S$$

Since π is a sequence of blow ups $\pi_i : S_i \to S_{i-1}$ with centre a point $p_i \in S_{i-1}$, we denote by E_i the (-1)-divisor in $\tilde{S}$ given by the full transform of p_i, and we denote by m_i the multiplicity in p_i of the proper transform of a general divisor in K_S, so that

Remark 1.1 a) $K_{\tilde{S}} \equiv H + \sum_i (m_i + 1) E_i$.

To simplify the notation, we set $\mathcal{E} := \sum_i m_i E_i$, $E := \sum_i E_i$.

b) If $\varphi_{|K_S|}$ is not composed with a pencil, then $\varphi_{|H|} : \tilde{S} \longrightarrow \Sigma_1 \subset \mathbb{P}^{p_g - 1}$ is a generically finite morphism from $\tilde{S}$ onto a surface Σ_1 in $\mathbb{P}^{p_g - 1}$ of degree $d = H^2 = K_{\tilde{S}}^2 - \sum_i m_i^2$ and

$$\varphi_{|H|}^*(\mathcal{O}_{\Sigma_1}(1)) = \mathcal{O}_{\tilde{S}}(H) = \mathcal{O}_{\tilde{S}}(\pi^* K_S - \mathcal{E}) = \mathcal{O}_{\tilde{S}}(K_{\tilde{S}} - \mathcal{E} - E) .$$

Definition 1.2 a) Let us denote by $\mathcal{F}_1$ the coherent sheaf of $\mathcal{O}_{\Sigma_1}$-modules $(\varphi_H)_* \mathcal{O}_{\tilde{S}}$.

Canonical Rings of Surfaces   43

b) We define $\tilde{\mathcal{R}}(S)$ as the graded ring associated to the divisor H on $\tilde{S}$, thus

$$\tilde{\mathcal{R}}(S) := \bigoplus_{m=0}^{\infty} H^0(\tilde{S}, \mathcal{O}_{\tilde{S}}(mH)) = \bigoplus_{m=0}^{\infty} H^0(\Sigma, \mathcal{F}_1(m)) \ .$$

We make the following easy observation:

Remark 1.3 $\tilde{\mathcal{R}}(S)$ is a (graded) subring of the canonical ring $\mathcal{R}(S) := \bigoplus_{m=0}^{\infty} H^0(S, \mathcal{O}_S(mK_S))$.

Proof. The claim follows from the fact that the natural homomorphism

$$H^0(\tilde{S}, \mathcal{O}_{\tilde{S}}(mH)) \longrightarrow H^0(\tilde{S}, \mathcal{O}_{\tilde{S}}(mK_{\tilde{S}})) \cong H^0(S, \mathcal{O}_S(mK_S))$$

is injective for all $m \geq 0$. $\square$

Remark 1.4 We have by our assumption

$$\tilde{\mathcal{R}}_1 = H^0(\tilde{S}, \mathcal{O}_{\tilde{S}}(H)) = H^0(\tilde{S}, \mathcal{O}_{\tilde{S}}(K_{\tilde{S}})) = \mathcal{R}_1 \ .$$

Consider first the Stein factorization of $\varphi_{|H|}$:

$$\tilde{S} \xrightarrow{\varphi_{|H|}} \Sigma_1 \subset \mathbb{P}^{p_g - 1}$$
$$\searrow^{\delta} \quad \nearrow_{\varepsilon_1}$$
$$Y$$

where in general Y is a normal algebraic surface, δ has connected fibres and ε_1 is a finite morphism.

We shall moreover from now on make the following assumption

B) $\varphi_{|H|}$ is a *birational morphism onto its image*, whence in particular $p_g(S) \geq 4$.

Under the above assumption we shall moreover consider a general projection of Σ_1 to a surface Σ in $\mathbb{P}^3$.

We have therefore the following diagram

$$\tilde{S} \xrightarrow{\varphi} \Sigma \subset \mathbb{P}^3$$
$$\searrow^{\delta} \quad \nearrow_{\varepsilon}$$
$$Y$$

and we may therefore assume

B') $\varphi : \tilde{S} \to \Sigma$ *is a birational morphism.*

44   I. C. Bauer et al.

We shall write the singular locus $\text{Sing}(\Sigma)$ of Σ as $\Gamma \cup Z$, where Γ is the subscheme of Σ corresponding to the conductor ideal $\mathcal{C}$ of the normalization morphism ε and where Z is the finite set $\varepsilon(\text{Sing}(Y)) \subset \Sigma$ (note that if the support of Γ is disjoint from Z, also Z has a natural subscheme structure given by the adjunction ideal).

Σ is Cohen-Macaulay, whence Γ is a pure subscheme of codimension 1.

We remark that our methods apply also if the degree of φ is equal to two, but we have then to make more complicated technical assumptions.

Defining $\mathcal{F} := (\varphi)_* \mathcal{O}_{\tilde{S}}$, we have

$$\tilde{\mathcal{R}}(S) := \bigoplus_{m=0}^{\infty} H^0(\tilde{S}, \mathcal{O}_{\tilde{S}}(mH)) = \bigoplus_{m=0}^{\infty} H^0(\Sigma, \mathcal{F}(m)) ,$$

whence we may observe that the graded ring $\tilde{\mathcal{R}}$ is a module over the polynomial ring $\mathcal{A} := \mathbb{C}[y_0, y_1, y_2, y_3]$ (the homogeneous coordinate ring of $\mathbb{P}^3$).

Since this module has a support of codimension 1, it has a graded free resolution of length equal to 1 if and only if it is a Cohen-Macaulay module.

The following result is essentially the same result as Theorem 2.5 of [Cil1].

Proposition 1.5 $\tilde{\mathcal{R}}$ *is a Cohen Macaulay $\mathcal{A}$-module if and only if the subscheme $\Gamma \subset \mathbb{P}^3$ is projectively normal.*

Proof. It is well known that the Cohen Macaulay property is equivalent to the vanishing of the cohomology groups

$$H^1(\Sigma, \mathcal{F}(m)) = H^1(\Sigma, \varepsilon_* \mathcal{O}_Y \otimes \mathcal{O}_\Sigma(m)) = H^1(Y, \mathcal{O}_Y(mH)) ,$$

for all m.

By Serre's theorem B(m) these groups obviously vanish for $m \gg 0$. Serre-Grothendieck duality tells us that these are the dual vector spaces of $\text{Ext}^1(\varepsilon_* \mathcal{O}_Y(m), \omega_\Sigma)$.

By the local-to-global spectral sequence of the Ext groups, there is an exact sequence

$$0 \to H^1(\mathcal{H}om(\varepsilon_* \mathcal{O}_Y(m), \omega_\Sigma)) \to \text{Ext}^1(\varepsilon_* \mathcal{O}_Y(m), \omega_\Sigma)$$
$$\to H^0(\mathcal{E}xt^1(\varepsilon_* \mathcal{O}_Y(m), \omega_\Sigma)) \to H^2(\mathcal{H}om(\varepsilon_* \mathcal{O}_Y(m), \omega_\Sigma)) .$$

But $\mathcal{E}xt^1(\varepsilon_* \mathcal{O}_Y(m), \omega_\Sigma)$ is zero because $\varepsilon_* \mathcal{O}_Y$ is a Cohen-Macaulay $\mathcal{O}_\Sigma$-module and ω_Σ is invertible.

Therefore it follows that $\text{Ext}^1(\varepsilon_* \mathcal{O}_Y(m), \omega_\Sigma) = 0$ if and only if

$$H^1(\mathcal{H}om(\varepsilon_* \mathcal{O}_Y(m), \omega_\Sigma)) = 0 .$$

In turn, since $\omega_\Sigma = \mathcal{O}_\Sigma(d - 4)$,

$$H^1(\mathcal{H}om(\varepsilon_* \mathcal{O}_Y(m), \omega_\Sigma)) = H^1(\mathcal{C}(-m + d - 4)) ,$$

Canonical Rings of Surfaces   45

where $\mathcal{C}$ is the conductor ideal of ε, and this last group, in view of the exact sequence

$$0 \to \mathcal{C} \to \mathcal{O}_\Sigma \to \mathcal{O}_\Gamma \to 0$$

is the cokernel of the map $H^0(\mathcal{O}_\Sigma(-m+d-4)) \to H^0(\mathcal{O}_\Gamma(-m+d-4))$. Since however $H^0(\mathcal{O}_\Sigma(n))$ is a quotient of $H^0(\mathcal{O}_{\mathbb{P}^3}(n))$ we conclude that our desired vanishing is equivalent to the projective normality of Γ. $\qquad\square$

Recalling that $\mathcal{O}_{\tilde{S}}(H) = \mathcal{O}_{\tilde{S}}(\pi^* K_S - \mathcal{E})$, we consider now the exact sequence

$$0 \longrightarrow \mathcal{O}_{\tilde{S}}(mH) \longrightarrow \mathcal{O}_{\tilde{S}}(m\pi^* K_S) \longrightarrow \mathcal{O}_{m\mathcal{E}} \longrightarrow 0 \,.$$

Observe moreover that, since S is minimal and of general type, for $m \geq 2$ we have

$$H^1(\tilde{S}, \mathcal{O}_{\tilde{S}}(m\pi^* K_S)) = H^1(S, \mathcal{O}_S(mK_S)) = 0 \,.$$

Whence, we arrive for each $m \geq 2$ to the following crucial exact sequence, which will be repeatedly used in the sequel

(i) $\quad 0 \longrightarrow \tilde{\mathcal{R}}_m \longrightarrow \mathcal{R}_m \longrightarrow H^0(\tilde{S}, \mathcal{O}_{m\mathcal{E}}) \longrightarrow H^1(\tilde{S}, \mathcal{O}_{\tilde{S}}(mH)) \longrightarrow 0 \,.$

The vanishing $H^1(\Sigma, \mathcal{F}(m)) = 0$ for all m, is clearly equivalent to the chain of equalities:

$$\dim H^1(\tilde{S}, \mathcal{O}_{\tilde{S}}(mH)) = \dim H^0(\Sigma, R^1(\varphi_H)_* \mathcal{O}_{\tilde{S}}(m))$$
$$= \operatorname{length}(R^1(\varphi_H)_* \mathcal{O}_{\tilde{S}}) := l \,.$$

Putting together the above considerations we obtain the following

Remark 1.6 $\tilde{\mathcal{R}}$ is a Cohen Macaulay module over the polynomial ring $\mathcal{A}$ if and only if the surface S is regular $(H^1(S, \mathcal{O}_S(K_S)) = 0)$ and
$$\dim \mathcal{R}_m - \dim \tilde{\mathcal{R}}_m = \dim H^0(\tilde{S}, \mathcal{O}_{m\mathcal{E}}) - l = \sum_i \frac{mm_i(mm_i+1)}{2} - l.$$

Proof. Assume $\tilde{\mathcal{R}}$ to be Cohen Macaulay: since we know that $H^1(\mathcal{O}_\mathcal{E}) = 0$, and that the map $\tilde{\mathcal{R}}_1 \to \mathcal{R}_1$ is an isomorphism (from the definition of $\tilde{\mathcal{R}}$), we get an exact sequence

$$0 \longrightarrow H^0(\tilde{S}, \mathcal{O}_\mathcal{E}) \longrightarrow H^1(\tilde{S}, \mathcal{O}_{\tilde{S}}(H)) \longrightarrow H^1(S, \mathcal{O}_S(K_S)) \longrightarrow 0$$

which shows that the surface must be regular.

If conversely the surface is regular, $H^1(S, \mathcal{O}_S) = H^1(S, \mathcal{O}_S(K_S)) = 0$, so the sequence (i) is exact also for $m = 0, 1$, then $\forall m \in \mathbb{Z}$.

In particular, $\dim \mathcal{R}_m - \dim \tilde{\mathcal{R}}_m = \dim H^0(\tilde{S}, \mathcal{O}_{m\mathcal{E}}) - \dim H^1(\tilde{S}, \mathcal{O}_{\tilde{S}}(mH))$, whence $\tilde{\mathcal{R}}$ is a Cohen Macaulay $\mathcal{A}$-module iff $H^1(\Sigma, \mathcal{F}(m)) = 0$, *i.e.*, iff $\dim H^1(\tilde{S}, \mathcal{O}_{\tilde{S}}(mH)) = l$, equivalently, iff

$$\dim \mathcal{R}_m - \dim \tilde{\mathcal{R}}_m = \dim H^0(\tilde{S}, \mathcal{O}_{m\mathcal{E}}) - l = \sum_i \frac{mm_i(mm_i+1)}{2} - l \,.$$

$\qquad\square$

46   I. C. Bauer et al.

Remark 1.7 $\tilde{\mathcal{R}} \subset \mathcal{R}$ is a subring, but it is easy to see that $\mathcal{R}$ is not a finitely generated $\tilde{\mathcal{R}}$-module.

We assume now that $\tilde{\mathcal{R}}$ is a Cohen-Macaulay $\mathcal{A}$-module, and we observe that it contains the coordinate ring of Σ and is contained in $\mathcal{R}$. Therefore, choosing a minimal system of generators $v_1 = 1, v_2, \ldots v_n$ of $\tilde{\mathcal{R}}$ as an $\mathcal{A}$-module, defining $l_i := \deg v_i$ we find (by Hilbert's syzygy theorem, as in [Cat1]), a resolution of the form

$$(\#) \qquad 0 \longrightarrow \oplus_{j=1}^{h} \mathcal{A}(-r_j) \xrightarrow{\alpha} \oplus_{i=1}^{h} \mathcal{A}(-l_i) \longrightarrow \tilde{\mathcal{R}} \longrightarrow 0 \ ;$$

Under the assumption that φ be birational follows that Σ has equation $f := \det \alpha = 0$.

As in [Cat1], being $\tilde{\mathcal{R}}$ a ring, the matrix has to fulfill the standard Rank Condition, which we will later recall.

In order to describe the ring $\mathcal{R}$, we first look for generators of $\mathcal{R}$ as an $\mathcal{A}$-module. On the other hand, when looking for generators of $\mathcal{R}$ as a ring, we may restrict ourselves to consider elements of low degree by virtue of the following result by M. Reid (cf. [Rei1], cf. also Ciliberto [Cil2]).

Theorem 1.8 *Let X be a canonical surface (i.e., the canonical model of a surface of general type). We suppose that*

(i) $p_g(X) \geq 2$, $K_X^2 \geq 3$,
(ii) $q(X) = 0$,
(iii) *X has an irreducible canonical curve $C \in |K_X|$.*

Then the canonical ring $\mathcal{R} = \mathcal{R}(X, K_X)$ of X is generated in degrees ≤ 3 and its relations are generated in degrees ≤ 6.

Now we define the $\tilde{\mathcal{R}}$-module

$$M := \Gamma_*(\mathcal{C}\omega_\Sigma) = \Gamma_*(\omega_Y) \ ,$$

where $\mathcal{C}$ is the conductor ideal of ε and $\Gamma_*(\mathcal{F})$ denotes as usual $\oplus_{n \in \mathbb{Z}} H^0(\mathcal{F}(n))$. We consider the following chain of inclusions of $\mathcal{A}$-modules

$$\mathcal{A}/(f) \subset \tilde{\mathcal{R}} \subset \Gamma_*(\varphi_* \omega_{\tilde{S}})[-1] \subset M[-1] \ .$$

We observed that $\tilde{\mathcal{R}}$ is Cohen Macaulay if and only if it has a free resolution as an $\mathcal{A}$-module of the form $(\#)$

$$0 \to L_1 \xrightarrow{\alpha} L_0 \to \tilde{\mathcal{R}} \to 0 \ .$$

Dualizing it, we get

$$0 \to L_0^\vee \otimes \Gamma_*(\omega_{\mathbb{P}^3}) \xrightarrow{\alpha^t} L_1^\vee \otimes \Gamma_*(\omega_{\mathbb{P}^3}) \to \mathrm{Ext}_{\mathcal{A}}^1(\tilde{\mathcal{R}}, \Gamma_*(\omega_{\mathbb{P}^3})) \to 0 \ .$$

By virtue of the exact sequence

$$0 \to \Gamma_*(\omega_{\mathbb{P}^3}) \xrightarrow{f} \Gamma_*(\omega_{\mathbb{P}^3})(\deg f) \to \Gamma_*(\omega_\Sigma) \to 0 \ ,$$

and since $\operatorname{Ext}^1_{\mathcal{A}}(\tilde{\mathcal{R}}, f) = 0$, we get

$$\begin{aligned}
\operatorname{Ext}^1_{\mathcal{A}}(\tilde{\mathcal{R}}, \Gamma_*(\omega_{\mathbb{P}^3})) &= \operatorname{Hom}_{\mathcal{A}}(\tilde{\mathcal{R}}, \Gamma_*(\omega_\Sigma)) \\
&= \Gamma_*(\mathcal{H}om_{\mathcal{O}_\Sigma}(\mathcal{F}, \omega_\Sigma)) = \Gamma_*(\mathcal{C}\omega_\Sigma) = M \ .
\end{aligned}$$

Moreover, M satisfies the Ring Condition (cf. [dJ-vS])

$$\tilde{\mathcal{R}} = \operatorname{Hom}(M, \mathcal{A}/(f)) = \operatorname{Hom}(M, M)$$

or, in other words, there is a bilinear pairing $\tilde{\mathcal{R}} \times M \to M$ which, in the given bases, is determined by the matrix $\beta = \Lambda^{h-1}(\alpha)$. In turn the Ring Condition is equivalent to the so called Rank Condition for (α): there exist elements λ^k_{jh} of $\mathcal{A}$ such that $\beta_{1k} = \sum \lambda^k_{jh}\beta_{jh}$.

Since we have the inclusion $\tilde{\mathcal{R}} \subset M[-1]$, we can fix bases $v_1 = 1, v_2, \ldots, v_h$, for $\tilde{\mathcal{R}}$, $z_1, z_2, \ldots, z_h$, for M, with $v_i z_j = \frac{\beta_{ij} z_1}{\beta_{11}}$ and such that z_1 is the image of $v_1 = 1$. For the same reason, $v_2, \ldots, v_n$ can be written as linear combinations $v_i = \sum \zeta_{ij} z_j / z_1$.

Now, as in [Cat1], the ring structure of $\tilde{\mathcal{R}}$ is equivalent to the Rank Condition which can also be phrased as follows:

R.C. the ideal of the $(h-1) \times (h-1)$ minors of α coincide with the ideal of the $(h-1) \times (h-1)$ minors of the matrix α' obtained by deleting the first row of α.

The polynomials λ^k_{jh} with $\beta_{1k} = \sum \lambda^k_{jh}\beta_{jh}$ determine therefore the ring structure of $\tilde{\mathcal{R}}$ by the following multiplication rule:

$$v_i v_h = \sum_{j,k} \zeta_{ij} \lambda^k_{jh} v_k \ .$$

We have now all the general ingredients at our disposal and we can explicitly describe our method in order to compute the canonical ring of the regular surfaces of general type with given values of the invariants K^2, p_g, canonical system with base points (and without fixed part), and birational canonical map.

Under the above assumptions, observe that obviously $p_g \geq 4$, moreover Castelnuovo's inequality $K^2 \geq 3p_g - 7$ holds, in particular the hypotheses of Theorem 1.8 are fullfilled.

We need first of all to assume that the subscheme Γ of $\mathbb{P}^3$ given by the conductor ideal of the normalization of Σ be projectively normal.

The last assumption, as we just saw, ensures that the ring $\tilde{\mathcal{R}}$ is a Cohen-Macaulay $\mathcal{A}$-module; whence, argueing as in [Cat1] we can find a length 1

48   I. C. Bauer et al.

presentation of $\tilde{\mathcal{R}}$ as an $\mathcal{A}$-module, given by a square matrix α fullfilling the Rank Condition .

Let $v_1, \ldots, v_h$ be the generators of $\tilde{\mathcal{R}}$ we used in order to write down α, and let $z_1, \ldots z_h$ be the dual generators of M (*i.e.*, the module M is generated by the z_i's and presented by the matrix α^t).

We have seen that there is an inclusion $\tilde{\mathcal{R}} \subset M[-1]$; assuming by sake of simplicity that ω_Y is Cartier, one can write explicit sections σ_d ($d \in \mathbb{N}$) of suitable line bundles $\mathcal{L}_d$ so that the above inclusion is obtained multiplying every element r_d homogeneous of degree d in $\tilde{\mathcal{R}}$, by σ_d. The σ_d's are of the form $e_d \cdot c$ where c is a section of the dual of the relative canonical bundle of the map $\delta : \tilde{S} \to Y$, and e_d is supported on the exceptional locus of $\pi : \tilde{S} \to S$.

It is not possible to construct a similar inclusion $\mathcal{R} \subset M$, but we can consider the module $\oplus_n H^0(K_{\tilde{S}} + nH)$. This is the submodule of M given by the elements divisible by c (in particular it contains $\tilde{R}$), and clearly it is a submodule of $\mathcal{R}$ so it is completely natural to denote it by $M \cap \mathcal{R}$.

The second step of our method is to study this module: first we compute the subset $\{w_1, \ldots w_r\}$ of a set of generators for $M \cap \mathcal{R}$ as an $\mathcal{A}$-module, consisting of the elements of degree ≤ 3 (with the grading of $\mathcal{R}$). Then we find the relations holding among them in degree ≤ 6. In fact, by Theorem 1.8, generators and relations in higher degrees will not be relevant for the canonical ring.

The elements $y_i's$, w_j's will not in general generate the canonical ring; the third step of our method consists in the research of the missing generators and relations.

It is clear that in every case $\mathcal{R}_1 \subset \mathcal{R} \cap M$. We assume now that the base points are simple (but a similar analysis can be carried out in every case), *i.e.*,

$$\mathcal{E} = E .$$

Directly by the definition follows the equality $K_{\tilde{S}} + H = 2\pi^* K_S$, so $H^0(K_{\tilde{S}} + H) = H^0(2K_S)$, *i.e.*, $\mathcal{R}_2 \subset \mathcal{R} \cap M$. Instead, if $\mathcal{E} \neq E$ the equality $h^0(K_{\tilde{S}} + H) = h^0(2K_S)$ cannot hold, otherwise $|2K_S|$ would not be base point free, as it has to be for $p_g > 0$ ([Fra]).

In degree 3 we have the exact sequence

$$0 \to H^0(K_{\tilde{S}} + 2H) \to H^0(3K_S) \to H^0(\mathcal{O}_\mathcal{E}) \to 0$$

since $H^1(K_{\tilde{S}} + 2H) = 0$ (by Mumford's vanishing theorem). Therefore, $\mathcal{R}_3 \cap M$ has codimension 3 in $\mathcal{R}_3$, *i.e.*, we need l elements $u_1, \ldots, u_l$, to complete a basis of $H^0(K_{\tilde{S}} + 2H)$ to a basis of $H^0(3K_S)$.

By Theorem 1.8 the generators $y_0, \ldots, y_3$ of $\mathcal{A}$ (seen as elements of $H^0(K_{\tilde{S}})$), together with $w_1, \ldots, w_r$ and $u_1, \ldots, u_l$ are a system of generators of $\mathcal{R}$ as a ring.

The relations as $\mathcal{A}$-module among the v_i's and among the w_j's are determined by the matrix α, and similarly the relations given by the products of type v_iv_j, and v_iw_j are also determined by α: these provide automatically a list of relations among the above generators of $\mathcal{R}$.

Some relation is still missing; in particular our method do not produce automatically any relation involving the u_k's. In order to complete the analysis one needs to find a way to espress the u_k's "in terms of M", so that the known relation among the elements of $\tilde{R}$ and M will produce also relations involving them. This should be possible case by case by "ad hoc" arguments, but we do not know a general argument.

We devote the next sections to the application of the above method to finding a description of the canonical ring of the stratum of the moduli space of surfaces of general type with $K^2 = 7$ and $p_g = 4$ corresponding to surfaces with birational canonical map and whose canonical system has base points.

2 The Canonical Ring of Surfaces with $K^2 = 7$, $p_g = 4$ Birational to a Sextic: From Algebra to Geometry

Let S be a minimal (smooth, connected) surface with $K^2 = 7$ and $p_g(S) = 4$. We remark that S is automatically regular (cf. [Deb]), *i.e.*, $q(S) = 0$.

In [Ba] the first author gave an exact description of minimal surfaces with $K^2 = 7$ and $p_g(S) = 4$, where the canonical system has base points, proving in particular that if moreover the canonical map is birational then the canonical system $|K_S|$ has exactly one simple base point $x \in S$.

Let $\pi : \tilde{S} \longrightarrow S$ be the blow up of S in x and let $E := \pi^{-1}(x)$ be the exceptional curve of π. Thus we have:

$$|K_{\tilde{S}}| = |\pi^*K_S| + E = |H| + 2E \ ,$$

where $|H|$ is base point free. Thus we will assume in this paragraph that $\varphi_{|H|} : \tilde{S} \longrightarrow \Sigma := \{F_6 = 0\}$ is a birational morphism (from $\tilde{S}$ onto a surface Σ of degree six in $\mathbb{P}^3$).

We recall now the description of the minimal surfaces with $K^2 = 7$ and $p_g(S) = 4$, whose canonical system has exactly one base point and whose canonical map is birational given in [Ba].

First we need the following definition.

Definition 2.1 A *generalized tacnode* is a two dimensional elliptic hypersurface singularity $(X, 0)$, such that the fundamental cycle has self intersection (-2). In particular, $(X, 0)$ is Gorenstein and by [Lau], Theorem (1.3), $(X, 0)$ is a double point singularity, whose local analytic equation is given by

$$z^2 = g(x, y) \ ,$$

where g vanishes of order four in 0. The normal cone of the singularity is given by the plane $\{z = 0\}$, called the tacnodal plane.

50 I. C. Bauer et al.

More precisely, a generalized tacnode $(X, 0)$ is the singularity obtained as the double cover branched along a curve with a quadruple point, which after a blow up decomposes in at most simple triple points or double points.

Theorem 2.2 ([Ba, Th. 3.11, 5.5]) 1) *Let S be a minimal surface with $K^2 = 7$ and $p_g(S) = 4$, whose canonical system has exactly one base point and whose canonical map is birational. Then the blow-up $\tilde{S}$ of the base point is the minimal desingularization of a surface $\Sigma \subset \mathbb{P}^3$ of degree six with the following properties:*
(a) *the double curve $\Gamma \subset \Sigma$ is a plane conic,*
(b) *if $\gamma \subset \mathbb{P}^3$ is the plane containing Γ, then Σ has a generalized tacnode $o \in \gamma \backslash \Gamma$ with tacnodal plane $\alpha \neq \gamma$,*
(c) *the image $\varphi_H(E)$ of the exceptional curve equals the line $\alpha \cap \gamma$.*
2) *The surfaces with $K^2 = 7$ and $p_g = 4$ such that the canonical system has exactly one base point and $\varphi_{|K|}$ is birational form an irreducible set $\mathfrak{M}_{(I.1)}$ of dimension 35 in their moduli space.*

Moreover, it was shown (ibidem) that for a general element of $\mathfrak{M}_{(I.1)}$ the canonical image Σ has an equation of the form

$$\alpha^2 Q^2 + \gamma F_5 = 0 \,,$$

where F_5 is an element of the linear subsystem $\Delta \subset |5H - \Gamma - 2o|$ in $\mathbb{P}^3$ consisting of quintics with tangent cone α^2 in o and where $Q \subset \mathbb{P}^3$ is an irreducible quadric containing Γ.

In this section we will study the ring $\tilde{\mathcal{R}}$ and the module M defined in the previous section in the case of the surfaces in the above class.

This study will allow us to compute the canonical ring and to give a purely algebraic proof of part 1) of Theorem 2.2, under a few generality assumptions.

For the convenience of the reader we will give here a list of notation (partly already introduced in the last section) which will be frequently used in the following.

Notation: We consider:

- $\pi : \tilde{S} \to S$, the blow-up of the base point of $|K_S|$;
- $\varphi : \tilde{S} \dashrightarrow \Sigma \subset \mathbb{P}^3$, the morphism induced by the canonical system;
- $\varepsilon : Y \to \Sigma$, the normalization of Σ;
- $\delta : \tilde{S} \to Y$, such that $\varphi = \varepsilon \circ \delta$;
- e, a generator of $H^0\left(\tilde{S}, \mathcal{O}_{\tilde{S}}\left(K_{\tilde{S}} - \pi^* K_S\right)\right)$ and E the corresponding divisor;
- c, a generator of $H^0\left(\tilde{S}, \mathcal{O}_{\tilde{S}}\left(\delta^* K_Y - K_{\tilde{S}}\right)\right)$ and C the corresponding divisor (this makes sense under the assumption 2) below);
- F_6 an equation of $\Sigma \subset \mathbb{P}^3$.

We have the following list of graded rings respectively modules:

Canonical Rings of Surfaces   51

- $\mathcal{A} := \bigoplus_{m=0}^{\infty} S^m \left(H^0 \left(\tilde{S}, \mathcal{O}_{\tilde{S}}(H) \right) \right) = \mathbb{C}[x_0, x_1, x_2, x_3],$
- $\tilde{\mathcal{R}} := \bigoplus_{m=0}^{\infty} H^0 \left(\Sigma, (\varphi_{|H|})_* \mathcal{O}_{\tilde{S}}(m) \right) = \bigoplus_{m=0}^{\infty} H^0 \left(\tilde{S}, \mathcal{O}_{\tilde{S}}(mH) \right),$
- $M := \bigoplus_{m \in \mathbb{Z}} H^0(\Sigma, \mathcal{C}\omega_{\Sigma}(m)) = \bigoplus_{m=-1}^{\infty} H^0(\tilde{S}, \mathcal{O}_{\tilde{S}}((m+1)H + 2E + C)),$
  where
- $\mathcal{C} := \mathcal{H}om_{\mathcal{O}_{\Sigma}} (\varepsilon_* \mathcal{O}_Y, \mathcal{O}_{\Sigma})$ is the conductor ideal;
- $\mathcal{R} = \bigoplus_{m=0}^{\infty} H^0(S, mK_S) = \bigoplus_{m=0}^{\infty} H^0 \left(\tilde{S}, \mathcal{O}_{\tilde{S}} (m(H + E)) \right),$ the canonical ring of S.

The assumptions we will make are the following:

1) The conductor ideal of ε defines a projectively normal subscheme of $\mathbb{P}^3$;
2) The singular points of Y do not lie in the preimage of the non normal locus of Σ; in particular it follows that ω_Y is Cartier.

These two assumptions give in fact no restriction, as it can be shown with geometrical arguments [Ba, Remark 3.1 and Prop. 3.6.(v)].

First, by the results of the previous section, we give a presentation of the ring $\tilde{\mathcal{R}}$ as $\mathcal{A}$-module.

Theorem 2.3 $\tilde{\mathcal{R}}$ *is a Cohen–Macaulay $\mathcal{A}$-module and has a resolution (as an $\mathcal{A}$-module) as follows*

$$0 \to \begin{matrix} \mathcal{A}(-5) \\ \oplus \\ \mathcal{A}(-4) \end{matrix} \xrightarrow{\ \alpha\ } \begin{matrix} \mathcal{A} \\ \oplus \\ \mathcal{A}(-3) \end{matrix} \to \tilde{\mathcal{R}} \to 0 \, .$$

Proof. (cf. [Cat1] for similar computations). By Remark 1.6, $\dim \tilde{\mathcal{R}}_m = \dim \mathcal{R}_m - \frac{m(m+1)}{2} + 1$, therefore by Riemann–Roch's Theorem, for $m \geq 2$, it equals $\chi(\mathcal{O}_S) + \frac{7}{2}m(m-1) - \frac{m(m+1)}{2} + 1 = 3m^2 - 4m + 6$. $\tilde{\mathcal{R}}$ is Cohen Macaulay by Theorem 1.5 (and Assumption 1)), whence it has a resolution as $\mathcal{A}$-module of length 1.

By definition the first generator of $\tilde{\mathcal{R}}$ has degree 0 ($\tilde{\mathcal{R}}_0 = H^0(\mathcal{O}_{\tilde{S}})$) and will be denoted by 1.

The above dimension formula gives us immediately that there are no other generators in degrees ≤ 2 and that one more generator (denoted by v) is needed in degree 3. Moreover, the relations live in degrees ≥ 4.

Again by the above dimension formula we get at least one relation in degree 4. If there were two independent relations in degree 4 they would have the form $x_0 \cdot v = f_4(x_i) \cdot 1$; $x_1 \cdot v = g_4(x_i) \cdot 1$ and this would force a nontrivial relation of the form $(x_1 f_4 - x_0 g_4) \cdot 1 = 0$ of degree 5, contradicting that obviously for any $f \in \mathcal{A}$ the equality $f \cdot 1 = 0$ implies that f is a multiple of F_6. By the dimension formula therefore there are no new generators in degree 4.

Again counting the dimensions we get a relation in degree 5; a straightforward computation shows that for all m, $\dim \tilde{\mathcal{R}}_m = \dim \mathcal{A}_m + \dim \mathcal{A}_{m-3} - \dim \mathcal{A}_{m-4} - \dim \mathcal{A}_{m-5}$; this shows that the resolution has the form

$$0 \to \begin{matrix} \mathcal{A}(-5) \\ \oplus \\ \mathcal{A}(-4) \\ \oplus \\ \mathcal{L} \end{matrix} \xrightarrow{\ \alpha\ } \begin{matrix} \mathcal{A} \\ \oplus \\ \mathcal{A}(-3) \\ \oplus \\ \mathcal{L} \end{matrix} \to \tilde{\mathcal{R}} \to 0$$

where $\mathcal{L}$ is a free module $\oplus \mathcal{A}(-s_i)$, with $s_i \geq 5$ for all i.

The minimality of the resolution ensures that $\mathcal{L} = 0$ (there are no non zero constants as coefficients in α; considering the row of α corresponding to the new generator of maximal degree (≥ 5) we get a row of zeroes, contradicting the injectivity of α). $\qquad\square$

Corollary 2.4 *M has a resolution as $\mathcal{A}$-module of the form*

$$0 \to \begin{matrix} \mathcal{A}(-4) \\ \oplus \\ \mathcal{A}(-1) \end{matrix} \xrightarrow{\ {}^t\alpha\ } \begin{matrix} \mathcal{A}(1) \\ \oplus \\ \mathcal{A} \end{matrix} \to M \to 0 \ .$$

We denote by z_{-1}, z_0 the generators of M in the respective degrees -1 and 0;

Remark 2.5 Notice that $z_{-1} = e^2 c$.

Corollary 2.6 *Up to a suitable choice of the generators of the $\mathcal{A}$-modules $\tilde{\mathcal{R}}, M$, we can write*

$$\alpha = \begin{pmatrix} QG + \gamma B & Qq \\ Q & \gamma \end{pmatrix};$$

where $\deg(\gamma, Q, q, G, B) = (1, 2, 2, 3, 4)$, *moreover*

$$vz_{-1} = qz_0$$
$$vz_0 = Bz_{-1} - Gz_0 \ .$$

Proof. By the above resolution of M we know the degrees of the entries of α. The Rank Condition for α means that the elements of the first row are in the ideal generated by the elements of the second row. To obtain now the desired form of α it suffices to add a suitable multiple of the second row to the first one. The two relations can be easily obtained writing explicitly the pairing $\tilde{\mathcal{R}} \times M \to M$. $\qquad\square$

In order to understand the structure of the canonical ring, we have now to investigate which elements of the module M can be divided by c. In fact, the

Canonical Rings of Surfaces 53

graded parts of our rings are related by the following (commutative) diagram:

$$(*) \qquad \begin{array}{ccc} \tilde{R}_n & \xrightarrow{e^2} & H^0(\tilde{S}, \mathcal{O}(nH + 2E)) \\ {\scriptstyle e^n}\downarrow & \swarrow & \downarrow{\scriptstyle c} \\ R_n & & M_{n-1} \end{array}$$

where the diagonal arrow is multication by e^{n-2} $(n \geq 2)$.

In order to write down explicitely the ring, we will fix a basis x_0, x_1, x_2, x_3 for $\mathcal{A}_1 = \tilde{R}_1$. We will denote by $y_i := ex_i$ the induced elements in $\mathcal{R}$.

Remark 2.7 First, since $\gamma \neq 0$ (or $\Sigma = \{\gamma F_5 = qQ^2\}$ would be reducible), we set $x_3 := \gamma$. Moreover we shall assume from now on that $G = G(x_0, x_1, x_2)$ (this is clearly possible without loss of generality by Corollary 2.6).

By the resolution of M, $\dim_{\mathbb{C}} M_1 = 13$ and a basis of M_1 is provided by the 10 elements of the form $x_i x_j z_{-1}$ and $x_0 z_0, x_1 z_0, x_2 z_0$ (the only relation in degree 1 has the form $x_3 z_0 = -Q z_{-1}$).

By Riemann–Roch, $P_2 = 12$; recalling that $c | z_{-1}$ diagram $(*)$ shows that we can fix the basis $x_0, x_1, x_2, x_3 = \gamma$ of $\mathcal{A}_1$ such that $c | x_i z_0$ for $i \geq 1$ but c does not divide $x_0 z_0$, and setting $w_0 := -\frac{x_1 z_0}{c}$, $w_1 := -\frac{x_2 z_0}{c}$, $w_0, w_1, y_i y_j$ is a basis of $\mathcal{R}_2$.

Lemma 2.8 $R = \mathbb{C}[y_0, y_1, y_2, y_3, w_0.w_1, u]/I$ where $\deg(y_i, w_j, u) = (1, 2, 3)$, *and the generators in degree ≤ 3 of the ideal I are given by the vanishing of the 2×2 minors of the following matrix:*

$$\begin{pmatrix} y_1 & y_2 & y_3 \\ w_0 & w_1 & Q(y_i) \end{pmatrix}$$

where $Q(y_i)$ is obtained from Q replacing every x_i by the corresponding y_i.

Proof. In the previous remark, we saw that in degree ≤ 2, $\mathcal{R}$ is generated by the y_i's and the w_j's. Every relation in degree ≤ 3 of $\mathcal{R}$ is of the form $w_0 l_0(y_i) + w_1 l_1(y_i) = g_3(y_i)$, where l_0 and l_1 are linear forms, g_3 is a cubic form. Thus we get from it the following relation in degree 2 for M as $\mathcal{A}$-module: $z_0(x_1 l_0(x_i) + x_2 l_1(x_i)) = z_{-1} g_3(x_i)$.

By Corollaries 2.4 and 2.6, must be a multiple of the relation $Q z_{-1} + x_3 z_0 = 0$. Thus we immediately see that this relation is a linear combination of the three given by the 2 by 2 minors of the matrix in the statement. We obtain therefore that the subspace of R_3 generated by the monomials in the y_i, w_i has dimension 25. Since $P_3 = 26$ we need to add a new generator u.

Finally, there are no more generators in degree ≥ 4 by Theorem 1.8. $\qquad\square$

We noticed in fact at the end of the previous section that if $E = \mathcal{E}$ we need exactly $l = h^0(\mathcal{O}_\mathcal{E})$ new generators of $\mathcal{R}$ in degree 3 which do not come from elements in M. This gives in our case $(l = 1)$ exactly one new generator in degree 3 (the only one not vanishing in E), that is, our generator u.

54   I. C. Bauer et al.

Remark 2.9 For later use we observe that c does not divide $x_0^2 z_0$. This holds since otherwise $\frac{x_0^2 z_0}{c}$ would yield (cf. diagram $(*)$) an element in $H^0(3H + 2E)$, whence $\frac{ex_0^2 z_0}{c}$ would be an element of R_3 linearly dependent upon the monomials in the y_i's, w_i's. This however contradicts Corollary 2.4.

As pointed out at the end of the previous section, the main problem in computing the canonical ring is given by the "additional" generators in degree 3 (in our case there is only one, namely u).

In the next lemma (part 5) we manage to express u "in terms of M"; this will allow us to compute the canonical ring.

Lemma 2.10 *Choosing suitable coordinates in $\mathbb{P}^3$ and suitable generators of M and $\tilde{\mathcal{R}}$ as $\mathcal{A}$-modules we can assume:*

1) *Q does not depend on the variable x_3;*
2) *$q = x_2^2$;*
3) *$(1, 0, 0, 0)$ is a double point, which is locally a double cover of the plane branched along a curve with a singularity of order at least 4.*
4) *$c^2 | x_2 z_0^2$;*
5) *$u = -\frac{w_1 z_0}{ec}$.*

Moreover $\varphi(E)$ is the line $x_2 = x_3 = 0$.

We would like to point out that the coordinates and generators in the previous lemma can be chosen such that Remark 2.5, Lemma 2.6, Remark 2.7 and Lemma 2.8 will remain valid as it can be traced in the following proof.

Proof. First, taking suitable linear combinations of rows and columns of α, we can assume that both q and Q do not depend on the variable x_3, and part 1) is proved.

In Corollary 2.6 we have seen that $vz_{-1} = qz_0$. As a matter of fact, for every quadric $q'(x_0, x_1, x_2)$ with the property

$$(**) \qquad\qquad z_{-1} | q' z_0 \ ,$$

$\frac{q' z_0}{z_{-1}}$ is an element in $\tilde{R}_3$; so if there were two independent quadrics with this property, we would get two independent elements of $\tilde{R}$ in degree 3, and by Theorem 2.3, for a suitable quadric q'' in the pencil generated by them, we would get a relation $g_3(x_i)z_{-1} = q''(x_0, x_1, x_2)z_0$, contradicting Corollary 2.4. Therefore there can be only one quadric with this property.

We already noticed (cf. Remark 2.9) that c does not divide $x_0^2 z_0$ but divides $x_i z_0$ for every $i \geq 1$; so q is the only quadric of the form $x_0 l(x_1, x_2) + l_1(x_1, x_2) l_2(x_1, x_2)$ such that qz_0 vanishes twice on E.

By definition $HE = 1$ whence $\varphi(E)$ is a line and there are two independent linear forms in $\mathbb{P}^3$ vanishing on E; in particular there is at least one linear form in the span of x_0, x_1, x_2 vanishing on E. Note that e does not divide z_0

Canonical Rings of Surfaces   55

(or we could easily find two different quadrics with the required property, a contradiction).

One of these linear forms belongs to the span of x_1, x_2: otherwise we could assume (up to a change of coordinates) that x_0 vanishes on E, and since q is divisible by e^2 we get $l_1 = l_2 = 0$ and $e^2|x_0$, contradicting again the unicity of q (we can take x_0x_1 and x_0x_2). Up to a change of coordinates we can then assume $e|x_2$; x_2^2 fulfills (**), hence $q = x_2^2$, and part 2) is proved.

Note that since $x_3z_0 = -Qz_{-1}$, and e does not divide z_0 (else the base point of the canonical system would also be a base point of the bicanonical system), $e^2|x_3$, and $\varphi(E) = \{x_2 = x_3 = 0\}$.

Let us write $C = C_1 + C_2$ (and accordingly $c = c_1c_2$) where C_2 is the greatest common divisor of C and the divisor of z_0, hence obviously $C_1 \neq 0$. Since c divides x_iz_0 for all $i \geq 1$, and not x_0z_0, so C_1 maps to the point $(1, 0, 0, 0)$.

By Assumption 2) $(1, 0, 0, 0)$ is an isolated singular point of Σ, and since the equation of Σ is given by the determinant of α, *i.e.*, by $Q^2x_2^2 = x_3QG + x_3^2B$, Q is invertible in a neighbourhood of it. Therefore Σ has a double point in $(1, 0, 0, 0)$, which is not a rational double point (otherwise $C = 0$). By the form of the equation of Σ we see immediately that the tangent cone has then an equation of the form $(x_2 + ax_3)^2 = 0$.

After a linear change of coordinates we can assume that the tangent cone is given by $x_2^2 = 0$. Notice that this coordinate change "corrupts" the previous choices, *i.e.*, statement 1) and 2) do not hold anymore, but we can easily act on the rows and columns of α in order to "recover" them.

We can now consider the double point as a (local analytic) double cover of the plane branched on a singular curve with a singularity of order at least 3. Assume that the singularity is a triple point. Then by [B-P-V] it has to be at least a $(3, 3)$-point (since otherwise we would have a rational double point). By our equations the tangent direction is $\{x_3 = 0\}$ and after a blowup there is again a triple point exactly on the intersection of the exceptional divisor with the strict transform of the "tangent" line $\{x_3 = 0\}$.

In particular the strict trasform of x_3 with respect to this blow up pulls back on $\tilde{S}$ to a divisor with some common component with C_1, while the one of x_1 has no common component with C_1. This however contradicts the equality $x_3z_0 = -Qe^2c$ since c divides x_1z_0 and Q is invertible at C. This shows that the branch curve has a singular point of order at least 4, and part 3) is proven. This in particular implies (looking at the equation of Σ) that $c_1^4|x_2^2$, so $c_1^2|x_2$ and $c^2|x_2z_0^2$; this proves part 4) of the statement.

It is now clear that $\frac{w_1z_0}{ec}$ is a holomorphic section in $3(H + E)$. Moreover, as we already observed, z_0 does not vanish on E (or S would have base points for the bicanonical system), and w_1 vanishes on E with multiplicity 1 (it vanishes there because it is multiple of x_2, with multiplicity 1 or w_1 would induce an element of $\tilde{R}$). So $\frac{w_1z_0}{ec}$ does not vanish on E; we can therefore choose $u = \frac{w_1z_0}{ec}$, and part 5) is proven. $\square$

56 I. C. Bauer et al.

The choice of u allows us immediately to write the matrix

$$A := \begin{pmatrix} y_1 & y_2 & y_3 & w_1 \\ w_0 & w_1 & Q(y_i) & u \end{pmatrix},$$

and to notice that the 2×2 minors of A are relations in $\mathcal{R}$.

The ring is in fact the canonical ring of a surface with 7 generators, so it has codimension 4. There is no structure theorem for rings of this codimension, but Reid noticed that most of them have 9 relations (joked by 16 syzygies) that can be expressed in some "formats" (cf. [Rei1], [Rei2]) that help in the study of the deformations.

An important format introduced by Reid is the "rolling factor" format; we try now to recall shortly how it is defined, referring to the above quoted papers by Reid for a more detailed treatement and other examples.

Definition 2.11 One says that a sequence of 9 equations $f_1, \ldots, f_9$ (usually joked by 16 syzygies and defining a Gorenstein ring of codimension 4, but we do not need this here) is in the "rolling factor" format if:

1) $f_1, \ldots f_6$ can be written as the (determinants of the) 2×2 minors of a 2×4 matrix A;
2) f_7 is in the ideal generated by the entries of the first row of A (for the matrix A above, it means that f_7 can be written as a linear combination $ay_1 + by_2 + cy_3 + dw_1 = 0$);
3) f_8 is obtained "rolling" f_7, *i.e.*, taking a linear combination with the same coefficients, but of the entries of the second row of A (in our case f_8 can be chosen as $aw_0 + bw_1 + cQ + du = 0$).
4) f_8 is in the ideal generated by the entries of the first rows of A and f_9 is obtained "rolling" f_8.

Notice that there can be different ways to "roll" the same equation, but all equivalent up to the equations given by the minors of A.

Remark 2.12 If $\mathcal{R} = \mathbb{C}[t_0, \ldots t_n]/I$ is an integral domain, I contains the 2×2 minors of a $2 \times n$ matrix, $n \geq 2$, and contains one equation f in the ideal generated by the entries of the first row of A, I must contain also the equation obtained "rolling" f.

This remark will be useful for the next theorem, where we will compute f_7 and prove that f_7 can be "rolled" twice, obtaining f_8 and f_9.

Philosophically, all the three relations come from a single relation in a bigger ring, that is, the last equation in Corollary 2.6.

Theorem 2.13 *The canonical ring of a surface with $p_g = 4$, $K^2 = 7$, such that $|K|$ has one simple base point and φ_K is birational, is of the form $\mathcal{R} := \mathbb{C}[y_0, y_1, y_2, y_3, w_0, w_1, u]/I$ where I is generated by the 2×2 minors of the matrix A above, and three more polynomials; one of degree 4 of the form $-w_1^2 + B(y_i) + \sum \mu_{ijk} y_i y_j w_k$, and the other 2 (of respective degrees 5,6) obtained rolling it twice (so they have the form $-w_1 u + \ldots, -u^2 + \ldots$).*

Canonical Rings of Surfaces   57

In the next section we will write explicitly these equations.

Proof. We know already all the generators and the relations in degree ≤ 3; moreover we know that all the minors of A are relations of $\mathcal{R}$. An easy dimension count shows that there is one relation missing in degree 4; this relation is in fact induced by the rank condition, as follows: in Corollary 2.6 we have seen that

$$v z_0 = B z_{-1} - G z_0 \, .$$

Using the fact that $v = \frac{q z_0}{z_{-1}} = \frac{x_2^2 z_0}{e^2 c}$ (cf. Corollary 2.6 and Lemma 2.10)

$$x_2^2 \frac{z_0^2}{e^2 c} = B z_{-1} - G z_0 \, .$$

Multipliying by $\frac{e^2}{c}$ we get the equality

$$w_1^2 = B(y_i) - \frac{G(y_i)}{ec} z_0 \, .$$

Using that $Q^2 x_2^2 = x_3 Q G + x_3^2 B$ is singular in $(1,0,0,0)$ (cf. proof of Lemma 2.10) and recalling that $G = G(x_0, x_1, x_2)$ (cf. Remark 2.7), we see that the coefficient of x_0^3 in G has to be zero and therefore $\frac{G(y_i)}{ec} z_0$ can be written as $-\sum \mu_{ijk} y_i y_j w_k$ for suitable coefficients μ_{ijk}: this provides a nontrivial relation in degree 4 of the form

$$(\#\#) \qquad\qquad w_1^2 = B(y_i) + \sum \mu_{ijk} y_i y_j w_k$$

which obviously is not in the ideal generated by the minors of A.

There are no further relations in degree 4 because they would force a new generator of $\mathcal{R}$ in degree 4, which is excluded by Theorem 1.8.

We showed in the proof of the last lemma that the singular point $(1,0,0,0)$ is locally a double cover of the plane branched along a curve with a singularity of order at least 4 and has tangent cone x_2^2.

It is easy to verify that this implies that $QG + x_3 B$ is contained in the ideal $(x_2^2, x_2 x_1^2, x_2 x_1 x_3, x_2 x_3^3, x_1^4, x_1 x_3^3, x_1^2 x_3^2, x_1 x_3^3, x_3^4)$, *i.e.*, the monomials $x_0^3, x_0^2 x_1, x_0^2 x_2, x_0 x_1^2$ do not appear in G, whence B is a quartic in $\mathbb{P}^3$ such that the monomials $x_0^4, x_0^3 x_1, x_0^3 x_3$ have coefficient zero.

Implementing this in $(\#\#)$ we find that the right side can be chosen (up to adding some element of the ideal generated by the minors of A) to be in the square of the ideal generated by the first row of A, hence it can be rolled twice.

By Theorem 1.8, we know that the ideal I of relations is generated in degree ≤ 6. We have three elements f_1, f_2, f_3 in I_3, f_4, f_5, f_6, f_7 in I_4, f_8 in I_5, f_9 in I_6.

Let I' be the ideal $(f_1, \ldots, f_9)$, R' be the quotient ring

$$\mathbb{C}[y_0, y_1, y_2, y_3, w_0, w_1, u] / I' \, .$$

58   I. C. Bauer et al.

To show $I' = I$ it suffices to show $I'_k = I_k\ \forall k \leq 6$, or equivalently, $\dim R'_k \leq \dim R_k$ for $k \leq 6$. This is a calculation done by Macaulay 2 (cf. Appendix 1, where the verification is done using the equations in Theorem 3.7).   □

3   The Canonical Ring of Surfaces with $K^2 = 7$, $p_g = 4$ Birational to a Sextic: Explicit Computations

Let S be as in the previous section, *i.e.*, a minimal surface with $K^2 = 7$, $p_g = 4$, such that the canonical system has one simple base point and the canonical map is birational.

In the last section we have shown that the image $\varphi(S) = \Sigma$ of the canonical map has an equation of the form

$$-x_2^2 Q^2 + x_3 QG + x_3^2 B$$

where Q is a quadric, G is a cubic, and both do not depend on x_3. Moreover, we have seen in the proof of Theorem 2.13 that the monomials $x_0^3, x_0^2 x_1, x_0^2 x_2$, $x_0 x_1^2$ do not appear in G, and that B is a quartic in $\mathbb{P}^3$ such that the monomials $x_0^4, x_0^3 x_1, x_0^3 x_3$ have coefficient zero.

In the following three subsections we write the canonical ring in three explicit ways, using different formats.

3.1   Rolling Factors Format

The proof of Theorem 2.13 provides immediately the equations in the "rolling factors" format that we have introduced in the previous section.

We have defined

$$A := \begin{pmatrix} y_1 & y_2 & y_3 & w_1 \\ w_0 & w_1 & Q(y_0, y_1, y_2) & u \end{pmatrix}.$$

We write

$$G = kx_1^3 + x_2 Q_0(x_1, x_2) + x_0 x_2 l(x_1, x_2)\,,$$
$$B = x_2 C(x_0, x_1, x_2, x_3) + x_3^2 Q_3(x_0, x_1, x_3) + x_3 x_1 Q_2(x_0, x_1) + x_1^2 Q_1(x_0, x_1)\,,$$

where $k \in \mathbb{C}$, l is linear, the Q_i's are quadratic and C is a cubic.

With the above notation, the proof of Theorem 2.13 gives the following description for the canonical ring:

Theorem 3.1 *Let S be a minimal (smooth, connected) surface with $K^2 = 7$ and $p_g(S) = 4$, such that the canonical system $|K_S|$ has one (simple) base point $x \in S$ and the canonical map is birational. Then the canonical ring $\mathcal{R}(S)$ of S can be written as*

$$\mathbb{C}[y_0, y_1, y_2, y_3, w_0, w_1, u]/\mathcal{I}\,,$$

Canonical Rings of Surfaces   59

where $\deg(y_i, w_j, u) = (1, 2, 3)$ *and* $\mathcal{I}$ *is generated by the* 2×2*-minors of the matrix*

$$\begin{pmatrix} y_1 & y_2 & y_3 & w_1 \\ w_0 & w_1 & Q(y_0, y_1, y_2) & u \end{pmatrix},$$

and by

$$-w_1^2 + kw_0y_1^2 + w_1Q_0(y_1, y_2) + w_1y_0l(y_1, y_2) + y_2C(y_0, y_1, y_2, y_3)$$
$$+y_3^2Q_3(y_0, y_1, y_3) + y_1y_3Q_2(y_0, y_1) + y_1^2Q_1(y_0, y_1) \, ,$$
$$-w_1u + kw_0^2y_1 + uQ_0(y_1, y_2) + uy_0l(y_1, y_2) + w_1C(y_0, y_1, y_2, y_3)$$
$$+y_3Q(y_0, y_1, y_2)Q_3(y_0, y_1, y_3) + w_0y_3Q_2(y_0, y_1) + y_1w_0Q_1(y_0, y_1) \, ,$$
$$-u^2 + kw_0^3 + w_1Q_0(w_0, w_1) + uy_0l(w_0, w_1)$$
$$+uC(y_0, y_1, y_2, y_3) + Q^2(y_0, y_1, y_2)Q_3(y_0, y_1, y_3)$$
$$+w_0Q(y_0, y_1, y_2)Q_2(y_0, y_1) + w_0^2Q_1(y_0, y_1) \, .$$

Proof. This is just the explicit expression of the computation in Theorem 2.13.
□

Remark 3.2 1) We remark that the above equations are not exactly "rolled", but only up to changing f_8 by a suitable combination of the 2×2-minors of A. Nevertheless we prefer to leave the equations in the above form because they are more readable.

2) Our goal is to show that this canonical ring can be deformed to a canonical ring of surfaces with the same invariants but with base point free canonical system.

Other ways of writing the same canonical ring might under this aspect be more convenient.

3.2   The $AM(^tA)$-format

Following the notation of M. Reid in [Rei2], the $AM(^tA)$-format is another way to describe the canonical ring of a Gorenstein variety of codimension 4 (defined by 9 relations). It was introduced by D. Dicks and M. Reid.

We briefly recall the definition of the $AM(^tA)$-format, referring for details to [Rei1], [Rei2], [Rei3].

Definition 3.3 Assume we are given a commutative ring

$$R = \mathbb{C}[x_1, \ldots, x_n]/(f_1, \ldots, f_9) \, .$$

Then we call this presentation in $AM(^tA)$-*format* iff there is a 2×4-matrix A such that $f_1, \ldots f_6$ are the 2×2-minors of A and there is a symmetric 4×4-matrix M such that

$$AM(^tA) = \begin{pmatrix} f_7 & f_8 \\ f_8 & f_9 \end{pmatrix} \, .$$

60   I. C. Bauer et al.

This format is slightly more restricted than the previous one. In fact, the equations given in the $AM(^tA)$ format are automatically in the "rolling factor" format, whereas equations given in the "rolling factor" format can be expressed in an $AM(^tA)$ format if and only if the 3 supplementary relations are "quadratic forms in the rows of A". This is equivalent to saying that the first one (the one we denoted by f_7) is in the square of the ideal generated by the entries of the first row of A, what happens to occur in our specific situation.

We are now ready to state our result.

Theorem 3.4 *Let S be a minimal (smooth, connected) surface with $K^2 = 7$ and $p_g(S) = 4$, such that the canonical system $|K_S|$ has one (simple) base point $x \in S$ and the canonical map is birational. Then the canonical ring $\mathcal{R}$ of S is of the form $\mathbb{C}[y_0, y_1, y_2, y_3, w_0, w_1, u]/\mathcal{I}$ ($\deg(y_i, w_j, u) = (1, 2, 3)$) and can be presented in the $AM(^tA)$-format with*

$$A = \begin{pmatrix} y_1 & y_2 & & y_3 & w_1 \\ w_0 & w_1 & Q(y_0, y_1, y_2) & u \end{pmatrix},$$

and

$$M = \begin{pmatrix} m_{11} & m_{12} & m_{13} & k_1 y_0 \\ & m_{22} & m_{23} & k_2 y_0 \\ & & m_{33} & L \\ \text{sym} & & & -1 \end{pmatrix}.$$

Here $k_1, k_2 \in \mathbb{C}$, $L = L(y_0, y_1, y_2, y_3)$ is linear and m_{ij} are quadratic forms depending on the variables $y_0, y_1, y_2, y_3, w_0, w_1$ as follows:

$$\begin{aligned}
m_{11} &= m_{11}(y_0, y_1, w_0, w_1) , \\
m_{12} &= a_1 w_1 + a_2 y_0 y_1 + a_3 y_0 y_2 + a_4 y_1^2 , \\
m_{13} &= m_{13}(y_0, y_1) , \\
m_{22} &= b_1 w_1 + b_2 y_0 y_2 + b_3 y_1^2 + b_4 y_1 y_2 + b_5 y_2^2 , \\
m_{23} &= m_{23}(y_0, y_1, y_2, y_3) , \\
m_{33} &= m_{33}(y_0, y_1, y_3) ,
\end{aligned}$$

where $a_i, b_j \in \mathbb{C}$.

As we will see in the proof, the coefficients of the entries of M can be explicitly determined from the equation of Σ.

Proof. First we note that A was already found in Theorem 3.1. In order to write down the matrix M, we have to set up some more notation.

We write explicitly (keeping the terminology which was introduced earlier)

$$\begin{aligned}
Q_0(y_1, y_2) &= q_{11} y_1^2 + q_{12} y_1 y_2 + q_{22} y_2^2 , \\
l(y_1, y_2) &= l_1 y_1 + l_2 y_2 ,
\end{aligned}$$

where $q_{ij}, l_k \in \mathbb{C}$.

We remark that the coefficient of y_0^2 in Q has to be different from zero because of the assumption that the generalized tacnode o is not contained in Γ. Therefore we can write

$$
\begin{aligned}
C(y_0, y_1, y_2, y_3) \\
= l'(y_0, y_1, y_2, y_3)Q + y_0(Q'_{11}y_1^2 + Q'_{12}y_1y_2 + Q'_{22}y_2^2) \\
+ y_0y_3l''(y_1, y_2, y_3) + sy_1^3 + y_2Q''(y_1, y_2) + y_3Q'''(y_1, y_2, y_3) \, ,
\end{aligned}
$$

where $s, Q'_{ij} \in \mathbb{C}$, l', l'' are linear forms and Q'', Q''' are quadratic forms.

Now, a rather lengthy, but straightforward calculation shows that for $M =$

$$
\begin{pmatrix}
kw_0 + Q_1 + q_{11}w_1 & \frac{1}{2}(q_{12}w_1 + y_0(Q'_{11}y_1 + Q'_{12}y_2) + sy_1^2) & \frac{Q_2}{2} & \frac{l_1y_0}{2} \\
& q_{22}w_1 + Q'_{22}y_0y_2 + Q'' & \frac{y_0l''+Q'''}{2} & \frac{l_2y_0}{2} \\
& & Q_3 & \frac{l'}{2} \\
\text{sym} & & & -1
\end{pmatrix} ,
$$

the conditions rank $A \le 1$ and $AM(^tA) = 0$ define the ideal $\mathcal{I}$ of Theorem 3.1.
$\square$

Remark 3.5 In [Rei2] M. Reid shows that, given a polynomial ring $\mathbb{C}[x_0, \ldots, x_n]$ and a Gorenstein $\mathbb{C}[x_0, \ldots, x_n]$–algebra $\mathcal{R}$ of codimension 4 and presented in $AM(^tA)$-format, all the syzygies are induced by A and M. In particular in this case the $AM(^tA)$-format is *flexible*, *i.e.*, every deformation of the matrices A and M preserving the symmetry of M induces a flat deformation of $\mathcal{R}$.

In order to find a flat family $\mathcal{S}/T$ of surfaces such that S_{t_0} is a surface with $K^2 = 7$, $p_g = 4$ such that the canonical system has one base point and the canonical map is birational, whereas S_t is a surface such that the canonical system has no base points for every $t \ne t_0$, it would be sufficient to find a deformation of the above matrices, which induces the right ring.

Unfortunately this does not work, since by [Cat1] the canonical ring of S_t (for $t \ne t_0$) is generated in degrees 1 and 2. In particular, if $\mathcal{R}_t$ is such a deformation of $\mathcal{R}$, for $t \ne 0$, one of the three relations in degree three has to eliminate the generator u. But, considering only deformations of $\mathcal{R}$ induced by deformations of A and M, the relations in degree three of $\mathcal{R}_t$ are given by the two by two minors of a matrix A_t with the following degrees in the entries:

$$
\begin{pmatrix} 1 & 1 & 1 \\ 2 & 2 & 2 \end{pmatrix} ,
$$

which obviously cannot eliminate an element u having degree 3.

62   I. C. Bauer et al.

3.3   The Antisymmetric and Extrasymmetric Format

The aim of this section is to give a third description of our canonical ring $\mathcal{R} = \mathbb{C}[y_0, y_1, y_2, y_3, w_0, w_1, u]/\mathcal{I}$.

Definition 3.6 1) A 6×6-matrix P is called *antisymmetric and extrasymmetric* if and only if it has the form

$$P = \begin{pmatrix} 0 & a & b & c & d & e \\ & 0 & f & g & h & d \\ & & 0 & i & pg & pc \\ & & & 0 & qf & qb \\ & & & & 0 & pqa \\ -\text{sym} & & & & & 0 \end{pmatrix}.$$

2) Assume we are given a commutative ring $\mathcal{R} = \mathbb{C}[x_1, \ldots, x_n]/\mathcal{I}$. If there is an antisymmetric and extrasymmetric matrix P whose 4×4 pfaffians generate the ideal $\mathcal{I}$, we say that the ring has an *antisymmetric and extrasymmetric format given by P*.

This format was first introduced by D. Dicks and M. Reid in a less general form [Rei1]. The above more general form appeared in [Rei3]. An easy computation shows that the ideal of the 15 (4×4) pfaffians of an antysimmetric and extrasymmetric matrix is in fact generated by nine of them.

Under suitable generality assumptions (e.g. that $\mathcal{R}$ be Gorenstein of codimension 4 and that the 9 equations be independent), it should be easy (but rather lengthy) to show, following the lines of Reid's argument used in [Rei1] to prove a special case, that the format is flexible (*i.e.*, that a deformation of the entries of P induces a flat deformation of the ring).

Since however we are only interested in our particular case, it does not pay off here to perform this calculation. Hence we will proceed as follows: We shall put our canonical ring in an antisymmetric and extrasymmetric format, deform the associated 6×6 antisymmetric and extrasymmetric matrix (preserving the extrasymmetry) and then verify the flatness of the induced family by the constancy of the Hilbert polynomial.

Using the same notation as in the previous section we obtain

Theorem 3.7 *Let S be a minimal (smooth, connected) surface with $K^2 = 7$ and $p_g(S) = 4$, whose canonical system $|K_S|$ has one (simple) base point $x \in S$ and yields a birational canonical map. Then the canonical ring of S can be presented as $\mathcal{R} = \mathbb{C}[y_0, y_1, y_2, y_3, w_0, w_1, u]/\mathcal{I}$, where $\deg(y_i, w_j, u) = (1, 2, 3)$ and the ideal of relations $\mathcal{I}$ of $\mathcal{R}$ is generated by the 4×4-pfaffians*

Canonical Rings of Surfaces   63

of the following antysimmetric and extrasymmetric matrix

$$P = \begin{pmatrix} 0 & 0 & w_0 & Q(y_0, y_1, y_2) & w_1 & u \\ & 0 & y_1 & y_3 & y_2 & w_1 \\ & & 0 & -u + \overline{C} & y_3\overline{Q}_3(w_0, y_0, y_1, y_3) & Q\overline{Q}_3 \\ & & & 0 & y_1\overline{Q}_1(w_0, w_1, y_0, y_1, y_3) & w_0\overline{Q}_1 \\ & & & & 0 & 0 \\ -\text{sym} & & & & & 0 \end{pmatrix},$$

where $Q, \overline{Q}_1, \overline{Q}_3$ *are quadratic forms of a subset of the given variables as indicated, and* $\overline{C}$ *is a cubic form. Moreover* $\overline{C}$ *does not depend on* u *and the* $w_i y_j$*'s for* $j \le 1$*,* $\overline{Q}_1$ *does not depend on* y_3^2*.*

Proof. Recall that the coefficient of x_0^2 in Q is different from zero. This allows us to choose $s \in \mathbb{C}$ and linear forms $\bar{l}_1(y_0, y_1), \bar{l}_2(y_0, y_1, y_2)$ such that

$$Q_2(y_0, y_1) = sQ(y_0, y_1, y_2) + y_1\bar{l}_1(y_0, y_1) + y_2\bar{l}_2(y_0, y_1, y_2) \, .$$

Therefore we can write

$$\overline{C} = q_{12}w_0y_2 + q_{22}w_1y_2 + y_0l(w_0, w_1) + C(y_0, y_1, y_2, y_3) + y_1y_3\bar{l}_2(y_0, y_1, y_2) \, ;$$
$$\overline{Q}_1 = kw_0 + q_{11}w_1 + Q_1(y_0, y_1) + y_3\bar{l}_1(y_0, y_1) \, ;$$
$$\overline{Q}_3 = -Q_3 - sw_0 \, .$$

Now the rest of the proof is a straightforward calculation. $\square$

4   An Explicit Family

In this section we will find an explicit deformation of the canonical ring $\mathcal{R}$ to the canonical ring $\mathcal{S}$ of a surface with $K^2 = 7$, $p_g = 4$, such that the canonical system has no base points. Before doing this we have to recall some of the results [Cat1] on surfaces with $K^2 = 7$, $p_g = 4$, whose canonical system is base point free.

Let X be a nonsingular surface with $K^2 = 7$, $p_g = 4$, such that the canonical system is base point free, hence the canonical map is a birational morphism onto a septic surface in $\mathbb{P}^3$. We denote by $\mathcal{S}$ the canonical ring of X. As in the previous case, we denote by y_i an appropriate basis of $H^0(X, \mathcal{O}_X(K))$. We set $\mathcal{A} := \mathbb{C}[y_0, y_1, y_2, y_3]$.

Theorem 4.1 1) $\mathcal{S}$ *has a minimal resolution as* $\mathcal{A}$*-module given by the matrix*

$$\alpha = \begin{pmatrix} d_1d_2y_0 + (d_3d_4 + d_2^2)y_1 + (d_2d_3 + d_1d_4)y_2 & d_4y_1 & d_1y_0 + d_2y_1 + d_3y_2 \\ d_4y_1 & y_0 & y_2 \\ d_1y_0 + d_2y_1 + d_3y_2 & y_2 & y_1 \end{pmatrix},$$

where d_1, d_2, d_3, d_4 *are arbitrary quadratic forms in* y_i*.*

64   I. C. Bauer et al.

2) α *satisfies the rank condition* $\Lambda^2(\alpha) = \Lambda^2(\alpha')$, *where* α' *is obtained by deleting the first row of* α, *and therefore induces a unique ring structure on* S *as quotient of* $\mathcal{B} = \mathcal{A}[w_0, w_1]$ *by the three relations given by*

$$\alpha \begin{pmatrix} 1 \\ w_0 \\ w_1 \end{pmatrix} = 0$$

and three more relations expressing $w_i w_j$ *as linear combination of the other monomials whose coefficients are determined by the adjoint matrix of* α.

3) *The surfaces with* $K^2 = 7$, $p_g = 4$ *such that the canonical system is base point free form an irreducible unirational component of dimension* 36 *in the moduli space* $\mathfrak{M}_{K^2=7, p_g=4}$ *of surfaces with* $K^2 = 7$, $p_g = 4$.

For a more precise formulation we refer to the original articles [Cat1] or [Cat3].

Remark 4.2 1) The previous theorem implies in particular that the canonical ring S is Gorenstein in codimension 3. Hence by the classical result of Buchsbaum and Eisenbud (cf. [B-E]) the ideal $J \subset \mathcal{A}[w_0, w_1] =: \mathcal{B}$ defining S can be minimally generated by the $2k \times 2k$-Pfaffians of a skewsymmetric $(2k + 1) \times (2k + 1)$-matrix. Writing down explicitly the above six defining equations, we can see that J is generated by only five of them, hence in our case $k = 2$.

2) More precisely, we have a selfdual resolution of S as $\mathcal{B}$-module as follows

$$0 \to \mathcal{B}(-9) \xrightarrow{f_2} \mathcal{B}(-5)^3 \oplus \mathcal{B}(-6)^2 \xrightarrow{f_1} \mathcal{B}(-4)^3 \oplus \mathcal{B}(-3)^2 \xrightarrow{f_0} \mathcal{B} \to S \to 0 ,$$

where f_1 is alternating, f_0 is given by the Pfaffians of f_1 and $f_2 = {}^t f_0$.

3) Vice versa, assume we have a $\mathbb{C}$-algebra S, which admits a resolution as above: then under suitable open condition S is the canonical ring of a surface X with $p_g = 4$, $K^2 = 7$ and free canonical system.

Our aim is now to take the matrix P in antisymmetric and extrasymmetric format and try to find a deformation P_t of P with the following properties:

0) For $t \neq 0$ in at least one of the Pfaffians of degree 3 the generator u appears with a non zero coefficient;
1) for $t \neq 0$ there is a skewsymmetric 5×5 -matrix Q_t, such that the ideal J_t generated by the 4×4-Pfaffians of Q_t coincides with $\mathcal{I}_t \cap \mathcal{B}$, where $\mathcal{I}_t$ is the ideal generated by the 4×4-Pfaffians of P_t, for every $t \neq 0$;
2) the entries of Q_t have the right degrees, *i.e.*, Q_t defines a map

$$\mathcal{B}(-5)^3 \oplus \mathcal{B}(-6)^2 \xrightarrow{Q_t} \mathcal{B}(-4)^3 \oplus \mathcal{B}(-3)^2 ;$$

3) $S_t := \mathbb{C}[y_0, y_1, y_2, y_3, w_0, w_1, u]/\mathcal{I}_t$ is a flat family.

From the previous remark we conclude that $\mathcal{S}_t$ (for $t \neq 0$), constructed as above, is the canonical ring of a surface X with $p_g = 4$, $K^2 = 7$ and free canonical system. Hence once we have found a deformation P_t as above, we have explicitly deformed the surfaces with $p_g = 4$, $K^2 = 7$, such that $|K_S|$ has one base point and induces a birational map to the surfaces with $p_g = 4$, $K^2 = 7$ and free canonical system.

The natural attempt now is to deform the entries of P preserving the extrasymmetry (this should give automatically the flatness), so that one of the relations in degree 3 eliminates u. At first glance we see a natural way: putting the deformation parameter t in the only entry of degree 0 (replacing the 0 therein) and replacing the symmetrical "zero" to preserve the extrasymmetry. This actually works and we have the following

Theorem 4.3 *Let P be an antisymmetric and extrasymmetric matrix as in Theorem 3.7. Consider the 1-parameter family of rings*

$$\mathcal{S}_t = \mathbb{C}[y_0, y_1, y_2, y_3, w_0.w_1, u]/\mathcal{I}_t$$

where the ideal $\mathcal{I}_t$ is given by the 4×4 pfaffians of the antysimmetric and extrasymmetric matrix

$$P_t = P + \begin{pmatrix} 0 & t & 0 & 0 & 0 & 0 \\ & 0 & 0 & 0 & 0 & 0 \\ & & 0 & 0 & 0 & 0 \\ & & & 0 & 0 & 0 \\ & & & & 0 & t\overline{Q}_1\overline{Q}_3 \\ -\text{sym} & & & & & 0 \end{pmatrix}.$$

This is a flat family and describes a flat deformation of the surface corresponding to the matrix P to surfaces with $p_g = 4$, $K^2 = 7$ and with $|K|$ base point free. For $t \neq 0$, $\mathcal{S}_t$ is isomorphic to $\mathbb{C}[y_0, y_1, y_2, y_3, w_0.w_1]/J_t$ where J_t is the ideal generated by the 4×4 pfaffians of the matrix

$$\begin{pmatrix} 0 & y_1 & -y_3 & -t^2 w_0 & -tQ \\ & 0 & y_2 & t^3\overline{Q}_3 & tw_1 \\ & & 0 & t^2 w_1 & t^2\overline{Q}_1 \\ & & & 0 & -t^3 c - t^2 y_1 Q + t^2 y_3 w_0 \\ -\text{sym} & & & & 0 \end{pmatrix}.$$

Proof. We have to check that the properties 0)-3) described above are fullfilled by our pair P_t, Q_t.

0): We immediately see that the 4×4-Pfaffian obtained eliminating the last two rows and columns eliminates the generator u;

1): once we eliminate u it is easy to write down explicitly $\mathcal{I}_t \cap \mathcal{B}$ for $t \neq 0$ and check that it coincides with the given ideal J_t (cf. Appendix 2);

2): obvious;

66   I. C. Bauer et al.

3): we expect that flatness holds in general once we preserve the extrasymmetry.

In our particular case, flatness follows auitomatically since:

a) $X_t := \mathrm{Proj}(\mathcal{B}/J_t)$ has dimension ≤ 2 by semicontinuity.
b) Then Q_t and its pfaffians give a resolution of J_t by [B-E].
c) Therefore the Hilbert polynomial of J_t is the same as the one of J_0, then the family is flat.

Finally, since the property that X_t has only R.D.P.'s as singularities is open, X_t is the canonical model of a surface of general type as required. $\quad\square$

Remark 4.4 As we have already said in the introduction, this particular degeneration was studied by Enriques in his book [Enr]. There Enriques states that such a degeneration should exist and suggest a way to construct such a family degenerating the canonical images. With the help of Macaulay 2 we have explicitly computed the degenerations of the canonical images corresponding to our family and we have found that the degeneration is not the one "predicted" by Enriques. In the following we recall briefly (see [Enr] for the details) Enriques' prediction and point out where is the difference.

The canonical image of a surface with $K^2 = 7$, $p_g = 4$ with base point free canonical system and birational canonical morphism is a surface in $\mathbb{P}^3$ of degree 7 with a singular curve of degree 7 and genus 4 having a triple point. Moreover the adjoint quadric is a quadric cone (with vertex in the singular point of the curve) whose intersection with the surface is given by the above curve of degree 7 counted twice.

As we have already seen, the canonical image of a surface with $K^2 = 7$, $p_g = 4$ with one simple base point for the canonical system and canonical map birational is a surface in $\mathbb{P}^3$ of degree 6 with a singular curve of degree 2 and a generalized tacnode. In this case there is an adjoint plane that is the plane through the conic ($\{x_3 = 0\}$ in our notation).

Enriques suggests to add to the sextic the adjoint plane. The intersection of this plane and the sextic is given by the singular conic and the line image of E both counted twice; this gives a reducible septic with a triple conic and a double line of "tacnodal type" (*i.e.*, near a general point of the line the surface has two branches tangent on the line).

Enriques states that it is possible to construct a family of septics with a singular curve of degree 7 and genus 4 having a triple point that degenerates to the above configuration so that the singular septic degenerates to the union of the conic (counted three times) and the line.

We wrote the canonical images of the family described by the 4×4 pfaffians of the 5×5 skewsymmetric matrix in Theorem 4.3, and, as anticipated, we did not get the situation predicted by Enriques. In fact the family of septics degenerates to the union of the sextic canonical image of the limit surface with a plane, but instead of the plane predicted by Enriques we have gotten the plane $\{x_2 = 0\}$ (the reduced tangent cone of the tacnode).

In fact, it is quite easy to compute also the degeneration of the adjoint quadric: consider the resolution of the canonical ring of a surface with base point free canonical system as $\mathcal{A}$-module as in Theorem 4.1. From this resolution one can immediately see that the adjoint quadric must be the determinant of the right-bottom 2 by 2 minor of the resolution matrix in Theorem 4.1 (*i.e.*, $y_0 y_1 - y_2^2$ in the coordinates chosen there). This minor depends only on the two relations in degree 3 of the canonical ring.

It is now easy to compute it for our family: we have just to write down the two relations in degree 3, write the 2 by 2 matrix of the coefficients of w_0 and w_1 in this two equations, and then compute the determinant.

In the notation of Theorem 4.3 the two relations in degree 3 (for $t \neq 0$) can be written as:

$$y_1 w_1 - w_0 y_2 + t y_3 \overline{Q}_3 \; ;$$
$$t y_1 \overline{Q}_1 + w_1 y_3 - Q y_2 \; .$$

The equation of the quadric cone depends clearly on the coefficients of w_0 and w_1 in $\overline{Q}_1$ and $\overline{Q}_1$ (but notice that it is in every case independent of y_0, so it cannot be a smooth quadric, as expected), but the two equations degenerate respectively to $y_1 w_1 - w_0 y_2$ and $w_1 y_3 - Q y_2$; the 2 by 2 minor degenerates then to

$$\begin{pmatrix} -y_2 & y_1 \\ 0 & y_3 \end{pmatrix}$$

and the quadric cone degenerate to the union of the adjoint plane and the tacnodal plane.

Geometrically we could say that the septic degenerates to the union of a sextic and a plane, the adjoint quadric to the union of the same plane and a different plane (the tacnodal plane); the two "identical" planes "simplify" and we are left with the sextic and his adjoint plane.

References

[B-P-V]   W. Barth, C. Peters, and A. Van de Ven, *Compact Complex Surfaces*, Springer-Verlag, 2001.

[Ba]   I. Bauer, Surfaces with $K^2 = 7$ and $p_g = 4$, *Mem. Am. Math. Soc.* **721** (2001), 79 p.

[Bom]   E. Bombieri, Canonical models of surfaces of general type, *Publ. Scient. I.H.E.S.* **42** (1973), 447–495.

[B-E]   D. A. Buchsbaum and D. Eisenbud, Algebra structures for finite free resolutions and some structure theorems for ideals of codimension 3, *Am. J. of Math.* **99** (1977), 447–485.

[Cat1]   F. Catanese, Commutative algebra methods and equations of regular surfaces, *Alg. Geom. – Bucharest 1982, Springer LNM* **1056** (1984), 68–111.

68   I. C. Bauer et al.

[Cat2]   F. Catanese, On the moduli spaces of surfaces of general type, *J. Diff. Geom.* **19** (1984), 483–515.

[Cat3]   F. Catanese, Homological algebra and algebraic surfaces, *Proceedings of Symposia in Pure Mathematics* **62**, 1 (1997), 3–56.

[Cil1]   C. Ciliberto, Canonical surfaces with $p_g = p_a = 4$ and $K^2 = 5, \ldots, 10$, *Duke Math. Journal* **48**(1) (1981), 121–157.

[Cil2]   C. Ciliberto, On the degree of the generators of the canonical ring of a surface of general type, *Rend. Semin. Mat., Torino* **41**(3) (1983), 83–111.

[Deb]   O. Debarre, Inégalités numériques pour les surfaces de type général, *Bull. Soc. Math. France* **110** (1982), 319–346.

[Dic1]   D. Dicks, Surfaces with $p_g = 3$, $K^2 = 4$ and extension-deformation theory, Warwick Ph. D. thesis, 1988.

[Dic2]   D. Dicks, Surfaces with $p_g = 3$, $K^2 = 4$, Warwick preprint, 1988.

[Enr]   F. Enriques, *Le Superficie Algebriche*, Zanichelli, Bologna, 1949.

[Fran]   A. Franchetta, Su alcuni esempi di superficie canoniche, *Rend. Sem. Mat. Roma* **3** (1939), 23–28.

[Fra]   P. Francia, On the base points of the bicanonical system, *Problems in the Theory of Surfaces and their Classification (Cortona, 1988), Sympos. Math.* **XXXII** (1991), 141–150, Academic Press, London.

[Gri]   E. Griffin, Families of quintic surfaces and curves, *Compositio Math.* **55** (1986) , 33–62.

[HorQ]   E. Horikawa, On deformations of quintic surfaces, *Inv. Math.* **31** (1975), 43–85.

[HorI-V]   E. Horikawa, Algebraic surfaces of general type with small c_1^2, I. *Ann. of Math.* **104** (1976), 357–387; II. *Inv. Math.* **37** (1976), 121–155; III. *Inv. Math.* **47** (1978), 209–248; IV. *Inv. Math.* **50** (1979), 103–128; V. *J. Fac. Sci. Univ. Tokyo, Sect. A. Math.* **283** (1981),745–755.

[Kod]   K. Kodaira, On characteristic systems of families of surfaces with ordinary singularities in a projective space, *Amer. J. of Math.* **87** (1965), 227–256.

[Kur]   M. Kuranishi, Deformations of compact complex manifolds, Presses de l'Universite de Montreal, 1969.

[dJ-vS]   T. de Jong and D. van Straten, Deformations of the normalization of hypersurfaces (English), *Math. Ann.* **288**(3) (1990), 527–547.

[Lau]   H. B. Laufer, On minimally elliptic singularities, *Am. J. of Math.* **99** (1977), 1257–1295.

[Max]   E. A. Maxwell, Regular canonical surfaces of genus three and four, *Proc. Cambridge Philosophical Soc.* **23** (1937), 306–310.

[Pal]   V. P. Palamodov, Deformations of complex spaces, *Russ. Math. Surveys* **31**(1) (1976), 129–197.

[Rei0]   M. Reid, Problem 12, Part 2, *Problemes ouverts, Séminaire de Mathématiques Supérieures*, Univ. Montréal, 1980.

[Rei1]   M. Reid, Infinitesimal view of extending a hyperplane section-deformation theory and computer algebra, *Alg. Geometry, L'Aquila, LNM* **1417** (1988), Springer Verlag, 214–286.

[Rei2]   M. Reid, Surfaces with $p_g = 3$, $K^2 = 4$ according to E. Horikawa and D. Dicks, *Proc. of Alg. Geometry mini Symposium, Tokyo Univ. Dec. 1989*, 1–22.

[Rei3]   M. Reid, Graded rings and birational geometry, *Proc. of algebraic geometry symposium (Kinosaki, Oct 2000)*, K. Ohno (Ed.), 1–72 .

Canonical Rings of Surfaces 69

[Zuc]   F. Zucconi, Surfaces with canonical map of degree three and $K^2 = 3p_g - 5$, *Osaka J. Math.* **34**(2), 411–428.

```
-- APPENDIX 1
-- This script checks that the relations we found in Theorem 2.13 are
-- all the relations of the canonical ring till degree 6, so they are
-- all the relations by Theorem 1.8.
-- First we write the ring: with all the variables and parameters we
-- need
R=QQ[u,w_1,w_0,y_0,y_1,y_2,y_3,
    a0,a1,a2,b0,d0,d1,t,Q,Q1,Q3,c,q00,q01,q02,q11,q12,q22,
    q100,q101,q103,q111,q113,q3,q300,q301,q303,q311,q313,q333,
    c000,c001,c002,c003,c011,c012,c013,c022,c023,c033,
    c111,c112,c113,c122,c123,c133,c222,c223,c333,
    MonomialOrder=>Lex
    ]
-- Now we write the matrix in Theorem 3.7
M=matrix{
{0,0,w_0,Q,w_1,u},
{0,0,y_1,y_3,y_2,w_1},
{-w_0,-y_1,0,-u+(c+a0*w_1*y_0-a1*w_1*y_1+a2*w_1*y_2+b0*w_0*y_0),
                                                  y_3*Q3,Q*Q3},
{-Q,-y_3,u-(c+a0*w_1*y_0-a1*w_1*y_1+a2*w_1*y_2+b0*w_0*y_0),0,
                (Q1+d0*w_0+d1*w_1)*y_1,w_0*(Q1+d0*w_0+d1*w_1)},
{-w_1,-y_2,-y_3*Q3,-(Q1+d0*w_0+d1*w_1)*y_1,0,0},
{-u,-w_1,-Q*Q3,-w_0*(Q1+d0*w_0+d1*w_1),0,0}
};
pfaff=pfaffians(4,M);

-- Here we restrict to 9 pfaffians and check that they are enough to
-- generate the whole pfaffian ideal (the second line gives ``true''
-- as output)
pfaff9=submatrix(gens(pfaff),,{0,1,5,2,6,9,3,4,7});
gens(pfaff) % ideal(pfaff9)==0
-- Then we write explicitly all the polynomials in the matrix
ourideal:=substitute(pfaff9,{

Q=>y_0*y_0+q01*y_0*y_1+q02*y_0*y_2+q11*y_1*y_1+q12*y_1*y_2+q22*y_2*y_2,

Q1=>q100*y_0*y_0+q101*y_0*y_1+q103*y_0*y_3+q111*y_1*y_1+q113*y_1*y_3,

Q3=>q3*w_0+q300*y_0*y_0+q301*y_0*y_1+q303*y_0*y_3+q311*y_1*y_1+
    q313*y_1*y_3+q333*y_3*y_3,

c=>c000*y_0*y_0*y_0+c001*y_0*y_0*y_1+c002*y_0*y_0*y_2+c003*y_0*y_0*y_3+
    c011*y_0*y_1*y_1+c012*y_0*y_1*y_2+c013*y_0*y_1*y_3+c022*y_0*y_2*y_2+
    c023*y_0*y_2*y_3+c033*y_0*y_3*y_3+c111*y_1*y_1*y_1+c112*y_1*y_1*y_2+
```

70   I. C. Bauer et al.

```
    c113*y_1*y_1*y_3+c122*y_1*y_2*y_2+c123*y_1*y_2*y_3+c133*y_1*y_3*y_3+
    c222*y_2*y_2*y_2+c223*y_2*y_2*y_3+c333*y_3*y_3*y_3
         })

-- Here we define the ideal of the monomials in all the degrees till 6
linear:=ideal(y_0,y_1,y_2,y_3);
quadrics:=ideal(w_0,w_1)+linear^2;
cubics:=ideal(mingens(ideal(u)+linear*quadrics));
quartics:=ideal(mingens(quadrics^2+linear*cubics));
quintics:=ideal(mingens(quadrics*cubics+linear*quartics));
sextics:=ideal(mingens(cubics*cubics+quadrics*quartics+
                                     linear*quintics));
-- finally we compute a system of generators, degree by degree, of the
-- resulting quotient. All of them turn out to be composed exactly by
-- P_n elements (resp. 4,12,26,47,75,110), that concludes the argument
-- in the proof of Theorem 2.13
K=mingens ideal((gens linear) % gb ourideal);
twoK=mingens ideal((gens quadrics) % gb ourideal);
threeK=mingens ideal((gens cubics) % gb ourideal);
fourK=mingens ideal((gens quartics) % gb ourideal);
fiveK=mingens ideal((gens quintics) % gb ourideal);
sixK=mingens ideal((gens sextics) % gb ourideal);

restart



-- APPENDIX 2

R=QQ[t,u,y_1..y_3,Q,Q1,Q3,w_0..w_1,c,
      Degrees=>{1,3,1,1,1,2,2,2,2,2,3}]

-- the 4 x 4 - Pfaffians of the matrix M define the relations
-- of the canonical ring of a surface with K^2 = 7, p_g = 4 such that
-- the canonical map has one base point and is birational.

M=matrix{
{0,0,w_0,Q,w_1,u},
{0,0,y_1,y_3,y_2,w_1},
{-w_0,-y_1,0,-u+c,y_3*Q3,Q*Q3},
{-Q,-y_3,u-c,0,Q1*y_1,w_0*Q1},
{-w_1,-y_2,-y_3*Q3,-Q1*y_1,0,0},
{-u,-w_1,-Q*Q3,-w_0*Q1,0,0}};
M=map(R^{-1,-2,3:0,1},R^{-2,-1,3:-3,-4},M)

-- we calculate the 4 x 4 - Pfaffians of M and extract the 9
-- ''important'' as above;
pfaff15=pfaffians(4,M);
pfaff9=submatrix(gens(pfaff15),,{0,1,5,2,6,9,3,4,7});
```

Canonical Rings of Surfaces 71

```
gens(pfaff15) % ideal(pfaff9)==0

-- we write the sixteen syzygies of them
syzs=syz pfaff9

-- we define the matrix $Mt = M + tM1$, which is the deformation
-- of $M$ whose pfaffians we want to understand;

M1=matrix(R,{
          {0,1,0,0,0,0},
          {-1,0,0,0,0,0},
          {0,0,0,0,0,0},
          {0,0,0,0,0,0},
          {0,0,0,0,0,Q3*Q1},
          {0,0,0,0,-Q3*Q1,0}
          });
M1=map(R^{-1,-2,3:0,1},R^{-2,-1,3:-3,-4},M1)
Mt=M+t*M1;
-- we calculate the 15 Pfaffians of Mt and verify that the same nine
-- Pfaffians of Pt again generate the whole ideal (the output of the
-- third line below is ''true'');
defpfaff=pfaffians(4,Mt);
defpfaff9=submatrix(gens(defpfaff),,{0,1,5,2,6,9,3,4,7});
gens(defpfaff) % ideal(defpfaff9)==0

-- If t in different from 0, one can eliminate the variable u using
-- the first equation in defpfaff9.

elimu=defpfaff9_(0,0)

-- we will use the following trick to eliminate u:
-- u appears only in degrees smaller than 2 in defpfaff9; we
-- multiply defpfaff9 by t^2, and reduce by elimu;
-- what we get is the same ideal defpfaff as before for every t
-- different from zero!
-- Finally we divide by t, wherever it is possible;
-- here it is crucial that we choose a monomial order such that tu
-- is the leading term of elimu: this forces the result to be
-- independent of u;

defpfaff9withoutu:= divideByVariable((t^2*defpfaff9) % elimu,t);

-- the following 5 generators are enough; we choose a strange order
-- in order to get a nicer result;
fiveequationsfortnotzero=submatrix(defpfaff9withoutu,,{5,4,2,3,1});
defpfaff9withoutu % ideal(fiveequationsfortnotzero)==0

-- now we look for the 5x5 matrix inducing these equations as
```

72   I. C. Bauer et al.

```
-- pfaffians:

lookforQ=syz fiveequationsfortnotzero;

-- among the 36 syzygies (Macaulay found a lot of them because he is
-- considering also the case t=0) one can easily find something that
-- looks interesting
almostQ=submatrix(lookforQ,,{1,3,2,5,9})
-- this matrix is not (yet) antisymmetric; we change coordinates in
-- the source and in the target in order to make it antisymmetric;
one:=matrix(R,{{1}})
diag1=one++one++one++(t*one)++(-1*one)
diag2=one++one++(-1*one)++(-t^2*one)++(-t*one)
Qt=diag1*almostQ*diag2

-- finally we check, whether for t different from 0, the Pfaffians
-- of Qt and the Pfaffians of Mt after having eliminated u generate
-- the same ideal;
pfaffQt=pfaffians(4,Qt);
pfaffQred=divideByVariable(gens pfaffQt,t);
fiveequationsfortnotzero % pfaffQred==0
pfaffQred % fiveequationsfortnotzero==0
```

Attractors

Araceli M. Bonifant[1] and John Erik Fornæss[2] *

[1] Instituto de Matemáticas UNAM, Campus Cuernavaca, Av. Universidad s/n, Col. Lomas de Chamilpa, 62210 Cuernavaca, Mor., MEXICO
   E-mail address: bonifant@matcuer.unam.mx
[2] Department of Mathematics, University of Michigan, East Hall 525, East University Avenue, Ann Arbor, MI 48109-1109, USA
   E-mail address: fornaess@math.lsa.umich.edu

Abstract In this paper we investigate attractors that are extended in space, but where the internal dynamics is ignored.

Table of Contents

1 Introduction 73
2 Endomorphisms 74
3 Hyperbolic Diffeomorphisms 77
4 Holomorphic Endomorphisms of $\mathbb{P}^k$ 79
   References 83

1 Introduction

Attractors play an important role in dynamics. The most basic ones are attracting fixed points (or periodic points). There has also been a lot of investigation about attractors that are extended in space. In that case, the focus has been on the investigation of detailed dynamical properties inside the attractor. See Lorenz [L], Tucker [T], Hénon [H], Bendedicks-Carleson [BC], Fornæss-Gavosto [FG], Fornæss-Weickert [FW], Jonsson-Weickert [JW] and Fornæss-Sibony [FS2]. In this paper we investigate attractors that are extended in space, but where the internal dynamics is ignored.

Let M be a topological space and $f : M \to M$ a continuous map. Let ρ be a continuous pseudometric on M and let μ be a probability measure on M. An *attractor*, in this paper, is a distinguished point p in M which is said to *absorb* all points $q \in M$ whose orbit get closer than a given number, called *radius*, $\varepsilon_p > 0$, *i.e.*, $\rho(f^n(p), f^n(q)) \leq \varepsilon_p$ for some integer $n \geq 0$. (An alternative terminology suggested by John Hubbard and John Milnor in private communications is to call such p, q, ϵ−friends.) The basin of attraction, $B(p)$,

---

* The second author is supported by an NSF grant.
  2000 AMS classification Primary: 32H50 ; Secondary 37A25
  Key words and phrases. attractors, mixing measures, complex dynamics.

74   Araceli M. Bonifant and John Erik Fornæss

of an attractor p consists of those q which gets absorbed by p. We will say that such q collide with p. We can also call such an attractor a collision-attractor, a sticky attractor or a planetary attractor. One might suggestively think of asteroids colliding with planets. Gravitational attraction makes the asteroid stick to the planet. More generally, structures grow by adding ingredients.

In Section 2 we consider the case of an invariant mixing ergodic measure and prove that all attractors absorb mass at the same rate. In Section 3, we show for hyperbolic maps with product structure that a dense set of points avoid any given attractor. In Section 4 we extend the result of Section 3 to arbitrary holomorphic maps of $\mathbb{P}^k$. We would like to thank the referee for many helpful comments.

2   Endomorphisms

Let (M, σ) be a compact metric space M, with metric σ and $f : M \to M$ a continuous map. Let ρ be a continuous pseudometric on M. We assume that M carries an invariant, mixing Borel probability measure μ. So $\mu(E) = \mu(f^{-1}(E))$ and

$$c_n(E, F) := \mu(f^{-n}(E) \cap F) \to_{n \to \infty} \mu(E)\mu(F)$$

for all measurable sets E, F. For $x \in M, \varepsilon > 0, n \geq 0$ let

$$C_\rho(x, \varepsilon, n) := \{y \in M; \rho(f^n(x), f^n(y)) \leq \varepsilon\}$$
$$C_\rho(x, \varepsilon, n) := \cup_{k \leq n} C_\rho(x, \varepsilon, k)$$
$$\mathcal{C}_\rho(x, \varepsilon) := \cup_n C_\rho(x, \varepsilon, n)$$
$$\mathcal{C}_\rho(x) := \cap_{\varepsilon > 0} \mathcal{C}_\rho(x, \varepsilon) .$$

So $\mathcal{C}_\rho(x, \varepsilon)$ is the basin $B(x)$ of x with the given radius, $\varepsilon_x = \varepsilon$.

Proposition 2.1. *Let $0 < \zeta < 1$. There exists an integer $N = N(\varepsilon, \zeta)$ so that if $x \in S_\mu$, the support of μ, i.e., the smallest compact set K for which $\mu(K^c) = 0$, then $\mu(C_\rho(x, \varepsilon, N)) \geq \zeta$.*

Proof. It suffices to prove the Proposition in the case when the pseudometric ρ is the given metric σ of M since those basins are smaller after changing ε. We cover S_μ by finitely many open balls $V_i = B(x_i, \varepsilon/2)$ each of which intersects S_μ. Set $\eta = \min_i \mu(V_i) > 0$. For $j \geq 0$, let P_j denote the following statement:

P_j: There exist an integer $n_j \geq 1$, finitely many open sets $\{U_\ell^j\}_\ell$ and compact sets $\{F_\ell^j\}_\ell$ in M such that
(i) $S_\mu \subset \cup_\ell U_\ell^j$
(ii) $\mu(F_\ell^j) \geq 1 - \left(1 - \frac{\eta}{2}\right)^j$
(iii) If $x \in U_\ell^j$ then $C_\rho(x, \varepsilon, n_j) \supset F_\ell^j$.

The statement P_0 is trivially satisfied with $U_1^0 = M$, $F_1^0 = \emptyset$ and $n_0 = 1$.

Suppose that we have proved $P_j, j \geq 0$. We prove P_{j+1}. For each i, ℓ we have from the mixing property of μ that

$$\mu(f^{-n}(V_i) \cap (M \setminus F_\ell^j)) \to_{n \to \infty} \mu(V_i)\mu(M \setminus F_\ell^j)$$
$$= \mu(V_i)(1 - \mu(F_\ell^j)) \ .$$

Hence we can find $n_{j+1} > n_j$ so that for all i, ℓ

$$\mu(f^{-n_{j+1}}(V_i) \cap (M \setminus F_\ell^j)) \geq \eta(1 - \mu(F_\ell^j)) - \delta$$

where $\delta > 0$ will be fixed later. (See formula $(*)$ below.)

Next let $K_{i,\ell}^j \subset M \setminus F_\ell^j$ be a compact subset of $f^{-n_{j+1}}(V_i) \cap (M \setminus F_\ell^j)$ so that

$$\mu(K_{i,\ell}^j) \geq \eta(1 - \mu(F_\ell^j)) - 2\delta \ .$$

Define next $F_{i,\ell}^{j+1} = F_\ell^j \cup K_{i,\ell}^j$. Also define $U_{i,\ell}^{j+1} = f^{-n_{j+1}}(V_i) \cap U_\ell^j$. We show that $\{U_{i,\ell}^{j+1}, F_{i,\ell}^{j+1}\}_{i,\ell}$ satisfy $(i), (ii), (iii)$ except for reordering.

(i) If $x \in S_\mu$, then by $P_j, x \in U_\ell^j$ for some ℓ. Also $f^{n_{j+1}}(x) \in S_\mu \subset \cup V_i$. Hence $f^{n_{j+1}}(x) \in V_i$ for some i. Therefore, $x \in U_{i,\ell}^{j+1}$. So $S_\mu \subset \cup U_{i,\ell}^{j+1}$.

(iii) If $x \in U_{i,\ell}^{j+1}$, then $x \in U_\ell^j$ so $C_\rho(x, \varepsilon, n_{j+1}) \supset C_\rho(x, \varepsilon, n_j) \supset F_\ell^j$. Also $x \in U_{i,\ell}^{j+1} \Rightarrow f^{n_{j+1}}(x) \in V_i$. If $y \in K_{i,\ell}^j$, then $y \in f^{-n_{j+1}}(V_i)$ so $f^{n_{j+1}}(y) \in V_i$ also. Hence $\rho(f^{n_{j+1}}(x), f^{n_{j+1}}(y)) < \varepsilon$ which implies that $y \in C_\rho(x, \varepsilon, n_{j+1})$. Hence $C_\rho(x, \varepsilon, n_{j+1}) \supset F_\ell^j \cup K_{i,\ell}^j = F_{i,\ell}^{j+1}$.

(ii) Since $K_{i,\ell}^j \subset M \setminus F_\ell^j$, we have

$$\mu(F_{i,\ell}^{j+1}) = \mu(F_\ell^j \cup K_{i,\ell}^j)$$
$$= \mu(F_\ell^j) + \mu(K_{i,\ell}^j)$$
$$\geq \mu(F_\ell^j) + \eta(1 - \mu(F_\ell^j)) - 2\delta$$
$$= (\eta - 2\delta) + \mu(F_\ell^j)(1 - \eta)$$
$$\geq \eta - 2\delta + (1 - (1 - \frac{\eta}{2})^j)(1 - \eta/2 - \eta/2)$$
$$= 1 - (1 - \frac{\eta}{2})^{j+1} + \frac{\eta}{2}(1 - \frac{\eta}{2})^j - 2\delta \ .$$

So we just have to choose

$$(*) \quad \delta = \frac{\eta}{4}(1 - \frac{\eta}{2})^j \ .$$

This proves the inductive hypothesis P_{j+1}.

To complete the proof of the proposition, choose $j \geq 0$ such that $(1 - \frac{\eta}{2})^j \leq 1 - \zeta$ and put $N = n_j$. Then, if $x \in S_\mu$, by $P_j(i)$, $x \in U_\ell^j$ for

76   Araceli M. Bonifant and John Erik Fornæss

some j, ℓ. Hence by $P_j(iii), \mu(C_\rho(x, \varepsilon, n_j)) \geq \mu(F_\ell^j)$ and by $P_j(ii), \mu(F_\ell^j) \geq 1 - (1 - \eta/2)^j \geq \zeta$. Therefore $\mu(C_\rho(x, \varepsilon, N)) = \mu(C_\rho(x, \varepsilon, n_j)) \geq \zeta$ as desired. $\square$

Corollary 2.2. *For every $x \in S_\mu$, $\mu(C_\rho(x)) = 1$.*

Proof. For every $\varepsilon > 0$, $C_\rho(x, \varepsilon) \supset C_\rho(x, \varepsilon, N(\varepsilon, \zeta))$ for all $0 < \zeta < 1$. Hence $\mu(C_\rho(x, \varepsilon)) = 1$. If $\varepsilon_1 < \varepsilon_2$, then $C_\rho(x, \varepsilon_1) \subset C_\rho(x, \varepsilon_2)$ and hence $C_\rho(x) = \cap_{n=1}^\infty C_\rho(x, 1/n)$ so $\mu(C_\rho(x)) = 1$. $\square$

Let μ^2 denote the product measure $\mu \times \mu$ on $M \times M$.

Corollary 2.3. *Let $\varepsilon > 0$. Under the conditions of Proposition 2.1, the set of $(p, q) \in M \times M$ which collide, i.e., $\rho(f^n(p), f^n(q)) \leq \varepsilon$ for some n, has full μ^2 measure.*

Proof. For any $p \in S_\mu$, $\mu(C_\rho(p, \varepsilon)) = 1$. All points $q \in C_\rho(p, \varepsilon)$ collide with p. Integrating over p, the result follows. $\square$

We finish this section by giving a few examples.

In the first two examples the map is ergodic but not mixing. The last three examples are all mixing, and attractors differ in whether their basins cover all of S_μ.

The irrational rotation of the unit circle, $f : T \to T$, $f(z) = ze^{i\psi}$ for a $e^{i\psi}$ not a root of unity, is ergodic for the arc-length measure $\mu = \frac{d\theta}{2\pi}$.

Example 2.4. We choose first the pseudometric $\rho_1(z, w) = |\Re(z - w)|$. For any attractor $p \in T$ and any $\varepsilon > 0$, $B(p) = T$. Physically one can think of the time-1 map of a harmonic oscillator where $\Re(z)$ represents position and $\Im(z)$ represents momentum. Any $q \in T$ eventually collides with p, i.e., gets closer than $\rho_1 = \varepsilon$, so the basin of p is the whole circle.

Example 2.5. However, for the metric $\rho_2(z, w) = |z - w|$, the basin is never the whole circle if ε is small. Physically this represents time-1 maps of particles in circular motion.

Example 2.6. If we use the pseudometric ρ_1 above for the mixing map $f = z^2$ on the circle, we also get that for small $\varepsilon > 0$, the basin of q will not contain all points.

Example 2.7. Let $N \geq 1, X = \{x = (x_n)_{n=0}^\infty, x_n \in \{0, 1\}; \prod_{k=0}^N x_{n+k} = 0 \ \forall n\}$. We define a metric on X, $\sigma(x, y) = \sum_{n=0}^\infty \frac{|x_n - y_n|}{2^n}$. The shift map $f : X \to X, f(x_0, x_1, \dots) = (x_1, x_2, \dots)$ is continuous and surjective. We choose a pseudometric $\rho(x, y) = |x_0 - y_0|$. Then we have that for any $0 < \varepsilon < 1, C_\rho((0, 0 \dots), \varepsilon, N + 1) = X$.

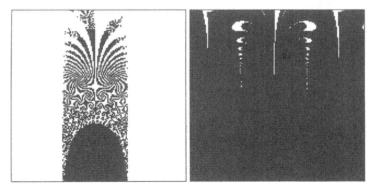

Figure 1. Attractor for $f(z) = \exp(i\psi z)$. The picture (left) shows the basin of 1 in white after 2 iterates. The picture (right) shows the attractor after 10 iterates. In both cases $\psi = .7$ and $\varepsilon = 0.5$. The pseudometric is $|\Re z - \Re w|$. The rectangles are [-8, 4]x [-14, -2]

Example 2.8. We can construct a compact set K in the plane and a continuous self-map f with an ergodic mixing probability measure μ with support K. Moreover, each point is an attractor which absorbs all points.

We proceed inductively. Suppose that K_n is a finite set $\{p_j\}_{j=1}^N$ and F_n is a permutation. Let $C_j, j = 1,\ldots, N$ be small circles centered at p_j. The radii will shrink very rapidly with n. Fix a number $\tilde{N} > N$ which is relatively prime with N and let $M \gg \tilde{N} \times N$. We put M equidistributed points $\{p_{j,k}\}_{k=1}^M$ around each p_j except around p_1 where we use $\{p_{1,k}\}_{k=1}^{M+\tilde{N}}$. Next we define the compact set $K_{n+1} := \{p_{j,k}\}$ and define F_{n+1} by letting

$$F_{n+1}(p_{j,k}) = p_{j+1,k}, \; j = 1,\ldots, N-1, k = 1,\ldots, M$$
$$F_{n+1}(p_{N,k}) = p_{1,k+1}, \; k = 1,\ldots, M$$
$$F_{n+1}(p_{1,k}) = p_{1,k+1}, \; k = M+1,\ldots, M+\tilde{N}-1$$
$$F_{n+1}(p_{1,M+\tilde{N}}) = p_{1,1} \, .$$

Any given two points on the limiting compact set have orbits that come arbitrarily close to each other.

In Fig. 1 we give pictures of an attractor after 2 and 10 iterations.

3 Hyperbolic Diffeomorphisms

Suppose that M is a smooth manifold with Riemannian metric $\rho = ds$. Let $f : M \to M$ be a smooth $\mathcal{C}^\infty$ diffeomorphism. Assume that K is a compact totally invariant perfect set. Suppose that f is hyperbolic on K with continuously varying stable subbundle E^s and unstable bundle E^u. We assume that $\dim E^s, \dim E^u > 0$ and that K has local product structure.

78   Araceli M. Bonifant and John Erik Fornæss

Proposition 3.1. *If $p \in K$, then $K \setminus C_\rho(p)$ is dense in K.*

Proof. Suppose $p \in K$. Pick $q \in K \setminus \{p\}$. We want to show that there are points in K arbitrarily close to q which doesn't belong to $C_\rho(p)$. Let $0 < \tau \ll 1$ be arbitrary and let $W^u_\tau(q)$ denote the local unstable manifold of q with radius τ. It suffices to find an integer n so that $W^u_\tau(q) \setminus C_\rho(p, 1/n) \neq \emptyset$.

We start by writing down estimates that follow from the hyperbolic local product structure. There exist constants $\delta, \eta > 0, \Lambda, \lambda > 1$ and an integer $N > 1$ so that whenever $x', y' \in W^u_\delta(z'), z' \in K$ and $f^j(x'), f^j(y') \in W^u_\delta(f^j(z')), j = 1, \ldots, m$ then

$$\rho(f^j(x'), f^j(y')) \geq \eta \rho(x', y') \; \forall \, j \leq N$$
$$\rho(f^j(x'), f^j(y')) > \lambda^j \rho(x', y'), \; N < j \leq m$$
$$\rho(f^j(x'), f^j(y')) \leq \Lambda^j \rho(x', y'), 1 \leq j \leq m \, .$$

(∗) Moreover, whenever $x'', z'' \in K$, $r > 0$ and $x'' \in W^u_{\delta/2}(z'')$ with $W^u_r(x'') \subset W^u_\delta(z'')$ there is a $y'' \in (W^u_r(x'') \setminus W^u_{\eta r}(x'')) \cap K$.(∗)

We will find an arbitrarily small $\tau > 0, \tau \gg \varepsilon > 0$ and sequences $\{q_k\}_{k=0}^\infty \subset W^u_\tau(q), \{n_k\}_{k=0}^\infty \subset \mathbb{Z}^+, 0 \leq n_0 < n_1 < \cdots$ so that $f^{n_k}(q_\ell) \in W^u_\tau(f^{n_k}(q_k)), \ell > k$ and $C_\rho(p, \varepsilon, n_k) \cap f^{-n_k}(W^u_\tau(f^{n_k}(q_k))) = \emptyset$. Then $\lim_{k \to \infty} q_k$ is in $\overline{W^u_\tau(q)}$ but is not in $C_\rho(p)$ and the proof will be complete.

The construction of the sequences $\{q_k\}, \{n_k\}$ goes by induction. The requirements needed for ε, τ will become evident during the proof. First of all we need ε, τ small enough that $p \notin W^u_\tau(q)$ to start the induction. Let $n_0 = 0, q_0 = q$. Suppose n_k, q_k have been chosen and that $C_\rho(p, \varepsilon, n_k) \cap f^{-n_k}(W^u_\tau(f^{n_k}(q_k))) = \emptyset$. There are two cases:

(I) $C_\rho(p, \varepsilon, n_k + 1) \cap f^{-n_k-1}(W^u_\tau(f^{n_k+1}(q_k))) = \emptyset$. Then we define $n_{k+1} = n_k + 1$ and set $q_{k+1} = q_k$.

(II) $C_\rho(p, \varepsilon, n_k + 1) \cap f^{-n_k-1}(W^u_\tau(f^{n_k+1}(q_k))) \neq \emptyset$. Then there is a point $x' \in W^u_{2\tau}(f^{n_k+1}(q_k))$ so that $x := f^{n_k+1}(p) \in W^s_{\text{loc}}(x')$ and $\rho(x, x') \leq C\varepsilon$ for some fixed constant C where C depends on the angle between the stable and unstable manifolds. Set $y := f^{n_k+1}(q_k)$. We consider two cases:

(IIA) $\rho(x', y) \geq \frac{C\varepsilon}{\eta}$

Then the distance between $f^j(x)$ or $f^j(x')$ and $f^j(y)$ remains strictly larger than ε for $j \leq N$ and for $j > N$, $\rho(f^j(x), f^j(y)), \rho(f^j(x'), f^j(y)) \geq \lambda^j \rho(x', y)$, as long as $\rho(x', y)\Lambda^j \leq \delta$ i.e., $j \log \Lambda \leq \log \frac{\delta}{\rho(x',y)}$ or $j \leq \frac{1}{\log \Lambda} \log \left(\frac{\delta}{\rho(x',y)}\right)$.

Assume first that $N \leq \frac{1}{\log \Lambda} \log \left(\frac{\delta}{\rho(x',y)}\right)$, i.e.,

(IIA1): $\rho(x', y) \leq \frac{\delta}{\Lambda^N}$.

For $j_0 = \left[\frac{1}{\log \Lambda} \log \left(\frac{\delta}{\rho(x',y)} \right) \right]$ we have

$$\rho(f^{j_0}(x'), f^{j_0}(y)) \geq \rho(x',y) \lambda^{\left[\frac{1}{\log \Lambda} \log \left(\frac{\delta}{\rho(x',y)} \right) \right]} .$$

In fact, we can assume that no point in

$$W^u \left(f^{j_0}(y), \frac{\rho(x',y)}{2} \lambda^{\left[\frac{1}{\log \Lambda} \log \left(\frac{\delta}{\rho(x',y)} \right) \right]} \right)$$

have been captured by p for $0 \leq j \leq j_0$, *i.e.*, points in

$$W^u \left(f^{j_0}(y), \frac{1}{2} \rho(x',y)^{1 - \frac{\log \lambda}{\log \Lambda}} \delta^{\frac{\log \lambda}{\log \Lambda}} \right)$$

have not been captured.

Notice that if $\varepsilon > 0$ is small enough,

$$\frac{1}{2} \left(\frac{C\varepsilon}{\eta} \right)^{1 - \frac{\log \lambda}{\log \Lambda}} \delta^{\frac{\log \lambda}{\log \Lambda}} > \varepsilon^\sigma, \sigma := 1 - \frac{1}{2} \frac{\log \lambda}{\log \Lambda}$$

and moreover, $\varepsilon^\sigma \ll \frac{\delta}{N\Lambda}$. We set $\tau = \varepsilon^\sigma, n_{k+1} = n_k + j_0$ and pick some $q_{k+1} \in W^u_{\tau/2}(f^{n_k+j_0}(q_k)) \cap K$.

(IIA2): $\rho(x',y) > \frac{\delta}{\Lambda^N}$. This can't happen since $\tau \ll \frac{\delta}{\Lambda^N}$ and $x' \in W^u_{2\tau}(y)$.

Next we deal with

(IIB): $\rho(x',y) < \frac{C\varepsilon}{\eta}$.

Using $(*)$ we can find a $y' \in W^u_\delta(y)$ with $\frac{2C\varepsilon}{\eta} \leq \rho(y',y) \leq \frac{2C\varepsilon}{\eta^2}$. Replacing y by y' we are back to case (IIA). $\qquad\square$

4 Holomorphic Endomorphisms of $\mathbb{P}^k$

Holomorphic endomorphisms on $\mathbb{P}^k$ carry a unique invariant measure μ, ergodic, mixing and of maximal entropy [FS1], [BD2].

Theorem 4.1. *Let $f : \mathbb{P}^k \to \mathbb{P}^k$ be a holomorphic map of degree d at least 2. Let ρ denote the Fubini-Study metric. Then for any $q \in S_\mu$ the complement of $\mathcal{C}_\rho(q)$ in S_μ is dense in S_μ.*

Proof. Fix $q \in S_\mu$. Let $p \in S_\mu$ and let $\varepsilon_1 > 0$. We need to find an $\varepsilon > 0$ and a point $p'' \in S_\mu \cap \Delta(p, \varepsilon_1) = \{x \in S_\mu; \rho(x,p) < \varepsilon_1\}$ with $p'' \in \mathcal{C}_\rho(q, \varepsilon)^c$.

(i) Suppose first that $\mathcal{O}(q)$ is a finite set S. By [BD1] the repelling periodic orbits are dense in S_μ, hence we can find a periodic point $p' \in S_\mu \cap \Delta(p, \varepsilon_1)$ not in S. Hence $\mathcal{O}(p') \cap \mathcal{O}(q) = \emptyset$ and since they are finite, we can choose $0 < \varepsilon < \rho(\mathcal{O}(q), \mathcal{O}(p'))$ and obtain that $p' \in \mathcal{C}_\rho(q, \varepsilon)^c$.

80   Araceli M. Bonifant and John Erik Fornæss

(ii) Suppose that $\mathcal{C}(q)$ is infinite. We argue using homoclinic orbits. Pick an integer $\ell \gg 1$. Let C denote the critical set of f. Set $C_\ell := C \cup f(C) \cup \cdots \cup f^\ell(C)$. For a generic point $x \in \mathbb{P}^k$ the measures $\mu_x^i := \frac{1}{d^{ki}} \sum_{y \in f^{-i}(x)} \delta_y$ converge weakly to μ. Here the points y are counted with multiplicity. In particular this holds for all x outside an algebraic subvariety of zero μ-mass. This subvariety is contained in C [BD2]. Since algebraic varieties carry no μ−mass [FS1], [BD2] and since repelling periodic orbits are dense in S_μ [BD1], we can find a point $p_0 \in [\Delta(p, \varepsilon_1) \cap S_\mu] \setminus C_\ell$ which is on a repelling periodic orbit $\{f^j(p_0) =: p_j\}_{j=0}^N, p_N = p_0$ contained in S_μ.

For each $i \geq 1$, let $T^i = \{p_j^i\}_{j=1}^{d^{ki}}$ denote all the f^{-1} preimages of p_0 counted with multiplicity. Since S_μ is totally invariant [FS1], all preimages are contained in S_μ. The measures $\mu^i := \frac{1}{d^{ki}} \sum_{j=1}^{d^{ki}} \delta_{p_j^i}$ converge weakly to μ.

Let p_1^1 denote a preimage of p_0 which is not on the periodic orbit of p_0 and let p_1^2, p_2^2 denote two distinct preimages of p_1^1. For $i \geq 2$, let $T_s^i \subset T^i, s = 1, 2$ denote the set of points p_j^i for which none of the points $p_j^i, f(p_j^i), \ldots, f^i(p_j^i)$ are on the critical set and $f^{i-2}(p_j^i) = p_s^2$.

Hence when $i \leq \ell$, T_s^i contains exactly $d^{k(i-2)}$ points, each of which has multiplicity one. Therefore, if $i \leq \ell, s = 1, 2$

$$\mu_s^i := \frac{1}{d^{ki}} \sum_{x \in T_s^i} \delta_x$$

has mass $\frac{1}{d^{2k}}$. However, for $i > \ell$ some preimages $f^{-i}(p)$ might be on the critical set. Hence the masses decrease, $\|\mu_s^{i+1}\| \leq \|\mu_s^i\|$.

By a counting argument as in [BD1], based on Bezouts theorem, we can, assuming that ℓ is large enough, have $\|\mu_s^i\| \geq \frac{1}{2d^{2k}}$ for all i, s.

Let $U = U(p_0)$ be a small neighborhood of p_0, $U \subset \Delta(p, \varepsilon_1)$, on which f^N is biholomorphic onto its image and $f^{-N}(U) \cap U \subset\subset U$. We can assume f^N is strictly expanding in some local coordinate system, i.e.,

$$\|f^N(x) - f^N(y)\| \geq \lambda \|x - y\| \; \forall \, x, y \in U, \lambda > 1$$

a constant. Since p_0 is in the support of μ, $\mu(U) =: \delta > 0$. From [FS1] it follows that for large enough r, $\mu(f^r(U)) \geq 1 - \frac{1}{4d^{2k}}$

Set $C^r := C \cup_{i=1}^r f^{-i}(C)$. Then C^r has no μ-mass and p_0 is not in C^r.

Set $V_0 = U \setminus C^r$ and $V_i = f^i(V_0), i = 1, \ldots, r$. Then $\{f^i\}$ are locally biholomorphic on V_0 and $\mu(V_r) = \mu(f^r(U)) \geq 1 - \frac{1}{4d^{2k}}$.

By the estimate of $\|\mu_s^i\|$, it follows that for large enough i_0, $\mu_s^{i_0}(V_r) > 0$, $s = 1, 2$. Therefore there exists two points $\tilde{p}_1, \tilde{p}_2 \in V_0 \cap S_\mu$ so that the orbits $\{f^i(\tilde{p}_s)\}_{i=0}^{r+i_0} \subset S_\mu$ contain no critical points and the orbits are disjoint except that $f^{r+i_0-1}(\tilde{p}_1) = f^{r+i_0-1}(\tilde{p}_2) = p_1^1$. Also $f^{r+i_0-2}(\tilde{p}_s) = p_s^2$.

Fix two small neighborhoods $W_s = W_s(\tilde{p}_s)$ so that all $f^i, i = 1, \ldots, r + i_0$ are biholomorphic on W_s, the images are all pairwise disjoint, except that $f^{r+i_0-1}(W_1) = f^{r+i_0-1}(W_2)$ and $f^{r+i_0}(W_s) = \Delta(p_0, \eta)$ for a small $\eta > 0$,

$s = 1, 2$. Define $W_s^{-1} = f^{-1}(W_s) \cap U$ and inductively, $W_s^{-(j+1)} = f^{-1}(W_s^{-j})$ $\cap U$. Fix a j_0 large enough that $W_s^{-j_0} \subset \Delta(p_0, \eta/2), s = 1, 2$. Increasing j_0 further, we can also arrange that the biholomorphic maps $f^{j_0+r+i_0} : W_s^{-j_0} \rightarrow \Delta(p_0, \eta)$ is strictly expanding.

We will define a sequence of open sets $\{U_i\}_{i \geq 0}, U_{i+1} \subset U_i$, and occasionally so that $U_{i+1} \subset\subset U_i$, $f^i : U_i \rightarrow W(i)$ is a biholomorphic map where $W(i)$ is one element of the following finite list of open sets:

$\Delta(p_0, \eta)$
$f^i(\Delta(p_0, \eta)), i = 1, \ldots, N - 1$
$W_s^{-i}, i = 1, \ldots, j_0, s = 1, 2$
$W_s, s = 1, 2$
$f^i(W_s), i = 1, \ldots, r + i_0 - 1, s = 1, 2.$

Since each of these open sets contain points in S_μ, each U_i will also. We moreover want $W(i)$ not to contain a point in $\Delta(f^i(q), \varepsilon)$, if $\varepsilon > 0$ is small enough.

To start we will define $U_0 = \Delta(p_0, \eta)$. Since q is not in U, we can take $W(0) = U_0$. Next we continue by setting $U_1, \ldots, U_{N-1} = U_0$ and $W(i) = f^i(U_i), i = 1, \ldots, N - 1$ unless $f^i(q)$ gets closer to this $W(i)$ than ε.

Assuming at first that $f^i(q)$ does not get closer than ε, we continue by setting $U_N = f^{-N}(U_0) \cap U_0$ and $W(N) = U_0$ and continue this process the same way. This procedure is only interrupted if $f^i(q)$ gets closer to these $W(i)$ than ε for some i. If this occurs for $i = i_1$, then we redefine U_i, call the new $U_i, \tilde{U}_i$ with $\tilde{U}_i \subset\subset U_i$ and if $i = jN + r, 0 \leq r \leq N$, then $f^i(\tilde{U}_i)$ is one of the sets $f^r(W_s^{-j_0}), s = 1, 2$. Since these two sets as well as their forward orbits until they reach p_s^2 are separated, choosing the right one of the two ensures that the orbit $f^j(q)$ stays at least an ε distance away until we return to p_s^2. But once we return there, we can repeat the process. Notice that by the strict expansion of the iterates f^N on U_0 and $f^{j_0+r+i_0}$ on $W_s^{-j_0}$, the diameters of the U_i shrink geometrically. Finally we set $p'' = \cap_{i>0} \overline{U}_i = \cap_{i>0} U_i$. Clearly $\rho(f^n(p''), f^n(q)) > \varepsilon \ \forall \ n \geq 0$ so $p'' \in \mathcal{C}_\rho(q, \varepsilon)^c$ as desired. $\qquad \square$

Since the measure μ is mixing, [FS1], we can combine the above result with Corollary 2.2 and obtain:

Corollary 4.2. *Let ρ denote the Fubini-Study metric on $\mathbb{P}^k$. Let $f : \mathbb{P}^k \rightarrow \mathbb{P}^k$ be a holomorphic map of degree at least 2. Then for every $z \in S_\mu$, $C_\rho(z)$ has full μ measure while the complement of $C_\rho(z)$ is dense in S_μ.*

The results here naturally suggest the following questions:

Question 4.3. Is it possible that all repelling periodic points of a holomorphic self-map of $\mathbb{P}^k, k \geq 2$, are in the critical orbit? In fact, given $C_\ell = C \cup f(C) \cup \cdots \cup C_\ell$, does there exist a repelling periodic orbit which does not intersect C_ℓ? (even for $\ell = 1$.)

82   Araceli M. Bonifant and John Erik Fornæss

Question 4.4. Is Proposition 3.1 valid for general hyperbolic endomorphisms in the case the pseudometric is a metric?

Question 4.5. When does Theorem 4.1 hold for rational self maps of $\mathbb{P}^2$?

We end with a partial result on Question 4.5, namely in the case of a complex Hénon maps.

If $f : \mathbb{C}^2 \to \mathbb{C}^2$ is a Hénon map of degree $d \geq 2$, then there is a unique invariant measure μ, ergodic, mixing and of maximal entropy [BLS]. The support of μ, S_μ, is a compact subset of $\mathbb{C}^2$. Let ρ be any continuous metric.

Proposition 4.6. *Let f be a Hénon map. For any $q \in S_\mu$, the complement $C_\rho(q)^c \cap S_\mu$ of $C_\rho(q)$ in S_μ is dense in S_μ.*

We state first a Lemma about behaviour near saddle points. Let p be a saddle periodic point for f. We choose a local holomorphic coordinate system (z, w) near p so that

$$W_{p,loc}^u = \{(z, w); w = 0, |z| < 1\}$$
$$W_{p,loc}^s = \{(z, w); z = 0, |w| < 1\}$$

are local unstable and stable manifolds respectively and $p = 0$. By [BLS] there is a transverse intersection of the global unstable and stable manifolds other than p.

Lemma 4.7. *Suppose $(z_0, 0), 0 < |z_0| < 1$ is a transverse homoclinic point, i.e., $(z_0, 0) \in W^s(p)$ and the tangent space of $W^s(p)$ at $(z_0, 0)$ is not the z-axis. Let $|z_0| < r < 1$, and let $0 < \delta < r - |z_0|$. Then there exists an arbitrarily small $\eta > 0$ and an integer $N > 1$ so that if $n \geq N$ and X is a holomorphic graph $w = g(z), z \in \Delta(z_0, \delta), |g| < \eta$ then $f^n(X)$ contains a relatively open set Y which is a graph $w = h(z), |z| < r, |w| < \eta$.*

Proof. Observe that the conclusion follows easily for the special case when $X = \{(z, 0); z \in \Delta(z_0, \delta)\}$. Finally we choose η small enough to complete the proof. □

Proof of the Proposition. Suppose $p, q \in S_\mu$ and $\varepsilon_1 > 0$. We will find an $\varepsilon > 0$ and a point $p'' \in S_\mu \cap \Delta(p, \varepsilon_1)$ such that $p'' \in C_\rho(q, \varepsilon)^c$.

If q is a periodic point, since periodic points are dense in S_μ [BLS] and points can carry no mass, we can let p'' be on another periodic orbit and we are done.

So we can suppose that the orbit of q is infinite. We choose a periodic saddle point $p' \in \Delta(p, \varepsilon_1)$ which then necessarily is not on the orbit of q. Use local coordinates (z, w) as in the above Lemma near p'. We can assume that the bidisc $\{|z|, |w| < 1\} \subset \Delta(p, \varepsilon_1)$. By [BLS] p' admits transverse homoclinic points. Hence we can find two distinct points $z_0^k \neq 0$ and disjoint discs $\overline{\Delta(z_0^0,\delta)}, \overline{\Delta(z_0^1,\delta)}$, with $z_0^k \in W^s(p'), k = 0,1$. Let (η^k, N^k) denote the numbers from the Lemma for the two points. We can choose $r > |z_0^k| + \delta, k = 0,1$ in both cases.

Letting $\eta = \min_k\{\eta^k\}$ and choosing $N \geq \max_k N^k$ large enough we can conclude:

Lemma 4.8. *Suppose X^k is a graph $w = g^k(z), z \in \Delta(z_0^k,\delta), |g^k| < \eta$, then $f^N(X^k)$ contains graphs $w = h^{k,\ell}(z), z \in \Delta(z_0^\ell,\delta), |h^{k,\ell}| < \eta, k,\ell = 0,1$.*

We fix

$$\varepsilon_2 = \min_{i \leq N} \rho\left(f^i\left(\overline{\Delta(z_0^0,\delta)} \times (|w| < \eta)\right) , \ f^i\left(\overline{\Delta(z_0^1,\delta)} \times (|w| < \eta)\right)\right).$$

Choose $\varepsilon < \varepsilon_2$ so that if $x \in S_\mu$ is any point and

$$\rho\left(f^i(x), f^i\left(\overline{\Delta(z_0^k,\delta)} \times (|w| \leq \eta)\right)\right) \leq \varepsilon$$

for some $0 \leq i \leq N$, then $\rho(f^i(x), f^i(\overline{\Delta(z_0^{1-k}\delta)} \times (|w| \leq \eta))) > \varepsilon$ for all $0 \leq i \leq N$.

We will find a point $p'' \in \cap_{i=0}^\infty U_i$ where $U_{i+1} \subset U_i \ \forall\ i, U_{i+1} \subset\subset U_i$ occasionally. Furthermore each U_i will contain a point in S_μ.

We define $U_0 = \Delta(z_0^k) \times \{0\}$ where $k \in \{0,1\}$ is such that $\rho(f^i(q), f^i(U_0)) > \varepsilon \ \forall\ i = 0,\ldots,N$. Set $U_i = U_0, i = 0,\ldots,N$. Since $(z_0^k,0) \in W^s(p')\cap W^u(p')$ is a transverse intersection, we know [BLS] that $(z_0^k,0) \in S_\mu$. Hence $U_i \cap S_\mu \neq \emptyset, i = 0,\ldots,N$.

Next, by Lemma 4.8, there are open subsets $U_N^k \subset\subset U_N, k = 0,1$ such that $f^N(U_N^k)$ are graphs in $\Delta(z_0^k,\delta) \times (|w| < \eta)$. In particular, $f^N(U_N^k) \subset f^N(W^u(p')) = W^u(p')$, it follows that $f^N(U_N^k)$ contains a transverse homoclinic intersection so $f^N(U_N^k) \cap S_\mu \neq \emptyset$. Since S_μ is totally invariant, U_N^k contains a point in S_μ.

We repeat the process starting with $X = X^k = f^N(U_N^k)$ for $k = 0,1$, chosen so that $\rho(f^{N+i}(q), f^i(X)) > \varepsilon \ \forall\ i = 0,\ldots,N$. Also set, for this k, $U_{N+1} = \cdots = U_{2N} = U_N^k$. Proceeding inductively we obtain a sequence $\{U_i\}, U_{i+1} \subset U_i, U_{i+N} \subset\subset U_i$ and $U_i \cap S_\mu \neq \emptyset$, and $\rho(f^i(U_i), f^i(q)) > \varepsilon$. We can finally choose $p'' \in (\cap \overline{U}_i) \cap S_\mu$. Then $\rho(f^i(q), f^i(p'')) > \varepsilon$ for all i, so $p'' \in \mathcal{C}(q,\varepsilon)^c \cap S_\mu$, as desired. $\qquad\square$

References

[BLS] Bedford, E., Lyubich, M., Smillie, J., Polynomial diffeomorphisms IV, *Invent. Math.* **112** (1993), 77–125.

[BC] Benedicks, M., Carleson, L., The dynamics of the Henon map, *Ann. Math.* **133** (1991), 73–170.

[BD1] Briend, J-Y., Duval, J., Exposants de Liapounoff et distribution des points periodiques d'un endomorphisme de $\mathbb{CP}^k$, *Acta Math.* **182** (1999), 143–157.

84   Araceli M. Bonifant and John Erik Fornæss

[BD2] Briend, J-Y., Duval, J., Deux caractérisations de la mesure d'equilibre d'un endomorphisme de $\mathbb{P}^k(\mathbb{C})$ preprint, 2000.

[FG] Fornæss, J.E., Gavosto, E.A., Existence of generic homoclinic tangencies for Hénon mappings, *Journal of Geometric Analysis* **2** (1992), 429–444.

[FS1] Fornæss, J. E., Sibony, N., Complex dynamics in higher dimension II. Modern methods in complex analysis, *Ann. Studies* **137** (1995), 135–182.

[FS2] Fornæss, J. E., Sibony, N., Dynamics of $\mathbb{P}^2$ (Examples), Stony Brook Conference on Laminations, *Contemp. Math.* **269** (2001), 47–86.

[FW] Fornæss, J.E, Weickert, B., Attractors on $\mathbb{P}^2$, several complex variables, (Berkeley, CA, 1995–1996), Cambridge Univ. Press, Cambridge, 297–307, 1999.

[H] Hénon, M., A two dimensional mapping with a strange attractor, *Comm. Math. Phys.* **50** (1976), 69–77.

[JW] Jonsson, M., Weickert, B., A nonalgebraic attractor in $\mathbb{P}^2$, *Proc. Amer. Math. Soc.* **128** (2000), 2999–3002.

[L] Lorenz, E., Deterministic non-periodic flow, *J. Atmos. Sci.* **20** (1963), 130–141.

[T] Tucker, W., The Lorenz attractor exists, *C. R. Acad. Sci. Paris Ser. I Math.* **328** (1999), 1197–1202.

A Bound on the Irregularity of Abelian Scrolls in Projective Space

Ciro Ciliberto[1] and Klaus Hulek[2]

[1] Università di Roma Tor Vergata, Dipartimento di Matematica, Via della Ricerca Scientifica, I 00173 Roma, Italy
E-mail address: `Cilibert@axp.mat.uniroma2.it`
[2] Institut für Mathematik, Universität Hannover, Postfach 6009, D 30060 Hannover, Germany
E-mail address: `Hulek@math.uni-hannover.de`

Abstract We prove that the irregularity of a smooth *abelian scroll* whose dimension is at least half of that of the surrounding projective space is bounded by 2. We also discuss some existence results and open problems.

Table of Contents

Introduction 85
1 Non-Existence of Scrolls 86
2 Existence of Scrolls 88
References 92

0 Introduction

Let $A \subset \mathbb{P}^{l-1}$ be an n-dimensional abelian variety and let $G \subset A$ be a subgroup of order k. Then one can define for every point $P \in A$ the linear space

$$S^G(P) := \operatorname{Span}\langle P + \rho; \rho \in G \rangle .$$

In general one expects $S^G(P)$ to have dimension $k - 1$ and the union of these spaces

$$Y = \bigcup_{P \in A} S^G(P)$$

is then a (possibly singular) scroll of expected dimension $n + k - 1$. The classical example of this construction is the case where E is an elliptic quintic normal curve in $\mathbb{P}^4$ and $G = \mathbb{Z}_2$. This leads to the quintic elliptic scroll, the only smooth irregular scroll in $\mathbb{P}^4$.

We shall refer to a smooth scroll Y constructed as above and having the expected dimension as the *abelian scroll* determined by the pair (A, G). In particular, if $\rho \in A_{(k)}$ is a point of order k, we can consider the cyclic

2000 *Mathematics Subject Classification*: 14K05, 14M07.

86   C. Ciliberto and K. Hulek

subgroup $G \simeq \mathbb{Z}_k$ generated by ρ. We call the corresponding abelian scroll the *cyclic scroll* determined by the pair (A, ρ).

Observe that for smooth abelian scrolls Y we have $2 \dim Y \leq l - 1$. This follows immediately from Barth's theorem which says that any smooth subvariety $Y \subset \mathbb{P}^{l-1}$ with $2 \dim Y > l - 1$ has irregularity $q(Y) = 0$. The main result of this paper is the following:

Theorem 0.1. *Let $Y \subset \mathbb{P}^{l-1}$ be a smooth abelian scroll with $2 \dim Y = l - 1$. Then $q(Y) \leq 2$, i.e. A is either an elliptic curve or an abelian surface. Moreover, in this case Y (or equivalently A) is linearly normally embedded.*

This non-existence result can be seen as an extension of a result of Van de Ven [V] which says that an abelian variety A of dimension n can be embedded in $\mathbb{P}^{2n}$ only if $n = 1$ or 2. We shall briefly discuss the existence of abelian scrolls and some open problems.

As we indicate at the end of section 2, the question considered here is related to the general interesting problem of classifying smooth projective varieties whose dimension equals their codimension, a problem motivated by questions of Griffiths and van de Ven concerning smooth surfaces in $\mathbb{P}^4$.

Acknowledgement

We warmly thank the referee for carefully reading the paper and, in particular, for having suggested an elegant simplification of the first version of this note which enabled us to avoid the use of some binomial identities. However, we thank M. Erné and S. Rao for having given us the right references for these identities in the first place. Both authors gratefully acknowledge partial support by the EU project EAGER (HPRN-CT-2000-00099). The second author was also partially supported by the DFG Schwerpunktprogramm "Globale Methoden in der komplexen Geometrie".

1   Non-Existence of Scrolls

This section is devoted to the proof of the non-existence theorem (0.1). The main ingredients in the proof are the double point formula and some estimates involving binomial coefficients.

Before we can prove our result we need some preparations. Consider the étale $k : 1$ quotient

$$\bar{\pi} : A \to \bar{A} := A/G$$

and let $\mathcal{L}$ be the line bundle on A which defines the embedding into $\mathbb{P}^{l-1}$. Then

$$\mathcal{E} := \bar{\pi}_* \mathcal{L}$$

is a rank k vector bundle on $\bar{A}$ and we set

$$X := \mathbb{P}(\mathcal{E}) \ .$$

The natural map $\bar{\pi}^* \mathcal{E} \to \mathcal{L}$ induces an inclusion of A into X as a multi-section and the linear system $|\mathcal{O}_X(1)| = |\mathcal{O}_{\mathbb{P}(\mathcal{E})}(1)|$ induces the complete linear system $|\mathcal{L}|$ on A and maps X onto a scroll Z (see [CH, Lemma 2.1]). If Y is embedded by a complete linear system, then $Z = Y$, otherwise Y is a projection of Z. Also as in [CH, p. 360] we have

$$\bar{\pi}^* \mathcal{E} = \mathcal{L} \oplus t^*_{\rho_1} \mathcal{L} \oplus \ldots \oplus t^*_{\rho_{k-1}} \mathcal{L}$$

where $G = \{1, \rho_1, \ldots, \rho_{k-1}\}$. Note that topologically $\tilde{X} := \mathbb{P}(\bar{\pi}^* \mathcal{E}) \cong A \times \mathbb{P}^{k-1}$ is trivial. We can also choose another covering $\hat{\pi} : \hat{A} \to A$ of order h where h divides k such that $\hat{X} := \mathbb{P}(\hat{\pi}^* \bar{\pi}^* \mathcal{E}) \cong \hat{A} \times \mathbb{P}^{k-1}$ as algebraic varieties.

Proof of Theorem 0.1. Let $m = \dim Y = n + k - 1$. Then the double point formula (see [F, Theorem (9.9.3)]) reads

$$[\mathbb{D}] = \varphi^* \varphi_*[X] - c_m(N_\varphi) \cap [X]$$

where $\varphi : X \to Y$ is the map given by $|\mathcal{O}_X(1)|$, possibly followed by a projection. We first determine the contribution $\varphi^* \varphi_*[X]$. We have already observed that topologically $\tilde{X}$ is a product and we can write for the hyperplane section on $\tilde{X}$:

$$\tilde{H} = c + h$$

where $c = c_1(\mathcal{L}) \in H^2(A, \mathbb{Z})$ and h is the positive generator of $H^2(\mathbb{P}^{k-1}, \mathbb{Z})$. Hence

$$\tilde{H}^m = \tilde{H}^{n+k-1} = (c + h)^{n+k-1} = \binom{n+k-1}{k-1} c^n \ .$$

Since $\tilde{X} \to X$ has degree k we find that

$$\varphi^* \varphi_*[X] = \frac{1}{k^2} \binom{n+k-1}{k-1}^2 (c^n)^2 \ .$$

Instead of the normal bundle of φ we compute the normal bundle of $\tilde{\varphi} := \varphi \circ f$ where $f : \tilde{X} \to X$ is the étale $k : 1$ map induced from $\bar{\pi} : A \to \bar{A}$. From the exact sequence

$$0 \to T_{\tilde{X}} \to \tilde{\varphi}^* T_{\mathbb{P}^{l-1}} \to N_{\tilde{\varphi}} \to 0$$

we find

$$c(N_{\tilde{\varphi}}) = c(\tilde{\varphi}^* T_{\mathbb{P}^{l-1}}) c(T_{\tilde{X}})^{-1}$$
$$= (1 + c + h)^l (1 + h)^{-k}$$

where we have again used that topologically $\tilde{X} = A \times \mathbb{P}^{k-1}$. Now we notice that $c^{n+1} = h^k = 0$ and, therefore, the top term of $c(N_{\tilde{\varphi}})$ is just the coefficient of the monomial $c^n h^{k-1}$, which is the same as the coefficient of the same monomial in

$$\binom{l}{n} c^n (1 + h)^{l-n} (1 + h)^{-k} = \binom{l}{n} c^n (1 + h)^{l-n-k} \ .$$

88   C. Ciliberto and K. Hulek

Since $l = 2n + 2k - 1$ we get

$$c_m(N_{\tilde{\varphi}}) = c^n \binom{n + k - 1}{n} \binom{2n + 2k - 1}{n}.$$

Altogether we find for the class of the double locus $\mathbb{D}$ that

$$[\mathbb{D}] = \frac{1}{k} c^n \left[\frac{1}{k} \binom{n + k - 1}{k - 1}^2 c^n - \binom{n + k - 1}{n} \binom{2n + 2k - 1}{n} \right].$$

Using Riemann-Roch on the abelian variety A we find

$$c^n \geq n!(2n + 2k - 1)$$

with equality if and only if the linear system is complete. Then the assertion of the theorem follows if one shows that

$$\binom{n + k - 1}{k - 1}(2n + 2k - 1)n! \geq k \binom{2n + 2k - 1}{n}$$

where equality holds if and only if $n = 1$ or 2. Indeed, it is a straightforward calculation to check equality for $n = 1$ and $n = 2$. Now assume $n \geq 3$. Evaluating the binomial coefficients and cancelling terms, the above inequality becomes

$$n! \prod_{l=2}^{n} (n + k - l + 1) \geq \prod_{l=2}^{n} (2n + 2k - l)$$

which can be rewritten as

$$\prod_{l=2}^{n} (n + k - l + 1)l \geq \prod_{l=2}^{n} (2n + 2k - l).$$

To check this inequality it is enough to check it termwise. The inequality

$$(n + k - l + 1)l \geq 2n + 2k - l \quad \text{for } l = 2, \ldots, n$$

is equivalent to

$$n + k \geq l \quad \text{for } l = 2, \ldots, n$$

which is trivially true. $\square$

2   Existence of Scrolls

In this section we shall discuss conditions under which a linearly normal abelian scroll Y determined by a pair (A, G) is smooth. We shall keep the notation as in section 1 and we shall start with the scroll $X = \mathbb{P}(\mathcal{E})$. Recall that A is contained in X as a k-section and that $\mathcal{O}_X(1)|_A \cong \mathcal{L}$, essentially by the definition of $\mathcal{O}_X(1)$. We shall always assume that $\mathcal{L}$ is very ample on A. The line bundle $\mathcal{L}$ is called $(m - 1)$-very ample, if for every cluster (i.e. 0-dimensional subscheme) $\zeta \subset A$ of length $\leq m$ the restriction map $H^0(A, \mathcal{L}) \to H^0(\zeta, \mathcal{L} \otimes \mathcal{O}_\zeta)$ is surjective and hence $\varphi_{|\mathcal{L}|}(\zeta)$ spans a $\mathbb{P}^{m-1}$.

A Bound on the Irregularity of Abelian Scrolls in Projective Space   89

Proposition 2.1. (i) *If $\mathcal{L}$ is $(k-1)$-very ample, then $|\mathcal{O}_X(1)|$ is base point free and $\varphi : X \to Y$ is a finite map.*
(ii) *If $\mathcal{L}$ is k-ample, then $\varphi : X \to Y$ is birational and an embedding near A.*

Proof. (i) A intersects each fibre of X in k independent points. This follows since by construction of X these k points are the image of points in $\tilde{X}$ given by the sections which come from the splitting of $\bar{\pi}^* \mathcal{E}$. Since these points are mapped to independent points in $\mathbb{P}^{l-1}$ it follows that $|\mathcal{O}_X(1)|$ restricted to a fibre of X cannot have base points. In order to show that φ is finite, it is enough to prove that $\mathcal{O}_X(1)$ has positive degree on each curve C in X. This can be checked by pulling back to the product $\hat{X} = \hat{A} \times \mathbb{P}^{k-1}$ where we have a product polarization.

(ii) It follows immediately from our assumption that every $\mathbb{P}^{k-1}$ of the scroll Y meets A only in the k points which span it. Hence for every point $P \in A$ we have that $\varphi^{-1}(P)$ consists of only one point. Similarly we prove that $d\varphi$ is injective along A. Now assume that $\varphi : X \to Y$ has degree $d \geq 2$. Over each point of $A \subset Y$ we have only one preimage. Hence φ is ramified along A, contradicting what we have just proved. $\qquad\square$

Proposition 2.2. *If $\mathcal{L}$ is $(2k-1)$-very ample, then $\varphi : X \to Y$ is an isomorphism and, in particular, Y is smooth.*

Proof. The assumption gives immediately that, for two points P and Q whose difference on A lies in the group G, we have $S^{k-1}(P) \cap S^{k-1}(Q) = \emptyset$. Hence the map φ is injective. It remains to prove that the differential $d\varphi$ is injective everywhere. The map $\varphi : X \to Y \subset \mathbb{P}^{l-1}$ embeds every fibre π_x of $\pi : X \to \bar{A}$ through a point $x \in X$ as a $(k-1)$-dimensional subspace which, by abuse of notation, we also denote by π_x. We consider the map

$$\gamma : A \to G(k-1, l-1) =: \mathrm{Gr}$$
$$x \mapsto \pi_x.$$

where π_x is the unique fibre of X containing the point x. This map factors through $\bar{A}$. Taking the projectivised differential of this map gives us a map

$$\omega = \mathbb{P}(d\gamma) : \mathbb{P}(T_{A,x}) \to \mathbb{P}(T_{\mathrm{Gr}, \pi_x}) = \mathrm{P}\, r(\pi_x, \pi^0)$$
$$t \mapsto \omega_t$$

where π^0 is a complementary subspace of π_x and $\mathrm{Pr}(\pi_x, \pi^0)$ is the projective space of projective transformations of π_x to π^0. (If $\pi_x = \mathbb{P}(U)$ and $\pi^0 = \mathbb{P}(W)$ then $\mathrm{Pr}(\pi_x, \pi^0) = \mathbb{P}(\mathrm{Hom}(U, W))$.) Here the identification $\mathbb{P}(T_{\mathrm{Gr}, \pi_x}) = \mathrm{Pr}(\pi_x, \pi^0)$ comes from the fact that there is a canonical isomorphism $T_{\mathrm{Gr}} = \mathrm{Hom}(\mathcal{U}, \mathcal{V})$ where $\mathcal{U}$ and $\mathcal{V}$ are the canonical sub bundle resp. the quotient bundle. We also have the Gauss map

$$\Gamma : X \dashrightarrow G(n+k-1, l-1)$$
$$P \longmapsto \bar{T}_{X,P}$$

90 C. Ciliberto and K. Hulek

where $\bar{T}_{X,P}$ is the projective closure of the image of the differential of φ at P, considered as a subspace through the point P. The relation between γ and Γ is the following. Let $P \in X$ and let $x \in \bar{\pi}^{-1}(\pi(P))$. Then

$$\bar{T}_{X,P} = \langle \pi_x, \omega_t(P); \ t \in \mathbb{P}(T_{A,x}) \rangle .$$

In order to check that $d\varphi$ is injective at P we have to prove that the projective map

$$\omega(P): \ \mathbb{P}(T_{A,x}) \dashrightarrow \pi^0$$
$$t \longmapsto \omega_t(P)$$

is injective, i.e. is well defined. If this is not the case, then we have a tangent direction $t \in \mathbb{P}(T_{A,x})$ such that the linear map associated to the map

$$\omega_t: \ \pi_x \dashrightarrow \pi^0$$
$$P \longmapsto \omega_t(P)$$

has a kernel. Assume that this is the case. Consider the germ of a holomorphic curve $(\mathbb{C}, 0) \to (A, x)$ which represents the tangent direction t at x. This determines a family of $(\mathbb{P}^{k-1})'s$ given by

$$\pi_s = \langle z_0(s), \dots, z_{k-1}(s) \rangle .$$

Choosing suitable coordinates in $\mathbb{P}^{l-1}$ we can assume that

$$z_i(0) = e_i \ ; \ i = 0, \dots, k-1$$

with $e_0, \dots, e_{l-1}$ the standard basis. For the complementary space π^0 we can choose $\pi^0 = \langle e_k, \dots, e_{l-1} \rangle$. It is an easy local computation to check that with respect to the coordinates on π_x and π^0 given by $e_0, \dots, e_{k-1}$ and $e_k, \dots, e_{l-1}$ the linear map associated to the map

$$\omega_0: \pi_x \to \pi^0$$

is given by the matrix

$$M = \begin{pmatrix} z'_{0k} & z'_{1k} & \cdots & z'_{k-1,k} \\ \vdots & \vdots & & \vdots \\ z'_{0,l-1} & z'_{1,l-1} & \cdots & z'_{k-1,l-1} \end{pmatrix} \ (s = 0) .$$

Hence ω_0 has a kernel if and only if $\operatorname{rank} M < k$. But this is equivalent to the assertion that $z_0(0), \dots, z_{k-1}(0), z'_0(0), \dots, z'_{k-1}(0)$ are dependent. In particular there is a cluster of length $2k$ on A which contradicts $(2k-1)$-very ampleness of $\mathcal{L}$. $\qquad \square$

We can apply this result immediately to elliptic normal curves $E \subset \mathbb{P}^{m-1}$. Recall that any $m-1$ points on E (possibly infinitely near) are linearly independent (otherwise we could find a hyperplane through these $m-1$ points and *any* other point on E.)

A Bound on the Irregularity of Abelian Scrolls in Projective Space 91

Corollary 2.3. *Let $m = 2n + 1$ be an odd integer. Then there exists an n-dimensional elliptic curve scroll Y in $\mathbb{P}^{2n}$.*

Proof. Let $E \subset \mathbb{P}^{m-1} = \mathbb{P}^{2n}$ be an elliptic normal curve of degree m and choose a subgroup G of order n of E, e.g. a cyclic subgroup. We want to prove that the abelian scroll Y determined by the pair (E, G) is smooth. This follows from Proposition 2.2 since the embedding line bundle $\mathcal{L}$ is $(m - 2 = 2n - 1)$-very ample. $\square$

Conversely we can also use our results to bound the very-ampleness of line bundles on abelian varieties.

Corollary 2.4. *Let $\mathcal{L}$ be a line bundle on an abelian variety A of dimension $\dim A = n \geq 3$ with $h^0(A, \mathcal{L}) = l$. Assume that $\mathcal{L}$ is k-very ample for some odd integer k. Then $k < l - 2n$.*

Proof. For $k = 1$ this is Van de Ven's result [V]. If $k \geq 3$ we can write $k = 2s - 1$ for some $s \geq 2$. The assertion then follows by combining Theorem (0.1) and Proposition (2.2), applied to the abelian scroll determined by a pair (A, G), where G can be chosen as any subgroup of order k of A, e.g. a cyclic subgroup. $\square$

The crucial open question which remains is the existence of abelian surface scrolls $Y \subset \mathbb{P}^{l-1}$ with $2 \dim Y = l - 1$. In [CH] we constructed one non-trivial such example, namely the following.

Example. Let $A \subset \mathbb{P}^6$ be a general (i.e. $\operatorname{rank} NS(A) = 1$) abelian surface embedded by a line bundle $\mathcal{L}$ of type $(1, 7)$ and let $\varepsilon \in A_{(2)}$ be a non-zero 2-torsion point. Then the cyclic scroll determined by the pair (A, ε) is smooth of dimension 3 and has degree 21.

The next step would be to investigate cyclic $\mathbb{P}^2$-scrolls determined by pairs (A, ρ) where $A \subset \mathbb{P}^8$ is an abelian surfaceof degree 18, embedded by a complete linear system of type $(1, 9)$ and $\rho \in A_{(3)}$ is a 3-torsion point. More generally we pose the

Problem. Let $A \subset \mathbb{P}^{2n}$, $n \geq 4$ be a general abelian surface of degree $4n$ and let $\rho \in A_{(n-1)}$ be a non-zero $(n - 1)$-torsion point. Is the cyclic scroll defined by the pair (A, ρ) smooth?

A positive answer to this question would provide us with an infinite series of smooth scrolls $Y \subset \mathbb{P}^{2n}$ whose dimension is half that of the surrounding space and with irregularity 2. Van de Ven has asked the question whether the irregularity of smooth surfaces in $\mathbb{P}^4$ is bounded by 2. As far as we know, this question is still open. In fact, we do not know of any smooth subvariety $Y \subset \mathbb{P}^{2n}$ with $\dim Y = n \geq 2$ and $q \geq 3$. Let us, therefore, pose the

Problem. Give examples of smooth subvarieties $Y \subset \mathbb{P}^{2n}$ with $\dim Y = n \geq 2$ and $q \geq 3$.

References

[CH]   C. Ciliberto and K. Hulek, A series of smooth irregular varieties in projective space, *Ann. Sc. Norm. Sup. di Pisa* - Serie IV, **XXVIII** (1999), 357–380.

[F]   W. Fulton, *Intersection Theory*, Second edition. Springer Verlag, Berlin, 1984.

[V]   A. Van de Ven, On the embedding of abelian varieties in projective spaces, *Ann. Mat. Pura Appl.* **103** (1975), 127–129.

On the Frobenius Integrability of Certain Holomorphic p-Forms

Jean-Pierre Demailly

Université de Grenoble I, Département de Mathématiques, Institut Fourier
38402 Saint-Martin d'Hères, France
E-mail address: `demailly@ujf-grenoble.fr`

Abstract The goal of this note is to exhibit the integrability properties (in the sense of the Frobenius theorem) of holomorphic p-forms with values in certain line bundles with semi-negative curvature on a compact Kähler manifold. There are in fact very strong restrictions, both on the holomorphic form and on the curvature of the semi-negative line bundle. In particular, these observations provide interesting information on the structure of projective manifolds which admit a contact structure: either they are Fano manifolds or, thanks to results of Kebekus-Peternell-Sommese-Wisniewski, they are biholomorphic to the projectivization of the cotangent bundle of another suitable projective manifold.

Table of Contents

1 Main Results 93
2 Proof of the Main Theorem 95
References 97

1 Main Results

Recall that a holomorphic line bundle L on a compact complex manifold is said to be *pseudo-effective* if $c_1(L)$ contains a closed positive $(1,1)$-current T, or equivalently, if L possesses a (possibly singular) hermitian metric h such that the curvature current $T = \Theta_h(L) = -i\partial\bar\partial \log h$ is nonnegative. If X is projective, L is pseudo-effective if and only if $c_1(L)$ belongs to the closure of the cone generated by classes of effective divisors in $H^{1,1}_{\mathbb{R}}(X)$ (see [Dem90], [Dem92]). Our main result is

Main Theorem. *Let X be a compact Kähler manifold. Assume that there exists a pseudo-effective line bundle L on X and a nonzero holomorphic section $\theta \in H^0(X, \Omega_X^p \otimes L^{-1})$, where $0 \leq p \leq n = \dim X$. Let $\mathcal{S}_\theta$ be the coherent subsheaf of germs of vector fields ξ in the tangent sheaf T_X, such that the contraction $i_\xi\theta$ vanishes. Then $\mathcal{S}_\theta$ is integrable, namely $[\mathcal{S}_\theta, \mathcal{S}_\theta] \subset \mathcal{S}_\theta$, and L has flat curvature along the leaves of the (possibly singular) foliation defined by $\mathcal{S}_\theta$.*

2000 *Mathematics Subject Classification*: 32Q15, 32J25.

94   Jean-Pierre Demailly

Before entering into the proof, we discuss several consequences. If $p = 0$ or $p = n$, the result is trivial (with $\mathcal{S}_\theta = T_X$ and $\mathcal{S}_\theta = 0$, respectively). The most interesting case is $p = 1$.

Corollary 1. *In the above situation, if the line bundle $L \to X$ is pseudo-effective and $\theta \in H^0(X, \Omega_X^1 \otimes L^{-1})$ is a nonzero section, the subsheaf $\mathcal{S}_\theta$ defines a holomorphic foliation of codimension 1 in X, that is, $\theta \wedge d\theta = 0$.*

We now concentrate ourselves on the case when X is a *contact manifold*, i.e. $\dim X = n = 2m+1$, $m \geq 1$, and there exists a form $\theta \in H^0(X, \Omega_X^1 \otimes L^{-1})$, called the *contact form*, such that $\theta \wedge (d\theta)^m \in H^0(X, K_X \otimes L^{-m-1})$ has no zeroes. Then $\mathcal{S}_\theta$ is a codimension 1 locally free subsheaf of T_X and there are dual exact sequences

$$0 \to L \to \Omega_X^1 \to \mathcal{S}_\theta^\star \to 0, \qquad 0 \to \mathcal{S}_\theta \to T_X \to L^\star \to 0 \ .$$

The subsheaf $\mathcal{S}_\theta \subset T_X$ is said to be the *contact structure* of X. The assumption that $\theta \wedge (d\theta)^m$ does not vanish implies that $K_X \simeq L^{m+1}$. In that case, the subsheaf is not integrable, hence L and K_X cannot be pseudo-effective.

Corollary 2. *If X is a compact Kähler manifold admitting a contact structure, then K_X is not pseudo-effective, in particular the Kodaira dimension $\kappa(X)$ is equal to $-\infty$.*

The fact that $\kappa(X) = -\infty$ had been observed previously by Stéphane Druel [Dru98]. In the projective context, the minimal model conjecture would imply (among many other things) that the conditions $\kappa(X) = -\infty$ and "K_X non pseudo-effective" are equivalent, but a priori the latter property is much stronger (and in large dimensions, the minimal model conjecture still seems far beyond reach!)

Corollary 3. *If X is a compact Kähler manifold with a contact structure and with second Betti number $b_2 = 1$, then K_X is negative, i.e., X is a Fano manifold.*

Actually the Kodaira embedding theorem shows that the Kähler manifold X is projective if $b_2 = 1$, and then every line bundle is either positive, flat or negative. As K_X is not pseudo-effective it must therefore be negative. In that direction, Boothby [Boo61], Wolf [Wol65] and Beauville [Bea98] have exhibited a natural construction of contact Fano manifolds. Each of the known examples is obtained as a homogeneous variety which is the unique closed orbit in the projectivized (co)adjoint representation of a simple algebraic Lie group. Beauville's work ([Bea98], [Bea99]) provides strong evidence that this is the complete classification in the case $b_2 = 1$.

We now come to the case $b_2 \geq 2$. If Y is an arbitrary compact Kähler manifold, the bundle $X = P(T_Y^\star)$ of hyperplanes of T_Y has a contact structure associated with the line bundle $L = \mathcal{O}_X(-1)$. Actually, if $\pi : X \to Y$ is the canonical projection, one can define a contact form $\theta \in H^0(X, \Omega_X^1 \otimes L^{-1})$ by setting

$$\theta(x) = \theta(y, [\xi]) = \xi^{-1}\pi^*\xi = \xi^{-1} \sum_{1 \le j \le p} \xi_j dy_j, \qquad p = \dim Y ,$$

at every point $x = (y, [\xi]) \in X$, $\xi \in T^*_{Y,y}\backslash\{0\}$ (observe that $\xi \in L_x = \mathcal{O}_X(-1)_x$). Morever $b_2(X) = 1 + b_2(Y) \ge 2$. Conversely, Kebekus, Peternell, Sommese and Wiśniewski [KPSW] have recently shown that every projective algebraic manifold X such that

(i) X has a contact structure,
(ii) $b_2 \ge 2$,
(iii) K_X is not nef (numerically effective)

is of the form $X = P(T^*_Y)$ for some projective algebraic manifold Y. However, the condition that K_X is not nef is implied by the fact that K_X is not pseudo-effective. Hence we get

Corollary 4. *If X is a contact projective manifold with $b_2 \ge 2$, then X is a projectivized hyperplane bundle $X = P(T^*_Y)$ associated with some projective manifold Y.*

The Kähler case of corollary 4 is still unsolved, as the proof of [KPSW] heavily relies on Mori theory (and, unfortunately, the extension of Mori theory to compact Kähler manifolds remains to be settled ...).

I would like to thank Arnaud Beauville, Frédéric Campana, Stefan Kebekus and Thomas Peternell for illuminating discussions on these subjects. The present work was written during a visit at Göttingen University, on the occasion of a colloquium in honor of Professor Hans Grauert for his 70th birthday.

2 Proof of the Main Theorem

In some sense, the proof is just a straightforward integration by parts, but there are slight technical difficulties due to the fact that we have to work with singular metrics.

Let X be a compact Kähler manifold, ω the Kähler metric, and let L be a pseudo-effective line bundle on X. We select a hermitian metric h on L with nonnegative curvature current $\Theta_h(L) \ge 0$, and let φ be the plurisubharmonic weight of the metric h in any local trivialisation $L_{|U} \simeq U \times \mathbb{C}$. In other words, we have

$$\|\xi\|^2_h = |\xi|^2 e^{-\varphi(x)}, \qquad \|\xi^\star\|^2_{h^\star} = |\xi^\star|^2 e^{\varphi(x)}$$

for all $x \in U$ and $\xi \in L_x$, $\xi^\star \in L^{-1}$. We then have a Chern connection $\nabla = \partial_{h^\star} + \overline{\partial}$ acting on all (p, q)-forms f with values in L^{-1}, given locally by

$$\partial_\varphi f = e^{-\varphi}\partial(e^\varphi f) = \partial f + \partial\varphi \wedge f$$

in every trivialization $L_{|U}$. Now, assume that there is a holomorphic section $\theta \in H^0(X, \Omega_X^p \otimes L^{-1})$, i.e., a $\overline{\partial}$-closed $(p, 0)$ form θ with values in L^{-1}. We compute the global L^2 norm

$$\int_X \{\partial_{h^*}\theta, \partial_{h^*}\theta\}_{h^*} \wedge \omega^{n-p-1} = \int_X e^\varphi \partial_\varphi\theta \wedge \overline{\partial_\varphi\theta} \wedge \omega^{n-p-1}$$

where $\{\ ,\ \}_{h^*}$ is the natural sesquilinear pairing sending pairs of L^{-1}-valued forms of type (p, q), (r, s) into $(p + s, q + r)$ complex valued forms. The right hand side is of course only locally defined, but it explains better how the forms are calculated, and also all local representatives glue together into a well defined global form; we will therefore use the latter notation as if it were global. As

$$d\left(e^\varphi\theta \wedge \overline{\partial_\varphi\theta} \wedge \omega^{n-p-1}\right) = e^\varphi \partial_\varphi\theta \wedge \overline{\partial_\varphi\theta} \wedge \omega^{n-p-1} + (-1)^p e^\varphi\theta \wedge \overline{\partial}\partial_\varphi\theta \wedge \omega^{n-p-1}$$

and $\overline{\partial}\partial_\varphi\theta = \overline{\partial}\partial\varphi \wedge \theta$, an integration by parts via Stokes theorem yields

$$\int_X e^\varphi \partial_\varphi\theta \wedge \overline{\partial_\varphi\theta} \wedge \omega^{n-p-1} = -(-1)^p \int_X e^\varphi \partial\overline{\partial}\varphi \wedge \theta \wedge \overline{\theta} \wedge \omega^{n-p-1}\ .$$

These calculations need a word of explanation, since φ is in general singular. However, it is well known that the $i\partial\overline{\partial}$ of a plurisubharmonic function is a closed positive current, in particular

$$i\partial\overline{\partial}(e^\varphi) = e^\varphi(i\partial\varphi \wedge \overline{\partial}\varphi + i\partial\overline{\partial}\varphi)$$

is positive and has measure coefficients. This shows that $\partial\varphi$ is L^2 with respect to the weight e^φ, and similarly that $e^\varphi\partial\overline{\partial}\varphi$ has locally finite measure coefficients. Moreover, the results of [Dem92] imply that there is a decreasing sequence of metrics h_ν^* and corresponding weights $\varphi_\nu \downarrow \varphi$, such that $\Theta_{h_\nu} \geq -C\omega$ with a fixed constant $C > 0$ (this claim is in fact much weaker than the results of [Dem92], and easy to prove e.g. by using convolutions in suitable coordinate patches and a standard gluing technique). Now, the results of Bedford-Taylor [BT76], [BT82] applied to the uniformly bounded functions $e^{c\varphi_\nu}$, $c > 0$, imply that we have local weak convergence

$$e^{\varphi_\nu}\partial\overline{\partial}\varphi_\nu \to e^\varphi\partial\overline{\partial}\varphi, \quad e^{\varphi_\nu}\partial\varphi_\nu \to e^\varphi\partial\varphi, \quad e^{\varphi_\nu}\partial\varphi_\nu \wedge \overline{\partial}\varphi_\nu \to e^\varphi\partial\varphi \wedge \overline{\partial}\varphi\ ,$$

possibly after adding $C'|z|^2$ to the φ_ν's to make them plurisubharmonic. This is enough to justify the calculations. Now, we take care of signs, using the fact that $i^{p^2}\theta \wedge \overline{\theta} \geq 0$ whenever θ is a $(p, 0)$-form. Our previous equality can be rewritten

$$\int_X e^\varphi\, i^{(p+1)^2} \partial_\varphi\theta \wedge \overline{\partial_\varphi\theta} \wedge \omega^{n-p-1} = -\int_X e^\varphi\, i\partial\overline{\partial}\varphi \wedge i^{p^2}\theta \wedge \overline{\theta} \wedge \omega^{n-p-1}\ .$$

Frobenius Integrability of Certain Holomorphic p-Forms 97

Since the left hand side is nonnegative and the right hand side is nonpositive, we conclude that $\partial_\varphi \theta = 0$ almost everywhere, i.e. $\partial \theta = -\partial \varphi \wedge \theta$ almost everywhere. The formula for the exterior derivative of a p-form reads

$$d\theta(\xi_0, \ldots, \xi_p) = \sum_{0 \le j \le p} (-1)^j \xi_j \cdot \theta(\xi_0, \ldots, \widehat{\xi_j}, \ldots, \xi_p)$$

$$(*) \qquad + \sum_{0 \le j < k \le p} (-1)^{j+k} \theta([\xi_j, \xi_k], \xi_0, \ldots, \widehat{\xi_j}, \ldots, \widehat{\xi_k}, \ldots, \xi_p) .$$

If two of the vector fields – say ξ_0 and ξ_1 – lie in S_θ, then

$$d\theta(\xi_0, \ldots, \xi_p) = -(\partial \varphi \wedge \theta)(\xi_0, \ldots, \xi_p) = 0$$

and all terms in the right hand side of $(\star)$ are also zero, except perhaps the term $\theta([\xi_0, \xi_1], \xi_2, \ldots, \xi_p)$. We infer that this term must vanish. Since this is true for arbitrary vector fields $\xi_2, \ldots, \xi_p$, we conclude that $[\xi_0, \xi_1] \in S_\theta$ and that S_θ is integrable.

The above arguments also yield strong restrictions on the hermitian metric h. In fact the equality $\partial \theta = -\partial \varphi \wedge \theta$ implies $\partial \overline{\partial} \varphi \wedge \theta = 0$ by taking the $\overline{\partial}$. Fix a smooth point in a leaf of the foliation, and local coordinates $(z_1, \ldots, z_n)$ such that the leaves are given by $z_1 = c_1, \ldots, z_r = c_r$ ($c_i = $ constant), in a neighborhood of that point. Then S_θ is generated by $\partial/\partial z_{r+1}, \ldots, \partial/\partial z_n$, and θ depends only on $dz_1, \ldots, dz_r$. This implies that $\partial^2 \varphi / \partial z_j \partial \overline{z}_k = 0$ for $j, k > r$, in other words (L, h) has flat curvature along the leaves of the foliation. The main theorem is proved.

References

[Bea98]   Beauville, A., Fano contact manifolds and nilpotent orbits, *Comm. Math. Helv.* **73**(4) (1998), 566–583.

[Bea99]   Beauville, A., Riemannian holonomy and algebraic geometry, Duke/alg-geom preprint 9902110, 1999.

[Boo61]   Boothby, W., Homogeneous complex contact manifolds, *Proc. Symp. Pure. Math. (Differential Geometry)* **3** (1961), 144–154.

[BT76]   Bedford, E., Taylor, B. A., The Dirichlet problem for a complex Monge-Ampère equation, *Invent. Math.* **37** (1976), 1–44.

[BT82]   Bedford, E., Taylor, B. A., A new capacity for plurisubharmonic functions, *Acta Math.* **149** (1982), 1–41.

[Dem90]   Demailly, J.-P., Singular Hermitian metrics on positive line bundles., in: Hulek K., Peternell T., Schneider M., Schreyer F. (Eds.), *Proceedings of the Bayreuth Conference "Complex Algebraic Varieties," April 2-6, 1990, Lecture Notes in Math.* **1507** (1992), Springer-Verlag.

[Dem92]   Demailly, J.-P., Regularization of closed positive currents and intersection theory, *J. Alg. Geom.* **1** (1992), 361–409.

[Dru98]   Druel, S., Contact structures on Algebraic 5-dimensional manifolds, *C.R. Acad. Sci. Paris* **327** (1998), 365–368.

[KPSW]   Kebekus, S., Peternell, Th., Sommese, A. J., Wiśniewski, J. A., Projective contact manifolds, 1999, *Invent. Math.*, to appear.

98   Jean-Pierre Demailly

[Wol65]   Wolf, J., Complex homogeneous contact manifolds and quaternionic symmetric spaces, *J. Math. Mech.* **14** (1965), 1033–1047.

Analytic Moduli Spaces
of Simple (Co)Framed Sheaves

Hubert Flenner[1] and Martin Lübke[2] *

[1] Fakultät für Mathematik der Ruhr-Universität, Universitätsstr. 150, Geb. NA 2/72, 44780 Bochum, Germany
E-mail address: Hubert.Flenner@ruhr-uni-bochum.de
[2] Mathematical Institute, Leiden University, PO box 9512, NL-2300 RA Leiden, The Netherlands
E-mail address: lubke@math.leidenuniv.nl

Abstract Let X be a complex space and $\mathcal{F}$ a coherent $\mathcal{O}_X$-module. A $\mathcal{F}$-*(co)framed* sheaf on X is a pair $(\mathcal{E}, \varphi)$ with a coherent $\mathcal{O}_X$-module $\mathcal{E}$ and a morphism of coherent sheaves $\varphi : \mathcal{F} \longrightarrow \mathcal{E}$ (resp. $\varphi : \mathcal{E} \longrightarrow \mathcal{F}$). Two such pairs $(\mathcal{E}, \varphi)$ and $(\mathcal{E}', \varphi')$ are said to be *isomorphic* if there exists an isomorphism of sheaves $\alpha : \mathcal{E} \longrightarrow \mathcal{E}'$ with $\alpha \circ \varphi = \varphi'$ (resp. $\varphi' \circ \alpha = \varphi$). A pair $(\mathcal{E}, \varphi)$ is called *simple* if its only automorphism is the identity on $\mathcal{E}$. In this note we prove a representability theorem in a relative framework, which implies in particular that there is a moduli space of simple $\mathcal{F}$-(co)framed sheaves on a given compact complex space X.

Table of Contents

1 Introduction 99
2 Preparations 101
3 Simple $\mathcal{F}$-Coframed Sheaves 104
4 Proof of Theorem 1.1 105
References 108

1 Introduction

Let X be a complex space and $\mathcal{F}$ a coherent $\mathcal{O}_X$-module. By a $\mathcal{F}$-*coframed* sheaf on X we mean a pair $(\mathcal{E}, \varphi)$ with

(a) $\mathcal{E}$ is a coherent $\mathcal{O}_X$-module,
(b) $\varphi : \mathcal{F} \longrightarrow \mathcal{E}$ is a morphism of coherent sheaves.

(Following [HL1], [HL2], a $\mathcal{F}$-*framed* sheaf is dually a pair $(\mathcal{E}, \varphi)$ with $\mathcal{E}$ as above and a morphism $\varphi : \mathcal{E} \longrightarrow \mathcal{F}$.) Two such pairs $(\mathcal{E}, \varphi)$ and $(\mathcal{E}', \varphi')$ are

---

* This paper was prepared during a visit of the second author to the University of Bochum which was financed by EAGER - European Algebraic Geometry Research Training Network, contract No. HPRN-CT-2000-00099 (BBW 99.0030). 2000 *Mathematics Subject Classification*: 32G13, 14D20.

100 Hubert Flenner and Martin Lübke

said to be *isomorphic* if there exists an isomorphism of sheaves $\alpha : \mathcal{E} \longrightarrow \mathcal{E}'$ with $\alpha \circ \varphi = \varphi'$. A pair $(\mathcal{E}, \varphi)$ is called *simple* if its only automorphism is the identity on $\mathcal{E}$. The purpose of this note is to show that there is a moduli space of simple $\mathcal{F}$-(co)framed sheaves on a given compact complex space X.

More generally, we will show the following relative result. Let $f : X \to S$ be a proper morphism of complex spaces. By a family of $\mathcal{F}$-coframed sheaves over S (or a $\mathcal{F}$-coframed sheaf on X/S in brief) we mean a $\mathcal{F}$-coframed sheaf $(\mathcal{E}, \varphi)$ on X that is S-flat. Such a family will be called *simple* if its restriction to each fibre $X(s) := f^{-1}(s)$ is simple.

We consider the set-valued functor $P : \mathfrak{An}_S \longrightarrow \mathfrak{sets}$ on the category of complex spaces over S such that $P(T)$ (for $T \in \mathfrak{An}_S$) is the set of all isomorphism classes of simple $\mathcal{F}_T$-coframed sheaves $(\mathcal{E}, \varphi)$ on X_T/T, where $X_T := X \times_S T$ and $\mathcal{F}_T$ is the pullback of $\mathcal{F}$ to X_T. The main result of this paper is

Theorem 1.1. *If X is cohomologically flat over S in dimension 0, then the functor P is representable by a (not necessarily separated) complex space.*

Thus, informally speaking there is a (relative) moduli space of $\mathcal{F}$-coframed sheaves on X/S. An inspection of the proof shows that the analogous result holds for simple $\mathcal{F}$-framed sheaves provided that $\mathcal{F}$ is flat over S. The reason that in the case of $\mathcal{F}$-framed sheaves we need this additional assumption is that only in this case the functor $\underline{\mathrm{Hom}}(\mathcal{E}, \mathcal{F})$ is known to be representable (see 2.3) whereas $\underline{\mathrm{Hom}}(\mathcal{F}, \mathcal{E})$ is representable as soon as $\mathcal{E}$ is S-flat.

Our main motivation for studying moduli spaces of $\mathcal{F}$-(co)framed sheaves is the following. In the case that $\mathcal{F}$ and $\mathcal{E}$ are locally free, a pair $(\mathcal{E}, \varphi)$ as above is called a *$\mathcal{F}$-(co)framed vector bundle* or *holomorphic pair*. Various types of holomorphic pairs over compact complex manifolds (e.g. the coframed ones with $\mathcal{F} = \mathcal{O}_X$ or the framed ones with arbitrary $\mathcal{F}$, see [OT1], [OT2]) can be identified with solutions of so-called *vortex equations* via a Kobayashi-Hitchin type correspondence. On complex surfaces, these solutions can further be identified with solutions of Seiberg-Witten equations, and moduli spaces $\mathcal{M}^{st}$ of *stable* holomorphic pairs (which are open subsets of the moduli spaces $\mathcal{M}^s$ of simple ones) can be used to effectively calculate Seiberg-Witten invariants in several cases.

It is important to notice that the set $\mathcal{M}^s$ a priori has two analytic structures. One is given by our result and makes it possible to determine $\mathcal{M}^s$ using complex-analytic deformation theory. The other one, which is the one relevant in Seiberg-Witten theory, is given by a gauge theoretical description as in [LL]. But the main result of that paper is in fact that these two structures are indeed the same.

Finally we mention that moduli spaces of stable $\mathcal{F}$-framed sheaves on *algebraic* manifolds have been constructed in [HL1].

Without a (co)framing, a moduli space of simple bundles was constructed in [KO] and in a more general context in [FS]. In this paper we follow closely the method of proof in the latter paper. The main difficulty is to verify the relative representability of the general criterion 4.3 for the functor P in 1.1. For this we will show in Sect. 2 that the functor of endomorphisms of $\mathcal{F}$-coframed sheaves is representable. In Sect. 3 we show the openness of the set of points where a coframed sheaf is simple. After these preparations it will be easy to give in Sect. 4 the proof of 1.1.

2 Preparations

We start with an algebraic lemma.

Lemma 2.1. *Let R be a ring, L, M, N R-modules, and $\alpha : L \longrightarrow M$, $\beta : L \longrightarrow N$ R-linear maps. If $K := \mathrm{coker}((\alpha, -\beta) : L \longrightarrow M \oplus N)$, then*

$$\mathbb{S}(K) \cong \mathbb{S}(M) \otimes_{\mathbb{S}(L)} \mathbb{S}(N)$$

where we consider the symmetric algebras $\mathbb{S}(M)$, $\mathbb{S}(N)$ as $\mathbb{S}(L)$-algebras via the maps $\mathbb{S}(\alpha)$, $\mathbb{S}(\beta)$.

Proof. There are canonical maps $M \longrightarrow K$ and $N \longrightarrow K$ given by

$$m \mapsto \text{residue class of} (m, 0) \ , \ n \mapsto \text{residue class of} (0, n).$$

These maps induce a commutative diagram

$$\begin{array}{ccc} \mathbb{S}(L) & \longrightarrow & \mathbb{S}(M) \\ \downarrow & & \downarrow \\ \mathbb{S}(N) & \longrightarrow & \mathbb{S}(K) \end{array}$$

and so, using the universal property of the tensor product, there is a natural map

$$\mathbb{S}(M) \otimes_{\mathbb{S}(L)} \mathbb{S}(N) \longrightarrow \mathbb{S}(K) \ .$$

Conversely, to construct a natural map in the other direction, note that the first graded piece of $\mathbb{S}(M) \otimes_{\mathbb{S}(L)} \mathbb{S}(N)$ is just $(M \oplus N)/L = K$, so there is an induced map

$$\mathbb{S}(K) \longrightarrow \mathbb{S}(M) \otimes_{\mathbb{S}(L)} \mathbb{S}(N) \ .$$

It is easy to check that these maps are inverse to each other. $\qquad\square$

Now let S be a complex analytic space, and let $\mathfrak{An}_S$ be the category of analytic spaces over S. Recall that for a coherent $\mathcal{O}_S$-module $\mathcal{F}$ over S the linear fibre space $\mathbb{V}(\mathcal{F})$ represents the functor

$$F : \mathfrak{An}_S \ni T \mapsto F(T) := \mathrm{Hom}(\mathcal{F}_T, \mathcal{O}_T)$$

(see [Fi] or [EGA, II 1.7]). Note that $F(T)$ has the structure of a $\Gamma(T, \mathcal{O}_T)$-module. Moreover, if $\mathcal{G}$ is another coherent $\mathcal{O}_S$-module and $T \mapsto G(T)$ is the

102   Hubert Flenner and Martin Lübke

associated functor as above, then a transformation of functors $F \longrightarrow G$ will be called *linear* if $F(T) \longrightarrow G(T)$ is $\Gamma(T, \mathcal{O}_T)$-linear for all $T \in \mathfrak{An}_S$. The reader may easily verify that there is a one-to-one correspondence between such linear transformations of functors and morphisms of sheaves $\mathcal{G} \longrightarrow \mathcal{F}$.

Proposition 2.2. *Let $H, F, G : \mathfrak{An}_S \longrightarrow \mathfrak{sets}$ be functors that are represented by linear fibre spaces $\mathbb{V}(\mathcal{H}), \mathbb{V}(\mathcal{F}), \mathbb{V}(\mathcal{G})$, respectively. Let $H \longrightarrow G$ and $F \longrightarrow G$ be linear morphisms of functors, and let $K := H \times_G F$ be the fibered product. Then K is represented by $\mathbb{V}(\mathcal{K})$ with*

$$\mathcal{K} := \mathrm{coker}((\alpha, -\beta) : \mathcal{G} \longrightarrow \mathcal{H} \times \mathcal{F}) \, ,$$

where $\alpha : \mathcal{G} \longrightarrow \mathcal{H}$ and $\beta : \mathcal{G} \longrightarrow \mathcal{F}$ are the morphisms of sheaves corresponding to $H \longrightarrow G$ and $F \longrightarrow G$.

Proof. The spaces $\mathbb{V}(\mathcal{H}), \mathbb{V}(\mathcal{F}), \mathbb{V}(\mathcal{G})$ are the analytic spectra associated to the symmetric algebras $\mathbb{S}(\mathcal{H}), \mathbb{S}(\mathcal{F}), \mathbb{V}(\mathcal{G})$, respectively, and $H \times_G F$ is represented by $\mathbb{V}(\mathcal{H}) \times_{\mathbb{V}(\mathcal{G})} \mathbb{V}(\mathcal{F})$ which is the analytic spectrum of $\mathbb{S}(\mathcal{H}) \otimes_{\mathbb{S}(\mathcal{G})} \mathbb{S}(\mathcal{F})$. Hence we need to verify that there is a natural isomorphism

$$\mathbb{S}(\mathcal{H}) \otimes_{\mathbb{S}(\mathcal{G})} \mathbb{S}(\mathcal{F}) \longrightarrow \mathbb{S}(\mathcal{K}) \, ,$$

but this is a consequence of Lemma 2.1. $\qquad\qquad\qquad\qquad\qquad\qquad\square$

Let $f : X \longrightarrow S$ be a fixed proper morphism of complex spaces, and let $\mathcal{E}, \mathcal{F}$ be coherent $\mathcal{O}_X$-modules, where $\mathcal{E}$ is flat over S. Let

$$H := \underline{\mathrm{Hom}}(\mathcal{F}, \mathcal{E}) : \mathfrak{An}_S \longrightarrow \mathfrak{sets}$$

be the functor given by

$$H(T) := \mathrm{Hom}_{X_T}(\mathcal{F}_T, \mathcal{E}_T) \, ,$$

where $X_T := X \times_S T$ and $\mathcal{E}_T, \mathcal{F}_T$ are the pullbacks of $\mathcal{E}, \mathcal{F}$ on X_T. We recall the following fact.

Theorem 2.3. *The functor H is representable by a linear fibre space over S.*

For a proof see e.g. [Fl2, 3.2] or [Bi].

Now let $\varphi : \mathcal{F} \longrightarrow \mathcal{E}$ and $\varphi' : \mathcal{F} \longrightarrow \mathcal{E}'$ be fixed morphisms of coherent sheaves on X. Let us consider the functor

$$M = \underline{\mathrm{Hom}}((\mathcal{E}, \varphi), (\mathcal{E}', \varphi')) : \mathfrak{An}_S \longrightarrow \mathfrak{sets}$$

defined as follows: For $T \in \mathfrak{An}_S$ the elements of $M(T)$ are the pairs

$$(c, \alpha) \in \Gamma(\mathcal{O}_{X_T}) \times \mathrm{Hom}_{X_T}(\mathcal{E}_T, \mathcal{E}_T')$$

such that the diagram

$$\begin{array}{ccc} \mathcal{F}_T & \xrightarrow{\varphi_T} & \mathcal{E}_T \\ {\scriptstyle c\cdot\mathrm{id}_{\mathcal{F}_T}}\downarrow & & \downarrow{\scriptstyle\alpha} \\ \mathcal{F}_T & \xrightarrow{\varphi_T'} & \mathcal{E}_T' \end{array}$$

commutes, i.e. such that $c \cdot \varphi_T' = \alpha \cdot \varphi_T$.

Proposition 2.4. *If $\mathcal{O}_X$, $\mathcal{E}$ and $\mathcal{E}'$ are flat over S, then M is representable by a linear fibre space $\mathbb{V}(\mathcal{M})$ over S.*

Proof. By Theorem 2.3 the functors

$$F := \underline{\mathrm{Hom}}(\mathcal{E}, \mathcal{E}') \ , \ \ H := \underline{\mathrm{Hom}}(\mathcal{O}_X, \mathcal{O}_X) \ , \ \ G := \underline{\mathrm{Hom}}(\mathcal{F}, \mathcal{E}')$$

are representable by linear fibre spaces over S. There are natural maps

$$H \longrightarrow G \ , \ c \mapsto c \cdot \varphi' \ ,$$

and

$$F \longrightarrow G \ , \ \alpha \mapsto \alpha \circ \varphi \ .$$

By definition we have

$$M = F \times_G H,$$

so the result follows from Proposition 2.2. $\qquad\qquad\square$

An important property of the sheaf $\mathcal{M}$ in Proposition 2.4 is given by

Lemma 2.5. *The following are equivalent.*

(a) *$\mathcal{M}$ is locally free.*
(b) *For every complex space $T \in \mathfrak{An}_S$ the canonical map*

$$f_*(\mathcal{H}om((\mathcal{E}, \varphi), (\mathcal{E}', \varphi'))) \otimes_{\mathcal{O}_S} \mathcal{O}_T \longrightarrow f_{T*}(\mathcal{H}om((\mathcal{E}_T, \varphi_T), (\mathcal{E}_T', \varphi_T')))$$

is an isomorphism.

Moreover, if one of these conditions holds then

(1) $$\mathcal{M} \cong [f_*(\mathcal{H}om((\mathcal{E}, \varphi), (\mathcal{E}', \varphi')))]^{\vee} \ .$$

Proof. First note that for every complex space $T \in \mathfrak{An}_S$ we have

(2) $$(\mathcal{M}_T)^{\vee} \cong f_{T*}(\mathcal{H}om((\mathcal{E}_T, \varphi_T), (\mathcal{E}_T', \varphi_T'))) \ .$$

Applying this to the case $T = S$, (1) follows immediately from the assumption that $\mathcal{M}$ is locally free. Moreover, if (a) is satisfied we have

$$\mathcal{M}_T^{\vee} \cong \mathcal{M}^{\vee} \otimes_{\mathcal{O}_S} \mathcal{O}_T \cong f_*(\mathcal{H}om((\mathcal{E}, \varphi), (\mathcal{E}', \varphi'))) \otimes_{\mathcal{O}_S} \mathcal{O}_T \ ,$$

104　Hubert Flenner and Martin Lübke

where for the last isomorphism we have used (1). Thus (b) follows.

Conversely, assume that (b) holds. Using (1) we infer from the isomorphism in (b) that

$$\mathcal{H}om(\mathcal{M}_T, \mathcal{O}_T) \cong \mathcal{H}om(\mathcal{M}, \mathcal{O}_S) \otimes_{\mathcal{O}_S} \mathcal{O}_T \ .$$

Applying this to $T = \{s\}$, $s \in S$ a reduced point, it follows that the map

$$\mathcal{H}om(\mathcal{M}, \mathcal{O}_S) \longrightarrow \mathcal{H}om(\mathcal{M}, \mathcal{O}_S/\mathfrak{m}_s)$$

is surjective for every point $s \in S$. Using standard arguments (see e.g. [EGA, 7.5.2]) we conclude that the functor $\mathcal{H}om(\mathcal{M}, -)$ is exact on the category of coherent $\mathcal{O}_S$-modules whence $\mathcal{M}$ is locally free, as required.　□

3　Simple $\mathcal{F}$-Coframed Sheaves

As before let $f : X \longrightarrow S$ be a proper morphism of complex spaces, and $\mathcal{F}$ a fixed $\mathcal{O}_X$-module. We consider $\mathcal{F}$-coframed sheaves $(\mathcal{E}, \varphi)$ on X/S, i.e. $\mathcal{E}$ is a S-flat coherent sheaf on X and $\varphi : \mathcal{F} \longrightarrow \mathcal{E}$ is a morphism of $\mathcal{O}_X$-modules.

Definition 3.1 $(\mathcal{E}, \varphi)$ *is called simple at* $s \in S$ *if its fibre* $(\mathcal{E}(s), \varphi(s))$ *is simple, i.e. if*

$$
\begin{array}{ccc}
\mathcal{F}(s) & \xrightarrow{\varphi(s)} & \mathcal{E}(s) \\
{\scriptstyle c \cdot \mathrm{id}_{\mathcal{F}(s)}} \downarrow & & \downarrow {\scriptstyle \alpha} \\
\mathcal{F}(s) & \xrightarrow{\varphi(s)} & \mathcal{E}(s)
\end{array}
$$

is a commutative diagram, then $\alpha = c \cdot \mathrm{id}_{\mathcal{E}(s)}$. *Moreover,* $(\mathcal{E}, \varphi)$ *is said to be simple over* S *if it is simple at every point.*

Notice that this definition of simpleness of $(\mathcal{E}(s), \varphi(s))$ coincides with the one given in the introduction. Later on we will will need that the points $s \in S$ at which $(\mathcal{E}, \varphi)$ is simple form an open set in S. For this we need the following considerations.

By Theorem 2.3 and Proposition 2.4 there are coherent $\mathcal{O}_S$-modules $\mathcal{H}$ and $\mathcal{G}$ such that

$$\underline{\mathrm{End}}(\mathcal{E}, \varphi) := \underline{\mathrm{Hom}}((\mathcal{E}, \varphi), (\mathcal{E}, \varphi)) \quad \text{and} \quad \underline{\mathrm{End}}(\mathcal{O}_X) := \underline{\mathrm{Hom}}(\mathcal{O}_X, \mathcal{O}_X)$$

are represented by $\mathbb{V}(\mathcal{H})$ resp. $\mathbb{V}(\mathcal{G})$. Let

$$\tilde{a} : \mathcal{G} \longrightarrow \mathcal{H} \quad \text{and} \quad \tilde{b} : \mathcal{H} \longrightarrow \mathcal{G}$$

be the $\mathcal{O}_S$-linear maps associated to the canonical morphisms of functors

$$a : \underline{\mathrm{End}}(\mathcal{E}, \varphi) \longrightarrow \underline{\mathrm{End}}(\mathcal{O}_X) \ , \text{ resp. } \ b : \underline{\mathrm{End}}(\mathcal{O}_X) \longrightarrow \underline{\mathrm{End}}(\mathcal{E}, \varphi) \ ,$$

$$(c, \alpha) \longmapsto c \qquad\qquad\qquad c \longmapsto (c, c \cdot \mathrm{id}_{\mathcal{E}}) \ .$$

As $a \circ b = \mathrm{id}_{\underline{\mathrm{End}}(\mathcal{O}_X)}$ we have $\tilde{b} \circ \tilde{a} = \mathrm{id}_{\mathcal{G}}$. In other words, $\mathcal{G}$ is a direct summand of $\mathcal{H}$ so that $\mathcal{H} \cong \mathcal{G} \oplus \mathcal{G}'$ for some coherent sheaf $\mathcal{G}'$ on S.

Analytic Moduli Spaces — 105

Lemma 3.2. *The following are equivalent.*

(1) $(\mathcal{E}, \varphi)$ *is simple on* S.
(2) $\mathcal{G}' = 0$.
(3) *The canonical morphism of functors* $b : \underline{\mathrm{End}}(\mathcal{O}_X) \longrightarrow \underline{\mathrm{End}}(\mathcal{E}, \varphi)$ *is an isomorphism.*

Proof. The functor $\underline{\mathrm{End}}(\mathcal{E}(s), \varphi(s))$ resp. $\underline{\mathrm{End}}(\mathcal{O}_{X(s)})$ on the category $\mathfrak{An}$ of all analytic spaces is represented by $\mathcal{H}(s)$ resp. $\mathcal{G}(s)$. Thus $(\mathcal{E}(s), \varphi(s))$ is simple if and only if $\mathcal{G}(s) \cong \mathcal{H}(s)$ which is equivalent to the vanishing of $\mathcal{G}'(s)$. Using Nakayama's lemma, the equivalence of (1) and (2) follows. Finally, the equivalence of (2) and (3) is immediate from the definition of $\mathcal{G}'$. $\qquad\square$

Corollary 3.3. *If $\mathcal{E}$ is S-flat, then the set of points $s \in S$ at which $(\mathcal{E}, \varphi)$ is simple, is an open subset of S.*

Proof. The set of points $s \in S$ for which $\mathcal{G}'/\mathfrak{m}_s \cdot \mathcal{G}' = 0$ is just the complement of the support of $\mathcal{G}'$ and hence Zariski-open in S. Using Lemma 3.2 we get the desired result. $\qquad\square$

Recall that a morphism $f : X \longrightarrow S$ of complex spaces is said to be *cohomologically flat in dimension* 0 if it is flat and if for every $s \in S$ the natural map $f_*(\mathcal{O}_X) \longrightarrow f_*(\mathcal{O}_{X(s)})$ is surjective.

Corollary 3.4. *If $(\mathcal{E}, \varphi)$ is simple and $f : X \longrightarrow S$ is cohomologically flat in dimension 0, then $f_*(\mathcal{O}_X)$ is a locally free $\mathcal{O}_S$-module, and the functor $\underline{\mathrm{End}}(\mathcal{E}, \varphi)$ is represented by $\mathbb{V}(f_*(\mathcal{O}_X)^{\vee})$.*

Proof. The fact that $f_*(\mathcal{O}_X)$ is locally free over S, is well known (see, e.g. [FS, 9.7]). Moreover, since $(\mathcal{E}, \varphi)$ is simple we have $\underline{\mathrm{End}}(\mathcal{E}, \varphi) = \underline{\mathrm{End}}(\mathcal{O}_X)$. As

$$(f_T)_*(\mathcal{O}_{X_T}) \cong f_*(\mathcal{O}_X) \otimes_{\mathcal{O}_S} \mathcal{O}_T$$

we get

$$\mathrm{Hom}_{X_T}(\mathcal{O}_{X_T}, \mathcal{O}_{X_T}) \cong \Gamma(X_T, \mathcal{O}_{X_T}) \cong \mathrm{Hom}_T(f_*(\mathcal{O}_X)^{\vee} \otimes_{\mathcal{O}_S} \mathcal{O}_T, \mathcal{O}_T) \,,$$

so the space $\mathbb{V}(f_*(\mathcal{O}_X)^{\vee})$ represents $\underline{\mathrm{End}}(\mathcal{E}, \varphi)$ as desired. $\qquad\square$

4 Proof of Theorem 1.1

An *isomorphism* of two $\mathcal{F}$-coframed sheaves $(\mathcal{E}, \varphi)$ and $(\mathcal{E}', \varphi')$ is an isomorphism $\alpha : \mathcal{E} \longrightarrow \mathcal{E}'$ such that

commutes. We note that $(\mathcal{E}, \varphi)$ and $(\mathcal{E}', \varphi')$ are isomorphic if and only if there is a pair

$$(c, \alpha) \in \Gamma(X, \mathcal{O}_X) \times \mathrm{Hom}_X(\mathcal{E}, \mathcal{E}')$$

such that

(a) c is a unit in $f_*(\mathcal{O}_X)$,
(b) α is an isomorphism,
(c) $\alpha \circ \varphi = c \cdot \varphi'$.

Notice that a simple pair $(\mathcal{E}, \varphi)$ has no automorphism besides $\mathrm{id}_\mathcal{E}$. If S is a reduced point then the converse also holds, i.e. $(\mathcal{E}, \varphi)$ is simple if and only if $\mathrm{id}_\mathcal{E}$ is its only automorphism.

Theorem 4.1. *Assume that X is cohomologically flat over S in dimension 0, and let $(\mathcal{E}, \varphi)$ and $(\mathcal{E}', \varphi')$ be simple pairs. Then the functor*

$$F : \mathfrak{An}_S \longrightarrow \mathfrak{sets}, \quad F(T) := \begin{cases} \{1\} & \text{if } (\mathcal{E}_T, \varphi_T) \cong (\mathcal{E}'_T, \varphi'_T), \\ \emptyset & \text{otherwise}, \end{cases}$$

is representable by a locally closed subspace of S.

Proof. As the sheaf $f_*(\mathcal{O}_X)$ is locally free over S we may assume that it has constant rank, say, r over $\mathcal{O}_S$. By Proposition 2.4 the functors

$$M := \underline{\mathrm{Hom}}((\mathcal{E}, \varphi), (\mathcal{E}', \varphi')), \quad M' := \underline{\mathrm{Hom}}((\mathcal{E}', \varphi'), (\mathcal{E}, \varphi))$$

are representable by linear fibre spaces $\mathbb{V}(\mathcal{M})$ resp. $\mathbb{V}(\mathcal{M}')$, where $\mathcal{M}$ and $\mathcal{M}'$ are coherent $f_*(\mathcal{O}_X)$-modules. If for some space $T \in \mathfrak{An}_S$ the pairs $(\mathcal{E}_T, \varphi_T)$ and $(\mathcal{E}'_T, \varphi'_T)$ are isomorphic, then by Corollary 3.4 $\mathcal{M}_T$ and $\mathcal{M}'_T$ are locally free $\mathcal{O}_T$-modules of rank r on T. Thus applying [FS, 9.10] as in the proof of [FS, 9.9], we are reduced to the case that $\mathcal{M}$ and $\mathcal{M}'$ are locally free $\mathcal{O}_S$-modules of rank r. Let us consider the pairings

$$M \times M' \longrightarrow \underline{\mathrm{End}}(\mathcal{E}, \varphi), \quad ((c, \alpha), (d, \beta)) \mapsto (cd, \beta \circ \alpha),$$
$$M' \times M \longrightarrow \underline{\mathrm{End}}(\mathcal{E}', \varphi'), \quad ((d, \beta), (c, \alpha)) \mapsto (cd, \alpha \circ \beta);$$

these correspond to pairings

$$\mathcal{M}^\vee \otimes \mathcal{M}'^\vee \xrightarrow{\gamma} f_*(\mathcal{O}_X),$$
$$\mathcal{M}'^\vee \otimes \mathcal{M}^\vee \xrightarrow{\gamma'} f_*(\mathcal{O}_X).$$

Using Lemma 2.5 it follows as in the proof of [FS, 9.9] that our functor F is represented by the open subset

$$S' := S \setminus \mathrm{supp}(\mathrm{coker}(\gamma) \oplus \mathrm{coker}(\gamma')) \,. \qquad \square$$

Now we consider the groupoid $\mathfrak{P} \longrightarrow \mathfrak{An}_S$, where for $T \in \mathfrak{An}_S$ the objects in $\mathfrak{P}(T)$ are the $\mathcal{F}_T$-coframed sheaves $(\mathcal{E}, \varphi)$ on X_T/T, where $\varphi : \mathcal{F}_T \longrightarrow \mathcal{E}$ is $\mathcal{O}_{X_T}$-linear. For $(\mathcal{E}, \varphi) \in \mathfrak{P}(T)$ and $(\mathcal{E}', \varphi') \in \mathfrak{P}(T')$, a morphism $(\mathcal{E}, \varphi) \longrightarrow (\mathcal{E}', \varphi')$ is a pair (f, α), where $f : T' \longrightarrow T$ is an S-morphism and $\alpha : f^*(\mathcal{E}) \longrightarrow \mathcal{E}'$ is an isomorphism of coherent sheaves such that the diagram

$$
\begin{array}{ccc}
f^*(\mathcal{F}_T) & \xrightarrow{f^*(\varphi)} & f^*(\mathcal{E}) \\
\| & & \downarrow \alpha \\
\mathcal{F}_{T'} & \xrightarrow{\varphi'} & \mathcal{E}'
\end{array}
$$

commutes.

Proposition 4.2.

(a) *Every object $(\mathcal{E}_0, \varphi_0)$ in $\mathfrak{P}(s)$, $s \in S$, admits a semiuniversal deformation.*
(b) *Versality is open in $\mathfrak{P}$.*

Proof. Let $\mathfrak{Q} \to \mathfrak{An}_S$ be the groupoid where the objects over a space $T \in \mathfrak{An}_S$ are the coherent $\mathcal{O}_{X_T}$-modules that are T-flat. As usual, given $\mathcal{E} \in \mathfrak{Q}(T)$ and $\mathcal{E}' \in \mathfrak{Q}(T')$, a morphism $\mathcal{E} \to \mathcal{E}'$ in $\mathfrak{Q}$ consists of a pair (f, α), where $f : T' \to T$ is an S-morphism and $\alpha : f^*(\mathcal{E}) \to \mathcal{E}'$ is an isomorphism of coherent sheaves. Assigning to a pair $(\mathcal{E}, \varphi)$ the sheaf $\mathcal{E}$ gives a functor $\mathfrak{P} \to \mathfrak{Q}$. It is well known that there are semiuniversal deformations in $\mathfrak{Q}$ (see [ST] or [BK]) and that versality is open is $\mathfrak{Q}$ (see e.g., [Fl1]).

The fibre of $\mathfrak{P} \to \mathfrak{Q}$ over a given object $\mathcal{E} \in \mathfrak{Q}(T)$ is the groupoid $\mathfrak{P}_\mathcal{E} \to \mathfrak{An}_T$ as explained in [Bi, Sect. 10]. More concretely, given a space $Z \in \mathfrak{An}_T$, an object in $\mathfrak{P}_\mathcal{E}$ over Z is a morphism

$$\varphi : \mathcal{F} \otimes_{\mathcal{O}_S} \mathcal{O}_Z \longrightarrow \mathcal{E} \otimes_{\mathcal{O}_T} \mathcal{O}_Z \,.$$

As the functor underlying $\mathfrak{P}_\mathcal{E}$ is representable by Theorem 2.3 we get that the objects in $\mathfrak{P}_\mathcal{E}(t)$, $t \in T$, admit semiuniversal deformations and that versality is open in $\mathfrak{P}_\mathcal{E}$. Applying [Bi, 10.12] gives the desired conclusion. $\qquad \square$

Before proving the main theorem we remind the reader of the following criterion for the representability of a functor which we present for our purposes in the form as given in [FS, 7.5]; see also [Bi, 3.1] or [KO, §2].

Theorem 4.3. *A functor $F : \mathfrak{An}_S \to \mathrm{sets}$ is representable by a complex space over S (resp. a separated complex space over S) if and only if the following conditions are satisfied.*

(1) *(Existence of semiuniversal deformations) Every $a_0 \in F(s)$, $s \in S$, admits a semiuniversal deformation.*

108   Hubert Flenner and Martin Lübke

(2) (Sheaf axiom) F is of local nature, i.e. for every complex space $T \in \mathfrak{An}_S$ the presheaf $T \supseteq U \mapsto F(U)$ on T is a sheaf.
(3) (Relative representability) For every $T \in \mathfrak{An}_S$ and a, $b \in F(T)$ the set-valued functor $\mathrm{Equ}(a, b)$ with

$$\mathrm{Equ}(a, b)(Z) := \begin{cases} \{1\} & if \ a_Z = b_Z, \\ \emptyset & otherwise, \end{cases}$$

is representable by a locally closed (resp. closed) subspace of T.
(4) (Openness of versality) For every $T \in \mathfrak{An}_S$ and $a \in F(T)$ the set of points $t \in T$ at which a is formally versal is open in T.

Proof (Proof of Theorem 1.1). We will verify that the conditions (1)–(4) in Theorem 4.3 are satisfied. (1) and (4) hold by Proposition 4.2. Moreover, (3) is just Theorem 4.1. Finally, (2) holds as simple pairs have no non-trivial automorphism. □

References

[Bi]   Bingener, J., Darstellbarkeitskriterien für analytische Funktoren, *Ann. Sci. École Norm. Sup.* **13**(4) (1980), 317–347.
[BK]   Bingener, J., Lokale Modulräume in der analytischen Geometrie I, II, with the cooperation of S. Kosarew, *Aspects of Mathematics* **D2**, **D3** (1987), Friedr. Vieweg & Sohn, Braunschweig.
[Fi]   Fischer, G., Complex analytic geometry, *Lecture Notes in Math.* **538** (1976), Springer-Verlag, Berlin-New York.
[Fl1]   Flenner, H., Ein Kriterium für die Offenheit der Versalität, *Math. Z.* **178** (1981), 449–473.
[Fl2]   Flenner, H., Eine Bemerkung über relative Ext-Garben, *Math. Ann.* **258** (1981), 175–182.
[FS]   Flenner, H., Sundararaman, D., Analytic geometry on complex superspaces, *Trans. AMS* **330** (1992), 1–40.
[EGA]   Grothendieck, A., Dieudonné, J., Éléments de géométrie algébrique, *Publ. Math. IHES* **4**, **8**, **11**, **17**, **20**, **24**, **28**, **32** (1961–1967).
[HL1]   Huybrechts, D., Lehn, M., Framed modules and their moduli, *Internat. J. Math.* **6** (1995), 297–324.
[HL2]   Huybrechts, D., Lehn, M., The geometry of moduli spaces of sheaves, *Aspects of Mathematics* **E31** (1997), Friedr. Vieweg & Sohn, Braunschweig.
[KO]   Kosarew, S., Okonek, C., Global moduli spaces and simple holomorphic bundles, *Publ. Res. Inst. Math. Sci.* **25** (1989), 1–19.
[LL]   Lübke, M., Lupascu, P., Isomorphy of the gauge theoretical and the deformation theoretical moduli space of simple holomorphic pairs, in preparation.
[OT1]   Okonek, Ch., Teleman, A., The coupled Seiberg-Witten equations, vortices, and moduli spaces of stable pairs, *Internat. J. Math.* **6** (1995), 893–910.
[OT2]   Okonek, Ch., Teleman, A., Gauge theoretical equivariant Gromov-Witten invariants and the full Seiberg-Witten invariants of ruled surfaces, preprint, 2000.

Analytic Moduli Spaces 109

[Ri] Rim, D.S., Formal deformation theory, in: *Séminaire de Géométrie Algébrique, SGA 7, Lecture Notes in Mathematics* **288** (1972), Springer Verlag Berlin-Heidelberg-New York.

[ST] Siu, Y.T., Trautmann, G., Deformations of coherent analytic sheaves with compact supports, *Mem. Amer. Math. Soc.* **29**(238) (1981).

[Su] Suyama, Y., The analytic moduli space of simple framed holomorphic pairs, *Kyushu J. Math.* **50** (1996), 65–82.

Cycle Spaces of Real Forms of $\mathrm{SL}_n(\mathbb{C})$

Alan T. Huckleberry[1] $\star$ and Joseph A. Wolf[2] $\star\star$

[1] Fakultät für Mathematik, Ruhr–Universität Bochum, D-44780 Bochum,
  Germany
  E-mail address: ahuck@cplx.ruhr-uni-bochum.de
[2] Department of Mathematics, University of California, Berkeley, California
  94720–3840, U.S.A.
  E-mail address: jawolf@math.berkeley.edu

Table of Contents

1. Background 111
2. Schubert Slices 114
   2.1. The Slice Theorem 115
   2.2. The Trace Transform Method 116
   2.3. Measurable Orbits 117
   2.4. The Transfer Lemma 120
3. Cycle Spaces of Open Orbits of $\mathrm{SL}_n(\mathbb{R})$ and $\mathrm{SL}_n(\mathbb{H})$ 121
   3.1. The Case of the Real Form $G_{\mathbb{R}} = \mathrm{SL}_n(\mathbb{R})$ 121
   3.2. The Case of the Real Form $G_{\mathbb{R}} = \mathrm{SL}_m(\mathbb{H})$ 126
   3.3. Example with $G_{\mathbb{R}} = U(n,1)$: Comparison of Transversal
       Varieties 130
   References 132

1 Background

Let us introduce the notation and goals of this paper in the context of an
example. For this let the complex Lie group $G_{\mathbb{C}} = \mathrm{SL}_3(\mathbb{C})$ act on the complex
projective space $Z = \mathbb{P}_2(\mathbb{C})$ in the usual way and consider the induced action
of the real form $G_{\mathbb{R}} = \mathrm{SL}_3(\mathbb{R})$. The latter has only two orbits on Z, the set
$M = \mathbb{P}_2(\mathbb{R})$ of real points, and its complement D.

This situation leads one to consider representations of $G_{\mathbb{R}}$ on linear spaces
that are defined by the complex geometry at hand. We focus our attention
on the open orbit D. Here M is totally real and has a basis of Stein neighborhoods. It follows that D is pseudoconcave. Consequently $\mathcal{O}(D) \cong \mathbb{C}$ and
we must look further for appropriate linear spaces.

$\star$ Research partially supported by the DFG Schwerpunkt "Global Methods in Complex Geometry".
$\star\star$ Research partially supported by the SFB–237 of the DFG and by NSF Grant
  DMS 99–88643.
2000 *Mathematics Subject Classification*: 32M05, 14C25.

If $\mathbb{E} \to D$ is a holomorphic $G_{\mathbb{R}}$-homogeneous vector bundle then, taking the pseudoconcavity into consideration, the theorem of Andreotti and Grauert [1] suggests that we consider the linear space $H^1(D; \mathcal{O}(\mathbb{E}))$. For simplicity consider the case where $\mathbb{E} \to D$ is the holomorphic cotangent bundle, dual to the holomorphic tangent bundle, and identify $H^1(D; \mathcal{O}(\mathbb{E}))$ with the space $H^{1,1}(D)$ of $\bar{\partial}$-closed $(1,1)$-forms modulo those that are exact.

The Andreotti–Norguet transform [2] allows one to represent such a Dolbeault cohomology space as a space of holomorphic functions on a space of cycles of appropriate dimension. For this, recall that a (compact) q-cycle C in a complex space X is a linear combination $C = n_1 C_1 + \cdots + n_k C_k$ where the n_j are positive integers and each C_j is an irreducible q-dimensional compact subvariety of X. Equipped with the topology of Hausdorff convergence, the space $\mathcal{C}^q(X)$ of all such cycles has a natural structure of complex space [3].

If X is a complex manifold, then the theorem of Andreotti and Norguet [2] states that the map defined by integration has values in $\mathcal{O}(\mathcal{C}^q(X))$:

$$AN : H^{q,q}(X) \to \mathcal{O}(\mathcal{C}^q(X)) \quad \text{where} \quad AN(\alpha)(C) = \int_C \alpha \,.$$

In general it is not a simple matter to explicitly describe either side of this correspondence. However, in the special case where D is an open $G_{\mathbb{R}}$-orbit such as the one in our example, there is at least a very natural space of cycles.

In the case of $D = \mathbb{P}_2(\mathbb{C}) \setminus \mathbb{P}_2(\mathbb{R})$ as above, choose $K_{\mathbb{R}} = SO_3(\mathbb{R})$ as a maximal compact subgroup of $G_{\mathbb{R}}$. Observe that $K_{\mathbb{R}}$ has a unique orbit in D that is a complex submanifold, namely the quadric curve $C_0 = \{z \in \mathbb{P}_2(\mathbb{C}) \mid z_0^2 + z_1^2 + z_2^2 = 0\}$. We may regard C_0 as a point in $\mathcal{C}^1(Z)$ and consider its orbit $\Omega := G_{\mathbb{C}} \cdot C_0$. The isotropy subgroup of $G_{\mathbb{C}}$ at C_0 is $K_{\mathbb{C}}\mathbb{Z}_3$, where $K_{\mathbb{C}}$ is the complexification $SO_3(\mathbb{C})$ of $K_{\mathbb{R}}$, and $\mathbb{Z}_3 = \{\omega I \mid \omega^3 = 1\}$ is the center of $G_{\mathbb{C}}$. So Ω may be regarded as the homogeneous space $G_{\mathbb{C}}/K_{\mathbb{C}}\mathbb{Z}_3$. That in turn can be identified with the complex symmetric 3×3 matrices of determinant 1, modulo $\mathbb{Z}_3$. So Ω is a very concrete, very familiar object.

It is another matter to give a concrete description of the space $\Omega(D) := \{C \in \Omega \mid C \subset D\}$, which is naturally associated to D. However we at least see that it is a $G_{\mathbb{R}}$-invariant open set in Ω and that, by restriction, we have the Andreotti–Norguet transform $AN : H^{1,1}(D) \to \mathcal{O}(\Omega(D))$.

Before looking more closely at the example, we introduce an appropriate general setting. For proofs and other basic facts we refer the reader to [14].

Let $Z = G_{\mathbb{C}}/Q$ be a projective algebraic variety, necessarily compact, viewed as a homogeneous manifold of a complex semisimple group $G_{\mathbb{C}}$. (Other terminology: Q is a parabolic subgroup of $G_{\mathbb{C}}$, or, equivalently, Z is a complex flag manifold.) Let $G_{\mathbb{R}}$ be a noncompact real form of $G_{\mathbb{C}}$. It can be shown that $G_{\mathbb{R}}$ has only finitely many orbits in Z; in particular at least one of them is open.

If D is such an open $G_{\mathbb{R}}$-orbit on Z, and $K_{\mathbb{R}}$ is a maximal compact subgroup of $G_{\mathbb{R}}$, then $K_{\mathbb{R}}$ has exactly one orbit in D that is a complex submani-

Cycle Spaces of Real Forms of $SL_n(\mathbb{C})$ 113

fold. We refer to it as the "base cycle" C_0 and regard it as a point $C_0 \in \mathcal{C}^q(Z)$ where $q = \dim_{\mathbb{C}} C_0$.

Since the action of $G_{\mathbb{C}}$ on $\mathcal{C}^q(Z)$ is algebraic, the orbit $\Omega := G_{\mathbb{C}} \cdot C_0$ is Zariski open in its closure. Define $\Omega(D)$ to be the connected component of $\{C \in \Omega \mid C \subset D\}$ that contains C_0.

In certain hermitian symmetric cases Ω is compact [15], but in those cases $\Omega(D)$ is just the associated bounded symmetric domain. There are also a few strange cases where $G_{\mathbb{R}}$ is transitive on Z [18]; in those cases $\Omega(D) = \Omega$ and it is reduced to a single point. In general, however, the isotropy subgroup of $G_{\mathbb{C}}$ at C_0 is a finite extension of the complexification of $K_{\mathbb{R}}$. By abuse of notation we write that finite extension as $K_{\mathbb{C}}$, so $\Omega = G_{\mathbb{C}}/K_{\mathbb{C}}$, and Ω is affine [15].

Just as in the example we have $AN : H^{q,q}(D) \to \mathcal{O}(\Omega(D))$, and it is of interest to understand the complex geometry of $\Omega(D)$, in particular with respect to functions in the image $AN(H^{q,q}(D))$. Recently Barlet and Magnusson developed some general methods involving "incidence varieties" [5], and Barlet and Koziarz developed a general "trace method" or "trace transform" for constructing holomorphic functions on cycle spaces [4]. These general results can be applied to our concrete situation; see [9, Appendix]. Here we discuss this only from the perspective of the trace transform, which produces functions in the image $\mathrm{Im}(AN)$ in a simple and elegant way.

In the example $Z = \mathbb{P}_2(\mathbb{C})$ above, let $S = \{z \in Z \mid z_2 = 0\}$. Note that $S \cap D$ can be regarded as the union of the upper and lower hemispheres in $S \cong \mathbb{P}_1(\mathbb{C})$, which are separated by $S \cap M \cong \mathbb{P}_1(\mathbb{R})$. The intersection $C_0 \cap S$ consists of exactly one point in each component, say p_0 and q_0. In fact, whenever $C \in \Omega(D)$ the intersection $C \cap S$ consists of two points $p(C)$ and $q(C)$, one in each component of $D \cap S$. Here the trace transform

$$\mathcal{T} : \mathcal{O}(S \cap D) \to \mathcal{O}(\Omega(D))$$

is given by $\mathcal{T}(f)(C) = f(p(C)) + f(q(C))$ whenever $f \in \mathcal{O}(S \cap D)$ and $C \in \Omega(D)$. In this case it is easy to see that $\mathcal{T}$ does have image in $\mathcal{O}(\Omega(D))$. It is also easy to see that, given C_∞ in the boundary $\mathrm{bd}\Omega(D)$ with $z \in C_\infty \cap \mathrm{bd}D$, there exists $f \in \mathcal{O}(S \cap D)$ with a pole at z and such that

$$\lim_{n \to \infty} |\mathcal{T}(f)(C_n)| = \infty \quad \text{whenever } \{C_n\} \text{ is a sequence in}$$
$$\Omega(D) \text{ that converges to } C_\infty.$$

This shows that $\Omega(D)$ is Stein. More precisely, given $z \in C_\infty$ as above, there are functions in the image of the trace transform $\mathcal{T}$ which display the holomorphic convexity of $\Omega(D)$ by having poles in the incidence variety $\{C \in \Omega \mid z \in C\}$.

In general this trace transform method, applied to a subvariety $S \subset Z$ of codimension q, transforms a function $f \in \mathcal{O}(S \cap D)$ to a certain function $\mathcal{T}(f) \in \mathcal{O}(\Omega(D))$. Here $C \cap S$ is finite for every $C \in \Omega(D)$, and $\mathcal{T}(f)$ is defined

114   Alan T. Huckleberry and Joseph A. Wolf

by $\mathcal{T}(f)(C) = \sum_{p \in C \cap S} f(p)$, counting intersection multiplicities. The trace transform method is an essential tool for the proof of the following result.

Theorem 1.1. *Let $G_\mathbb{R}$ be a real form of $G_\mathbb{C} = \mathrm{SL}_n(\mathbb{C})$. Let D be an open $G_\mathbb{R}$-orbit in a complex flag manifold $Z = G_\mathbb{C}/Q$. Then either $G_\mathbb{R}/K_\mathbb{R}$ is a bounded symmetric domain and $\Omega(D) = G_\mathbb{R}/K_\mathbb{R}$, or $\Omega(D)$ is a Stein domain in $\Omega = G_\mathbb{C}/K_\mathbb{C}$.*

This is proved for $G_\mathbb{R} = \mathrm{SL}_n(\mathbb{R})$ in [9] using ad hoc generalizations of the transversal Schubert variety $S = \{z \in \mathbb{P}_2(\mathbb{C}) \mid z_2 = 0\}$ above. The method of transversal Schubert varieties has since been systematized [8] and may very well lead to a general proof, without conditions on Z or $G_\mathbb{R}$, that $\Omega(D)$ is Stein. This approach is reviewed in Sect. 2. There we also show that it suffices to understand Schubert variety intersections $\Sigma = S \cap D$ in the measurable case, where it is known that $\Omega(D)$ is Stein [15].

In Sect. 3 we go to the case case $G_\mathbb{C} = \mathrm{SL}_n(\mathbb{C})$, obtaining an immense simplification of the combinatorial aspects of [9] and a relatively elementary proof of the theorem stated above.

In [17] transversal varieties T are constructed in open $G_\mathbb{R}$-orbits D in a hermitian symmetric flag manifold $Z = G_\mathbb{C}/Q$. Thus $D = G_\mathbb{R}/K_\mathbb{R}$ is a bounded symmetric domain and Z, as a homogeneous space of the compact real form of $G_\mathbb{C}$, is the compact dual hermitian symmetric space. This is done with the partial Cayley transforms intrinsic to the $G_\mathbb{R}$-orbit structure of Z, and for each open orbit D it produces a precisely described bounded symmetric domain $\Sigma_T = T \cap D$ of Z-codimension equal to the dimension of the base cycle C_0 in D, and such that T intersects C_0 transversally. Just as in the case of the Schubert slices above, the closure $c\ell(D)$ meets every $G_\mathbb{R}$-orbit in $c\ell(D)$. The trace transform method can therefore be used to transfer the Stein property from Σ_T to $\Omega(D)$, thus proving the above theorem in the hermitian symmetric space case.

The transversal varieties T of [17] have the advantage that they are constructed at an explicit base point in C_0. This leads to a concrete description of the slice Σ_T mentioned above. The Schubert slices Σ_S have the advantage that they exist in general, but the disadvantage that no distinguished base point is given in the construction. (So far, this has meant that the Stein property must be proved by ad hoc considerations.) At the end of Sect. 3 we give an example which shows that Σ_T and Σ_S can be very different: $Z = \mathbb{P}_n(\mathbb{C})$, D is the complement of the closure of the unit ball $B_n \subset \mathbb{C}^n \subset Z$, Σ_T is a disk whose closure $c\ell(\Sigma_T)$ is transversal to the boundary $\mathrm{bd}(B_n)$, and $\Sigma_S \cong \mathbb{C}$ in such a way that its closure $c\ell(\Sigma_S) \cong \mathbb{P}_1(\mathbb{C})$ is the projective tangent line to a point on $\mathrm{bd}(B_n)$.

2   Schubert Slices

As in Sect. 1, $G_\mathbb{C}$ denotes a complex connected semisimple linear algebraic group with a given noncompact real form $G_\mathbb{R}$. Let Q be a (complex) parabolic

Cycle Spaces of Real Forms of $SL_n(\mathbb{C})$ 115

subgroup of $G_{\mathbb{C}}$, so the homogeneous space $G_{\mathbb{C}}/Q$ is a projective algebraic variety. Let D denote an open $G_{\mathbb{R}}$-orbit on Z, and fix a maximal compact subgroup $K_{\mathbb{R}}$ of $G_{\mathbb{R}}$. Let $C_0 = K_{\mathbb{R}}(z_0)$ denote the unique $K_{\mathbb{R}}$-orbit in D that is a complex submanifold, so of course $D = G_{\mathbb{R}}(z_0)$, and define $q := \dim_{\mathbb{C}} C_0$.

We write $\mathfrak{g}_{\mathbb{C}}$, $\mathfrak{g}_{\mathbb{R}}$, $\mathfrak{q}$ and $\mathfrak{k}_{\mathbb{R}}$ for the respective Lie algebras of $G_{\mathbb{C}}$, $G_{\mathbb{R}}$, Q and $K_{\mathbb{R}}$, and we write $\mathfrak{k}_{\mathbb{C}}$ for the complexification of $\mathfrak{k}_{\mathbb{R}}$.

The action of $G_{\mathbb{C}}$ on the cycle space $\mathcal{C}^q(Z)$ is algebraic, so the orbit $\Omega := G_{\mathbb{C}} \cdot C_0$ is Zariski open in its closure. If $G_{\mathbb{R}}$ is of hermitian type, then for certain special orbits D (called "holomorphic type" in [19]) it is in fact closed. There are also a few strange cases where $D = Z$, so $C_0 = Z$ and $\mathcal{C}^q(Z)$ is reduced to a point; see [18]. But in general the $G_{\mathbb{C}}$-stabilizer of C_0 is a finite extension of the analytic subgroup of $G_{\mathbb{C}}$ with Lie algebra $\mathfrak{k}_{\mathbb{C}}$; see [15]. By abuse of notation, we write $K_{\mathbb{C}}$ for that stabilizer, so $\Omega = G_{\mathbb{C}}/K_{\mathbb{C}}$ and Ω is an affine, spherical homogeneous space.

Let $\Omega(D)$ denote the connected component of $\{C \in \Omega \mid C \subset D\}$ that contains C_0. It contains the Riemannian symmetric space $G_{\mathbb{R}} \cdot C_0 = G_{\mathbb{R}}/K_{\mathbb{R}}$ as a closed totally real submanifold of real dimension equal to $\dim_{\mathbb{C}} \Omega(D)$. In the special hermitian cases mentioned above, where Ω is compact, our $\Omega(D)$ is the bounded symmetric domain $G_{\mathbb{R}}/K_{\mathbb{R}}$ [15].

2.1   The Slice Theorem

A Borel subgroup of $G_{\mathbb{C}}$ is, by definition, a maximal connected solvable subgroup. Borel subgroups are complex algebraic subgroups, and any two are conjugate in $G_{\mathbb{C}}$. Let B be a Borel subgroup of $G_{\mathbb{C}}$. If $Z = G_{\mathbb{C}}/Q$ as above, then B has only finitely many orbits on Z, and each orbit O is algebraic–geometrically equivalent to a complex affine space $\mathbb{A}^{\dim O}$. The covering of Z by the closures $S = c\ell(O)$ of B-orbits realizes Z as a CW complex. In fact the "Schubert varieties" form a free set of generators of the integral homology $H_*(Z;\mathbb{Z})$. See [6] for these and other basic facts.

If $G_{\mathbb{R}} = K_{\mathbb{R}} A_{\mathbb{R}} N_{\mathbb{R}}$ is an Iwasawa decomposition (see, for example, [7], [13] or [12]), then we refer to the connected solvable group $A_{\mathbb{R}} N_{\mathbb{R}}$ as an *Iwasawa component*. The important Borel subgroups, for our considerations of Schubert slices transversal to the base cycle C_0, will be those that contain an Iwasawa component.

In most cases there is no Borel subgroup defined over $\mathbb{R}$. See [14] for an analysis of this. So instead one considers "minimal parabolic subgroups." They are the $P_{\mathbb{R}} = M_{\mathbb{R}} A_{\mathbb{R}} N_{\mathbb{R}}$ where $G_{\mathbb{R}} = K_{\mathbb{R}} A_{\mathbb{R}} N_{\mathbb{R}}$ is an Iwasawa decomposition and $M_{\mathbb{R}}$ is the centralizer of $A_{\mathbb{R}}$ in $K_{\mathbb{R}}$. Any two minimal parabolic subgroups of $G_{\mathbb{R}}$ are conjugate. If $P_{\mathbb{R}} = M_{\mathbb{R}} A_{\mathbb{R}} N_{\mathbb{R}}$ is a minimal parabolic subgroup of $G_{\mathbb{R}}$, then its complexification $P_{\mathbb{C}} = M_{\mathbb{C}} A_{\mathbb{C}} N_{\mathbb{C}}$ specifies the class of Borel subgroups that contain the Iwasawa component $A_{\mathbb{R}} N_{\mathbb{R}}$. Those are the $B = B_M A_{\mathbb{C}} N_{\mathbb{C}}$ where B_M is a Borel subgroup of $M_{\mathbb{C}}$.

Choose a Borel subgroup that contains an Iwasawa component, as above. Let $S = c\ell(O)$ be a B-Schubert variety with $\mathrm{codim}_Z(S) = q$, and let Σ

be a connected component of $S \cap D$. We refer to such a component Σ as a "Schubert slice." The main properties of Schubert slices, formulated and proved in [8], can be summarized as follows.

Theorem. *The intersection $\Sigma \cap C_0$ is non–empty and is transversal at each of its points. Suppose $z_0 \in \Sigma \cap C_0$. Then B_M fixes z_0, $\Sigma = A_{\mathbb{R}} N_{\mathbb{R}}(z_0)$, and $K_{\mathbb{R}} \cdot c\ell(\Sigma) = c\ell(D)$.*

Remarks. (1) $c\ell(\Sigma)$ meets every $G_{\mathbb{R}}$-orbit in $c\ell(D)$ because $K_{\mathbb{R}} \cdot c\ell(\Sigma) = c\ell(D)$.

(2) It would be extremely interesting to explicitly compute homology class $[C_0] \in H_{2q}(Z; \mathbb{Z})$.

2.2 The Trace Transform Method

Let $\Sigma = S \cap D$ as above and $f \in \mathcal{O}(\Sigma)$. If $C \in \Omega(D)$, then $\Sigma \cap C$ is finite because it is a compact subvariety of the Stein manifold Σ. Define $\mathcal{T}(f)(C) := \sum_{p \in \Sigma \cap C} f(p)$, counting multiplicities. This defines a holomorphic function on $\Omega(D)$, and as a result we have the *trace transform*

$$\mathcal{T} = \mathcal{T}_{\Sigma} : \mathcal{O}(\Sigma) \to \mathcal{O}(\Omega(D)) . \tag{2.1}$$

See [4] and [9, Appendix].

Corollary 2.2. *The cycle space $\Omega(D)$ is holomorphically separable. More precisely, if $C_1 \neq C_2$ in $\Omega(D)$ then there exist a Schubert slice Σ and a function $f \in \mathcal{O}(\Sigma)$ with $\mathcal{T}_{\Sigma}(f)(C_1) \neq \mathcal{T}_{\Sigma}(f)(C_2)$.*

Proof. Suppose that we have $C_1, C_2 \in \Omega(D)$ such that $\mathcal{T}_{\Sigma}(f)(C_1) = \mathcal{T}_{\Sigma}(f)(C_2)$ for every Schubert slice Σ and every $f \in \mathcal{O}(\Sigma)$. As the orbit O is affine, holomorphic functions separate points on Σ. Now, as we vary Σ, its generic intersections with C_1 and C_2 coincide. It follows that $C_1 \cap C_2$ contains interior points of C_1 or C_2, and each is the algebraic hull of that set of interior points. Thus $C_1 = C_2$. $\square$

Theorem 2.3. *If Σ is a Stein manifold, then so is $\Omega(D)$.*

Proof. Let $\{C_n\}$ be a sequence in $\Omega(D)$, $\{C_n\} \to C_{\infty} \in \text{bd}(\Sigma(D))$. Each $\Sigma \cap C_n$ is finite. Choose $p \in C_{\infty} \cap \text{bd}(D)$. Since $c\ell(D) = K_{\mathbb{R}} \cdot c\ell(\Sigma)$ for any choice of $K_{\mathbb{R}}$, we may choose the Schubert slice Σ, in other words choose the Iwasawa component $A_{\mathbb{R}} N_{\mathbb{R}}$, so that $p \in c\ell(\Sigma)$. Choose $p_n \in \Sigma \cap C_n$ such that $\{p_n\} \to p$ and choose $f \in \mathcal{O}(\Sigma)$ such that (i) $\limsup |f(p_n)| = \infty$ and (ii) $f(q_n) = 0$ for all other intersection points $q_n \in \Sigma \cap C_n$. For (ii) use finiteness of the $\Sigma \cap C_n$. Thus the trace transform satisfies $\limsup |\mathcal{T}(f)(C_n)| = \infty$. Consequently $\Omega(D)$ is holomorphically convex. As it is holomorphically separable, it is Stein. $\square$

Cycle Spaces of Real Forms of $SL_n(\mathbb{C})$ 117

Remarks. (1) See [4] for trace transform proofs of more general results on holomorphic convexity of cycle spaces.

(2) We emphasize that, if some Schubert slice Σ is Stein, then we display the Stein property for $\Omega(D)$ using only functions on $\Omega(D)$ that are in the image of the trace transform. In many cases those functions can be chosen so that their polar sets are contained in incidence divisors $\mathcal{H}_Y$, where Y is a B–invariant divisor on S. Such functions are in fact rational functions on the closure $X = c\ell(\Omega)$ in the cycle space. See [9].

2.3 Measurable Orbits

An open $G_\mathbb{R}$-orbit on Z is called *measurable* if it carries a $G_\mathbb{R}$-invariant pseudo–Kähler structure. There are a number of equivalent conditions, e.g. that the isotropy subgroups of $G_\mathbb{R}$ in D are reductive. This is always the case for $Z = G_\mathbb{C}/B$ where B is a Borel subgroup of $G_\mathbb{C}$. Also, if one open $G_\mathbb{R}$-orbit on Z is measurable, then every open $G_\mathbb{R}$-orbit on Z is measurable. In other words, measurability of open orbits is a property of the pair $(G_\mathbb{R}, Z)$. See [14] for details.

It is known that if D is measurable then $\Omega(D)$ is Stein [15]. The proof is not constructive in the sense that it goes via the solution to the Levi Problem. In particular it is not at all clear whether the functions on $\Omega(D)$ that display the Stein property have anything to do with the cohomology of D. Thus, even in the measurable case, constructive methods such as those used in our Corollaries 2.2 and 2.3 are of interest.

Let us discuss the real form $G_\mathbb{R} = SL_n(\mathbb{R})$ of $G_\mathbb{C} = SL_n(\mathbb{C})$ with respect to the concept of measurability. The manifolds $Z = G_\mathbb{C}/Q$ are the classical flag manifolds Z_δ. Here $\delta = (d_1, \ldots, d_k)$ is a "dimension symbol" of integers with $0 < d_1 < \ldots < d_k < n$, and Z_δ consists of the flags

$$z = (\{0\} \subset \Lambda_{d_1} \subset \cdots \subset \Lambda_{d_k} \subset \mathbb{C}^n),$$

where Λ_d is a d-dimensional linear subspace of $\mathbb{C}^n$.

Let τ be the standard antiholomorphic involution of $\mathbb{C}^n$, complex conjugation of $\mathbb{C}^n$ over $\mathbb{R}^n$. A flag $z \in Z_\delta$ is called τ-*generic* if, for each $\{i, j\}$, $\dim(\Lambda_{d_i} \cap \tau(\Lambda_{d_j}))$ is minimal, i.e. is equal to $\max\{d_i + d_j - n, 0\}$. See [9] for a proof that z is τ-generic if and only if $G_\mathbb{R}(z)$ is open in Z_δ.

Note that if $n = 2m$ and $\Lambda = \mathrm{Span}(v_1, v_2, \ldots, v_m)$ in such a way that $\Lambda \oplus \tau(\Lambda) = \mathbb{C}^n$, then the ordered basis $\{\mathrm{Re}(v_1), \mathrm{Im}(v_1), \ldots, \mathrm{Re}(v_m), \mathrm{Im}(v_m)\}$ defines an orientation on $\mathbb{R}^n$ that depends only on Λ. Comparing this to the standard orientation we may speak of Λ as being positively or negatively oriented.

If $\delta = (d_1, \ldots, d_k)$ is a dimension symbol with some $d_i = m$, then we say that a τ-generic flag $z = (\{0\} \subset \Lambda_{d_1} \subset \cdots \subset \Lambda_{d_k} \subset \mathbb{C}^n) \in Z_\delta$ is *positively* (resp. *negatively*) oriented if Λ_m is positively (resp. negatively) oriented. Since $G_\mathbb{R}$ preserves this notion of orientation it has at least two open orbits in Z_δ.

118   Alan T. Huckleberry and Joseph A. Wolf

In fact $G_\mathbb{R}$ has exactly two open orbits on Z_δ in this case, and otherwise has only one open orbit; see [9, §2.2].

We say that $\delta = (d_1, \ldots, d_k)$ is *symmetric* if $\delta = \delta'$ where $\delta' = (n - d_k, \ldots, n - d_1)$.

Proposition 2.4. *An open* $\mathrm{SL}_n(\mathbb{R})$*-orbit on* Z_δ *is measurable if and only if* δ *is symmetric.*

Proof. Consider the open orbit $D = G_\mathbb{R}(z_0)$. We may assume that Q is the $G_\mathbb{C}$-stabilizer of z_0. The involution τ extends from $\mathbb{C}^n$ to $G_\mathbb{C}$ by $\tau(g)(v) = \tau(g(\tau(v)))$. Now $L = Q \cap \tau(Q)$ is the complexification of the $G_\mathbb{R}$-stabilizer of z_0, and [14] D is measurable if and only if L is a Levi factor of Q, which, by a dimension count, is equivalent to L being reductive. To check that this condition is equivalent to δ being symmetric it is convenient to use a particular basis. Also, for notational simplicity, we only describe the even–dimensional case.

Let $\{e_1, \ldots, e_{2m}\}$ be the standard basis of $\mathbb{C}^{2m}$, define $h_j = e_j + \sqrt{-1}\, e_{m+j}$ for $1 \le j \le m$, and use the basis $\mathbf{h} = \{h_1, \ldots, h_m; \tau(h_1), \ldots, \tau(h_m)\}$ to define a base point $z_0 \in Z_\delta$ as follows. If $d_i \le m$ then $\Lambda_{d_i} = \mathrm{Span}(h_1, \ldots, h_{d_i})$; if $d_i > m$ then $\Lambda_{d_i} = \mathrm{Span}(h_1, \ldots, h_m, \tau(h_m), \ldots, \tau(h_{m+(m-d_i+1)}))$.

The action of τ on $G_\mathbb{C}$ is given, in matrices relative to the basis $\mathbf{h}$, in $m \times m$ blocks, by $\tau(g) = \overline{gJ}$ where $J = \begin{pmatrix} 0 & I \\ I & 0 \end{pmatrix}$ in $m \times m$ blocks and the bar is complex conjugation of matrix entries.

In the basis $\mathbf{h}$, Q consists of the block form upper triangular matrices with block sizes given by δ, and $\tau(Q)$ consists of the lower triangular matrices with block sizes given by δ'. Now $\delta = \delta'$ if and only if L consists of all block diagonal matrices with block size given by δ, and the latter is equivalent to reductivity of L. $\square$

Remarks. (1) The correspondence $\delta \mapsto \delta'$ implements an instance of the flag duality of [10].

(2) The condition $\delta = \delta'$ of Proposition 2.4 is analogous to the tube domain criterion of [11].

The following result, along with Proposition 2.4, will lead to a description of the measurable flag manifolds Z_δ for any real form of $G_\mathbb{C} = \mathrm{SL}_n(\mathbb{C})$.

Proposition 2.5. *Let* $G_\mathbb{R}^1$ *and* $G_\mathbb{R}^2$ *be two real forms of a connected complex semisimple Lie group* $G_\mathbb{C}$. *Let* τ_1 *and* τ_2 *be the antiholomorphic involutions of* $G_\mathbb{C}$ *with respective fixed point sets* $G_\mathbb{R}^1$ *and* $G_\mathbb{R}^2$, *and suppose that* $\beta := \tau_1 \cdot \tau_2^{-1}$ *is an inner automorphism of* $G_\mathbb{C}$. *Fix a complex flag manifold* $Z = G_\mathbb{C}/Q$. *Then the open* $G_\mathbb{R}^1$*-orbits on* Z *are measurable if and only if the open* $G_\mathbb{R}^2$*-orbits on* Z *are measurable.*

Proof. Let $\mathfrak{h} \subset \mathfrak{q}$ be a Cartan subalgebra of $\mathfrak{g}_\mathbb{C}$. Here lower case Gothic letters denote Lie algebras of groups denoted by the corresponding upper

Cycle Spaces of Real Forms of $SL_n(\mathbb{C})$ 119

case Roman letters. Then the nilradical $\mathfrak{q}^{nil}$ of $\mathfrak{q}$ is a sum $\sum_{\alpha \in R} \mathfrak{g}_{\mathbb{C}}^{\alpha}$ of $\mathfrak{h}$-root spaces, and $\mathfrak{h}$ determines a choice $\mathfrak{q}^{red} = \mathfrak{h} + \sum_{\alpha \in L} \mathfrak{g}_{\mathbb{C}}^{\alpha}$ of Levi component of $\mathfrak{q}$. The *opposite* (of the $G_{\mathbb{C}}$-conjugacy class) of $\mathfrak{q}$ is the ($G_{\mathbb{C}}$-conjugacy class of the) parabolic subalgebra $\mathfrak{q}^- = \mathfrak{q}^{red} + \sum_{\alpha \in R} \mathfrak{g}_{\mathbb{C}}^{-\alpha}$. This "opposition" is a well defined relation between conjugacy classes of parabolic subalgebras. The point here, for us, is the fact [14, Theorem 6.7] that the $G_{\mathbb{R}}^i$-orbits on Z are measurable if and only if $\tau_i(\mathfrak{q})$ is opposite to $\mathfrak{q}$, i.e. is $G_{\mathbb{C}}$-conjugate to $\mathfrak{q}^-$.

Let $\text{Int}(G_{\mathbb{C}})$ denote the group of inner automorphisms of $G_{\mathbb{C}}$. Now the $G_{\mathbb{R}}^1$-orbits on Z are measurable, if and only if $\mathfrak{q}^- = \alpha\tau_1\mathfrak{q}$ for some $\alpha \in \text{Int}(G_{\mathbb{C}})$, if and only if $\mathfrak{q}^- = \gamma\tau_2\mathfrak{q}$ for some $\gamma \in \text{Int}(G_{\mathbb{C}})$ where $\gamma = \alpha\beta$, if and only if the $G_{\mathbb{R}}^2$-orbits on Z are measurable. $\qquad \square$

The real forms of $SL_n(\mathbb{C})$ are the real special linear group $SL_n(\mathbb{R})$, the quaternion special linear group $SL_m(\mathbb{H})$ defined for $n = 2m$, and the special unitary groups $SU(p,q)$ with $p + q = n$. The quaternion special linear group is defined as follows. We have $\mathbb{R}$-linear transformations of $\mathbb{C}^{2m}$ given by

$$\mathbf{i} : v \mapsto \sqrt{-1}v, \quad \mathbf{j} : v \mapsto \left(\begin{smallmatrix} 0 & I_m \\ -I_m & 0 \end{smallmatrix} \right)\overline{v}, \quad \mathbf{k} = \mathbf{ij} : v \mapsto \sqrt{-1}\left(\begin{smallmatrix} 0 & I_m \\ -I_m & 0 \end{smallmatrix} \right)\overline{v} \quad (2.6)$$

where $v \mapsto \overline{v}$ is complex conjugation of $\mathbb{C}^{2m}$ over $\mathbb{R}^{2m}$. Then $\mathbf{i}^2 = \mathbf{j}^2 = \mathbf{k}^2 = -I_{2m}$, and any different ones of $\mathbf{i}, \mathbf{j}$ and $\mathbf{k}$ anticommute. So they generate a quaternion algebra $\mathbb{H}$ of linear transformations of $\mathbb{C}^{2m}$, and we have

$$\mathbb{H}^m : \text{quaternionic vector space structure on } \mathbb{C}^{2m} \text{ defined by (2.6).} \quad (2.7)$$

An $\mathbb{R}$-linear transformation of $\mathbb{C}^{2m}$ is *quaternion–linear* if it commutes with every element of $\mathbb{H}$. The group $SL_m(\mathbb{H})$ is defined to be the group of all quaternion–linear transformations of $\mathbb{C}^{2m}$ of determinant 1, in other words all volume preserving linear transformations of $\mathbb{H}^n$. Thus $SL_m(\mathbb{H})$ is the centralizer of $\mathbf{j}$ in $SL_{2m}(\mathbb{C})$, and we have

Lemma 2.8. $SL_m(\mathbb{H})$ *is the real form of* $SL_{2m}(\mathbb{C})$ *such that* $\tau(g) = \mathbf{j}g\mathbf{j}^{-1}$ *is complex conjugation of* $SL_{2m}(\mathbb{C})$ *over* $SL_m(\mathbb{H})$.

Combining Lemma 2.8 with Propositions 2.4 and 2.5 we have

Corollary 2.9. *Consider a flag manifold* Z_δ *of* $SL_{2m}(\mathbb{C})$. *Then the following are equivalent.*
(i) *The open* $SL_{2m}(\mathbb{R})$-*orbits on* Z_δ *are measurable.*
(ii) *The open* $SL_m(\mathbb{H})$-*orbits on* Z_δ *are measurable.*
(iii) *The dimension symbol* δ *is symmetric.*

Since $SU(p,q)$ has a compact Cartan subgroup T, complex conjugation of $SL_{p+q}(\mathbb{C})$ over $SU(p,q)$ sends every $\mathfrak{t}_{\mathbb{C}}$-root to its negative, so the discussion of opposition in the proof of Proposition 2.5 shows that complex conjugation of $SL_{p+q}(\mathbb{C})$ over $SU(p,q)$ sends $\mathfrak{q}$ to its opposite. Thus every open $SU(p,q)$-orbit on every Z_δ is measurable. Now Corollary 2.9 gives us

120   Alan T. Huckleberry and Joseph A. Wolf

Corollary 2.10. *Let Z_δ be a flag manifold of $G_\mathbb{C} = \mathrm{SL}_n(\mathbb{C})$. Let $G_\mathbb{R}$ be a real form of $G_\mathbb{C}$. Then the open $G_\mathbb{R}$-orbits on Z_δ are measurable except when (i) δ is not symmetric and (ii) $G_\mathbb{R} = \mathrm{SL}_n(\mathbb{R})$, or $G_\mathbb{R} = \mathrm{SL}_n(\mathbb{H})$ with $n = 2m$ even.*

2.4   The Transfer Lemma

As we already indicated, one of our main goals here is to show that, whenever D is an open orbit of a real form of $G_\mathbb{C} = \mathrm{SL}_n(\mathbb{C})$ in an arbitrary flag manifold Z_δ, the cycle space $\Omega(D)$ is a Stein domain. Since this is known for measurable orbits, by Corollary 2.10 it is enough to consider Z_δ for δ non-symmetric and either $G_\mathbb{R} = \mathrm{SL}_n(\mathbb{R})$ or $G_\mathbb{R} = \mathrm{SL}_m(\mathbb{H})$ with $2m = n$. We carry this out by our Schubert slice method. For any given Z this is related to an analysis of a certain associated measurable flag manifold $\widetilde{Z}$.

Given a complex flag manifold $Z = G_\mathbb{C}/Q$ and a real form $G_\mathbb{R} \subset G_\mathbb{C}$, there exists a root–theoretically canonically associated parabolic group $\widetilde{Q} \subset Q$ such that (i) the open $G_\mathbb{R}$-orbits in $\widetilde{Z} = G_\mathbb{C}/\widetilde{Q}$ are measurable and (ii) $\widetilde{Q}$ is maximal for this. Let $\pi : \widetilde{Z} \to Z$ be the holomorphic bundle defined by $g\widetilde{Q} \mapsto gQ$, using $\widetilde{Q} \subset Q$. Its (k-dimensional) typical fiber is $F = Q/\widetilde{Q}$. If z_0 is the neutral point in Z associated to Q and $D = G_\mathbb{R}(z_0)$ is open, then there is a unique open orbit $\widetilde{D} = G_\mathbb{R}(\widetilde{z}_0)$ in $\pi^{-1}(D)$. The fiber of the $G_\mathbb{R}$-homogeneous fibration $\pi|_{\widetilde{D}} : \widetilde{D} \to D$ is Zariski open in F and isomorphic to an affine space $\mathbb{A}^k$. See [16] or [10] for the details and for related information.

Now K has unique orbits in D and $\widetilde{D}$ which are complex submanifolds, i.e. the base cycles C_0 and $\widetilde{C}_0$ in the respective spaces. Since $\mathbb{A}^k$ is affine, $\pi|_{\widetilde{C}_0} : \widetilde{C}_0 \to C_0$ has finite fibers and, since the base is simply connected, it is in fact biholomorphic. Thus in a very natural way we have an induced $G_\mathbb{R}$-equivariant open immersion $\pi_* : \Omega(\widetilde{D}) \to \Omega(D)$.

It can happen that the isotropy subgroup of $G_\mathbb{C}$ at $C_0 \in \Omega$ is a finite extension of the isotropy subgroup of $G_\mathbb{C}$ and $\widetilde{C}_0 \in \widetilde{\Omega}$, but in general this is not the case. Nevertheless, we may think of $\Omega(\widetilde{D})$ as an open subset of $\Omega(D)$. In certain cases it has been shown that $\Omega(\widetilde{D}) = \Omega(D)$, for example when $G_\mathbb{R} = \mathrm{SL}_n(\mathbb{R})$ (see [9]). In general this a very interesting open problem.

We now compare the Schubert slices in D and $\widetilde{D}$. For this we fix an Iwasawa decomposition $G_\mathbb{R} = K_\mathbb{R} A_\mathbb{R} N_\mathbb{R}$, let B be a Borel subgroup of $G_\mathbb{C}$ which contains the Iwasawa component $A_\mathbb{R} N_\mathbb{R}$ and let S be a q-codimensional Schubert variety in Z. We may assume that the base point $z_0 \in S \cap C_0$. Recall that $(S \cap C_0) \subset (O \cap C_0)$, where $O = B(z_0)$ is the open B-orbit in S.

The cycle $\widetilde{C}_0$ is likewise q-dimensional. Thus we restrict our attention to the sets $\mathcal{S}$ and $\widetilde{\mathcal{S}}$ of q-codimensional B-Schubert varieties in the respective spaces, and we note that the projection induces a natural injective map $\pi^* : \mathcal{S} \to \widetilde{\mathcal{S}}$.

Proposition 2.11. *Let* $S \in \mathcal{S}$ *and* $\widetilde{S} = \pi^*(S)$. *Choose the neutral point* $\widetilde{z}_0 \in \widetilde{S} \cap \widetilde{C}_0$, *let* $z_0 := \pi(\widetilde{z}_0)$, *let* $\widetilde{\Sigma}$ *denote the Schubert slice* $A_\mathbb{R} N_\mathbb{R}(\widetilde{z}_0)$ *in* $\widetilde{D}$ *and let* $\Sigma = \pi(\widetilde{\Sigma})$. *Then* $\Sigma = A_\mathbb{R} N_\mathbb{R}(z_0)$ *is a Schubert slice in* D, *and the map* $\pi|_{\widetilde{\Sigma}} : \widetilde{\Sigma} \to \Sigma$ *has the same fibers as* $\pi|_{\widetilde{D}}$.

Proof. Equivariance and the basic properties of Schubert slices immediately imply all but the last statement. A dimension count shows that the fibers of $\pi|_{\widetilde{\Sigma}}$ are open in those of $\pi|_{\widetilde{D}}$. Since $\widetilde{\Sigma}$ is closed in $\widetilde{D}$, they must therefore agree. Thus we have the last assertion. $\qquad\square$

We refer to the following as the "Transfer Lemma". The interesting aspect of the transfer is given, of course, by combining it with Corollary 2.3.

Lemma 2.12. *The Schubert slice* Σ *is a Stein manifold if and only if* $\widetilde{\Sigma}$ *is a Stein manifold.*

Proof. The fiber of $\pi|_{\widetilde{D}}$ is a Zariski open orbit of a solvable group in the π-fiber F, so it is isomorphic to $\mathbb{A}^k$. It is therefore an open Schubert cell in F; see [9]. If $\widetilde{\Sigma} = A_\mathbb{R} N_\mathbb{R}(\widetilde{z}_0)$ as above, which is open in $\widetilde{O} = B(\widetilde{z}_0)$, now Proposition 2.11 shows that $\pi|_{\widetilde{\Sigma}}$ and $\pi|_{\widetilde{O}}$ have the same fibers. Since O is equivalent to an affine space, the holomorphic bundle $\pi|_{\widetilde{O}} : \widetilde{O} \to O$ is trivial. Therefore $\widetilde{\Sigma} \cong \Sigma \times \mathbb{A}^k$ and the assertion follows. $\qquad\square$

3 Cycle Spaces of Open Orbits of $SL_n(\mathbb{R})$ and $SL_n(\mathbb{H})$

In this section we consider cycle spaces $\Omega(D)$ of open orbits D of $SL_n(\mathbb{R})$, and of $SL_m(\mathbb{H})$ where $n = 2m$, on flag manifolds Z_δ of $G_\mathbb{C} = SL_n(\mathbb{C})$. In particular, using Schubert slices and the trace transform method, we prove that $\Omega(D)$ is a Stein domain in the affine homogeneous space $\Omega = G_\mathbb{C}/K_\mathbb{C}$. This was shown for $SL_n(\mathbb{R})$ in [9], but the proof here, which relies on Theorem 2.3, is essentially simpler.

3.1 The Case of the Real Form $G_\mathbb{R} = SL_n(\mathbb{R})$

For notational simplicity we restrict our attention to the even dimensional case, $n = 2m$, and, if the choice arises, to the open orbit D of positively oriented flags in Z_δ. Recall that the dimension symbol $\delta = (d_1, \dots, d_k)$ is called symmetric if $\delta = \delta'$ where $\delta' = (n - d_k, \dots, n - d_1)$. Proposition 2.4 says that D is measurable if and only if δ is symmetric. In that measurable case it is known that $\Omega(D)$ is Stein [15]. The proof in [15] is not constructive: functions displaying the Stein properties are not given. Thus the independent constructive proof given here can be of use even in the measurable case.

Let $V = \mathbb{C}^n$, let $\{e_1, \dots, e_{2m}\}$ denote its standard basis, define $f_j = e_{2j-1} + \sqrt{-1}\, e_{2j}$ for $1 \leq j \leq m$, and consider the basis $\mathbf{b} = \{f_1, \dots, f_m; \tau(f_m), \dots, \tau(f_1)\}$ of V where τ is the complex conjugation $v \mapsto \overline{v}$ of V that leaves the

122   Alan T. Huckleberry and Joseph A. Wolf

e_i fixed. We refer to **b** as the standard ordered basis of isotropic vectors with respect to the complex bilinear form $b(v, w) = {}^t v \cdot w$. As usual $K_{\mathbb{C}} = SO_n(\mathbb{C})$ and $K_{\mathbb{R}} = SO_n(\mathbb{R})$ relative to b. Let B denote the Borel subgroup of $G_{\mathbb{C}} = SL_n(\mathbb{C})$ that fixes the full flag defined by the standard basis $\{e_1, \ldots, e_{2m}\}$. Then $B = A_{\mathbb{C}} N_{\mathbb{C}}$ where $G_{\mathbb{R}} = K_{\mathbb{R}} A_{\mathbb{R}} N_{\mathbb{R}}$ is a fixed Iwasawa decomposition such that $A_{\mathbb{R}} N_{\mathbb{R}} = (B \cap G_{\mathbb{R}})^0$.

If $z = (\{0\} \subset \Lambda_{d_1} \subset \cdots \subset \Lambda_{d_k} \subset \mathbb{C}^n) \in Z_\delta$, then we define the *lower part* $L(z)$ to be the flag $(\{0\} \subset \Lambda_{d_1} \subset \cdots \subset \Lambda_{d_\ell} \subset \mathbb{C}^n) \in Z_{L(\delta)}$, defined by the conditions $d_\ell \leqq m$ and $d_{\ell+1} > m$ and by $L(\delta) = (d_1, \ldots, d_\ell)$. (The *upper part* is the flag $(\{0\} \subset \Lambda_{d_{\ell+1}} \subset \cdots \subset \Lambda_{d_k} \subset \mathbb{C}^n)$.) Let $L : Z_\delta \to Z_{L(\delta)}$ denote the associated map of flag manifolds.

The open orbit $D \subset Z_\delta$ under discussion is the set of all (positively oriented in case some $d_i = m$) τ-*generic* flags. Those are the ones such that, for each $\{i, j\}$, $\dim(\Lambda_{d_i} \cap \tau(\Lambda_{d_j}))$ is minimal, i.e. is equal to $\max\{d_i + d_j - n, 0\}$. See [9, §2.2]

Lemma 3.1. *The base cycle C_0 is the set of all maximally isotropic flags in D. In other words C_0 consists of the flags*

$$z = (\{0\} \subset \Lambda_{d_1} \subset \cdots \subset \Lambda_{d_k} \subset \mathbb{C}^n) \in Z_\delta$$

in D such that, for each $\{i, j\}$, $\dim(\Lambda_{d_i} \cap \Lambda_{d_j}^\perp)$ is maximal. If δ is symmetric, then the flag $z \in C_0$ if and only if all the subspaces Λ_d in $L(z)$ are isotropic. In that case, $\Lambda_d^\perp = \Lambda_{n-d}$ for all Λ_d in $L(z)$.

Remark. This result is contained in [9, §2.3], but our argument is more direct.

Proof. It suffices to prove Lemma 3.1 when Q is a Borel subgroup of $G_{\mathbb{C}}$. That is the case where $\delta = (1, 2, \ldots, n)$. Then the set C_0' of maximally isotropic flags $z \in Z_\delta$ is given by: the subspaces Λ_d in $L(z)$ are isotropic and satisfy $\Lambda_d^\perp = \Lambda_{n-d}$. The orthogonal groups $O_n(\mathbb{R})$ and $O_n(\mathbb{C})$ act transitively on C_0', with action on $k(z)$ determined by $k(L(z))$ for $z \in C_0'$. The action of $O_n(\mathbb{C})$ is transitive by Witt's Theorem, and C_0' is a complex flag manifold of $O_n(\mathbb{C})$. Thus also the maximal compact subgroup $O_n(\mathbb{R})$ is transitive on C_0'. Passing to identity components, now $C_0' \cap D$ is an orbit of $K_{\mathbb{R}} = SO_n(\mathbb{R})$ that is a complex flag manifold of $K_{\mathbb{C}} = O_n(\mathbb{C})$, and thus is the base cycle C_0. $\square$

The standard ordered basis $\{f_1, \ldots, f_m; \tau(f_m), \ldots, \tau(f_1)\}$ of isotropic vectors will be used to determine base points of Schubert slices, i.e. points of intersection of Σ with C_0. For δ nonsymmetric we will re-order this basis in a simple way.

If $g \in B$ then, up to a nonzero scalar multiplication, which has no effect on Z_δ,

$$g(f_j) = f_j + z_j \tau(f_j) + \sum_{i<j}(\xi_{i,j} f_i + \eta_{i,j} \tau(f_i)) . \tag{3.2}$$

Here z_j, the $\xi_{i,j}$ and the $\eta_{i,j}$ are arbitrary complex numbers as g ranges over B.

Lemma 3.3. *If $g \in A_{\mathbb{R}}N_{\mathbb{R}}$, then (again up to a nonzero scalar multiplication) $g(f_j)$ has the same form (3.2) where z_j ranges over the unit disk ($|z| < 1$), and where the $\xi_{i,j}$ and the $\eta_{i,j}$ are arbitrary complex numbers, as g ranges over B.*

Proof.

$$g(f_j) = g(e_{2j-1}) + \sqrt{-1}\, g(e_{2j}) = \left(\lambda_{2j-1}\, e_{2j-1} + \sum\nolimits_{i<2j-1} \beta_{i,2j-1}\, e_i \right)$$
$$+ \sqrt{-1}(\lambda_{2j}\, e_{2j} + \alpha_j\, e_{2j-1}) + \sum\nolimits_{i<2j-1} (\beta_{i,2j}\, e_i)$$

with λ_i positive real and $\alpha_j, \beta_{i,k}$ arbitrary real. It follows immediately that the $\xi_{i,j}$ and the $\eta_{i,j}$ range over all of $\mathbb{C}$ as g ranges over B.

We can't normalize the leading coefficient to 1 in (3.2) without losing track of the fact $g \in A_{\mathbb{R}}N_{\mathbb{R}}$, but without that normalization we use (3.2) to express

$$g(f_j) = x_j f_j + y_j \tau(f_j) + x_j \sum\nolimits_{i<j} (\xi_{i,j} f_i + \eta_{i,j} \tau(f_i))$$
$$= (x_j + y_j)e_{2j-1} + \sqrt{-1}(x_j - y_j)e_{2j} + x_j \sum\nolimits_{i<j} (\xi_{i,j} f_i + \eta_{i,j} \tau(f_i)) \ .$$

Equating coefficients of e_{2j-1} we have $x_j + y_j = \lambda_{2j-1} + \sqrt{-1}\, \alpha_j$. Equating coefficients of e_{2j} we have $x_j - y_j = \lambda_{2j}$. In other words $x_j = 1/2(\lambda_{2j-1} + \lambda_{2j} + \sqrt{-1}\, \alpha_j)$ and $y_j = 1/2(\lambda_{2j-1} - \lambda_{2j} + \sqrt{-1}\, \alpha_j)$. Now

$$z_j = y_j/z_j = (\lambda_{2j-1} - \lambda_{2j} + \sqrt{-1}\, \alpha_j)/(\lambda_{2j-1} + \lambda_{2j} + \sqrt{-1}\, \alpha_j) \ .$$

As the λ_i are positive real and α_j is arbitrary real, the only restriction as g varies over $A_{\mathbb{R}}N_{\mathbb{R}}$ is $|z_j| < 1$. $\qquad\square$

Since $g \in A_{\mathbb{R}}N_{\mathbb{R}}$, and in particular $\tau(g) = g$, the descriptions of Lemma 3.3 also describe the $g(\tau(f_j))$.

If δ is symmetric with $L(\delta) = (d_1, \dots, d_\ell)$ we define our base point

$$z_0 = (\{0\} \subset \Lambda_{d_1} \subset \dots \subset \Lambda_{d_\ell} \subset \Lambda_{n-d_\ell} \subset \dots \subset \Lambda_{n-d_1} \subset V) \tag{3.4}$$

by

$$\Lambda_{d_i} = \mathrm{Span}(f_1, \dots f_{d_i}) \text{ and } \Lambda_{n-d_i} = \Lambda_{d_i}^{\perp} \text{ for } i \leqq \ell \ . \tag{3.5}$$

This is the flag associated with the standard ordered basis. Each Λ_{d_i} is isotropic, so $z_0 \in D$.

Lemma 3.6. *Suppose that δ is symmetric and that z_0 is defined as in (3.4) and (3.5). Then $B(z_0) \cap C_0 = \{z_0\}$. In particular $(A_{\mathbb{R}}N_{\mathbb{R}})(z_0) = \Sigma$ is a Schubert slice,*

124   Alan T. Huckleberry and Joseph A. Wolf

Proof. Let $g \in B$. According to (3.2), $g(f_1)$ is isotropic if and only if $z_1 = 0$, in other words if and only if it is a multiple of f_1. Recursively in k one uses (3.2) to see that $g(\mathrm{Span}(f_1, \ldots, f_k))$ is isotropic if and only if g fixes $\mathrm{Span}(f_1, \ldots, f_k)$. Thus $z \in (B(z_0) \cap C_0)$ implies $L(z) = L(z_0)$. As $z \in C_0$, and by symmetry of δ, this implies $z = z_0$. $\qquad\square$

In dealing with nonsymmetric δ it is necessary to change the ordering of the basis. Keep in mind here that, in the standard ordered basis $\mathbf{b}$, the only difference between the f_j and the $\tau(f_j)$ is that the f_j are in increasing order and the $\tau(f_j)$ are in decreasing order. If δ is not symmetric it is necessary to exchange the roles of certain of the f_j with the corresponding $\tau(f_j)$. This will amount to changing the ordering of certain matrix blocks.

The ordering must be changed if there is a "gap" in $L(\delta)$ in the sense that, for some $d_a < d_e$ adjacent in $L(\delta)$, we have $n - d_e$, $n - d$ and $n - d_a$ in δ where $d_a < d < d_e$. In the gaps, here for indices $d_a + 1$ through $d_e - 1$, we change the order of the f_j to be decreasing with j and the order of the $\tau(f_j)$ to be increasing with j. Thus in the lower part $L(z_0)$ of the base point we will have

$$\Lambda_{d_a} = \mathrm{Span}(f_1, \ldots, f_{d_a}) \subset \mathrm{Span}(f_1, \ldots, f_{d_e}) = \Lambda_{d_e}$$

as usual, and in the upper part of the base point we will have

$$\begin{aligned}
\Lambda_{n-d_e} &= \mathrm{Span}\left(f_1, \ldots, f_m, \tau(f_m), \ldots, \tau(f_{d_e+1})\right) \\
&\subset \mathrm{Span}\left(f_1, \ldots, f_m, \tau(f_m), \ldots, \tau(f_{d_a+1})\right) = \Lambda_{n-d_a}
\end{aligned}$$

as usual, but

$$\begin{aligned}
\Lambda_{n-d} = \mathrm{Span}\big(&f_1, \ldots, f_m, \tau(f_m), \\
&\ldots, \tau(f_{d_e+1}), \tau(f_{d_a+1}), \tau(f_{d_a+2}), \ldots, \tau(f_{d_a+(d_e-d)})\big) .
\end{aligned}$$

Gaps may occur in the upper part of δ as well, but they will not require any reordering.

Lemma 3.7. *Let δ be any dimension symbol, and z_0 the base point in Z_δ defined by the basis reordered as above. Then $z_0 \in C_0$, $B(z_0) \cap C_0 = \{z_0\}$, and $\Sigma := (A_{\mathbb{R}} N_{\mathbb{R}})(z_0)$ is a Schubert slice.*

Proof. The spaces $\Lambda_{d_j} = \mathrm{Span}(f_1, \ldots, f_{d_j})$ in $L(z_0)$ are isotropic, and if $n - d_j \in \delta$ then $\Lambda_{n-d_j} = \Lambda_{d_j}^\perp$ as before. The intermediate spaces Λ_{n-d} satisfy $\Lambda_{d_a} \subset \Lambda_{n-d}^\perp \subset \Lambda_{d_e}$. Thus z_0 is maximally isotropic, so $z_0 \in C_0$.

Fix $g \in B$ so that $g(z_0)$ is maximally isotropic. As in Lemma 3.6, the fact that $g(\mathrm{Span}(f_1, \ldots, f_j))$ is isotropic implies

$$g(\mathrm{Span}(f_1, \ldots, f_j)) = \mathrm{Span}(f_1, \ldots, f_j) .$$

If in addition $n - d_j \in \delta$, then, since $g(z_0)$ is maximally isotropic, $g(\Lambda_{n-d_j}) = \Lambda_{n-d_j}$. Thus, to show that $g(z_0) = z_0$ we need only discuss $g(\Lambda_{n-d})$ for the

Cycle Spaces of Real Forms of $\mathrm{SL}_n(\mathbb{C})$ 125

intermediate spaces in a gap, i.e. for $d_a < d < d_e$ as in the discussion just before the statement of Lemma 3.7.

As in (3.2) we express (up to a scalar that fixes the base point in Z_δ)

$$g(\tau(f_{d_a+1})) = \tau(f_{d_a+1}) + zf_{d_a+1} + \sum_{i \leq d_a}(\xi_i f_i + \eta_i \tau(f_i)). \qquad (3.8)$$

Since g fixes $\mathrm{Span}(f_1, \ldots, f_{d_a})$, we have $\mathrm{Span}(f_1, \ldots, f_{d_a}) \subset g(\Lambda_{n-d})$. As $g(\Lambda_{n-d})$ is maximally isotropic we also have $\mathrm{Span}(f_1, \ldots, f_{d_a}) \subset g(\Lambda_{n-d})^\perp$. Therefore all the η_i vanish in (3.8). In other words,

$$g(\tau(f_{d_a+1})) \in \mathrm{Span}(\tau(f_{d_a+1}), f_1, \ldots, f_m).$$

Proceeding recursively in k for the $g(\tau(f_{d_a+k}))$ we obtain $g(\Lambda_{n-d}) = \Lambda_{n-d}$. Now $g \in B$ with $g(z_0) \in C_0$ implies $g(z_0) = z_0$. $\qquad \square$

Our final goal in this section is to give an explicit description of the Schubert slice Σ determined by the base point z_0. For this let Δ denote the open unit disc in $\mathbb{C}$. The result for real special linear groups is

Proposition 3.9. *Let δ be any dimension symbol, let $z_0 \in C_0 \subset D \subset Z_\delta$ denote the base point defined above, and consider the associated Schubert slice $\Sigma = (A_\mathbb{R} N_\mathbb{R})(z_0)$. Then Σ is biholomorphic to $\Delta^p \times \mathbb{C}^q$ where $p = d_\ell$ if $\delta = L(\delta)$, $p = m$ if $\delta \neq L(\delta)$, and $p + q = \mathrm{codim}_\mathbb{C}(C_0)$.*

Since $\Delta^p \times \mathbb{C}^q$ is Stein, Theorem 2.3 gives us

Corollary 3.10. *The cycle space $\Omega(D)$ of an open $\mathrm{SL}_n(\mathbb{R})$-orbit D in a flag manifold Z_δ is a Stein domain.*

Proof of Proposition 3.9. Let $g \in B$. Applying (3.2) one has

$$g(f_j) = f_j + z_j \tau(f_j) + \sum_{i<j} \eta_{i,j} \tau(f_i) \text{ modulo } \mathrm{Span}(g(f_1), \ldots, g(f_{j-1})).$$
$$(3.11)$$

As mentioned above, the only restriction imposed if $g \in A_\mathbb{R} N_\mathbb{R}$ is $|z_j| < 1$. Also,

$$g(\tau(f_j)) = \tau(f_j) + \sum_{i<j} \xi_{i,j} \tau(f_i) \text{ modulo } \mathrm{Span}(g(f_1), \ldots, g(f_m)), \quad (3.12)$$

where the $\xi_{i,j}$ can be chosen without restriction, even if $g \in A_\mathbb{R} N_\mathbb{R}$. Compare with Lemma 3.3.

Now let z_0 be the base point in Z_δ defined by the re-ordered basis of isotropic vectors. We may suppose that Q is the isotropy subgroup Q_{z_0} of $G_\mathbb{C}$ at z_0. Let $\mathfrak{q}^{\mathrm{nil}}$ denote the nilradical $\mathfrak{q}^{\mathrm{nil}} = \sum_{\alpha \in R} \mathfrak{g}_\mathbb{C}^{-\alpha}$ of $\mathfrak{q}$, where R is the appropriate set of positive roots, let $\mathfrak{u}^+ = \sum_{\alpha \in R} \mathfrak{g}_\mathbb{C}^\alpha$, opposite to $\mathfrak{q}^{\mathrm{nil}}$, which represents the holomorphic tangent space to Z_δ at z_0, and let U^+ denote the corresponding unipotent subgroup of $G_\mathbb{C}$. We will describe $B(z_0)$ in the coordinate chart $U^+(z_0)$.

126   Alan T. Huckleberry and Joseph A. Wolf

Consider the most complicated case, where the upper part of δ is not empty. Given $1 < j \leqq m$, let I_j denote the set of indices $i < j$ such that the 1-parameter group

$$g_{i,j}(t) : \tau(f_j) \mapsto \tau(f_j) + t\tau(f_i), \ \tau(f_k) \mapsto \tau(f_k) \text{ for } k \neq j,$$
$$f_k \mapsto f_k \text{ for } 1 \leqq k \leqq m$$

belongs to U^+. Then the $g(z_0), g \in B$, are described by (3.11) and by

$$g(\tau(f_j)) = \tau(f_j) + \sum\nolimits_{i \in I_j} \xi_{i,j}\tau(f_i) \ . \tag{3.13}$$

Here, as g runs over B all coefficients run over $\mathbb{C}$ independently, and if g is constrained to run over $A_\mathbb{R} N_\mathbb{R}$ the only restrictions are $|z_j| < 1$ for $1 \leqq j \leqq m$. In this case $p = m$.

Finally consider the case where the upper part of δ is empty. Then the considerations of (3.12) and (3.13) are not needed, and the $g(z_0)$, $g \in A_\mathbb{R} N_\mathbb{R}$, are described by

$$g(f_j) = f_j + \tau(f_j) + \sum\nolimits_{i<j} \eta_{i,j}\tau(f_i), \ 1 \leqq j \leqq d_\ell \ ,$$

where, as g runs over $A_\mathbb{R} N_\mathbb{R}$, the $\eta_{i,j}$ run over $\mathbb{C}$, and the z_j run over Δ, independently. In this case $p = d_\ell$. $\qquad\qquad\qquad\qquad\square$

3.2   The Case of the Real Form $G_\mathbb{R} = \mathrm{SL}_m(\mathbb{H})$

As in (2.7), $n = 2m$ and $V = \mathbb{C}^{2m}$ carries a quaternionic vector space structure $\mathbb{H}^m$ defined by $\mathbf{j} : v \mapsto J\bar{v}$ where $J = \left(\begin{smallmatrix} 0 & I_m \\ -I_m & 0 \end{smallmatrix}\right)$. Lemma 2.8 exhibits the quaternion linear group $G_\mathbb{R} = \mathrm{SL}_m(\mathbb{H})$ as the real form of $G_\mathbb{C} = \mathrm{SL}_{2m}(\mathbb{C})$ that is the centralizer of $\mathbf{j}$ in $\mathrm{SL}_{2m}(\mathbb{C})$. The Cartan involution $\theta : g \mapsto {}^t\bar{g}^{-1}$ of $\mathrm{SL}_{2m}(\mathbb{C})$ commutes with the complex conjugation $\tau : g \mapsto \mathbf{j}g\mathbf{j}^{-1}$ of $\mathrm{SL}_{2m}(\mathbb{C})$ over $\mathrm{SL}_m(\mathbb{H})$, so it restricts to a Cartan involution (which we also call θ) of $\mathrm{SL}_m(\mathbb{H})$. Thus $K_\mathbb{R} := G_\mathbb{R} \cap SU_{2m} = Sp_m$, the unitary symplectic group, whose complexification $K_\mathbb{C}$ is the complex symplectic group $Sp_m(\mathbb{C})$. Now $K_\mathbb{C} = \{g \in G_\mathbb{C} \mid g^*\omega = \omega\}$ where ω is the standard complex symplectic structure on V defined by $\omega(v,w) = {}^tv \cdot Jw$.

We refer to a flag $z = (\Lambda_j) \in Z_\delta$ as $\mathbf{j}$-generic if the dimensions $\dim(\Lambda_i \cap \mathbf{j}(\Lambda_j))$ are minimal for all Λ_i, Λ_j in z. A flag $z \in Z_\delta$ is maximally ω-isotropic if the $\dim(\Lambda_i \cap \Lambda_j^\perp)$ are maximal for all Λ_i, Λ_j in z. Here $\perp$ refers to ω. Maximally ω-isotropic flags are $\mathbf{j}$-generic.

Proposition 3.14. *There is just one open $G_\mathbb{R}$-orbit D in Z_δ ; it is the set of all $\mathbf{j}$-generic flags. The base cycle C_0 in D, in other words the unique closed $K_\mathbb{C}$-orbit in D, is the set of all maximally ω-isotropic flags in Z_δ .*

Cycle Spaces of Real Forms of $\mathrm{SL}_n(\mathbb{C})$ 127

Proof. Define D to be the set of all **j**-generic flags in Z_δ. To show that D is the unique open $G_\mathbb{R}$-orbit in Z_δ it suffices to show that $G_\mathbb{R}$ is transitive on D. Let Z_ϵ denote the manifold of full flags, corresponding to the dimension symbol $\epsilon = (1, 2, \ldots, 2m - 1)$ Every **j**-generic flag $z \in Z_\delta$ can be filled out to a **j**-generic flag $\widetilde{z} \in Z_\epsilon$. To show that $G_\mathbb{R}$ is transitive on D, now it suffices to prove transitivity on the set D_ϵ of all **j**-generic flags in Z_ϵ. We proceed to do that.

Let z_ϵ be the base point in Z_ϵ associated to the ordered basis

$$\{e_1, \ldots, e_m, \mathbf{j}(e_m), \ldots, \mathbf{j}(e_1)\}$$

where $\{e_1, \ldots, e_{2m}\}$ is the standard basis of V. Let z be any **j**-generic flag in Z_ϵ. Then z is defined by an ordered basis $\{v_1, \ldots, v_m, w_m, \ldots, w_1\}$. By **j**-genericity,

$$\mathrm{Span}(v_1, \ldots, v_m) \cap \mathbf{j}(\mathrm{Span}(v_1, \ldots, v_m)) = 0 \,,$$

so the set $\{v_1, \ldots, v_m\}$ is linearly independent over $\mathbb{H}$, and we can define $g \in G_\mathbb{R}$ by $g(v_j) = e_j$ for $1 \leqq j \leqq m$. It follows that $g(\mathbf{j}(v_j)) = \mathbf{j}(e_j)$ for $1 \leqq j \leqq m$. In other words we may assume that z is defined by

$$\{e_1, \ldots, e_m, w_m, \ldots, w_1\} \,.$$

Since we may redefine each w_j modulo $\mathrm{Span}(e_1, \ldots, e_m)$ we may also assume that each $w_j \in \mathbf{j}(\mathrm{Span}(e_1, \ldots, e_m))$. By **j**-genericity,

$$w_m \notin \mathbf{j}(\mathrm{Span}(e_1, \ldots, e_{m-1})) \,,$$

so we may suppose $w_m = \mathbf{j}(e_m) + \sum_{i<m} a_i \, \mathbf{j}(e_i)$. Now define $g \in G_\mathbb{R}$ by $g(e_i) = e_i$ for $i < m$ and $g(e_m) = e_m - \sum_{i<m} a_i \, \mathbf{j}(e_i)$. In other words, we may assume $w_m = \mathbf{j}(e_m)$. Continuing this procedure we may assume that each $w_j = \mathbf{j}(e_j)$. Thus we have $g \in G_\mathbb{R}$ with $g(z) = z_\epsilon$. This completes the proof that the set of all **j**-generic flags in Z_δ forms the unique open $G_\mathbb{R}$-orbit D there.

For the second statement, assume first that the dimension symbol δ is symmetric. Let C_1 denote the set of all maximally ω-isotropic flags in Z_δ. Then C_1 is closed in Z_δ because it is defined by equations, and it contains the base point of $z_0 \in C_0 \subset Z_\delta$ defined by the ordered basis

$$\{e_1, \ldots, e_m, \mathbf{j}(e_m), \ldots, \mathbf{j}(e_1)\} \,.$$

As C_0 is the unique closed $K_\mathbb{C}$-orbit in D, we need only check that $K_\mathbb{C}$ acts transitively on C_1.

Let $z \in C_1$ and denote

$$L(z) = (0 \subset \mathrm{Span}(v_1, \ldots, v_{d_1}) \subset \ldots \subset \mathrm{Span}(v_1, \ldots, v_{d_\ell}) \subset V) \,.$$

128   Alan T. Huckleberry and Joseph A. Wolf

We may recursively normalize so that $\omega(v_i, \mathbf{j}(v_j)) = \delta_{i,j}$ for $1 \leqq i, j \leqq d_\ell$. By Witt's Theorem, the map $v_1 \mapsto e_i$, $\mathbf{j}(v_i) \mapsto \mathbf{j}(e_i)$, for $1 \leqq i \leqq d_\ell$, extends to an element $k \in K_{\mathbb{C}}$. Since $L(k(z)) = L(z_0)$, δ is symmetric, and $k(z)$ and z_0 are maximally ω-isotropic, it follows that $k(z) = z_0$. The second statement is therefore proved for symmetric δ.

To handle the nonsymmetric case, let $\widetilde{\delta}$ denote the symmetrized dimension symbol $\delta \cup \delta'$ consisting of all the d_j and all the $n - d_j$. The set $\widetilde{C}_1$ of all maximally isotropic flags in $Z_{\widetilde{\delta}}$ maps onto C_1 under the natural projection $Z_{\widetilde{\delta}} \to Z_\delta$. As $\widetilde{C}_1$ is a closed $K_{\mathbb{C}}$-orbit, C_1 also is a closed $K_{\mathbb{C}}$-orbit. That completes the proof of the second statement. $\qquad\square$

Exactly as in Sect. 3.1 the main goal here is to give a concrete description of a Schubert slice for a given open $G_{\mathbb{R}}$-orbit D in the flag manifold Z_δ for an arbitrary dimension symbol δ. For this it is first necessary to determine an appropriate Iwasawa decomposition $G_{\mathbb{R}} = K_{\mathbb{R}} A_{\mathbb{R}} N_{\mathbb{R}}$.

Regard $V = \mathbb{H}^m$, the direct sum of quaternionic lines $\mathbb{H}(e_j)$, $1 \leqq j \leqq m$. Our $A_{\mathbb{R}}$ consists of the hermitian operators $v = \sum v_j \mapsto \sum a_j v_j$ where $v_j \in \mathbb{H}(e_j)$ and $a_j > 0$ with $a_1 a_2 \ldots a_m = 1$. Evidently this group is commutative and is contained in $G_{\mathbb{R}}$, and its elements are semisimple with all eigenvalues positive real. It is maximal for this: any such group containing the group $A_{\mathbb{R}}$ just defined, preserves each line $\mathbb{H}(e_i)$, acts on $\mathbb{H}(e_i)$ by positive real scalars, and preserves volume, hence is equal to $A_{\mathbb{R}}$. Thus our $A_{\mathbb{R}}$ is the split component of an Iwasawa decomposition of $G_{\mathbb{R}}$.

Consider $\delta = (2, 4, \ldots, 2m - 2)$ and let $z_0 \in Z_\delta$ be the flag associated to the ordered $\mathbb{C}$-basis $\{e_1, \mathbf{j}(e_1), e_2, \mathbf{j}(e_2), \ldots, e_m, \mathbf{j}(e_m)\}$. The isotropy subgroup $P_{\mathbb{C}}$ of $G_{\mathbb{C}}$ at z_0 is normalized by $\mathbf{j}$, in other words invariant under complex conjugation of $G_{\mathbb{C}}$ over $G_{\mathbb{R}}$, so $P_{\mathbb{C}}$ is the complexification of $P_{\mathbb{R}} = P_{\mathbb{C}} \cap G_{\mathbb{R}}$, and a moment's thought shows that $P_{\mathbb{R}} = M_{\mathbb{R}} A_{\mathbb{R}} N_{\mathbb{R}}$ where (i) $M_{\mathbb{R}} = Z_{K_{\mathbb{R}}}(A_{\mathbb{R}}) = (\mathrm{SL}_1(\mathbb{H}))^m$, the product of the quaternion special linear groups of the $\mathbb{H}(e_i)$, (ii) $A_{\mathbb{R}}$ is the group we defined above, and (iii) $N_{\mathbb{R}}$ is a real form of the unipotent radical of $P_{\mathbb{C}}$. So $P_{\mathbb{R}}$ is a minimal parabolic subgroup of $G_{\mathbb{R}}$ and we have the Iwasawa decomposition $G_{\mathbb{R}} = K_{\mathbb{R}} A_{\mathbb{R}} N_{\mathbb{R}}$.

B will be the Borel subgroup of $G_{\mathbb{C}}$ defined by the ordered $\mathbb{C}$-basis $\{e_1, \mathbf{j}(e_1), e_2, \mathbf{j}(e_2), \ldots, e_m, \mathbf{j}(e_m)\}$ of V. It contains $A_{\mathbb{R}} N_{\mathbb{R}}$ and is contained in $P_{\mathbb{C}}$.

We proceed as in Sect. 3.1. The main points are (i) to determine a base point $z_0 \in C_0 \subset D \subset Z_\delta$ such that $B(z_0) \cap C_0 = \{z_0\}$ and (ii) to explicitly compute $\Sigma = (A_{\mathbb{R}} N_{\mathbb{R}})(z_0)$. We summarize this as follows.

Proposition 3.15. *There exists $z_0 \in C_0$ such that $B(z_0) \cap C_0 = \{z_0\}$ and such that $(A_{\mathbb{R}} N_{\mathbb{R}})(z_0) = B(z_0) \cong \mathbb{C}^p$ where $p = \mathrm{codim}_{\mathbb{C}}(C_0)$.*

Proof. First consider the case of a symmetric dimension symbol δ. Here let $z_0 \in Z_\delta$ be associated to the standard ordered basis $\{e_1, \ldots, e_m, \mathbf{j}(e_m), \ldots, \mathbf{j}(e_1)\}$. Normalizing the leading coefficients, for $g \in B$ we have

$$g(e_j) = e_j + \sum\nolimits_{i < j} (z_{i,j} e_i + w_{i,j} \, \mathbf{j}(e_i)) \text{ for } 1 \leqq j \leqq m . \qquad (3.16)$$

By induction on k now

$$g(\mathrm{Span}(e_1,\ldots,e_k)) \text{ is } \omega\text{-isotropic if and only if}$$
$$g(\mathrm{Span}(e_1,\ldots,e_k)) = \mathrm{Span}(e_1,\ldots,e_k) \ .$$

Thus, if $g(z_0) \in C_0$ then $L(g(z_0)) = L(z_0)$. Since δ is symmetric, $g(z_0) \in C_0$ further implies $g(z_0) = z_0$. Now $B(z_0) \cap C_0 = \{z_0\}$ and $\Sigma = (A_{\mathbb{R}} N_{\mathbb{R}})(z_0)$ is a Schubert slice.

We parameterize $B(z_0)$ in the case of symmetric δ. As in Sect. 3.1 we use the coordinate chart $U^+(z_0)$ where U^+ is the unipotent subgroup of $G_{\mathbb{C}}$ whose Lie algebra $\mathfrak{u}^+$ is opposite to the nilradical $\mathfrak{q}_{z_0}^{\mathrm{nil}}$ of the isotropy subalgebra of $\mathfrak{g}_{\mathbb{C}}$ at z_0. We have $\delta = (d_1,\ldots,d_k)$ and symmetry implies $d_k = n - d_1$. Let $L(\delta) = (d_1,\ldots,d_\ell)$. If $g \in B$ and $j \leqq m$ then, modulo linear combinations of the $g(e_i)$ for $1 \leqq i < j$, and normalizing the leading coefficients,

$$g(e_j) = e_j + \sum\nolimits_{i<j} z_{i,j}\, \mathbf{j}(e_i) \ . \tag{3.17}$$

Similarly, modulo linear combinations of the $g(e_i)$ for $1 \leqq i \leqq m$, normalizing leading coefficients,

$$g(\mathbf{j}(e_b)) = \mathbf{j}(e_b) + \sum\nolimits_{a<b} w_{a,b}\, \mathbf{j}(e_a) \ . \tag{3.18}$$

Thus $B(z_0)$ is parameterized by (3.17) and (3.18) for $1 \leqq j \leqq d_\ell$ and $d_{\ell+1} \leqq b \leqq m$. There are no restrictions on the coefficients $z_{i,j}, w_{a,b} \in \mathbb{C}$.

If $g \in A_{\mathbb{R}} N_{\mathbb{R}}$ then we have, again normalizing leading coefficients,

$$g(e_j) = e_j + \sum\nolimits_{i<j} (\xi_{i,j} e_i + \eta_{i,j}(j)(e_i)) \text{ for } 1 \leqq j \leqq m \tag{3.19}$$

where there are no restrictions on the $\xi_{i,j}, \eta_{i,j} \in \mathbb{C}$. If we set $\xi_{a,b} = \overline{w}_{a,b}$ and $\eta_{i,j} = z_{i,j}$ and compare (3.18) with (3.19) we see that $(A_{\mathbb{R}} N_{\mathbb{R}})(z_0) = B(z_0)$.

We just proved Proposition 3.15 in the case of symmetric dimension symbol, using the base point $z_0 \in Z_\delta$ associated to the standard ordered basis $\{e_1,\ldots,e_m,\mathbf{j}(e_m),\ldots,\mathbf{j}(e_1)\}$. For the general case we must re-order this basis in order to manage gaps in $L(\delta)$. This goes essentially as in the case of $\mathrm{SL}_n(\mathbb{R})$.

As in Sect. 3.1, suppose that there is a gap in $L(\delta)$ between adjacent entries $d_a < d_e$ there. We reorder the standard basis so that, in the range of the gap, the ordered subset

$$\{\mathbf{j}(e_{d_e}),\mathbf{j}(e_{d_e-1}),\ldots,\mathbf{j}(e_{d_a+2}),\mathbf{j}(e_{d_a+1})\}$$

is reversed to

$$\{\mathbf{j}(e_{d_a+1}),\mathbf{j}(e_{d_a+2}),\ldots,\mathbf{j}(e_{d_e-1}),\mathbf{j}(e_{d_e})\} \ .$$

The base point $z_0 \in Z_\delta$ is the flag associated to this reordered basis.

130   Alan T. Huckleberry and Joseph A. Wolf

As in the symmetric case, if $g \in B$ and $g(z_0) \in C_0$, then $g(\Lambda_{d_i}) = \Lambda_{d_i}$ for all d_i except possibly when $d_i = n - d$ and d is in a gap, $d_a < d < d_e$ as above. In that case the reordered basis yields

$$\Lambda_{n-d} = \text{Span}(e_1, \ldots, e_m, \mathbf{j}(e_m), \ldots$$
$$\ldots, \mathbf{j}(e_{d_e+1}), \mathbf{j}(e_{d_a+1}), \mathbf{j}(e_{d_a+2}), \ldots, \mathbf{j}(e_{d_a+(d_e-d)})) .$$

Now

$$g(\mathbf{j}(e_{d_a+j})) = \mathbf{j}(e_{d_a+j}) + \sum_{u \leq d_a+j} z_{u,j} e_u + \sum_{u < d_a+j} w_{u,j} \mathbf{j}(e_u) .$$

But

$$g(\text{Span}(e_1, \ldots, e_m, \mathbf{j}(e_m), \ldots, \mathbf{j}(e_{d_e+1})))$$
$$= \text{Span}(e_1, \ldots, e_m, \mathbf{j}(e_m), \ldots, \mathbf{j}(e_{d_e+1})) .$$

In particular $\text{Span}(e_1, \ldots, e_m) \subset g(\Lambda_{n-d})$. Thus, modulo elements of $g(\Lambda_{n-d})$,

$$g(\mathbf{j}(e_{d_a+j})) = \mathbf{j}(e_{d_a+j}) + \sum_{u < d_a+j} w_{u,j} \mathbf{j}(e_u) . \tag{3.20}$$

Because of the "maximally isotropic" condition,

$$\text{Span}(e_1, \ldots, e_{d_a}) \subset g(\Lambda_{n-d})^\perp ,$$

and that implies vanishing of the $w_{u,j}$ for all u, j. Thus $g(\Lambda_{n-d}) = \Lambda_{n-d}$ for gap flag entries as well. We have proved $B(z_0) \cap C_0 = \{z_0\}$.

Finally, we show that $B(z_0) = (A_{\mathbb{R}} N_{\mathbb{R}})(z_0)$ as in the symmetric case. For this note that there was no change in the ordering in $L(z_0)$ and, as we saw in (3.20), in the upper part of z_0 the point is that only terms involving $\mathbf{j}(e_u)$ appear. So the choice $\xi_{u,j} = \overline{w_{u,j}}$ is possible as in the symmetric case. This completes the proof of Proposition 3.15.   □

Corollary 3.21. *For an arbitrary flag manifold Z_δ of $G_{\mathbb{C}} = \text{SL}_{2m}(\mathbb{C})$, and an open orbit $D \subset Z_\delta$ of $G_{\mathbb{R}} = \text{SL}_m(\mathbb{H})$, the cycle space $\Omega(D)$ is a Stein domain in the affine homogeneous space $\Omega = G_{\mathbb{C}}/K_{\mathbb{C}}$.*

Proof. This is immediate from Theorem 2.3 and the fact that $\Sigma \cong \mathbb{C}^p$ is Stein.   □

Corollaries 3.10 and 3.21, and the fact that open $SU(k, \ell)$-orbits are measurable, combine to prove Theorem 1.1.

3.3   Example with $G_{\mathbb{R}} = U(n, 1)$: Comparison of Transversal Varieties

In this section $G_{\mathbb{R}}$ is the unitary group $U(n, 1)$ acting on the complex projective space $Z = \mathbb{P}_n(\mathbb{C})$, and $G_{\mathbb{C}}$ is its complexification $GL_{n+1}(\mathbb{C})$. We use the unitary group $U(n,1)$ rather than special unitary group $SU(n,1)$ for notational convenience; this has no effect on the spaces we consider. Fix a $G_{\mathbb{R}}$-orthonormal basis: the hermitian form h that defines $G_{\mathbb{R}}$ is given by $h(e_i, e_j) = \delta_{i,j}$ if $1 \leq i \leq n$, by $-\delta_{i,j}$ if $i = n+1$. There are 3 orbits: the open ball $D_0 = G_{\mathbb{R}}(z_0)$, $z_0 = e_{n+1}\mathbb{C}$ consisting of negative definite lines; its boundary $S = G_{\mathbb{R}}(cz_0)$, $cz_0 = (e_n + e_{n+1})\mathbb{C}$, consisting of isotropic lines; and the complement $D_1 = G_{\mathbb{R}}(z_1)$, $z_1 = c^2 z_0 = e_n\mathbb{C}$ of $D_0 \cup S$ consisting of positive definite lines. Here c is a certain Cayley transform.

Let $P_{\mathbb{R}}$ denote the $G_{\mathbb{R}}$-stabilizer of the point $cz_0 \in S$. Its Lie algebra $\mathfrak{p}_{\mathbb{R}}$ is the sum of the non–positive eigenspaces of $\mathrm{ad}(x_0)$, so $\mathfrak{p}_{\mathbb{R}} = \mathfrak{p}_{\mathbb{R}}^0 + \mathfrak{p}_{\mathbb{R}}^{-1} + \mathfrak{p}_{\mathbb{R}}^{-2}$ where $\mathfrak{p}_{\mathbb{R}}^s$ is the s-eigenspace. Calculate

$$
\mathfrak{p}_{\mathbb{R}}^0 = \left\{ \begin{pmatrix} \mathfrak{u}(n-1) & 0 & 0 \\ 0 & a & b \\ 0 & b & a \end{pmatrix} \right\}, \quad \mathfrak{p}_{\mathbb{R}}^{-1} = \left\{ \begin{pmatrix} 0_{n-1} & u & u \\ v & 0 & 0 \\ -v & 0 & 0 \end{pmatrix} \right\},
$$
$$
\mathfrak{p}_{\mathbb{R}}^{-2} = \left\{ \begin{pmatrix} 0_{n-1} & 0 & 0 \\ 0 & d & d \\ 0 & -d & -d \end{pmatrix} \right\}
$$

(3.22)

where $a, b, d \in \mathbb{R}$, $u \in \mathbb{R}^{(n-1)\times 1}$ and $v \in \mathbb{R}^{1\times(n-1)}$. The real parabolic subalgebra $\mathfrak{p}_{\mathbb{R}}$ and its complexification $\mathfrak{p}_{\mathbb{C}} = \mathfrak{p}_{\mathbb{C}}^0 + \mathfrak{p}_{\mathbb{C}}^{-1} + \mathfrak{p}_{\mathbb{C}}^{-2}$ are given by

$$
\mathfrak{p}_{\mathbb{R}} = \left\{ \begin{pmatrix} \mathfrak{u}(n-1) & u & u \\ v & a+d & b+d \\ -v & b-d & a-d \end{pmatrix} \right\} \quad \text{and}
$$

(3.23)

$$
\mathfrak{p}_{\mathbb{C}} = \left\{ \begin{pmatrix} \mathfrak{gl}(n-1;\mathbb{C}) & u & u \\ v & a+d & b+d \\ -v & b-d & a-d \end{pmatrix} \right\}.
$$

The base cycle in D_1 is

$$
C_1 = K_0(z_1) = (U(n) \times U(1))(z_1) \tag{3.24}
$$

$$
= \{\mathrm{Span}(v) \in Z \mid v \in \mathrm{Span}(e_1 \wedge \cdots \wedge e_n)\} \cong \mathbb{P}_{n-1}(\mathbb{C}).
$$

The complexification $P_{\mathbb{C}}$ of $P_{\mathbb{R}}$ has Levi factor $P_{\mathbb{C}}^0$ with Lie algebra $\mathfrak{p}_{\mathbb{C}}^0$. The Borel subalgebras of $\mathfrak{p}_{\mathbb{C}}^0$ are just the

$$
\mathfrak{b}_1 = \left\{ \begin{pmatrix} p_1' & 0 & 0 \\ 0 & a & b \\ 0 & b & a \end{pmatrix} \,\middle|\, p_1' \in \mathfrak{b}_1' \text{ and } a, b \in \mathbb{C} \right\} \tag{3.25}
$$

where $\mathfrak{b}_1'$ is a Borel subalgebra of $\mathfrak{gl}_{n-1}(\mathbb{C})$. Let B_1 denote the Borel subgroup of P^0 with Lie algebra $\mathfrak{b}_1$. Then the fixed points of B_1 on Z are $z_1 = (e_n + e_{n+1})\mathbb{C}$, $(e_n - e_{n+1})\mathbb{C}$, and a unique point on Y_1. We choose $\mathfrak{b}_1'$ to be the lower triangular matrices in $\mathfrak{gl}_{(n-1)}(\mathbb{C})$. Then $e_{n-1}\mathbb{C}$ is the unique B_1-fixed point in Y_1.

132   Alan T. Huckleberry and Joseph A. Wolf

The unipotent radical $P_{\mathbb{C}}^{\text{unip}}$ of $P_{\mathbb{C}}$ has Lie algebra $\mathfrak{p}_{\mathbb{C}}^{\text{nil}} = \mathfrak{p}_{\mathbb{C}}^{-1} + \mathfrak{p}_{\mathbb{C}}^{-2}$. The Borel subgroup $B_1 \in P_{\mathbb{C}}^0$ determines the Borel subgroup $B = B_1 P_{\mathbb{C}}^{\text{unip}} \subset G_{\mathbb{C}}$ with Lie algebra $\mathfrak{b} = \mathfrak{b}_1 + \mathfrak{p}_{\mathbb{C}}^{-1} + \mathfrak{p}_{\mathbb{C}}^{-2}$,

$$\mathfrak{b} = \left\{ \begin{pmatrix} p'_1 & u & u \\ v & a+d & b+d \\ -v & b-d & a-d \end{pmatrix} \middle| \begin{array}{l} p'_1 \text{ is lower triangular}, \\ u \in \mathbb{C}^{(n-1)\times 1}, v \in \mathbb{C}^{1\times(n-1)}, \\ \text{and } a,b,d \in \mathbb{C}. \end{array} \right\}. \tag{3.26}$$

Compute

$$\exp \begin{pmatrix} 0 & u & u \\ v & d & d \\ -v & -d & -d \end{pmatrix} = \begin{pmatrix} I & u & u \\ v & 1+d+\frac{1}{2}vu & d+\frac{1}{2}vu \\ -v & -d+\frac{1}{2}vu & 1-d+\frac{1}{2}vu \end{pmatrix}$$

to see that the orbit $B(e_{n-1}\mathbb{C})$ satisfies

$$B(e_{n-1}\mathbb{C}) = P^-(e_{n-1}\mathbb{C}) = \{(e_{n-1} + te_n - te_{n+1})\mathbb{C} \mid t \in \mathbb{C}\}. \tag{3.27}$$

Here in effect t is the last entry of the row vector v in the matrix exponential just above. That comes from $\mathfrak{p}_{\mathbb{C}}^{-1}$. It comes from $\mathfrak{p}_{\mathbb{R}}^{-1}$ if and only if $u^* + v = 0$, which does not effect the range of possibilities for v. Thus $B(e_{n-1}\mathbb{C}) = P_{\mathbb{R}}^{\text{unip}}(e_{n-1}\mathbb{C})$.

We have an Iwasawa decomposition $G_{\mathbb{R}} = K_{\mathbb{R}}A_{\mathbb{R}}N_{\mathbb{R}} = N_{\mathbb{R}}A_{\mathbb{R}}K_{\mathbb{R}}$ given by $K_{\mathbb{R}} = U(n) \times U(1)$, $A_{\mathbb{R}} = \exp(\mathfrak{a}_{\mathbb{R}})$, and $\mathfrak{n}_{\mathbb{R}} = (\mathfrak{n}_{\mathbb{R}} \cap \mathfrak{b}_1) + \mathfrak{p}_{\mathbb{R}}^{\text{unip}}$. Since $B(e_{n-1}\mathbb{C}) = P_{\mathbb{R}}^{\text{unip}}(e_{n-1}\mathbb{C})$ we have the slice

$$\Sigma_S := (A_{\mathbb{R}}N_{\mathbb{R}})(e_{n-1}\mathbb{C}) = B(e_{n-1}\mathbb{C}). \tag{3.28}$$

This explicitly shows that the slice (3.28) is biholomorphic to $\mathbb{C}$, with closure in Z that is the projective line based on the subspace $\text{Span}(e_{n-1}, (e_n - e_{n+1}))$ of $\mathbb{C}^{n+1}$. Now it is transversal to C_1, with complementary dimension, so it meets every cycle C transversally and at just one point, and thus is a Schubert slice. In sharp contrast, the transversal variety Σ_T constructed in [17] is, as described at the end of Sect. 1, holomorphically equivalent to the unit disk Δ in $\mathbb{C}$.

References

[1]   A. Andreotti and H. Grauert, Théorème de finitude pour la cohomologie des espaces complexes, *Bull. Soc. Math. France* **90** (1962), 193–259.

[2]   A. Andreotti and F. Norguet, La convexité holomorphe dans l'espace analytique des cycles d'une variété algébrique, *Ann. Scuola Norm. Sup. Pisa* **21** (1967), 31–82.

[3]   D. Barlet, Familles analytiques de cycles paramétrées par un espace analytique réduit, *Lecture Notes Math.* **482** (1975), 1–158, Springer.

[4] D. Barlet and V. Koziarz, Fonctions holomorphes sur l'espace des cycles: la méthode d'intersection, *Math. Research Letters* **7** (2000), 537–549.

[5] D. Barlet and J. Magnusson, Intégration de classes de cohomologie méromorphes et diviseurs d'incidence, *Ann. Sci. École Norm. Sup.* **31** (1998), 811–842.

[6] A. Borel, *Linear algebraic groups*, Second edition, Graduate Texts in Math. **126**, Springer-Verlag, New York, 1991.

[7] S. Helgason, *Differential Geometry and Symmetric Spaces*, Academic Press, 1962.

[8] A. T. Huckleberry, On certain domains in cycle spaces of flag manifolds, to appear in *Math. Ann.*

[9] A. T. Huckleberry and A. Simon, On cycle spaces of flag domains of $SL_n(\mathbb{R})$, to appear in *Crelle's Journal.*

[10] A. T. Huckleberry and J. A. Wolf, Flag duality, *Ann. of Global Anal. and Geometry* **18** (2000), 331–340.

[11] R. L. Lipsman and J. A. Wolf, Canonical semi–invariants and the Plancherel formula for parabolic groups, *Trans. Amer. Math. Soc.* **269** (1982), 111–131.

[12] N. R. Wallach, *Real Reductive Groups*, I, II, Academic Press, 1988.

[13] G. Warner, Harmonic Analysis on Semisimple Lie Groups, I, II, Springer–Verlag, 1972.

[14] J. A. Wolf, The action of a real semisimple group on a complex manifold, I: Orbit structure and holomorphic arc components, *Bull. Amer. Math. Soc.* **75** (1969), 1121–1237.

[15] J. A. Wolf, The Stein condition for cycle spaces of open orbits on complex flag manifolds, *Annals of Math.* **136** (1992), 541–555.

[16] J. A. Wolf, Exhaustion functions and cohomology vanishing theorems for open orbits on complex flag manifolds, *Math. Res. Letters* **2** (1995), 179–191.

[17] J. A. Wolf, Hermitian symmetric spaces, cycle spaces, and the Barlet–Koziarz intersection method for the construction of holomorphic functions, *Math. Research Letters* **7** (2000), 551–564.

[18] J. A. Wolf, Real groups transitive on complex flag manifolds, *Proc. Amer. Math. Soc.* **129** (2001), 2483–2487.

[19] J. A. Wolf and R. Zierau, Cayley transforms and orbit structure in complex flag manifolds, *Transformation Groups* **2** (1997), 391–405.

On a Relative Version of Fujita's Freeness Conjecture

Yujiro Kawamata

Department of Mathematical Sciences, University of Tokyo, Komaba, Meguro, Tokyo, 153-8914, Japan
E-mail address: kawamata@ms.u-tokyo.ac.jp

Table of Contents

1 Introduction 135
2 Review on the Hodge Bundles 137
3 Parabolic Structure in Several Variables 138
4 Base Change and a Relative Vanishing Theorem 141
5 Proof of Theorem 1.7 144
  References 146

1 Introduction

The following is Fujita's freeness conjecture:

Conjecture 1.1. Let X be a smooth projective variety of dimension n and H an ample divisor. Then the invertible sheaf $\mathcal{O}_X(K_X + mH)$ is generated by global sections if $m \geq n + 1$, or $m = n$ and $H^n \geq 2$.

We have a stronger local version of Conjecture 1.1 (cf. [4]):

Conjecture 1.2. Let X be a smooth projective variety of dimension n, L a nef and big invertible sheaf on X, and $x \in X$ a point. Assume that $L^n > n^n$ and $L^d Z \geq n^d$ for any irreducible subvariety Z of X of dimension d which contains x. Then the natural homomorphism

$$H^0(X, \omega_X \otimes L) \to \omega_X \otimes L \otimes \kappa(x)$$

is surjective.

We shall extend the above conjecture to a relative setting.

Let $f : Y \to X$ be a surjective morphism of smooth projective varieties. We note that the geometric fibers of f are not necessarily connected. Assume that there exists a normal crossing divisor $B = \sum_{i=1}^{h} B_i$ on X such that f is smooth over $X_0 = X \setminus B$. Then the sheaves $R^q f_* \omega_{Y/X}$ are locally free for $q \geq 0$ ([2] for $q = 0$ and [6] in general). We note that even if we change the birational model of Y, the sheaf $R^q f_* \omega_{Y/X}$ does not change.

The relative version is the following:

2000 *Mathematics Subject Classification*: 14J60, 14C20, 14J17, 14Q20.

136   Yujiro Kawamata

Conjecture 1.3. Let $f : Y \to X$ be a surjective morphism from a smooth projective variety to a smooth projective variety of dimension n such that f is smooth over $X_0 = X \setminus B$ for a normal crossing divisor B on X. Let H be an ample divisor on X. Then the locally free sheaf $R^q f_* \omega_Y \otimes \mathcal{O}_X(mH)$ is generated by global sections if $m \geq n+1$, or $m = n$ and $H^n \geq 2$.

We have again a stronger local version:

Conjecture 1.4. Let $f : Y \to X$ be a surjective morphism from a smooth projective variety to a smooth projective variety of dimension n such that f is smooth over $X_0 = X \setminus B$ for a normal crossing divisor B on X. Let L be a nef and big invertible sheaf on X, and $x \in X$ a point. Assume that $L^n > n^n$ and $L^d Z \geq n^d$ for any irreducible subvariety Z of X of dimension d which contains x. Then the natural homomorphism

$$H^0(X, R^q f_* \omega_Y \otimes L) \to R^q f_* \omega_Y \otimes L \otimes \kappa(x)$$

is surjective for any $q \geq 0$.

In [4], the following strategy toward Conjectures 1.1 and 1.2 was developed:

Theorem 1.5 ([4]). *Let X be a smooth projective variety of dimension n, L a nef and big invertible sheaf, and $x \in X$ a point. Assume the followig condition: for any effective $\mathbb{Q}$-divisor D_0 on X such that (X, D_0) is KLT, there exists an effective $\mathbb{Q}$-divisor D on X such that*

(1) *$D \equiv \lambda L$ for some $0 < \lambda < 1$,*
(2) *The pair $(X, D_0 + D)$ is properly log canonical at x, and*
(3) *$\{x\}$ is a log canonical center for $(X, D_0 + D)$.*

Then the natural homomorphism

$$H^0(X, \omega_X \otimes L) \to \omega_X \otimes L \otimes \kappa(x)$$

is surjective.

Theorem 1.6 ([4]).

(1) *In the situation of Conjecture 1.2, assume that $\dim X \leq 3$. Then the condition of Theorem 1.5 is satisfied for L.*
(2) *In the situation of Conjecture 1.1, assume that $\dim X = 4$. Then the condition of Theorem 1.5 is satisfied for $L = mH$ with $m \geq 5$ or $m = 4$ and $H^4 \geq 2$.*

Our main result is the relative version of Theorem 1.5:

Theorem 1.7. *Let* $f : Y \to X$ *be a surjective morphism from a smooth projective variety to a smooth projective variety of dimension n such that f is smooth over $X_0 = X \setminus B$ for a normal crossing divisor B on X. Let L be a nef and big invertible sheaf on X, and $x \in X$ a point. Assume the followig condition: for any effective $\mathbb{Q}$-divisor D_0 on X such that (X, D_0) is KLT, there exists an effective $\mathbb{Q}$-divisor D on X such that*

(1) $D \equiv \lambda L$ *for some* $0 < \lambda < 1$,
(2) *The pair* $(X, D_0 + D)$ *is properly log canonical at x, and*
(3) $\{x\}$ *is a log canonical center for* $(X, D_0 + D)$.

Then the natural homomorphism

$$H^0(X, R^q f_* \omega_Y \otimes L) \to R^q f_* \omega_Y \otimes L \otimes \kappa(x)$$

is surjective for any $q \geq 0$. In particular, Conjecture 1.4 for $\dim X \leq 3$ and Conjecture 1.3 for $\dim X = 4$ hold.

The main tool for the proof of Theorem 1.5 was the $\mathbb{Q}$-divisor version of the Kodaira vanishing theorem, so-called Kawamata-Viehweg vanishing theorem. In order to prove Theorem 1.7, we shall extend the Kollár vanishing theorem, the relative version of the Kodaira vanishing theorem, to the $\mathbb{Q}$-divisor version. We had a similar result already in [3] Theorem 3.3, but we need more precise version.

In Sect. 2, we review the construction of the canonical extension for a variation of Hodge structures. In order to describe the behavior of the Hodge bundles at infinity, we shall introduce the notion of parabolic structures over arbitrary dimensional base in Sect. 3. In Sect. 4, we shall prove the base change theorem of the parabolic structures (Lemma 4.1) and derive the correct $\mathbb{Q}$-divisor version of the Kollár vanishing theorem (Theorem 4.2). The main result will be proved in Sect. 5.

Remark 1.8. (1) It is easy to prove Conjecture 1.1 in the case L is very ample. [6] proved Conjecture 1.3 in the case L is very ample.

(2) Let $\pi : P = \mathbb{P}(\mathcal{F}) \to X$ be the projective space bundle associated to our locally free sheaf $\mathcal{F} = R^q f_* \omega_{Y/X}$, and $\mathcal{O}_P(H)$ the tautological invertible sheaf. Then the sheaf $\mathcal{F} \otimes \mathcal{O}_X(K_X + L)$ is generated by global sections, if and only if $\mathcal{O}_P(H + \pi^*(K_X + L))$ is so. Since $K_P = -rH + \pi^*(K_X + \det(\mathcal{F}))$ if $r = \text{rank}(\mathcal{F})$, Conjecture 1.3 is related to, but different from, Conjecture 1.1.

2 Review on the Hodge Bundles

Let X be a smooth projective variety, and B a normal crossing divisor. Let $H_{\mathbb{Z}}$ be a local system on $X_0 = X \setminus B$. A *variation of Hodge structures* on X_0 is defined as a decreasing filtration $\{\mathcal{F}_0^p\}$ of locally free subsheaves on $\mathcal{H}_0 = H_{\mathbb{Z}} \otimes \mathcal{O}_{X_0}$ which satisfy certain axioms ([1]). Assume in addition that all the local monodromies of $H_{\mathbb{Z}}$ around the branches of B are unipotent. Then we define a locally free sheaf $\mathcal{H}$ on X called the *canonical extension* of $\mathcal{H}_0$ as follows. Let $\{z_1, \cdots, z_n\}$ be local coordinates at a given point $x \in X$ such that B is defined by an equation $z_1 \cdots z_r$ near x. Let T_i be the monodromies of $H_{\mathbb{Z}}$ around the branches of B defined by $z_i = 0$. Let v be a multi-valued flat section of $H_{\mathbb{Z}}$. Then the expression

$$s = \exp(-\frac{1}{2\pi\sqrt{-1}} \sum_{i=1}^{r} \log T_i \log z_i)v$$

is single-valued and gives a holomorphic section of $\mathcal{H}_0$, where $\log T_i$ is defined by a finite power series of $T_i - 1$. The canonical extension $\mathcal{H}$ is defined as a locally free sheaf on X whose local basis near x consists of the s when v making a basis of $H_{\mathbb{Z}}$. The filtration $\{\mathcal{F}_0^p\}$ extends to a filtration $\{\mathcal{F}^p\}$ of locally free subsheaves on $\mathcal{H}$ such that the quotients $\mathcal{H}/\mathcal{F}^p$ are also locally free for all p ([8]).

Let $f : Y \to X$ be a surjective morphism from another smooth projective variety which is smooth over X_0. Let $d = \dim Y - \dim X$, $Y_0 = f^{-1}(X_0)$ and $f_0 = f|_{Y_0}$. Then $H_{\mathbb{Z}} = R^{d+q}f_{0*}\mathbb{Z}_{Y_0}$ is a variation of Hodge structures on X_0 for any $q \geq 0$. We know that the canonical extension $\mathcal{F}^d$ in this case coincides with the direct image sheaf $R^q f_*\omega_{Y/X}$ ([2] and [6]).

The following lemma is obvious:

Lemma 2.1. *Let X be a smooth projective variety, and B a normal crossing divisor. Let $H_{\mathbb{Z}}$ be a variation of Hodge structures on $X_0 = X \setminus B$ whose local monodromies around the branches of B are unipotent, and $\mathcal{H}$ the canonical extension of $\mathcal{H}_0 = H_{\mathbb{Z}} \otimes \mathcal{O}_{X_0}$ on X. Let $\pi : X' \to X$ be a generically finite and surjective morphism from a smoth projective variety such that $B' = (\pi^*B)_{\mathrm{red}}$ is a normal crossing divisor. Let $H'_{\mathbb{Z}} = \pi^*H_{\mathbb{Z}}$ be the induced variation of Hodge structures on $X'_0 = X' \setminus B'$, and $\mathcal{H}'$ the canonical extension of $\mathcal{H}'_0 = H'_{\mathbb{Z}} \otimes \mathcal{O}_{X'_0}$ on X'. Then $\mathcal{H}' = \pi^*\mathcal{H}$.*

3 Parabolic Structure in Several Variables

We generalize the notion of parabolic structures on vector bundles ([7]) over higher dimensional base space:

Definition 3.1. Let $f : Y \to X$ be a surjective morphism of smooth projective varieties. Assume that there exists a normal crossing divisor $B = \sum_{i=1}^{h} B_i$ on X such that f is smooth over $X_0 = X \setminus B$. Fixing a nonnegative integer q, we define a *parabolic structure* on the sheaf $\mathcal{F} = R^q f_*\omega_{Y/X}$. It is a decreasing filtration of subsheaves $F^{t_1,\dots,t_h} = F^{t_1,\dots,t_h}(\mathcal{F}) \subset \mathcal{F}$ with multi-indices $t = (t_1, \ldots, t_h)$ $(t_i \in \mathbb{R}_{\geq 0})$ defined by

$$\Gamma\left(U, F^{t_1, \ldots, t_h}(\mathcal{F})\right) = \left\{ s \in \Gamma(U, \mathcal{F}) \mid \left(\prod_i z_i^{-t_i}\right) \text{sis } L^2 \right.$$
$$\left. \text{with respect to the Hodge metric} \right\},$$

where z_i is a local equation of the branch B_i on an open subset $U \subset X$.

Lemma 3.2. (1) $F^t \supset F^{t'}$ for $t \leq t'$, i.e., $t_i \leq t_i'$ for all i.
(2) $F^{t_1, \ldots, t_i + \varepsilon, \ldots, t_h} = F^{t_1, \ldots, t_i, \ldots, t_h}$ for $0 < \varepsilon \ll 1$.
(3) $F^{t_1, \ldots, t_i + 1, \ldots, t_h} = F^{t_1, \ldots, t_i, \ldots, t_h} \otimes \mathcal{O}_X(-B_i)$.
(4) Let $Y_0 = f^{-1}(X_0)$, $f_0 = f|_{Y_0}$ and $d = \dim Y - \dim X$. If all the local monodromies of $R^{d+q} f_{0*} \mathbb{Z}_{Y_0}$ around the branches of B are unipotent, then $F^t = F^0$ for any $t = (t_1, \ldots, t_h)$ with $0 \leq t_i < 1$.

Proof. (1) through (3) are obvious. (4) follows from the fact that the growth of the Hodge metric is logarithmic in this case ([2]). $\square$

Remark 3.3. For negative values of the t_i, we can also define F^t as subsheaves of $\mathcal{F} \otimes \mathcal{O}_X(mB)$ for sufficiently large m by using Lemma 3.2 (3). We shall also write $F^{\sum_i t_i B_i}$ instead of $F^{t_1, \ldots, t_h}$.

Definition 3.4. For a local section $s \in \Gamma(U, \mathcal{F})$, we define its *order of growth* along B by

$$\text{ord}(s) = \sum_i \text{ord}_i(s) B_i = \inf \left\{ \sum_i (1 - t_i) B_i \mid s \in \Gamma(U, F^{t_1, \ldots, t_h}(\mathcal{F})) \right\}.$$

We note that $s \notin \Gamma(U, F^{B - \text{ord}(s)}(\mathcal{F}))$, and

$$\Gamma(U, F^{t_1, \ldots, t_h}(\mathcal{F})) = \left\{ s \in \Gamma(U, \mathcal{F}) \mid \text{ord}(s) + \sum_i t_i B_i < B \right\}.$$

There is a nice local basis of the sheaf $\mathcal{F} = R^q f_* \omega_{Y/X}$:

Lemma 3.5. *At any point $x \in X$, there exists an open neighborhood U in the classical topology and a $\Gamma(U, \mathcal{O}_U)$-free basis $\{s_1, \ldots, s_k\}$ of $\Gamma(U, \mathcal{F})$ such that*

$$(\prod_i z_i^{\lfloor t_i + \text{ord}_i(s_1) \rfloor}) s_1, \ldots, (\prod_i z_i^{\lfloor t_i + \text{ord}_i(s_k) \rfloor}) s_k$$

generates $\Gamma(U, F^{t_1, \ldots, t_h}(\mathcal{F}))$ for any t, where the z_i are local equations of the B_i on U. In particular, the sheaf $F^{t_1, \ldots, t_h}(\mathcal{F})$ is locally free for any t.

140   Yujiro Kawamata

Proof. We shall prove that the filtration F^t is determined by the local monodromies of the cohomology sheaf $R^{d+q} f_{0*} \mathbb{Z}_{Y_0}$ around the branches of B which are known to be quasi-unipotent, where $d = \dim Y - \dim X$.

We take an open neighborhood U of $x \in X$ in the classical topology which is isomorphic to a polydisk with coordinates $\{z_1, \ldots, z_n\}$ such that $B \cap U$ is defined by $z_1 \cdots z_r = 0$. To simplify the notation, we write X instead of U. There exists a finite surjective and Galois morphism $\pi : X' \to X$ from a smooth variety which is etale over X_0 such that, for the induced morphism $f' : Y' \to X'$ from a desingularization Y' of the fiber product $Y \times_X X'$, the local system $R^{d+q} f'_{0*} \mathbb{Z}_{Y'_0}$ has unipotent local monodromies around the branches of $B' = \pi^{-1}(B)$, where we set $X'_0 = \pi^{-1}(X_0)$, $Y'_0 = f'^{-1}(X'_0)$ and $f'_0 = f'|_{Y'_0}$. Let $\sigma : Y' \to Y$ be the induced morphism.

We may assume that X' is isomorphic to a polydisk centered at a point $x' = \pi^{-1}(x)$ with coordinates $\{z'_1, \ldots, z'_n\}$, and the morphism $\pi : X' \to X$ is given by $\pi^* z_i = z_i'^{m_i}$ for some positive integers m_i, where $m_i = 1$ for $i > r$. The Galois group $G = \mathrm{Gal}(X'/X)$ is isomorphic to $\prod_i \mathbb{Z}/(m_i)$. Let $g_1, \ldots, g_r$ be generators of G such that $g_i^* z'_j = \zeta_{m_i}^{\delta_{ij}} z'_j$ for some roots of unity ζ_{m_i} of order m_i.

The group G acts on the sheaves $R^q f'_* \omega_{Y'/X'}$ and $\omega_{X'}$ equivariantly such that the invariant part $(\pi_* (R^q f'_* \omega_{Y'/X'} \otimes \omega_{X'}))^G$ is isomorphic to $R^q f_* \omega_{Y/X} \otimes \omega_X$, because $(\sigma_* \omega_{Y'})^G = \omega_Y$ and $R^p \sigma_* \omega_{Y'} = 0$ for $p > 0$.

The vector space $R^q f'_* \omega_{Y'/X'} \otimes \kappa(x')$ is decomposed into simultaneous eigenspaces with respect to the action of G. Let $s_{x'}$ be a simultaneous eigenvector such that $g_i^* s_{x'} = \zeta_{m_i}^{a_i} s_{x'}$ for some a_i with $0 \le a_i < m_i$. Let $\bar{s}'$ be a section of $R^q f'_* \omega_{Y'/X'}$ which extends $s_{x'}$. Then the section

$$ s' = \frac{1}{\prod_i m_i} \sum_{i=1}^r \sum_{k_i=0}^{m_i-1} \frac{(\prod_i g_i^{k_i})^* \bar{s}'}{\prod_i \zeta_{m_i}^{a_i k_i}} $$

satisfies that $s'(x') = s_{x'}$ and $g_i^* s' = \zeta_{m_i}^{a_i} s'$.

On the other hand, $dz'_1 \wedge \cdots \wedge dz'_n = (\prod_i m_i^{-1} z_i'^{1-m_i}) dz_1 \wedge \cdots \wedge dz_n$ is a generating section of $\omega_{X'}$. Therefore, $(\prod_i z_i'^{-a_i}) s'$ descends to a section s of $R^q f_* \omega_{Y/X}$. If the $s_{x'}$ varies among a basis of $R^q f'_* \omega_{Y'/X'} \otimes \kappa(x')$, then the corresponding sections s make a basis of the locally free sheaf $R^q f_* \omega_{Y/X}$.

We have $\mathrm{ord}_i(s) = a_i/m_i$, since the Hodge metric on the sheaf $R^q f'_* \omega_{Y'/X'}$ has logarithmic growth along B'. Therefore, the sections $(\prod_i z_i^{\lfloor t_i + a_i/m_i \rfloor}) s$ form a basis of a locally free sheaf $F^{t_1, \ldots, t_h}(R^q f_* \omega_{Y/X})$. $\square$

Remark 3.6. The Hodge metric and the flat metric on the canonical extension of the variation of Hodge structures $R^{d+q} f_{0*} \mathbb{Z}_{Y_0} \otimes \mathcal{O}_{X_0}$ coincide when restricted to the subsheaf $R^q f_* \omega_{Y/X}$. Therefore, the statement that the Hodge metric on the canonical extension has logarithmic growth is easily proved for the sheaf $R^q f_* \omega_{Y/X}$.

4 Base Change and a Relative Vanishing Theorem

By using the basis obtained in Lemma 3.5, we can study the base change property of the sheaf $R^q f_* \omega_{Y/X}$:

Lemma 4.1. *Let* $\pi : X' \to X$ *be a generically finite and surjective morphism from a smooth projective variety such that* $B' = (\pi^* B)_{\mathrm{red}} = \sum_{i'=1}^{h'} B'_{i'}$ *is a normal crossing divisor. Let* $\mu : Y' \to Y \times_X X'$ *be a birational morphism from a smooth projective variety such that the induced morphism* $f' : Y' \to X'$ *is smooth over* $X'_0 = X' \setminus B'$. *Let* $\sigma : Y' \to Y$ *be the induced morphism. Then the following hold.*

(1) Let $\{s_1, \ldots, s_k\}$ *be the basis of* $\Gamma(U, R^q f_* \omega_{Y/X})$ *in Lemma 3.5, and let* U' *be an open subset of* X' *in the classical topology such that* $\pi(U') \subset U$. *Then the equality* $\mathrm{ord}(\pi^* s_j) = \pi^* \mathrm{ord}(s_j)$ *holds, and the basis* $\{\pi^* s_1, \ldots, \pi^* s_k\}$ *of* $\Gamma(U', \pi^* R^q f_* \omega_{Y/X})$ *satisfies the conclusion of Lemma 3.5 in the sense that sections*

$$
\left(\prod_{i'} z_{i'}^{\prime \llcorner t_{i'} + \mathrm{ord}_{i'}(\pi^* s_1) \lrcorner} \right) \pi^* s_1, \ldots, \left(\prod_{i'} z_{i'}^{\prime \llcorner t_{i'} + \mathrm{ord}_{i'}(\pi^* s_k) \lrcorner} \right) \pi^* s_k
$$

form a basis of $\Gamma(U', F^{t'_1, \ldots, t'_{h'}}(R^q f'_* \omega_{Y'/X'}))$ *for any* $t' = (t'_1, \ldots, t'_{h'})$, *where the* $z'_{i'}$ *are local equations of the* $B'_{i'}$ *on* U'. *In particular, the sections*

$$
\left(\prod_{i'} z_{i'}^{\prime \llcorner \mathrm{ord}_{i'}(\pi^* s_1) \lrcorner} \right) \pi^* s_1, \ldots, \left(\prod_{i'} z_{i'}^{\prime \llcorner \mathrm{ord}_{i'}(\pi^* s_k) \lrcorner} \right) \pi^* s_k .
$$

form a basis of $\Gamma(U', R^q f'_* \omega_{Y'/X'})$.

(2) There is an equality of subsheaves of $\pi^* R^q f_* \omega_{Y/X}$:

$$
F^{t'_1, \ldots, t'_{h'}}(R^q f'_* \omega_{Y'/X'}) = \sum_t \pi^* F^{t_1, \ldots, t_h}(R^q f_* \omega_{Y/X})
$$

$$
\otimes \mathcal{O}_{X'} \left(- \llcorner \sum_{i'} t'_{i'} B'_{i'} + \sum_i (1 - t_i) \pi^* B_i \lrcorner \right),
$$

where the summation is taken for all $t = (t_1, \ldots, t_h)$ *with* $0 \le t_i < 1$ *inside the sheaf* $\pi^* R^q f_* \omega_{Y/X}$. *In particular,*

$$
R^q f'_* \omega_{Y'/X'} = \sum_t \pi^* F^{t_1, \ldots, t_h}(R^q f_* \omega_{Y/X}) \otimes \mathcal{O}_{X'} \left(- \llcorner \sum_i (1 - t_i) \pi^* B_i \lrcorner \right).
$$

Proof. (1) Since the s_j are derived from the basis in the case of unipotent monodromies, we obtain our assertion by Lemma 2.1.

142   Yujiro Kawamata

(2) We can check the assertion locally. We write $\pi^* B_i = \sum_{i'} m_{ii'} B_{i'}$ for some nonnegative integers $m_{ii'}$. Then the left hand side is generated by the sections

$$\left(\prod_{i'} z_{i'}^{\llcorner t'_{i'} + \mathrm{ord}_{i'}(\pi^* s_j) \lrcorner} \right) \pi^* s_j$$

for $1 \leq j \leq k$, while each component of the right hand side is by

$$\left(\prod_{i'} z_{i'}^{\sum_i (\llcorner t_i + \mathrm{ord}_i(s_j) \lrcorner) m_{ii'} + \llcorner t'_{i'} + \sum_i (1 - t_i) m_{ii'} \lrcorner} \right) \pi^* s_j \ .$$

Since $\mathrm{ord}_{i'}(\pi^* s_j) = \sum_i \mathrm{ord}_i(s_j) m_{ii'}$, we should compare

$$\llcorner t'_{i'} + \sum_i \mathrm{ord}_i(s_j) m_{ii'} \lrcorner$$

and

$$\min_t \left\{ \sum_i (\llcorner t_i + \mathrm{ord}_i(s_j) \lrcorner) m_{ii'} + \llcorner t'_{i'} + \sum_i (1 - t_i) m_{ii'} \lrcorner \right\} \ .$$

We have

$$\sum_i (\llcorner t_i + \mathrm{ord}_i(s_j) \lrcorner) m_{ii'} + \llcorner t'_{i'} + \sum_i (1 - t_i) m_{ii'} \lrcorner - \llcorner t'_{i'} + \sum_i \mathrm{ord}_i(s_j) m_{ii'} \lrcorner$$
$$> \sum_i (\llcorner t_i + \mathrm{ord}_i(s_j) \lrcorner) m_{ii'} + \sum_i (1 - t_i - \mathrm{ord}_i(s_j)) m_{ii'} - 1 > -1,$$

hence

$$\min_t \left\{ \sum_i (\llcorner t_i + \mathrm{ord}_i(s_j) \lrcorner) m_{ii'} + \llcorner t'_{i'} + \sum_i (1 - t_i) m_{ii'} \lrcorner \right\}$$
$$\geq \llcorner t'_{i'} + \sum_i \mathrm{ord}_i(s_j) m_{ii'} \lrcorner \ .$$

On the other hand, if we set $t_i = 1 - \mathrm{ord}_i(s_j) - \varepsilon_i$ for $0 < \varepsilon_i \ll 1$, then

$$\sum_i \left(\llcorner t_i + \mathrm{ord}_i(s_j) \lrcorner \right) m_{ii'} + \llcorner t'_{i'} + \sum_i (1 - t_i) m_{ii'} \lrcorner \} = \llcorner t'_{i'} + \sum_i \mathrm{ord}_i(s_j) m_{ii'} \lrcorner \ .$$

Therefore, they are equal. Since the minimum is attained at a value of t which does not depend on i' but only on j, we obtain the equality. $\qquad \square$

Now we state the vanishing theorem for $\mathbb{Q}$-divisors in the relative setting. This theorem will be used as an essential tool in the proof of the main theorem.

Theorem 4.2. *In the situation of Definition 3.1, let L be a nef and big $\mathbb{Q}$-divisor on X whose fractional part is supported on B. Then*

$$H^p(X, \sum_t F^{t_1, \dots, t_h}(R^q f_* \omega_{Y/X}) \otimes \omega_X(\ulcorner L - \sum_i (1 - t_i) B_i \urcorner)) = 0$$

for $p > 0$ and $q \geq 0$, where the summation is taken for all $t = (t_1, \dots, t_h)$ with $0 \leq t_i < 1$ inside the sheaf $R^q f_* \omega_Y \otimes \mathcal{O}_X(\ulcorner L \urcorner)$.

On a Relative Version of Fujita's Freeness Conjecture   143

Proof. By [2], there exists a normal crossing divisor $\bar{B}$ such that $B \leq \bar{B}$ which satisfies the following: there exists a finite surjective and Galois morphism $\pi : X' \to X$ from a smooth projective variety which is etale over $\bar{X}_0 = X \setminus \bar{B}$ and such that π^*L has integral coefficients and all the local monodromies of $R^{d+q} f'_{0*} \mathbb{Z}_{Y'_0}$ are unipotent under the notation of Lemma 4.1. We replace B by $\bar{B}$ and let G be the Galois group of π. By [6] Theorem 2.1 and [3] Theorem 3.3, we have

$$H^p(X', R^q f'_* \omega_{Y'} \otimes \mathcal{O}_{X'}(\pi^*L)) = 0$$

for $p > 0$. Since $\pi^*(K_X + B) = K_{X'} + B'$, we have by Lemma 4.1

$$H^p \left(X', \sum_t \pi^* F^{t_1, \ldots, t_h} \left(R^q f_* \omega_{Y/X} \right) \right.$$

$$\left. \otimes \mathcal{O}_{X'} \left(- \lfloor \sum_{i'} B'_{i'} + \sum_i (1 - t_i) \pi^* B_i \rfloor + \pi^*(K_X + B + L) \right) \right) = 0 \, .$$

We want to calculate the G-invariant part of the locally free sheaf

$$\mathcal{G} = F^{t_1, \ldots, t_h} \left(R^q f_* \omega_{Y/X} \right)$$

$$\otimes \, \pi_* \left(\mathcal{O}_{X'} \left(- \lfloor \sum_{i'} B'_{i'} + \sum_i (1 - t_i) \pi^* B_i \rfloor + \pi^*(K_X + B + L) \right) \right) \, .$$

For this purpose, let A be the largest divisor on X such that

$$\pi^* A \leq \lceil - \sum_{i'} B'_{i'} - \sum_i (1 - t_i) \pi^* B_i \rceil + \pi^*(B + L) \, .$$

This is equivalent to the condition

$$\pi^* A < \pi^* \left(L - \sum_i (1 - t_i) B_i \right) + \pi^* B \, .$$

Hence we obtain

$$A = \lceil L - \sum_i (1 - t_i) B_i \rceil \, .$$

Since

$$0 = H^p(X, \mathcal{G})^G = H^p \left(X, \sum_t F^{t_1, \ldots, t_h} (R^q f_* \omega_{Y/X}) \otimes \omega_X(A) \right) \, ,$$

our assertion is proved.   □

Remark 4.3. We note that the sum

$$\sum_t F^{t_1, \ldots, t_h} (R^q f_* \omega_{Y/X}) \otimes \omega_X \left(\lceil L - \sum_i (1 - t_i) B_i \rceil \right)$$

144   Yujiro Kawamata

is a locally free sheaf because it is the G-invariant part of a locally free sheaf as shown in the above proof, though the expression looks complicated. It coincides with the subsheaf of L^2-sections of the locally free sheaf $R^q f_* \omega_{Y/X} \otimes \omega_X(\ulcorner L \urcorner)$. We can consider the non-vanishing problem for this sheaf.

5   Proof of Theorem 1.7

Let $\{s_1, \cdots, s_k\}$ be the basis of $R^q f_* \omega_{Y/X}$ on a neighborhood U of x which is obtained in Lemma 3.5. We shall prove that the image of the homomorphism

$$H^0(X, R^q f_* \omega_Y \otimes L) \to R^q f_* \omega_Y \otimes L \otimes \kappa(x)$$

contains $s_j \otimes \omega_X \otimes L \otimes \kappa(x)$ for any j. Let us consider $\mathrm{ord}(s_j)$ as an effective $\mathbb{Q}$-divisor on X by setting the coefficients of the irreducible components of B which do not intersect U to be 0. By the assumption of the theorem, there exists an effective $\mathbb{Q}$-divisor D such that $D \sim_{\mathbb{Q}} \lambda L$ with $0 < \lambda < 1$, $(X, \mathrm{ord}(s_j) + D)$ is properly log canonical at x, and that $\{x\}$ is a minimal log canonical center. By the perturbation of D, we may assume that $\{x\}$ is the only log canonical center which contains x, and there exists only one log canonical place E above the center $\{x\}$. Let $\mu : X' \to X$ be a birational morphism from a smooth projective variety such that E appears as a smooth divisor on X'. We write

$$\mu^*(K_X + \mathrm{ord}(s_j) + D) = K_{X'} + E + F ,$$

where the coefficients of F are less than 1. We may assume that the union of the exceptional locus of μ and the support of $\mu^{-1}(B+D)$ is a normal crossing divisor. Let $B' = \mu^* B_{\mathrm{red}} = \sum_{i'} B'_{i'}$. Since

$$K_{X'} + (1 - \lambda)\mu^* L = \mu^*(K_X + L) - E - F + \mu^* \mathrm{ord}(s_j) ,$$

we obtain by Theorem 4.2

$$H^1 \left(X', \sum_{t'} F^{t'_1, \dots, t'_{h'}}(R^q f'_* \omega_{Y'/X'}) \otimes \mathcal{O}_{X'} \left(\mu^*(K_X + L) \right. \right.$$

$$\left. \left. -E + \ulcorner -F + \mu^* \mathrm{ord}(s_j) - \sum_{i'}(1 - t'_{i'})B'_{i'} \urcorner \right) \right) = 0 .$$

Since

$$\sum_{t'} F^{t'_1, \dots, t'_{h'}}(R^q f'_* \omega_{Y'/X'})$$

$$\otimes \mathcal{O}_{X'} \left(\mu^*(K_X + L) + \ulcorner -F + \mu^* \mathrm{ord}(s_j) - \sum_{i'}(1 - t'_{i'})B'_{i'} \urcorner \right)$$

On a Relative Version of Fujita's Freeness Conjecture   145

is a locally free sheaf on X', we have a surjective homomorphism

$$H^0\left(X', \sum_{t'} F^{t'_1,\ldots,t'_{h'}}(R^q f'_* \omega_{Y'/X'})\right.$$

$$\left.\otimes \mathcal{O}_{X'}\left(\mu^*(K_X + L) + \ulcorner -F + \mu^* \mathrm{ord}(s_j) - \sum_{i'}(1 - t'_{i'})B'_{i'}\urcorner\right)\right)$$

$$\to H^0\left(E, \sum_{t'} F^{t'_1,\ldots,t'_{h'}}(R^q f'_* \omega_{Y'/X'})\right.$$

$$\left.\otimes \mathcal{O}_E\left(\mu^*(K_X + L) + \ulcorner -F + \mu^* \mathrm{ord}(s_j) - \sum_{i'}(1 - t'_{i'})B'_{i'}\urcorner\right)\right).$$

We have

$$\mu_*\left(\ulcorner -F + \mu^* \mathrm{ord}(s_j) - \sum_{i'}(1 - t'_{i'})B'_{i'}\urcorner\right) \le 0$$

if $0 \le t'_{i'} < 1$. Hence

$$H^0\left(X', \sum_{t'} F^{t'_1,\ldots,t'_{h'}}(R^q f'_* \omega_{Y'/X'})\right.$$

$$\left.\otimes \mathcal{O}_{X'}\left(\mu^*(K_X + L) + \ulcorner -F + \mu^* \mathrm{ord}(s_j) - \sum_{i'}(1 - t'_{i'})B'_{i'}\urcorner\right)\right)$$

$$\subset H^0(X, R^q f_* \omega_{Y/X} \otimes \mathcal{O}_X(K_X + L)).$$

We note that the $t'_{i'}$ need not be contained in the interval $[0, 1)$ in the above sum. On the other hand, if we define the $t'_{i'}$ by

$$\sum_{i'} t'_{i'} B'_{i'} = \sum_{i'}(1 - \varepsilon'_{i'})B'_{i'} - \mu^* \mathrm{ord}(s_j)$$

for sufficiently small and positive numbers $\varepsilon'_{i'}$, then the divisor

$$\ulcorner -F + \mu^* \mathrm{ord}(s_j) - \sum_{i'}(1 - t'_{i'})B'_{i'}\urcorner$$

is effective and its support does not contain E, even if E is contained in B'. Since $\mathrm{ord}(\mu^* s_j) = \mu^* \mathrm{ord}(s_j)$, we have $\mu^* s_j \in F^{t'_1,\ldots,t'_{h'}}(R^q f'_* \omega_{Y'/X'})$ for such $t'_{i'}$. Hence

$$\mu^* s_j \otimes \mathcal{O}_E(\mu^*(K_X + L)) \in H^0\left(E, \sum_{t'} F^{t'_1,\ldots,t'_{h'}}(R^q f'_* \omega_{Y'/X'})\right.$$

$$\left.\otimes \mathcal{O}_E\left(\mu^*(K_X + L) + \ulcorner -F + \mu^* \mathrm{ord}(s_j) - \sum_{i'}(1 - t'_{i'})B'_{i'}\urcorner\right)\right).$$

Therefore, $s_j \otimes \omega_X \otimes L \otimes \kappa(x)$ is contained in the image of $H^0(X, R^q f_* \omega_Y \otimes L)$.

146   Yujiro Kawamata

References

[1] P. Griffiths, Period of integrals on algebraic manifolds III, *Publ. Math. I.H.E.S.* **38** (1970), 125–180.

[2] Y. Kawamata, Characterization of abelian varieties, *Compositio Math.* **43** (1981), 253–276.

[3] Y. Kawamata, Pluricanonical systems on minimal algebraic varieties, *Invent. Math.* **79** (1985), 567–588.

[4] Y. Kawamata, On Fujita's freeness conjecture for 3-folds and 4-folds, alg-geom/9510004, *Math. Ann.* **308** (1997), 491–505.

[5] Y. Kawamata, K. Matsuda, and K. Matsuki, Introduction to the minimal model problem, *Adv. Study Pure Math.* **10** (1987), 283–360.

[6] J. Kollár, Higher direct images of dualizing sheaves I, *Ann. of Math.* **123** (1986), 11–42.

[7] V. B. Mehta and C. S. Seshadri, Moduli of vector bundles over curves with parabolic structures, *Math. Ann.* **248** (1980), 205–239.

[8] W. Schmid, Variation of Hodge structure: the singularities of period mapping, *Invent. Math.* **22** (1973), 211–319.

Characterizing the Projective Space after Cho, Miyaoka and Shepherd-Barron

Stefan Kebekus*

Math. Institut, Universität Bayreuth, 95440 Bayreuth, Germany
E-mail address: stefan.kebekus@uni-bayreuth.de

Table of Contents

1 Introduction..147
2 Setup...148
3 Proof of the Characterization Theorem......................150
  References...154

1 Introduction

The aim of this paper is to give a short proof of the following characterization of the projective space.

Theorem 1.1. *Let X be a projective manifold of dimension $n \geq 3$, defined over the field $\mathbb{C}$ of complex numbers. Assume that for every curve $C \subset X$, we have $-K_X.C \geq n + 1$. Then X is isomorphic to the projective space.*

A proof was first given in the preprint [CMSB00] by K. Cho, Y. Miyaoka and N. Shepherd-Barron. While our proof here is shorter, involves substantial technical simplifications and is perhaps more transparent, the essential ideas are taken from that preprint – see Sect. 3.3.

This paper aims at simplicity, not at completeness. The methods also yield other, more involved characterization results which we do not discuss here. The preprint [CMSB00] discusses these thoroughly.

Acknowledgement. This paper was worked out while the author enjoyed the hospitality of the University of Washington at Seattle, the University of British Columbia at Vancouver and Princeton University. The author would like to thank K. Behrend, J. Kollár and S. Kovács for the invitations and for numerous discussions.

---

* The author gratefully acknowledges support by the Forschungsschwerpunkt "Globale Methoden in der komplexen Analysis" of the Deutsche Forschungsgemeinschaft.

2000 *Mathematics Subject Classification*: 14M20, 32M15, 14E08, 14E30.

148   Stefan Kebekus

2 Setup

2.1 The Space of Rational Curves

For the benefit of the reader who is not entirely familiar with the deformation
theory of rational curves on projective manifolds, we will briefly recall the
basic facts about the parameter space of rational curves. Our basic reference
is Kollárs book [Kol96] on rational curves. The reader might also wish to
consider the less technical overview in [Keb01].

If a point $x \in X$ is given, then there exists a scheme $\mathrm{Hom}_{\mathrm{bir}}(\mathbb{P}_1, X,$
$[0:1] \mapsto x)$ whose geometric points correspond to generically injective morphisms from $\mathbb{P}_1$ to X which map the point $[0:1] \in \mathbb{P}_1$ to x. There exists a
universal morphism

$$\mu : \mathrm{Hom}_{\mathrm{bir}}(\mathbb{P}_1, X, [0:1] \mapsto x) \times \mathbb{P}_1 \to X$$
$$(f, p) \mapsto f(p)$$

If $\mathbb{B}$ denotes the group of automorphisms of $\mathbb{P}_1$ which fix the point $[0:1]$,
then $\mathbb{B}$ acts by composition naturally on the space $\mathrm{Hom}_{\mathrm{bir}}(\mathbb{P}_1, X, [0:1] \mapsto x)$.
The quotient in the sense of Mumford exists. We obtain a diagram as follows.

$$(2.1) \quad \mathrm{Hom}^n_{\mathrm{bir}}(\mathbb{P}_1, X, [0:1] \mapsto x) \times \mathbb{P}_1 \xrightarrow{\quad U_x \quad} \overset{\mu}{\overbrace{\mathrm{Univ}^{rc}(x, X)}} \underset{\iota_x}{\xrightarrow{\quad\quad}} X$$

$$\downarrow \qquad\qquad\qquad\qquad \pi_x \downarrow$$

$$\mathrm{Hom}^n_{\mathrm{bir}}(\mathbb{P}_1, X, [0:1] \mapsto x) \xrightarrow{\quad u_x \quad} \mathrm{RatCurves}^n(x, X)$$

Here $\mathrm{Hom}^n_{\mathrm{bir}}(\ldots)$ is the normalization of $\mathrm{Hom}_{\mathrm{bir}}(\ldots)$, the morphisms U_x and
u_x have the structure of principal $\mathbb{B}$-bundles and π_x is a $\mathbb{P}_1$-bundle. The
restriction of ι_x to any fiber of π_x is generically injective, i.e. birational onto
its image.

The space $\mathrm{RatCurves}^n(x, X)$ is called the "space of rational curves
through x". This name is perhaps a bit misleading because the correspondence

$$e : \mathrm{RatCurves}^n(x, X) \to \{\text{rational curves in } X \text{ which contain } x\}$$
$$h \mapsto \iota_x(\pi^{-1}(h))$$

is not bijective in general. Although e is surjective, it may happen that the
restriction of e to an irreducible component $H \subset \mathrm{RatCurves}^n(x, X)$ is only
generically injective: several points in H may correspond to the same rational
curve.

2.2 Results of Mori's Theory of Rational Curves

The following theorem summarizes some of the classic results of Mori theory,
in particular Mori's famous existence theorem for rational curves on manifolds

Characterizing the Projective Space   149

where K_X is not nef. While most statements can been found explicitly or implicitly in the papers [Mor79] and [KMM92], some results found their final formulation only years later. We refer to [Kol96] for proper attributions.

Theorem 2.1 (Classic results on families of rational curves). *Under the assumptions of Theorem 1.1, let x be an arbitrary closed point of X. Then [Kol96, Thm. II.5.14] there exists a rational curve $\ell \subset X$ which contains x and satisfies $-K_X.\ell = n + 1$.*

Let $H_x \subset \mathrm{RatCurves}^n(x, X)$ be the irreducible component which contains the point corresponding to ℓ and consider the restriction of the diagram (2.1) above:

(2.2)
$$\begin{array}{ccc} U_x & \xrightarrow{\iota_x} & X \\ {\scriptstyle \mathbb{P}_1 \text{ -bundle}}\Big\downarrow {\scriptstyle \pi_x} & & \\ H_x & & \end{array}$$

Then the following holds.

1. *The variety H_x is compact [Kol96, Prop. II.2.14] and has dimension $\dim H_x = n - 1$ [Kol96, thms. II.3.11 and II.1.7].*
2. *If $x \in X$ is a general point, then the variety H_x is smooth [Kol96, Cor. II.3.11.5]. Note that we do not have to assume that $x \in X$ is very general since we are working with an irreducibly family*

$$H \subset \mathrm{RatCurves}^n(x, X)$$

   rather than the space $\mathrm{RatCurves}^n(x, X)$ which has countably many components.
3. *The evaluation morphism ι_x is finite away from $\iota_x^{-1}(x)$ (Mori's Bend-and-Break, [Kol96, Cor. II.5.5]). In particular, ι_x is surjective.*
4. *If $\ell \subset X$ is curve corresponding to a general point of H_x, then ℓ is smooth [Kol96, Thm. II.3.14] and the restriction $T_X|_\ell$ is an ample vector bundle on ℓ [Kol96, Cor. II.3.10.1].*

2.3   Singular Rational Curves

It was realized very early by Miyaoka ([Miy92], see also [Kol96, Prop. V.3.7.5]) that the singular curves in the family H_x play a pivotal rÿle in the characterization problem. A systematic study of families of singular curves, however, was not carried out before the paper [Keb00]. In that paper, the author gave a sharp bound on the dimension of the subvariety $H_x^{\mathrm{Sing}} \subset H_x$ whose points correspond to singular rational curves and described the singularities of those curves which are singular at a general point $x \in X$.

The following theorem summarizes the results of [Keb00] which form the centerpiece of our argumentation. A singular curve is called "immersed" if the normalization morphism has constant rank one. A singular curve which is not immersed is often said to be "cuspidal".

Theorem 2.2 (Singular curves in H_x, [Keb00, Thm. 3.3]). *Let $x \in X$ be a general point. Then the closed subfamily $H_x^{\text{Sing}} \subset H_x$ of singular curves has dimension at most one. The subfamily $H_x^{\text{Sing},x} \subset H_x^{\text{Sing}}$ of curves which are singular at x is at most finite. If $H_x^{\text{Sing},x}$ is not empty, then the associated curves are immersed.*

In our setup, we obtain a good description of $\iota_x^{-1}(x)$ as an immediate corollary.

Corollary 2.3. *The preimage $\iota_x^{-1}(x)$ contains a section, which we call σ_∞, and at most a finite number of further points, called z_i. The tangential morphism $T\iota_x$ has rank one along σ_∞.*

The universal property of the blow-up [Har77, Prop. II.7.13] therefore allows us to extend diagram (2.2) as follows:

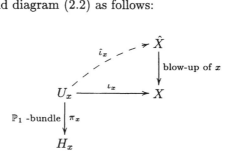

where the rational map $\hat{\iota}_x$ is well-defined away from the points z_i.

We end with a further description of $\hat{\iota}_x$.

Proposition 2.4 ([Keb00, Thm. 3.4]). *Let $x \in X$ be a general point. If $E \cong \mathbb{P}(T_X^*|_x)$ is the exceptional divisor of the blow-up[1], then the restricted morphism*

$$\hat{\iota}_x|_{\sigma_\infty} : \sigma_\infty \to E$$

is finite. In particular, since $\dim \sigma_\infty = \dim H_x = n - 1$, the morphism $\hat{\iota}_x|_{\sigma_\infty}$ is surjective.

3 Proof of the Characterization Theorem

Throughout the present chapter, let $x \in X$ be a general point. We will keep the notation introduced in Sect. 2.

---
[1] We use Grothendieck's notation: if V is a vector space, then $\mathbb{P}(V^*) = V \setminus \{0\}/\mathbb{C}^*$.

3.1 The Neighborhood of σ_∞

As a first step towards the proof of Theorem 1.1, we need to study the neighborhood of the section $\sigma_\infty \subset U_x$.

Proposition 3.1. *If $E \cong \mathbb{P}(T_X^*|_x)$ is the exceptional divisor of the blow-up, then*

(1) *The restricted morphism $\hat{\imath}_x|_{\sigma_\infty}$ is an embedding. In particular, $H_x \cong \sigma_\infty \cong \mathbb{P}_{n-1}$.*

(2) *The tangent map $T\hat{\imath}_x$ has maximal rank along σ_∞. In particular,*

$$N_{\sigma_\infty, U_x} \cong N_{E,\hat{X}} \cong \mathcal{O}_{\mathbb{P}_{n-1}}(-1) .$$

The remaining part of the present section is devoted to a proof of Proposition 3.1. Note that statement (2) follows immediately from statement (1) and from Corollary 2.3. To show statement (1) requires some work.

By Proposition 2.4 and by Zariski's main theorem, we are done if we show that $\hat{\imath}_x|_{\sigma_\infty}$ is birational. For this, we argue by contradiction and assume for the moment that $\hat{\imath}_x|_{\sigma_\infty}$ is *not* birational, let $\ell \subset X$ be a general curve associated with H_x and let $F \subset U_x$ be the corresponding fiber of π_x. The subvariety $\iota_x^{-1}(\ell) \subset U_x$ will then contain a curve B such that

(1) B is not contained in σ_∞.
(2) B is not a fiber of the projection π_x.
(3) $B \cap \sigma_\infty$ contains a point y_1 which is different from $y_0 := F \cap \sigma_\infty$.

In order to see that we can find a curve B which is not a fiber of the projection π_x, recall that the correspondence between points in H_x and curves in X is generically injective and that ℓ was generically chosen.

Summing up, in order to show Proposition 3.1, it suffices to show the following claim which contradicts the assumption that $\hat{\imath}_x|_{\sigma_\infty}$ was not birational.

Claim 3.2 *Let $\ell \subset X$ be a general curve associated with H_x and let $B \subset \iota_x^{-1}(\ell)$ be any curve which satisfies items (1) and (2). Then B is disjoint from σ_∞.*

A proof will be given in the next few subsections.

3.1.1. The Normal Bundle of ℓ. Since $\dim H_x = n - 1 > 1$, it is a direct consequence of Theorem 2.2 that ℓ is smooth and therefore isomorphic to the projective line. A standard theorem, which is attributed to Grothendieck, but probably much older, asserts that a vector bundle on $\mathbb{P}_1$ always decomposes into a sum of line bundles. For the restriction of the tangent bundle T_X to ℓ, all summands must be positive by Theorem 2.1(3). The splitting type is therefore known:

$$T_X|_\ell \cong \mathcal{O}(2) \oplus \mathcal{O}(1)^{\oplus n-1} .$$

The normal bundle of ℓ in X is thus isomorphic to

(3.1) $$N_{\ell/X} \cong \mathcal{O}(1)^{\oplus n-1}.$$

We will use this splitting later in Sect. 3.1.3 to give an estimate on certain self-intersection numbers.

3.1.2. Reduction to a Ruled Surface. As a next step let $\tilde{B}$ be the normalization of B and perform a base change via the natural morphism $\tilde{B} \to H_x$. We obtain a diagram as follows:

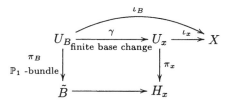

The bundle U_B will now contain two distinct distinguished sections. Let $\sigma_{B,\infty} \subset \iota_B^{-1}(x)$ be the section which is contracted to a point and choose a component $\sigma_{B,0} \subset \gamma^{-1}(B)$. In order to prove Claim 3.2 we have to show that these sections are disjoint.

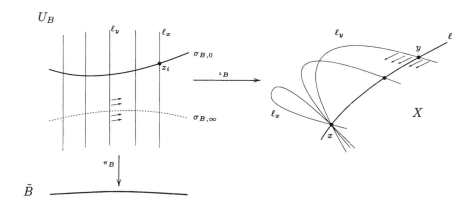

Figure 1. Reduction to a ruled surface

3.1.3. Estimate for the Self-intersection of $\sigma_{B,0}$. Let d be the mapping degree of the restricted evaluation $\iota_B|_\sigma$. We will now show that the self-intersection number of the distinguished section $\sigma_{B,0}$ in the ruled surface U_B is at most d.

Characterizing the Projective Space 153

Claim 3.3 *The natural map*

$$T\iota_B : N_{\sigma_{B,0},U_B} \to N_{\ell,X}$$

between the normal bundles is not trivial.

Proof. Let $\hat{H}_x \subset H_x$ be the closed proper subvariety whose points correspond to curves which are either not smooth or whose normal bundle is not of the form (3.1). Since ℓ was generically chosen, ℓ is not contained in the proper subvariety $\iota_x(\pi_x^{-1}(\hat{H}_x)) \subset X$. Consequence: if $F \subset U_B$ is a general fiber of the morphism π_B, then $\iota_B(F)$ is a smooth curve with normal bundle $N_{\iota_B(F),X} \cong \mathcal{O}(1)^{\oplus n-1}$, and the tangent map $T\iota_B$ has rank two along $F \setminus (F \cap \sigma_{B,\infty})$. In particular, ι_B has maximal rank at $F \cap \sigma_{B,0}$, and the claim is shown.

We obtain the estimate

$$(3.2) \qquad \sigma_{B,0}^2 = \deg N_{\sigma_{B,0},U_B}$$

$$\leq d \cdot \underbrace{(\text{max. degree of sub-linebundles in } N_{\ell,X})}_{=1 \text{ by } (3.1)} = d .$$

3.1.4. Intersection Numbers on the Ruled Surface. Let F be a fiber of the projection π_B and $H \in \operatorname{Pic}X$ be any ample line bundle. We obtain the following list of intersection numbers.

$$\iota_B^*(H).\sigma_{B,\infty} = 0 \qquad \text{because } \sigma_{B,\infty} \text{ is contracted to a point}$$
$$\iota_B^*(H).\sigma_{B,0} = d \cdot \iota_B^*(H).F \qquad \iota|_F : F \to \ell \text{ is birational and } \iota|_{\sigma_{B,0}} \text{ is } d:1$$
$$\sigma_{B,0}.F = 1 \qquad \text{because } \sigma_{B,0} \text{ is a section .}$$

Consequence: we may write the following numerical equivalence of divisors on U_B:

$$\sigma_{B,0} \equiv \sigma_{B,\infty} + d \cdot F.$$

We end the proof of Claim 3.2 and of Proposition 3.1 with the calculation

$$\sigma_{B,0}^2 = \sigma_{B,0} \cdot (\sigma_{B,\infty} + d \cdot F)$$
$$= \sigma_{B,0} \cdot \sigma_{B,\infty} + d .$$

The inequality (3.2) shows that $\sigma_{B,0} \cdot \sigma_{B,\infty} = 0$. The distinguished sections are therefore disjoint. The proof of Proposition 3.1 is finished.

3.2 Factorization of ι_x

To end the proof of Theorem 1.1, consider the Stein-factorization of the morphism ι_x. We obtain a sequence of morphisms

$$U_x \xrightarrow[\alpha]{\iota_x} Y \xrightarrow{\beta} X$$

154   Stefan Kebekus

where α contracts the divisor σ_∞, and β is a finite map.

Since $R^1\pi_{x*}(\mathcal{O}_{U_x}) = 0$, the push-forward of the twisted ideal sheaf sequence

$$0 \to \mathcal{O}_{U_x} \to \mathcal{O}_{U_x}(\sigma_\infty) \to \underbrace{\mathcal{O}_{U_x}(\sigma_\infty)|_{\sigma_\infty}}_{\cong \mathcal{O}_{\mathbb{P}_{n-1}}(-1)} \to 0$$

gives a sequence

$$0 \to \mathcal{O}_{\mathbb{P}_{n-1}} \to \mathcal{E} \to \mathcal{O}_{\mathbb{P}_{n-1}}(-1) \to 0$$

on $H_x \cong \mathbb{P}_{n-1}$ where $\mathcal{E}$ is a vector bundle of rank two and $U_x \cong \mathbb{P}(\mathcal{E}^*)$. Since $\mathrm{Ext}^1_{\mathbb{P}_{n-1}}(\mathcal{O}_{\mathbb{P}_{n-1}}(-1), \mathcal{O}_{\mathbb{P}_{n-1}}) = 0$, the bundle U_x is thus isomorphic to

$$U_x \cong \mathbb{P}\left(\mathcal{O}_{\mathbb{P}_{n-1}}(-1) \oplus \mathcal{O}_{\mathbb{P}_{n-1}}\right).$$

Consequence: there exists a morphism $\alpha' : U_x \to \mathbb{P}_n$ which contracts σ_∞. An application of Zariski's main theorem shows that $\alpha = \alpha'$. In particular, $Y \cong \mathbb{P}_n$.

The fact that β is an isomorphism now follows from [Laz84]: note that the β-images of the lines through $\alpha(x)$ are the curves associated with H_x. Recall the adjunction formula for a finite, surjective morphism:

$$-K_{\mathbb{P}_n} = \beta^*(-K_X) + (\text{branch divisor}) .$$

To see that β is birational, and thus isomorphic, it is therefore sufficient to realize that for a curve $\ell \in H_x$, we have

$$-K_X.\ell = -K_{\mathbb{P}_n}.(\text{line}) = n + 1 .$$

This ends the proof of Theorem 1.1.

3.3   Attributions

The reduction to a ruled surface and the calculation of the intersection numbery have already been used in [Miy92] (see also [Kol96, Prop. V.3.7.5]) to give a criterion for the existence of singular rational curves. The estimate (3.2) is taken from [CMSB00] where a similar estimate is used in a more complex and technically involved situation to prove a statement similar to Proposition 3.1.

The calculation that $U_x \cong \mathbb{P}(\mathcal{O}(-1) \oplus \mathcal{O})$ is modelled after [Mor79].

References

[CMSB00] K. Cho, Y. Miyaoka, and N.I. Shepherd-Barron, Characterizations of projective spaces and applications. Preprint, October–December 2000

[Har77]	R. Hartshorne, *Algebraic Geometry. Graduate Texts in Mathematics* **52** (1977), Springer.
[Keb00]	S. Kebekus, Families of singular rational curves. LANL-Preprint math.AG/0004023, to appear in *J. Alg. Geom.* (2000).
[Keb01]	S. Kebekus, Rationale Kurven auf projektiven Mannigfaltigkeiten (German). Habilitationsschrift. Available from the author's home page at http://btm8x5.mat.uni-bayreuth.de/~kebekus, February 2001.
[KMM92]	J. Kollár, Y. Miyaoka, and S. Mori, Rational connectedness and boundedness of fano manifolds, *J. Diff. Geom.* (1992), 765–769.
[Kol96]	J. Kollár. *Rational Curves on Algebraic Varieties, Ergebnisse der Mathematik und ihrer Grenzgebiete 3. Folge* **32**, Springer, 1996.
[Laz84]	R. Lazarsfeld. Some applications to the theory of positive vector bundles, Proceedings Acireale, in S. Greco and R. Strano, editors, *Complete Intersections, Lecture Notes in Math.* **1092** (1984), Springer, 29–61.
[Miy92]	Y. Miyaoka, Lecture at The University of Utah, unpublished, 1992.
[Mor79]	S. Mori, Projective manifolds with ample tangent bundles, *Ann. of Math.* **110** (1979), 593–606.

Manifolds with Nef Rank 1 Subsheaves in Ω^1_X

Stefan Kebekus[1], Thomas Peternell[2], and Andrew J. Sommese[3]

[1] Lehrstuhl Mathematik VIII, Universität Bayreuth, 95440 Bayreuth, Germany
 E-mail address: stefan.kebekus@uni-bayreuth.de
[2] Lehrstuhl Mathematik I, Universität Bayreuth, 95440 Bayreuth, Germany
 E-mail address: thomas.peternell@uni-bayreuth.de
[3] Department of Mathematics, University of Notre Dame, Notre Dame, Indiana
 46556-5683, USA
 E-mail address: sommese@nd.edu

Table of Contents

1. Introduction 157
2. Generalities 158
3. The case where $\kappa(X) = 1$ 158
4. The case where $\kappa(X) = 0$ 159
  References 163

1 Introduction

This paper is concerned with the question how a subbundle or subsheaf L of the cotangent bundle Ω^1_X of an algebraic variety X impacts the global geometry of X. There is clearly no hope to answer this question in general without further assumptions on L. For reasonable assumptions, one might think, for instance, of L being a maximal destabilizing subsheaf or the dual of the cokernel of a foliation of X.

Here we consider the case that $L \subset \Omega^1_X$ is a locally free subsheaf of rank one and assume that there exists a positive rational number α such that $\alpha L = K_X$. This setup appeared naturally in the study [KPSW00] of projective manifolds which carry a contact structure and is of course related to the stability problem for the cotangent bundle. Examples for this situation are provided by manifolds X whose Iitaka fibration $f : X \to C$ maps onto a smooth curve C of genus $g(C) \geq 1$ —set $L = f^*\Omega^1_C$.

Recall that it follows immediately from the well-known result of Bogomolov [Bog79] (see also [Mou98]) that $\kappa(X) \leq 1$. The case where $\kappa(X) = -\infty$ should not occur because it is expected that manifolds with negative Kodaira dimension are uniruled, and because then K_X could not be nef. In dimension 3 this has already been shown by Miyaoka and Mori. If $\kappa(X) = 1$, then K_X is semi-ample and we obtain information on L via the Iitaka fibration.

2000 *Mathematics Subject Classification*: 14J60, 14J40.

158   Stefan Kebekus et al.

If $\kappa(X) = 0$ we conjecture that X is a quotient of a product $A \times Y$ where A is Abelian and Y is simply connected. We prove this conjecture if $\dim X \leq 4$ or if Ueno's Conjecture K holds.

2 Generalities

In the situation explained in the introduction which we will keep for all of this paper – X a projective manifold, $L \subset \Omega^1_X$ a nef locally free subsheaf of rank 1 with $\alpha L = K_X$, we obtain as a first result:

Proposition 2.1. *If the Kodaira dimension $\kappa(X) \geq 0$, then $K_X \equiv 0$ or $\alpha \geq 1$.*

Proof. We argue by absurdity: assume that $K_X \not\equiv 0$ and $\alpha < 1$. The inclusion $L \to \Omega^1_X$ gives a non-zero element

$$\theta \in H^0(\Omega^1_X \otimes L^*) = H^0\left(\wedge^{\dim X - 1}T_X \otimes \left(1 - \frac{1}{\alpha}\right)K_X\right).$$

Suppose that $\alpha < 1$. Then $(1 - \frac{1}{\alpha}) < 0$ and thus we can find positive integers m, p such that

$$H^0(\underbrace{S^m(\wedge^{\dim X - 1}T_X) \otimes \mathcal{O}(-pK_X)}_{=:E}) \neq 0 .$$

In order to derive a contradiction, recall the result of Miyaoka ([Miy87, Cor. 8.6]; note that X cannot be uniruled) that $\Omega^1_X|_C$ is nef for a sufficiently general curve $C \subset X$ cut out by general hyperplane sections of large degree. Remark that $E^*|_C$ is nef, too. We may assume without loss of generality that p is big enough so that pK_X has a section with zeroes, and so does $E|_C$. See [CP91, Prop. 1.2] for a list of basic properties of nef vector bundles which shows that this is impossible.   □

3 The Case Where $\kappa(X) = 1$

In this case K_X is semi-ample and we give a description of L in terms of the Kodaira-Iitaka map.

Theorem 3.1. *If $\kappa(X) = 1$, then K_X is semi-ample, i.e. some multiple is generated by global sections. Let $f : X \to C$ be the Iitaka fibration and B denote the divisor part of the zeroes of the natural map $f^*(\Omega^1_C) \to \Omega^1_X$. Then there exists an effective divisor $D \in \mathrm{Div}(X)$ such that*

$$L = f^*(\Omega^1_C) \otimes \mathcal{O}_X(B - D) .$$

Furthermore, if p is chosen such that $pK_X \in f^*(\mathrm{Pic}(C))$ and such that $p\alpha$ is an integer, then

$$p(\alpha - 1)K_C = f_*(pK_{X|C}) + f_*(p\alpha(D - B)) .$$

Proof. Since $K_X \not\equiv 0$ and $K_X^2 = 0$ by [KPSW00, Prop. 3.2], the numerical dimension $\nu(X) = 1$. In this setting [Kaw85, Thm. 1.1] proves that K_X is semi-ample, i.e. that a sufficiently high multiple of K_X is globally generated. By [Bog79, Lem. 12.7], the canonical morphism

$$L \to \Omega^1_{X/C}$$

is generically 0. Let B denote the divisor part of the zeroes of $f^*(\Omega^1_C) \to \Omega^1_X$. Then we obtain an exact sequence

$$0 \to f^*(\Omega^1_C) \otimes \mathcal{O}_X(B) \to \Omega^1_X \to \tilde{\Omega}^1_{X/C} \to 0$$

where the cokernel $\tilde{\Omega}^1_{X/C}$ is torsion free. Since $\Omega^1_{X/C} = \tilde{\Omega}^1_{X/C}$ away from a closed subvariety, the induced map

$$L \to \tilde{\Omega}^1_{X/C}$$

vanishes everywhere. Hence there exists an effective divisor D such that

$$L \subset f^*(\Omega^1_C) \otimes \mathcal{O}_X(B) \quad \text{i.e.} \quad L = f^*(\Omega^1_C) \otimes \mathcal{O}_X(B - D) \,.$$

The last equation is obvious. $\qquad\square$

4 The Case Where $\kappa(X) = 0$

We now investigate the more subtle case $\kappa(X) = 0$. We pose the following

Conjecture 4.1. Let X be a projective manifold. If $\kappa(X) = 0$, then $K_X \equiv 0$. Hence (see [Bea83]) there exists a finite étale cover $\gamma : \tilde{X} \to X$ such that $\gamma^*(L) = \mathcal{O}_{\tilde{X}}$ and $\tilde{X} = A \times Y$ where A is Abelian and Y is simply connected.

We will prove the conjecture in a number of cases, in particular if $\dim X \leq 4$ or if the well-known Conjecture K holds (see [Mor87, §10] for a detailed discussion). Recall that the Albanese map of a projective manifold X with $\kappa(X) = 0$ is surjective and has connected fibers [Kaw81].

Conjecture 4.2 (Ueno's Conjecture K). If X is a nonsingular projective variety with $\kappa(X) = 0$, then the Albanese map is birational to an étale fiber bundle over $\text{Alb}(X)$ which is trivialized by an étale base change.

It is known that Conjecture K holds if $q(X) \geq \dim X - 2$.

The proof requires two technical lemmata. The first is a characterization of pull-back divisors. It appears implicitly in [Kaw85, p. 571].

Lemma 4.3 (Kawamata's pull-back lemma). *Let $f : X \to Y$ be an equidimensional surjective projective morphism with connected fibers between normal quasi-projective $\mathbb{Q}$-factorial varieties and let $D \in \mathbb{Q}\text{Div}(X)$ be an f-nef $\mathbb{Q}$-divisor such that $f(\text{Supp}(D)) \neq Y$. Then there exists a number $m \in \mathbb{N}^+$ and a divisor $A \in \text{Div}(Y)$ such that $mD = f^*(A)$.*

160   Stefan Kebekus et al.

Here is a sketch of the argument. By taking hyperplane sections on Y and on X, we reduce ourselves to Zariski's lemma for normal surfaces. Note that the normal version follows immediately from the smooth case [BPV84, III.8.2] by passing to a desingularisation.

The next lemma is very similar in nature. We include it here for lack of an adequate reference.

Lemma 4.4. *Let $f : X \to Y$ be a surjective projective morphism between quasi-projective $\mathbb{Q}$-factorial varieties and let $D \in \mathbb{Q}\operatorname{Div}(X)$ be an f-nef $\mathbb{Q}$-divisor which is of the form*

$$D = f^*(H) + \sum \lambda_i E_i$$

where $H \in \mathbb{Q}\operatorname{Div}(Y)$ and $\operatorname{codim}_Y f(E_i) \geq 2$. If X is normal, then $\lambda_i \leq 0$ for all i.

Proof. Choose a general surface section S of X. By Seidenberg's theorem, S will then be normal. Note that $\dim f(S) = 2$; each $E_i \cap S$ is an irreducible curve on S; any of the $S \cap E_i$ map to a point under f. Thus perhaps by passing to a connected component of $\sum \lambda_i (S \cap E_i)$, we have a sum of compact curves $\sum \lambda_i (S \cap E_i)$ on a surface S over a point y. If some of the λ_i were positive we would have a curve $A - B$ over a point y with A and B effective, and $A \cdot B \geq 0$. This would give $(A - B) \cdot A \geq 0$ by the f-nefness, or $A \cdot A \geq B \cdot A \geq 0$. But this is absurd since $A \cdot A < 0$ by Grauert's theorem on the negative definiteness of the intersection matrix of curves that get collapsed on a surface.

Note the negative definiteness is true even for the rational intersections of arbitrary divisors on a normal surface [Sak84], and therefore applies to our situation. $\qquad\square$

We will now show that Conjecture K implies our Conjecture 4.1.

Theorem 4.5. *If Conjecture K holds, then Conjecture 4.1 holds as well.*

In order to clarify the structure of the proof, we single out the case where $H^0(X, L) \neq 0$. More precisely, we consider the weakened

Assumptions 4.6. Let X be a projective manifold and $L \subset \Omega_X^1$ be a locally free nef subsheaf of rank one. Assume that $\kappa(X) = 0$ and that there is a positive rational number α such that $\alpha L \subset K_X$ is a subsheaf. Assume furthermore that $H^0(X, L) \neq 0$.

Lemma 4.7. *Under the weakened Assumptions 4.6, if Conjecture K holds for X, then Conjecture 4.1 holds as well, i.e. $L \equiv 0$.*

Proof. In inequality $H^0(X, L) \neq 0$ directly implies that $q(X) > 0$, i.e. that the Albanese map $f : X \to A = \operatorname{Alb}(X)$ is not trivial.

In this setting, Conjecture K yields a diagram as follows:

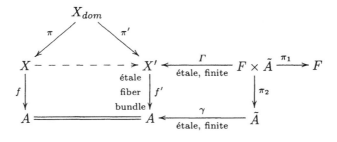

where π and π' are birational morphisms. Let E be the zero-set of a section of L; i.e. $L = \mathcal{O}(E)$. We will show that $E = 0$.

Claim 1. The divisor E is supported on f-fibers, i.e. $f(\mathrm{Supp}(E)) \neq A$.

Proof of Claim 1. the inclusion $L \subset \Omega_X^1$ yields an element

$$\theta \in H^0(\Omega_X^1(-E)) .$$

Since $\theta = f^*(\eta)$ for some one-form η on A, we conclude that

$$f(E) \subset \mathrm{Sing}(f) ,$$

where $\mathrm{Sing}(f)$ is the set of all $y \in A$ such that $f^{-1}(y)$ is singular. In particular, E does not meet the general fiber of f. This shows claim 1.

As a next step in the proof of Lemma 4.7, we set

$$A^0 := \{a \in A | \dim f^{-1}(a) = \dim X - \dim A\} \quad \text{and} \quad X^0 := f^{-1}(A^0)$$

and show

Claim 2. the divisor E does not intersect X^0, i.e. $\mathrm{Supp}(E) \cap X^0 = \emptyset$.

Proof of Claim 2. by Kawamata's pull-back Lemma 4.3 there exists a number $m \in \mathbb{N}^+$ such that $mE|_{X^0} \in f^*(D|_{A^0})$ for some $D \in \mathrm{Div}(A)$. This implies

$$\overline{f^{-1}(D \cap A^0)} \subset \mathrm{Supp}(E) .$$

Now consider X_{dom}. Since X is smooth, we can find a $\mathbb{Q}$-divisor for the canonical bundle $\omega_{X_{\mathrm{dom}}}$ of the form

$$K_{X_{\mathrm{dom}}} = \pi^*(\underbrace{K_X}_{\alpha E + (\text{effective})}) + \sum \lambda_i B_i$$

where the B_i are π-exceptional divisors and $\lambda_i \in \mathbb{Q}^+$. In particular, we have

$$\overline{(\pi \circ f)^{-1}(D \cap X^0)} \subset \mathrm{Supp}(\pi^*(E)) \subset \mathrm{Supp}(K_{X_{\mathrm{dom}}}) .$$

Since $\pi \circ f = \pi' \circ f'$ and $K_{X'} := (\pi')_*(K_{X_{\mathrm{dom}}})$ is a $\mathbb{Q}$-divisor for the anti-canonical bundle $\omega_{X'}$, we find

$$\overline{(f')^{-1}(D \cap X^0)} \subset \mathrm{Supp}(K_{X'}).$$

We derive a contradiction from the last inequality by noting that

$$(\pi_2 \circ \gamma)^{-1}(D) \subset \mathrm{Supp}(\underbrace{K_{F \times \tilde{A}}}_{=\Gamma^*(K_{X'})}).$$

On the other hand, it follows from the adjunction formula that every effective $\mathbb{Q}$-divisor for the canonical bundle $\omega_{F \times \tilde{A}}$ must be contained in $\pi_1^*(\mathbb{Q}\,\mathrm{Div}(F))$. This shows Claim 2.

Application of Claims 1 and 2. we know that E is an effective and nef divisor satisfying $\mathrm{codim}_A(f(\mathrm{Supp}(E))) \geq 2$. By Lemma 4.4 this is possible if and only if $E = 0$. $\qquad\square$

The preceding lemma enables us to finish the proof of Theorem 4.5 by reducing to the modified weakened Assumptions 4.6. Since we wish to re-use the same argumentation later, we formulate a technical reduction lemma:

Lemma 4.8 (Reduction Lemma). *If $L \equiv 0$ holds true for all varieties of a given dimension d satisfying the weakened Assumptions 4.6 , then Conjecture 4.1 holds for all varieties of dimension d.*

Proof. Let X be a variety as in Conjecture 4.1 and assume that X is of dimension d. If $H^0(X, L) \neq 0$, then we can stop here.

Otherwise, we find a minimal positive integer m and an effective divisor E such that $(L)^m = \mathcal{O}_X(E)$. We build a sequence of morphisms

$$\tilde{Y} \xrightarrow[\text{desing.}]{\sigma} Y \xrightarrow[\text{cyclic cover}]{\gamma} \hat{X} \xrightarrow[\text{desing. of } E]{\pi} X$$

as follows: Let $\pi : \hat{X} \to X$ be a sequence of blow-ups with smooth centers such that $\mathrm{Supp}(\pi^*(E))$ has only normal crossings. Set $\hat{E} = \pi^*(E)$ and $\hat{L}^* = \pi^*(L^*)$ and let $f : Y \to \hat{X}$ be the cyclic covering (followed by normalization) associated with the section

$$s \in H^0(X, L^m)$$

which defines $\hat{E}$; see e.g. [EV92, 3.5] for a more detailed description of this construction. By [EV92, 3.15], Y is irreducible, étale over $\hat{X} \setminus \mathrm{Supp}(\hat{E})$ and smooth over $\hat{X} \setminus \mathrm{Sing}(\mathrm{Supp}(\hat{E}))$. Moreover

$$H^0(Y, \gamma^*(\hat{L})) \neq 0.$$

Let $\sigma : \tilde{Y} \to Y$ be a desingularization. We will show:

Claim: $\kappa(\tilde{Y}) = 0$.

Application of the claim. If the claim holds true, we conclude as follows: let $\tilde{L}^* = (\sigma \circ \gamma)^*(\hat{L}^*)$. Then $H^0(\tilde{X}, \tilde{L}) \neq 0$, and furthermore $\tilde{L} \subset \Omega^1_{\tilde{Y}}$ by virtue of the canonical morphism

$$\underbrace{(\sigma \circ \gamma \circ \pi)^*}_{=: \Gamma}(\Omega^1_X) \to \Omega^1_{\tilde{Y}} \ .$$

Similarly, since X is smooth, $\Gamma^* K_X \subset K_{\tilde{Y}}$. Thus, Lemma 4.7 applies to $\tilde{Y}$, showing that $\Gamma^*(L) \equiv 0$ so that L was numerically trivial in the first place. This shows the lemma and thus finishes the proof of Theorem 4.5.

Proof of the claim. Since Γ is an étale cover away from E, we can find a divisor for the canonical bundle $\omega_{\tilde{Y}}$ which is of the form

$$K_{\tilde{Y}} = \Gamma^*(\alpha E) + D$$

where D is effective and $\Gamma(D) \subset E$. This already implies that there is a number $k \in \mathbb{N}^+$ such that $(k\Gamma^*(E) - K_{\tilde{Y}})$ is effective. Thus $\kappa(K_{\tilde{Y}}) \leq \kappa(\Gamma^*(E)) = 0$, and the claim is shown. $\qquad\Box$

Thus ends the proof of Theorem 4.5.

Finally we show Conjecture 4.1 in the case where $\dim X \leq 4$.

Theorem 4.9. *If $\kappa(X) = 0$ and $\dim X \leq 4$, then $K_X \equiv 0$.*

Proof. Using the reduction Lemma 4.8, it is sufficient to consider the weakened setting 4: let X be as in 4. At first we argue exactly as in the proof of lemma 4.7: the Albanese map $f : X \to A = \mathrm{Alb}(X)$ is surjective and has connected fibers and we have $f(\mathrm{Supp}(E)) \neq A$, where E is the zero-divisor of the section in L. Furthermore we have $\kappa(X, E) = 0$.

Recall that Conjecture K holds if $q(X) \geq \dim X - 2$; see [Mor87, p. 316]. By Lemma 4.7 we are finished in these cases. It remains to consider the case where $q(X) = 1$.

Because f is equidimensional in this case, Kawamata's pull-back lemma 4.3 applies: there is a number $m \in \mathbb{N}^+$ such that $mE = f^*(D)$. It follows immediately that $\kappa(A, D) = 0$. Since A is a torus, this is possible if and only if $D = 0$. $\qquad\Box$

Remark 4.10. Actually, the proof of Lemma 4.8 and Theorem 4.9 show that Conjecture 4.1 holds if $q(X) \geq \dim X - 2$ or if $H^0(X, L) \neq 0$ and $q(X) = 1$. For the first statement, note that $q(X)$ increases when passing to a cover.

References

[Bea83] A. Beauville, Variétés Kählériennes dont la première classe de Chern est nulle, *J. Diff. Geom.* **18** (1983), 755–782.

164   Stefan Kebekus et al.

[Bog79] F. Bogomolov, Holomorphic tensors and vector bundles on projective varieties, *Math. USSR Izv.* **13** (1979), 499–555.

[BPV84] W. Barth, C. Peters and A. Van de Ven, Compact complex surfaces, Springer, 1984.

[CP91] F. Campana and T. Peternell, Projective manifolds whose tangent bundles are numerically effective, *Math. Ann.* **289** (1991), 169–187.

[EV92] H. Esnault and E. Viehweg, *Lectures on Vanishing Theorems*, Birkhäuser Verlag, 1992.

[Kaw81] Y. Kawamata, Characterization of Abelian varieties, *Compositio Math.* **43** (1981), 253–276.

[Kaw85] Y. Kawamata, Pluricanonical systems on minimal algebraic varieties, *Inv. Math.* **79** (1985), 567–588.

[KPSW00] S. Kebekus, T. Peternell, A. Sommese, and J. Wiśniewski, Projective contact manifolds, *Invent. Math.* **142**(1) (2000), 1–15.

[Miy87] Y. Miyaoka, Deformation of a morphism along a foliation, In S. Bloch, editor, *Algebraic Geometry* **46**(1) (1987), *Proceedings of Symposia in Pure Mathematics*, pages 245–268, Providence, Rhode Island, American Mathematical Society.

[Mor87] S. Mori, Classification of higher-dimensional varieties, In *Proceedings of Symposia in Pure Mathematics* **46** (1987), *Proceedings of Symposia in Pure Mathematics*, pages 269–331, Providence, Rhode Island, American Mathematical Society.

[Mou98] C. Mourougane, Version kählériennes du théorème d'annulation de Bogomolov, *Collect. Math.* **49** (1998), 433–445.

[Sak84] F. Sakai, Weil divisors on normal surfaces, *Duke Math. J.* **51** (1984), 877–887.

The Simple Group of Order 168
and K3 Surfaces

Keiji Oguiso[1] and De-Qi Zhang[2]

[1] Department of Mathematical Sciences, University of Tokyo, Komaba, Meguro, Tokyo Japan
E-mail address: oguiso@ms.u-tokyo.ac.jp
[2] Department of Mathematics, National University of Singapore, 2 Science Drive 2, Singapore 117543
E-mail address: matzdq@math.nus.edu.sg

Abstract The aim of this note is to characterize a K3 surface of Klein-Mukai type in terms of its symmetry.

Table of Contents

0 Introduction 165
1 The Niemeier Lattices 168
2 Proof of the Main Theorem 170
   References 180

0 Introduction

The group $L_2(7)$ is by the definition the projectivized special linear group $\mathrm{PSL}(2, \mathbf{F}_7)$ and is generated by the three projective transformations of $\mathbf{P}^1(\mathbf{F}_7)$ of order 7, 3, 2:

$$\alpha : x \mapsto x + 1; \beta : x \mapsto 2x; \gamma : x \mapsto -x^{-1} ,$$

where the coefficient 2 in β is a generator of the cyclic group $(\mathbf{F}_7^\times)^2 (\simeq \mu_3)$. (See for instance [CS, Chap. 10].) As well-known, this group is of order 168 and is characterized as the second smallest non-commutative simple group.

One of interesting connections between $L_2(7)$ and complex algebraic geometry goes back to the result of the great German mathematicians Hurwitz and Klein in Göttingen: $|L_2(7)| = 84(3 - 1)$ is the largest possible order of a group acting on a genus-three curve and the so called Klein quartic curve

$$C_{168} = \left\{ x_1 x_2^3 + x_2 x_3^3 + x_3 x_1^3 = 0 \right\} \subset \mathbf{P}^2$$

2000 *Mathematics Subject Classification*: 14J28.

is the unique genus-three curve admitting an $L_2(7)$-action. The action of $L_2(7)$ on C_{168} is the projective transformation induced by (one of two essentially the same) 3-dimensional irreducible representation V_3 of $L_2(7)$ given by

$$\alpha \mapsto \begin{pmatrix} \zeta_7 & 0 & 0 \\ 0 & \zeta_7^2 & 0 \\ 0 & 0 & \zeta_7^4 \end{pmatrix}; \quad \beta \mapsto \begin{pmatrix} 0 & 0 & 1 \\ 1 & 0 & 0 \\ 0 & 1 & 0 \end{pmatrix}; \quad \gamma \mapsto \frac{-1}{\sqrt{-7}} \begin{pmatrix} a & b & c \\ b & c & a \\ c & a & b \end{pmatrix},$$

where $\zeta_7 = \exp(2\pi\sqrt{-1}/7)$, $a = \zeta_7^2 - \zeta_7^5$, $b = \zeta_7 - \zeta_7^6$, $c = \zeta_7^4 - \zeta_7^3$, and the branch of $\sqrt{-7}$ is chosen so that $\sqrt{-7} = \zeta_7 + \zeta_7^2 + \zeta_7^4 - \zeta_7^3 - \zeta_7^5 - \zeta_7^6$. The other 3-dimensional irreducible representation is the composition of the representation V_3 with the outer automorphism of $L_2(7)$ given by $\alpha \mapsto \alpha^{-1}$, $\beta \mapsto \beta$ and $\gamma \mapsto \gamma$ (see [ATLAS] and [Bu, §267]). The Klein curve C_{168} together with $L_2(7)$-action also appears in the McKay correspondece problem [Ma] and a classification of Calabi-Yau threefolds [Og].

Our interest in this note is a relation between $L_2(7)$ and K3 surfaces.

Throughout this note, a *K3 surface* means a simply-connected smooth complex *algebraic* surface X with a nowhere vanishing holomorphic 2-form ω_X. We call an automorphism $g \in \mathrm{Aut}(X)$ *symplectic* if $g^*\omega_X = \omega_X$. According to Mukai's classification [Mu1], there are eleven maximum finite groups acting on K3 surfaces symplectically, and among them, there appear two simple groups: the group $L_2(7)$ and the alternating group A_6 of degree 6. In the same paper, Mukai also gives a beautiful example of K3 surface X_{168} with $L_2(7)$-action, where

$$X_{168} = \{x_0^4 + x_1 x_2^3 + x_2 x_3^3 + x_3 x_1^3 = 0\} \subset \mathbf{P}^3 .$$

This is the cyclic cover of $\mathbf{P}^2$ of degree 4 branched along the Klein quartic curve C_{168}. Here, the action of $L_2(7)$ on X_{168} is naturally induced by the action on C_{168}. Note that this X_{168} now admits a larger group action of $L_2(7) \times \mu_4$, where μ_4 is the Galois group of the covering. On the other hand, the smooth plane curve H_{168} of degree 6 defined by

$$\{5x_1^2 x_2^2 x_3^2 - x_1^5 x_2 - x_2^5 x_3 - x_3^5 x_1 = 0\} \subset \mathbf{P}^2$$

– the zero locus of the Hessian of the Klein quartic curve – is also invariant under the same $L_2(7)$-action on $\mathbf{P}^2$. So, the K3 surface

$$X'_{168} = \{y^2 = 5x_1^2 x_2^2 x_3^2 - x_1^5 x_2 - x_2^5 x_3 - x_3^5 x_1\} \subset \mathbf{P}(1, 1, 1, 3) ,$$

i.e. the double cover of $\mathbf{P}^2$ branched along H_{168}, also admits an $L_2(7)$-action. However, it will turn out that these two K3 surfaces X_{168} and X'_{168} are not isomorphic to each other (see Remark (2.12) and Proposition 2 in Added in Proof). Therefore, K3 surfaces having $L_2(7)$-action are no more unique and it is of interest to characterize the Klein-curve-like K3 surface X_{168} in a flavour similar to that of Hurwitz and Klein. This is the aim of this short note.

Throughout this note, we set $G := L_2(7)$. Our main observation is as follows:

Main Theorem. *Let X be a K3 surface. Assume that $G \subset \mathrm{Aut}(X)$. Let $\tilde{G}$ be a finite subgroup of $\mathrm{Aut}(X)$ such that $G \subset \tilde{G}$. Then,*

(1) *$\tilde{G}/G$ is a cyclic group of order 1, 2, 3, or 4; and*

(2) *if $\tilde{G}/G$ is of the maximum order 4, then $(X, \tilde{G})$ is isomorphic to the Klein-Mukai pair $(X_{168}, L_2(7) \times \mu_4)$.*

Here an isomorphism means an equivariant isomorphism with respect to group actions. See also Proposition 1 in Added in Proof for an improvement of the statement (1).

The main difference between genus-three curves and K3 surfaces is that there are no canonical polarizations on K3 surfaces. In other words, we do not know *a priori* which K3 surfaces are quartic K3 surfaces or which polarizations are invariant under the group action. Indeed, the determination of the invariant polarization for $\tilde{G}$ – this will turn out to be of degree four if $|\tilde{G}/G| = 4$ (Claim 2.10) – is the most crucial part in this note. Besides Mukai's pioneering work, we are much inspired by a series of Kondo's work [Ko1], [Ko2] on a lattice theoretic proof of Mukai's classification and the determination of the K3 surface with the largest finite group action as well as the action. Especially we will fully exploit his brilliant idea of studying invariant lattices through an embedding of their orthogonal complements into some Niemeier lattices. This enables us to relate the problem with the Mathieu group M_{24} and the binary Golay code $\mathcal{C}_{24}$ (Sect. 1) and provides a very powerful tool in calculating the discriminants of the invariant lattices $H^2(X, \mathbf{Z})^G$ also in our setting. Combining this with the additional group action $\tilde{G}/G$, we shall determine the invariant polarization in the maximum case $|\tilde{G}/G| = 4$. Once we find the invariant polarization in a lattice-theoretic way, we can continue the proof by coming back to more algebro-geometric arguments. One of the advantages of the algebro-geometric argument is perhaps that we can then express the K3 surface and the group action in a very concrete way as in the theorem.

Acknowledgement

Both authors would like to express their thanks to Professor M. L. Lang for his help on Mathieu groups, to Professor Dolgachev for his critical reading of the first draft and pointing out the reference [Ed] for the use in the proof of Claim 2.11, and to Professors S. Mukai and S. Kondo for their valuable comments. The main idea of this work was found during the first named author's stay at the National University of Singapore in August 2000. He would like to express his thanks to the National University of Singapore for financial support and to staff members there for the warm hospitality.

168   Keiji Oguiso and De-Qi Zhang

1   The Niemeier Lattices

In this section, we recall some basic facts on the Niemeier lattices needed in our arguments. Our main reference concerning Niemeier lattices and their relations with Mathieu groups is [CS, Chaps. 10, 11, 16, 18].

(1.1). In this note, the even *negative* definite unimodular lattices of rank 24 are called Niemeier lattices. (We changed the sign from positive into negative.) There are exactly 24 isomorphism classes of the Niemeier lattices and each isomorphism class is uniquely determined by its root lattice N_2, i.e. the sublattice generated by all the roots, the elements x with $x^2 = -2$. Except the so called Leech lattice which contains no roots, the other 23 lattices are the over-lattices of their root lattices, which are:

$$A_1^{\oplus 24}, A_2^{\oplus 12}, A_3^{\oplus 8}, A_4^{\oplus 6}, A_6^{\oplus 4}, A_8^{\oplus 3}, A_{12}^{\oplus 2}, A_{24},$$

$$D_4^{\oplus 6}, D_6^{\oplus 4}, D_8^{\oplus 3}, D_{12}^{\oplus 2}, D_{24}, E_6^{\oplus 4}, E_8^{\oplus 3}, A_5^{\oplus 4} \oplus D_4, A_7^{\oplus 2} \oplus D_5^{\oplus 2},$$

$$A_9^{\oplus 2} \oplus D_6, A_{15} \oplus D_9, E_8 \oplus D_{16}, E_7^{\oplus 2} \oplus D_{10}, E_7 \oplus A_{17}, E_5 \oplus D_7 \oplus A_{11}.$$

We denote the Niemeier lattices N whose root lattices are $A_1^{\oplus 24}$, $A_2^{\oplus 12}$ and so on by $N(A_1^{\oplus 24})$, $N(A_2^{\oplus 12})$ and so on.

(1.2). In what follows, we regard the set of roots $R := \{r_i | 1 \le i \le 24\}$ corresponding to the vertices of the Dynkin diagram, as the set of the simple roots of N. Denote by $O(N)$ (resp. by $O(N_2)$) the group of isometries of N (resp. of N_2) and by $W(N)$ the Weyl group generated by the reflections given by the roots of N. Here $O(N) \subset O(N_2)$ and $W(N)$ is a normal subgroup of both $O(N)$ and $O(N_2)$. The invariant hyperplanes of the reflections divide $N \otimes \mathbf{R}$ into (finitely many) chambers. Then, each chamber is a fundamental domain of the action of $W(N)$ and the quotient group $S(N) := O(N)/W(N)$ is identified with a subgroup of symmetry of the distinguished chamber $\mathcal{C} := \{x \in N \otimes \mathbf{R} | (x, r) > 0 r \in R\}$ and also a subgroup of a larger group $S_{24} = \text{Aut}_{\text{set}}(R)$.

The groups $S(N)$ are very explicitly calculated in [CS, Chaps. 18, 16]. (See also [Ko1].) The following is a part of the results there:

Lemma (1.3) [CS, Chaps. 18, 16]. *Let N be a non-Leech Niemeier lattice. Then,*

(1) $S(N) = M_{24}$ *if* $N = N(A_1^{\oplus 24})$;
(2) $S(N) = C_2 \ltimes (C_2^{\oplus 3} \rtimes L_3(2))$ *if* $N = N(A_3^{\oplus 8})$; *and*
(3) *for other* N, *the order* $|S(N)|$ *is not divisible by 7.*

Let us add a few remarks about the groups appearing in the Lemma above. The next (1.4) and (1.5) are concerned with the first case (1) and (1.6) is for the second case (2).

The Simple Group of Order 168 and K3 Surfaces   169

(1.4). Observe that $A_1^{\oplus 24} \subset N(A_1^{\oplus 24}) \subset (A_1^{\oplus 24})^*$ and that

$$(A_1^{\oplus 24})^*/A_1^{\oplus 24} = \oplus_{i=1}^{24} \mathbf{F}_2 \bar{r}_i \simeq \mathbf{F}_2^{\oplus 24} .$$

Here $\bar{r}_i := r_i/2 \, \mathrm{mod} \mathbf{Z} r_i$ is the standard basis of the i-th factor $(A_1)^*/A_1$. We also identify

$$S_{24} = \mathrm{Aut}_{\mathrm{set}}(R) = \mathrm{Aut}_{\mathrm{set}}(\{\bar{r}_i\}_{i=1}^{24}) .$$

The linear subspace of $(A_1^{\oplus 24})^*/A_1^{\oplus 24}$

$$\mathcal{C}_{24} := N(A_1^{\oplus 24})/A_1^{\oplus 24} \simeq \mathbf{F}_2^{\oplus 12}$$

encodes the information which elements of $(A_1^{\oplus 24})^*$ lie in $N(A_1^{\oplus 24})$. Besides this role, this subspace $\mathcal{C}_{24}$ carries the structure of the binary self-dual code of Type II with minimal distance 8, called the (extended) binary Golay code.

Among many equivalent definitions, the Mathieu group M_{24} of degree 24 is defined to be the subgroup of S_{24} preserving $\mathcal{C}_{24}$, i.e.

$$M_{24} := \{\sigma \in S_{24} | \sigma(\mathcal{C}_{24}) = \mathcal{C}_{24}\} .$$

As well-known, M_{24} is a simple group of order $24 \cdot 23 \cdot 22 \cdot 21 \cdot 20 \cdot 16 \cdot 3$ and acts on the set $\{\bar{r}_i\}_{i=1}^{24}$ as well as on R quintuply transitively.

Let $\mathcal{P}(R)$ be the power set of R, i.e. the set consisting of the subsets of R. Then, $\mathcal{P}(R)$ bijectively corresponds to the set $(A_1^{\oplus 24})^*/A_1^{\oplus 24}$ by:

$$\iota : \mathcal{P}(R) \ni A \mapsto \bar{r}_A := \frac{1}{2} \sum_{r_j \in A} r_j \, \mathrm{mod} \, A_1^{\oplus 24} \in \left(A_1^{\oplus 24}\right)^* / A_1^{\oplus 24} .$$

Set $\mathcal{E} := \iota^{-1}(\mathcal{C}_{24})$. Then $A \in \mathcal{E}$ if and only if $\frac{1}{2} \sum_{r_j \in A} r_j$ is in $N(A_1^{\oplus 24})$. Moreover, it is known that $\emptyset, R \in \mathcal{E}$ and that if $A \in \mathcal{E}$ $(A \neq R, \emptyset)$ then $|A|$ is either 8, 12, or 16. We call $A \in \mathcal{E}$ an Octad (resp. a Dodecad) if $|A| = 8$ (resp. 12). Note that $B \in \mathcal{E}$ with $|B| = 16$ is then the complement of an Octad (in R), i.e. B is of the form $R - A$ for some Octad A. There are exactly 759 Octads.

The following fact called the Steiner property $St(5, 8, 24)$ and its proof are both needed in the proof of our main result:

Fact (1.5) (the Steiner property). *For each 5-elemet subset S of R, there exists exactly one Octad O such that $S \subset O$.*

Proof. Since M_{24} is quintuply transitive on R, there exists an Octad O such that $S \subset O$. Let O_1 and O_2 be two Octads. Then, by the definition, their symmetric difference $(O_1 - O_2) \cup (O_2 - O_1)$ is also an element of $\mathcal{E}$. Thus, $|(O_1 - O_2) \cup (O_2 - O_1)|$ is either 0, 8, 12, or 16 and we have that $|O_1 \cap O_2|$ is either 8, 4, 2 or 0. Therefore, if $S \subset O_1$ and $S \subset O_2$, then one has $|O_1 \cap O_2| \geq 5$, whence $|O_1 \cap O_2| = 8$. This means that $O_1 = O_2$. $\qquad \square$

(1.6). In the second case, we identify (non-canonically) the set of eight connected components of the Dynkin diagram $A_3^{\oplus 8}$ with the three-dimensional linear space $\mathbf{F}_2^{\oplus 3}$ over $\mathbf{F}_2$ by letting one connected component to be 0. The group $C_2 \rtimes (C_2^{\oplus 3} \rtimes L_3(2))$ is the semi-direct product, where C_2 interchanges the two edges of all the components, $C_2^{\oplus 3}$ is the group of the parallel transformations of the affine space $\mathbf{F}_2^{\oplus 3}$ and $L_3(2)(\simeq L_2(7))$ is the linear transformation group of $\mathbf{F}_2^{\oplus 3}$ which fixes (point wise) the three simple roots in the identity component.

2 Proof of the Main Theorem

In what follows, we set $L := H^2(X, \mathbf{Z})$, $L^G := \{x \in L | g^* x = x$ for all $g \in G\}$, and $L_G := (L^G)^\perp$ in L. This L is the unique even unimodular lattice of index $(3, 19)$. We also denote by S_X the Néron-Severi lattice and by T_X $(:= S_X^\perp$ in $L)$, the transcendental lattice. Since G is simple and non-commutative, we have $G = [G, G]$. In particular, G acts on X symplectically. Therefore L^G contains both T_X and the invariant ample classes under G, namely the pull back of ample classes of X/G. In addition, since G is maximum among finite symplectic group actions [Mu1], G is normal in $\tilde{G}$ and the quotient group $\tilde{G}/G$ acts faithfully on $H^{2,0}(X) = \mathbf{C}\omega_X$. In particular, $\tilde{G}/G$ is a cyclic group of order I such that the Euler function $\varphi(I)$ divides rank T_X [Ni].

Claim (2.1).

(1) rank $L^G = 3$. In particular, rank $T_X = 2$ and (up to scalar) there is exactly one G-invariant algebraic cycle class H. Moreover, this class H is ample and is also invariant under $\tilde{G}$.

(2) $|\tilde{G}/G|$ is either 1, 2, 3, 4 or 6.

Proof. The equality rank $L^G = 3$ is a special case of a general formula of Mukai. Here for the convenience of readers, we shall give a direct argument along [Mu1, Prop. 3.4]. Let us consider the natural representaion ρ of G on the cohomology ring of X

$$\tilde{L} := \oplus_{i=0}^4 H^i(X, \mathbf{Z}) = H^0(X, \mathbf{Z}) \oplus L \oplus H^4(X, \mathbf{Z}) .$$

Then, by the representation theory of finite groups and by the Lefschetz $(1, 1)$-Theorem, one has

$$2 + \operatorname{rank} L^G = \operatorname{rank} \tilde{L}^G = \frac{1}{|G|} \sum_{g \in G} \operatorname{tr}(\rho(g)) = \frac{1}{|G|} \sum_{g \in G} \chi_{\mathrm{top}}(X^g) .$$

Here, the terms in the last sum are calculated by Nikulin [Ni] as follows:

$$\chi_{\mathrm{top}}(X^g) = 24, 8, 6, 4, 4, 2, 3, 2$$

if

$$\mathrm{ord}(g) = 1, 2, 3, 4, 5, 6, 7, 8 .$$

Observe also that $n_1 = 1$, $n_2 = 21$, $n_3 = 56$, $n_4 = 42$, $n_7 = 48$ and $n_j = 0$ for other j if $G = L_2(7)$, where n_d denotes the cardinality of the elements of order d in G. Now combining all of these together, we obtain

$$2 + \mathrm{rank}\, L^G = \frac{1}{168}(24 + 8 \times 21 + 6 \times 56 + 4 \times 42 + 3 \times 48) = 5 .$$

The remaining assertions now follow from the facts summarized before Claim (2.1). $\qquad\square$

Remark (2.2). By (2.1)(1), K3 surfaces with $L_2(7)$-action are of the maximum Picard number 20. By a similar case-by-case calculation, one can also show that the invariant lattices are positive definite and of rank 3 if a K3 surface admits one of the eleven maximum symplectic group actions listed in [Mu1]. In particular, one has that

(1) the invariant lattices (tensorized by $\mathbf{R}$) have the hyperkähler three-space structure;
(2) such *algebraic* K3 surfaces are of the maximum Picard number 20 and are then at most countably many by [SI].

It would be very interesting to describe all of such (algebraic) K3 surfaces as rational points of the twister spaces corresponding to the invariant lattices (tensorized by $\mathbf{R}$).

Next we determine the discriminant of L^G.

Key Lemma. $|\mathrm{det}\, L^G| = 196$.

The proof of Key Lemma will be given after Claim (2.6). Technically, this is the most crucial step and the next embedding Theorem due to Kondo is the most important ingredient in our proof of Key Lemma:

Theorem (2.3) [Ko1]. *Under the notation explained in Sect. 1, one has the following:*

(1) *For a given finite symplectic action H on a K3 surface, there exists a non-Leech Niemeier lattice N such that $L_H \subset N$. Moreover, the action of H extends to an action on N so that $L_H \simeq N_H$ and that N^H contains a simple root.*
(2) *This group action of H on N preserves the distinguished Weyl chamber C and the natural homomorphism $H \to S(N)$ is injective.*

Corollary (2.4). *Under the notation of Theorem (2.3), one has:*

(1) $\mathrm{rank}\, N^H = \mathrm{rank}\, L^H + 2$. *In particular,* $\mathrm{rank}\, N^G = \mathrm{rank}\, L^G + 2 = 5$.
(2) $|\mathrm{det}\, N^H| = |\mathrm{det}\, L^H|$.

172   Keiji Oguiso and De-Qi Zhang

Proof. Since rank $N^H = 24 -$ rank N_H and rank $L^H = 22 -$ rank L_H, the first part of the assertion (1) follows from $N_H \simeq L_H$. Now the last part of (1) follows from (2.1). Recall that L and N are unimodular and the embeddings $L^H \subset L$ and $N^H \subset N$ are primitive. Then $|\det L^H| = |\det L_H|$ and $|\det N^H| = |\det N_H|$. Combining these with $N_H \simeq L_H$, one obtains $|\det N^H| = |\det L^H|$.

$\square$

Let us return back to our original situation and determine the Niemeier lattice N for our G. Note that N is not the Leech lattice by (2.3)(1).

Claim (2.5). *The Niemeier lattice N in Theorem (2.3) for $G = L_2(7)$ is* $N(A_1^{\oplus 24})$.

Proof. By Theorem (2.3)(2), $168 = |G|$ divides $|S(N)|$. Thus, N is either $N(A_1^{\oplus 24})$ or $N(A_3^{\oplus 8})$ by (1.3). Suppose that the latter case occurs. Then, by (1.3), we have $G \subset C_2 \times (C_2^{\oplus 3} \rtimes L_3(2))$. Since G is simple, a normal subgroup $G \cap C_2$ is trivial, i.e. $G \subset C_2^{\oplus 3} \rtimes L_3(2)$. Again, for the same reason, one has $G \subset L_3(2)$ (and in fact equal). The Dynkin diagram $A_3^{\oplus 7}$, the complement of the identity component in $A_3^{\oplus 8}$, consists of the 21 simple roots r_{i1}, r_{i2}, r_{i3} ($1 \leq i \leq 7$) such that $(r_{i1}, r_{i2}) = (r_{i2}, r_{i3}) = 1$ but $(r_{i1}, r_{i3}) = 0$, $(r_{ik}, r_{jl}) = 0$ if $i \neq j$. Therefore the action G on these 21 simple roots satisfies $g(r_{i2}) = r_{g(i)2}$, where g in the right hand side is regarded as an element of the permutation of the seven components. Therefore $g(r_{i1})$ is either $r_{g(i)1}$ or $r_{g(i)3}$. Thus, G is embedded into a subgroup of the permutation subgroup $C_2^{\oplus 7} \rtimes S_7$ of the 14 simple roots r_{i1}, r_{i3}. Here, the indices $1, 3$ are so labelled that $\sigma \in S_7$ acts as $\sigma(r_{i1}) = r_{\sigma(i)1}$ and $\sigma(r_{i3}) = r_{\sigma(i)3}$ and the i-th factor of $C_2^{\oplus 7}$ acts as a permutation of r_{i1} and r_{i3}. Since G is simple and can not be embedded in $C_2^{\oplus 7}$, we have $G \subset S_7$. Therefore, $g(r_{i1}) = r_{g(i)1}$ and $g(r_{i3}) = r_{g(i)3}$. In conclusion, the orbits of the action G on the 24 simple roots are $\{r_{01}\}$, $\{r_{02}\}$, $\{r_{03}\}$, $\{r_{i1} | 1 \leq i \leq 7\}$, $\{r_{i2} | 1 \leq i \leq 7\}$, $\{r_{i3} | 1 \leq i \leq 7\}$. In particular, the 24 simple roots are divided into exactly 6 G-orbits. Since these 24 roots generate the Niemeier lattice $N = N(A_3^{\oplus 8})$ over $\mathbf{Q}$, we have then rank $N^G = 6$, a contradiction to (2.4)(1). Hence the Niemeier lattice for our G is $N(A_1^{\oplus 24})$.

$\square$

From now we set $N := N(A_1^{\oplus 24})$. By (2.5) and (1.4), we have

$$G \subset M_{24} \subset S_{24} = \mathrm{Aut}_{\mathrm{set}}(R) .$$

Here $R := \{r_i\}_{i=1}^{24}$ is the set of the simple roots of N and the last inclusion is the natural one explained in (1.4). This allows us to use the table of the cyclic types of elements of M_{24} given in [EDM] for its action on R. One may also talk about the orbit decomposition type of the action of G on R. Although we donot know much about how G is embedded in M_{24}, we can say at least the following:

Claim (2.6) (cf. [Mu2]). *The orbit decomposition type of R by G is either*

$$[14, 1, 1, 7, 1] \text{ or } [8, 7, 1, 7, 1] .$$

The Simple Group of Order 168 and K3 Surfaces   173

Proof. Since $N^G = 5$ by (2.4)(1), the 24 simple roots of N are divided into exactly 5 G-orbits. (See the last argument of the Claim (2.5).) Set the orbit decomposition type as $[a, b, c, d, e]$. Then $a + b + c + d + e = 24$ and each entry is less than 21. In addition, since G is simple and contains an element of order 7, if $a \leq 6$ then $a = 1$, for otherwise the natural non-trivial representation $G \to S_a$ would have a non-trivial kernel. Moreover, if $a \geq 7$, then a divides $168 = |G|$. This is because the action of G on each orbit is, by the definition, transitive. Therefore a is either 1, 7, 8, 12 or 14. If $a = 12$, then an element of order 7 in G has already 5 fixed points in this orbit. However, by [EDM], the cycle type of order 7 element in M_{24} is $(7)^3(1)^3$ and therefore has only 3 fixed points, a contradiction. Hence a is either 1, 7, 8 or 14. Clearly the same holds for b, c, d, e. Now by combining these, together with the equality $a + b + c + d + e = 24$, we obtain the result. $\square$

Proof of Key Lemma. By (2.4)(2), we may calculate $|\det N^G|$ instead. Let us renumber the 24 simple roots according to the orbit decompositions found in (2.6):

$$\{r_1, \cdots, r_{14}\} \cup \{r_{15}\} \cup \{r_{16}\} \cup \{r_{17}, \cdots, r_{23}\} \cup \{r_{24}\} \qquad (*)$$

or

$$\{r_1, \cdots, r_8\} \cup \{r_9, \cdots, r_{15}\} \cup \{r_{16}\} \cup \{r_{17}, \cdots, r_{23}\} \cup \{r_{24}\} . \qquad (**)$$

Consider the case $(*)$ first. Recall that rank $N^G = 5$ and $N^G = N \cap (A_1^{\oplus 24})^G \otimes \mathbf{Q}$. Then $b_1 = \sum_{i=1}^{14} r_i$, $b_2 = r_{15}$, $b_3 = r_{16}$, $b_4 = \sum_{i=17}^{23} r_i$ and $b_5 = r_{24}$ form the basis of $(A_1^{\oplus 24})^G$. Moreover, by (1.4), we see that $N^G/(A_1^{\oplus 24})^G$ consists of the elements of the form $\sum_{i \in I} b_i/2$, where the set of the simple roots $\{r_j\}$ appearing in the sum $\sum_{i \in I} b_i/2$ is either R, $\emptyset$, an Octad, complement of an Octad, or a Dodecad. However, by the shape of the orbit decomposition, there are no cases where a Dodecad appears. Therefore, in order to get an integral basis of N^G we may find out all the Octads and their complements appearing in the forms above.

Claim (2.7). *By reordering the three 1-element orbits if necessary, the union of the fourth and fifth orbits $\{r_{17}, r_{18}, \cdots, r_{23}, r_{24}\}$ forms an Octad.*

Proof. Let $\alpha \in G$ be an element of order 7. Then the cycle type of α (on R) is $(7)^3(1)^3$ [EDM]. In particular, a simple root x forms a 1-element orbit if $\alpha^k(x) = x$ for some k with $1 \leq k \leq 6$. Moreover, one can adjust the numbering of the roots in the fourth orbit $\{r_{17}, r_{18}, \cdots, r_{23}\}$ so as to be that $\alpha(r_i) = r_{i+1}$ $(17 \leq i \leq 22)$ and $\alpha(r_{23}) = r_{17}$. Let us consider the 5-element set $S := \{r_{17}, r_{18}, \cdots, r_{21}\}$. Then by (1.5), there is an Octad O such that $S \subset O$. We shall show that $O = \{r_{17}, r_{18}, \cdots, r_{23}, r_{24}\}$. For this purpose, assuming first that $r_{22}, r_{23} \notin O$, we shall derive a contradiction. Under this assumption, one has $\{r_{19}, r_{20}, r_{21}\} \subset O \cap \alpha^2(O)$ and $(O \cap \alpha^2(O)) \cap \{r_{17}, r_{18}, r_{22}, r_{23}\} = \emptyset$. (Here for the last equality we used the fact that $\alpha^2(r_{22}) = r_{17}$ and $\alpha^2(r_{23})$

174   Keiji Oguiso and De-Qi Zhang

$= r_{18}$.) The last equality also implies that $O \neq \alpha^2(O)$. Let us consider the symmetric difference $D = (O - \alpha^2(O)) \cup (\alpha^2(O) - O)$. Then $|D|$ must be either 8, 12, 16 whence $|O \cap \alpha^2(O)| = 4$ (See the proof of (1.5)). Thus, there exists an $x \in R$ such that $O \cap \alpha^2(O) = \{r_{19}, r_{20}, r_{21}, x\}$ and one has

$$O = \{r_{17}, r_{18}, r_{19}, r_{20}, r_{21}, x, y, z\} .$$

Here none of x, y, z lies in the fourth orbit. Since $x \in \alpha^2(O)$, one has either $x = \alpha^2(y)$ or $x = \alpha^2(x)$ (by changing the role of y and z if neccesary). In each case, we have $\alpha^2(z) \neq z$, whence $\alpha^k(z) \neq z$ for all k with $1 \leq k \leq 6$. In particular, z is in the first orbit.

Consider first the case where $x = \alpha^2(y)$. In this case, both x and y belong to the first orbit. Let us rename the elements in the first orbit so as to be that $\alpha(r_i) = r_{i+1}$ $(1 \leq i \leq 6)$, $\alpha(r_7) = r_1$; $\alpha(r_{7+i}) = r_{i+8}$ $(1 \leq i \leq 6)$, $\alpha(r_{14}) = r_8$ and that $y = r_1$ and $x = r_3$. Then, we have

$$O = \{r_{17}, r_{18}, r_{19}, r_{20}, r_{21}, r_3, r_1, z\} ,$$

and

$$\alpha^3(O) = \{r_{20}, r_{21}, r_{22}, r_{23}, r_{17}, r_6, r_4, \alpha^3(z)\} .$$

Considering the symmetric difference of O and $\alpha^3(O)$ as before, one finds that $O \cap \alpha^3(O)$ is a 4-element set. Therefore $|\{r_1, r_3, z\} \cap \{r_4, r_6, \alpha^3(z)\}| = 1$. Combining this with $\alpha^3(z) \neq z$, one has either $r_1 = \alpha^3(z)$, $r_3 = \alpha^3(z)$, $z = r_4$, or $z = r_6$. Hence O satisfies one of the following four:

$$O = \{r_{17}, r_{18}, r_{19}, r_{20}, r_{21}, r_1, r_3, r_5\} \tag{1}$$

$$O = \{r_{17}, r_{18}, r_{19}, r_{20}, r_{21}, r_1, r_3, r_7\} \tag{2}$$

$$O = \{r_{17}, r_{18}, r_{19}, r_{20}, r_{21}, r_1, r_3, r_4\} \tag{3}$$

$$O = \{r_{17}, r_{18}, r_{19}, r_{20}, r_{21}, r_1, r_3, r_6\} . \tag{4}$$

In the case (1), one calculates

$$\alpha^2(O) = \{r_{19}, r_{20}, r_{21}, r_{22}, r_{23}, r_3, r_5, r_7\}$$

and has then $O \cap \alpha^2(O) = \{r_{19}, r_{20}, r_{21}, r_3, r_5\}$. In particular, the two Octads O and $\alpha^2(O)$ share 5 elements in common. Then, by the Steiner property (1.5), we would have $O = \alpha^2(O)$, a contradiction. By considering $O \cap \alpha(O)$ in the cases (2), (3) and $O \cap \alpha^2(O)$ in the case (4), we can derive a contradiction in the same manner, too. Thus, the case $x = \alpha^2(y)$ is imposible.

Next we consider the case where $\alpha^2(x) = x$. In this case, this x forms a 1-element orbit and satisfies $\{r_{18}, r_{19}, r_{20}, r_{21}, x\} \subset O \cap \alpha(O)$. However, the Steiner property would then imply $O = \alpha(O)$, whence $O = \alpha^2(O)$, a contradiction.

Therefore, the Octad O satisfies either $r_{22} \in O$ or $r_{23} \in O$, i.e.

$$\{r_{17}, r_{18}, r_{19}, r_{20}, r_{21}, r_{22}\} \subset O ,$$

The Simple Group of Order 168 and K3 Surfaces   175

or
$$\{r_{23}, r_{17}, r_{18}, r_{19}, r_{20}, r_{21}\} \subset O .$$

Then one has either
$$\{r_{18}, r_{19}, r_{20}, r_{21}, r_{22}\} \subset O \cap \alpha(O) ,$$

or
$$\{r_{17}, r_{18}, r_{19}, r_{20}, r_{21}\} \subset O \cap \alpha(O) .$$

Hence, by the Steiner property, we have $O = \alpha(O)$, whence $O = \alpha^k(O)$ for all k. This implies that the Octad O is of the form

$$O = \{r_{17}, r_{18}, r_{19}, r_{20}, r_{21}, r_{22}, r_{23}, x\}$$

for some root x. Since $O = \alpha(O)$, we have also $\alpha(x) = x$. Hence this x forms a 1-element orbit set. $\qquad\square$

By this Claim, one has
$$b_6 := (b_4 + b_5)/2 \in N^G$$

and also
$$b_7 := (b_1 + b_2 + b_3)/2 \in N^G .$$

By the remark before Claim (2.7) and the Steiner property (1.5), we also see that there are no other Octads appearing in the sum $\sum_{i \in I} b_i/2$. Since $\sum_{i=1}^5 b_i/2 = b_6 + b_7$, the seven elements $b_1, \cdots, b_7$ then generate N^G over $\mathbf{Z}$. Moreover, since $b_1 = 2b_7 - b_2 - b_3$ and $b_4 = 2b_6 - b_5$, we finally see that b_7, b_2, b_3, b_6, b_5 form an integral basis of N^G. Using $(r_i, r_j) = -2\delta_{ij}$, we find that the intersection matrix of N^G under this basis is given as A below:

$$A = \begin{pmatrix} -8 & -1 & -1 & 0 & 0 \\ -1 & -2 & 0 & 0 & 0 \\ -1 & 0 & -2 & 0 & 0 \\ 0 & 0 & 0 & -4 & -1 \\ 0 & 0 & 0 & -1 & -2 \end{pmatrix}, \quad B = \begin{pmatrix} -4 & 0 & 0 & 0 & 0 \\ 0 & -4 & -1 & 0 & 0 \\ 0 & -1 & -2 & 0 & 0 \\ 0 & 0 & 0 & -4 & -1 \\ 0 & 0 & 0 & -1 & -2 \end{pmatrix} .$$

Let us consider next the case $(**)$. Since G acts on the first 8-element orbit transitively, for each root r_i in the first orbit, one can find an element $\alpha' \in G$ of order 7 such that $\alpha'(r_i) \neq r_i$. Now, by the same argument based on the fact that the cycle type of order 7 element is $(7)^3(1)^3$ and the Steiner property (1.5) (together with the remark above), one finds that (after reordering the two 1-element orbits) the union of the second and third orbits and the union of the fourth and fifth orbits are both Octads. This also implies that the first orbit is an Octad. Then again by the same argument as in the previous case, one can easily see that the five elements $b_1 := \sum_{i=1}^8 r_i/2$, $b_2 := \sum_{i=9}^{16} r_i/2$, $b_3 = r_{16}$, $b_4 = \sum_{i=17}^{24} r_i/2$, and $b_5 = r_{24}$ form an integral basis of N^G in the second case, and that the intersection matrix of N^G under this basis is given as B above. Clearly $|\det N^G| = 196$ in both cases. This proves Key Lemma. $\qquad\square$

176   Keiji Oguiso and De-Qi Zhang

Next we shall study possible extensions $G \subset \tilde{G}$. Recall that we have already shown that $\tilde{G}/G \simeq \mu_I$, where I is either 1, 2, 3, 4 or 6, and that $\tilde{G}/G$ acts faithfully on T_X. Set $\tilde{G}/G = \langle \tau \rangle$.

The next Lemma is valid for any $G \subset \tilde{G}$ if $\tilde{G}/G$ acts faithfully on T_X and if rank $T_X = 2$.

Lemma (2.8).

(1) *Assume that* $\mathrm{ord}(\tau) = 3$. *Then, as* $\mathbf{Z}[\tau^*]$-*modules, one has*

$$T_X \simeq \mathbf{Z}[x]/(x^2 + x + 1) \, ,$$

where τ^* *acts on the right hand side by the multiplication by* x. *In particular, one can take an integral basis* e_1, e_2 *of* T_X *such that* $\tau^*(e_1) = e_2$ *and* $\tau^*(e_2) = -(e_1 + e_2)$. *Moreover, under this basis, the intersection matrix of* T *is of the form* $((e_i, e_j)) = \begin{pmatrix} 2m & -m \\ -m & 2m \end{pmatrix}$.

(2) *Assume that* $\mathrm{ord}(\tau) = 4$. *Then, as* $\mathbf{Z}[\tau^*]$-*modules, one has*

$$T_X \simeq \mathbf{Z}[x]/(x^2 + 1) \, ,$$

where τ^* *again acts on the right hand side by the multiplication by* x. *In particular, one can take an integral basis* e_1, e_2 *of* T_X *such that* $\tau^*(e_1) = e_2$ *and* $\tau^*(e_2) = -e_1$. *Moreover, under this basis, the intersection matrix of* T *is of the form* $((e_i, e_j)) = \begin{pmatrix} 2m & 0 \\ 0 & 2m \end{pmatrix}$.

Proof. The first part of the two assertions is due to the fact that $\mathbf{Z}[\zeta_3]$ and $\mathbf{Z}[\zeta_4]$ are both PID. (For more detail, see for example [MO].) By taking an integral basis of T_X corresponding to 1 and x (in the right hand side), one obtains the desired representation of the action of τ^*. Now combining this with $(\tau^*(a), \tau^*(b)) = (a, b)$, we get the intersection matrix as claimed.   □

The next Claim completes the first assertion of the main Theorem:

Claim (2.9). $I \neq 6$.

A similar method is exploited in [Ko2] and [OZ] in other settings with somewhat different flavours and will be also adopted in the next Claim (2.10).

Proof. Assuming to the contrary that $\tilde{G}/G = \langle g \rangle \simeq \mu_6$, we shall derive a contradiction. By (2.1), one has $L^G \supset T_X \oplus \mathbf{Z}H$, where H is the primitive ample class invariant under G. Set $(H^2) = 2n$. Since T_X is primitive in L^G, we can choose an integral basis of L^G as e_1, e_2 and $e_3 = (aH + be_1 + ce_2)/\ell$, where e_1 and e_2 are the integral basis of T_X found in (2.8)(1) applied for $\tau := g^2$ and ℓ and a, b, c are integers such that $(\ell, a) = 1$. Then

$$L^G/(T_X \oplus \mathbf{Z}H) = \langle \overline{e_3} \rangle \simeq C_\ell \, ,$$

The Simple Group of Order 168 and K3 Surfaces   177

where $\overline{e_3} = e_3 \bmod (T_X \oplus \mathbf{Z}H)$. Since H is also stable under $\tilde{G}$, we have $\tau^*(\overline{e_3}) = \overline{e_3}$ and

$$\tau^*(be_1 + ce_2)/\ell \equiv (be_1 + ce_2)/\ell \bmod T_X .$$

On the other hand, by the choice of e_1, e_2, we calculate

$$\tau^*(be_1 + ce_2)/\ell = (-ce_1 + (b - c)e_2)/\ell .$$

Therefore, $b \equiv -c$ and $c \equiv b - c \bmod \ell$. In particular, $b \equiv -c$ and $3b \equiv 3c \equiv 0$ $\bmod \ell$. This, together with the primitivity of $\mathbf{Z}H$ in L^G, implies that $\ell = 1$ or 3, that is, $[L^G : T_X \oplus \mathbf{Z}H] = 1$ or 3.

If $\ell = 1$, we have $L^G = T_X \oplus \mathbf{Z}H$ and $196 = 6m^2n$. However 6 is not a divisor of 196, a contradiction.

Consider the case $\ell = 3$. Then, (by using the primitivity of H in L^G and by adding an element of T_X to e_3 if necessary), we can take one of $(\pm H \pm (e_2 - e_3))/3$ as e_3. Put $\sigma := g^3$. Then $\sigma^*H = H$ and $\sigma^*|T_X = -\mathrm{id}$. Using these two equalities, we calculate

$$\sigma^*(e_3) = \sigma^*((\pm H \pm (e_2 - e_3))/3) = (\pm H \mp (e_2 - e_3))/3 .$$

However, one would then have

$$\pm 2(e_2 - e_3)/3 = e_3 - \sigma^*(e_3) \in L^G ,$$

a contradiction to the primitivity of T_X in L^G. Hence $I \neq 6$.   □

From now, we consider the maximum case $\tilde{G}/G = \langle \tau \rangle \simeq \mu_4$.

Claim (2.10). $(H^2) = 4$.

Proof. As in (2.9), one has $L^G \supset T_X \oplus \mathbf{Z}H$, where H is the primitive ample class invariant under G. Set $(H^2) = 2n$. Since T_X is primitive in L^G, we can choose an integral basis of L^G as e_1, e_2 and $e_3 = (aH + be_1 + ce_2)/\ell$, where e_1 and e_2 are the integral basis of T_X found in (2.8)(2) and ℓ and a, b, c are integers such that $(\ell, a) = 1$. Then, as in (2.9), we have

$$L^G/(T_X \oplus \mathbf{Z}H) = \langle \overline{e_3} \rangle \simeq C_\ell ,$$

where $\overline{e_3} = e_3 \bmod (T_X \oplus \mathbf{Z}H)$. Since H is also stable under $\tilde{G}$, we have $\tau^*(\overline{e_3}) = \overline{e_3}$ and

$$\tau^*(be_1 + ce_2)/\ell \equiv (be_1 + ce_2)/\ell \bmod T_X .$$

On the other hand, by the choice of e_1, e_2, we calculate

$$\tau^*(be_1 + ce_2)/\ell = (be_2 - ce_1)/\ell .$$

178   Keiji Oguiso and De-Qi Zhang

Therefore, $b \equiv c$ and $c \equiv -b \bmod \ell$. In particular, $b \equiv c$ and $2b \equiv 2c \equiv 0$ $\bmod \ell$. This, together with the primitivity of $\mathbf{Z}H$ in L^G, implies that $\ell = 1$ or 2, that is, $[L^G : T_X \oplus \mathbf{Z}H] = 1$ or 2.

In the first case, we have $L^G = T_X \oplus \mathbf{Z}H$ and $196 = 8m^2n$. However 8 is not a divisor of 196, a contradiction.

In the second case, we have $2^2 \cdot 196 = 8m^2n$, i.e. $m^2n = 2 \cdot 7^2$. Then (m, n) is either $(1, 2 \cdot 7^2)$ or $(7, 2)$. In the first case we have $X = X_4$ by the result of Shioda and Inose [SI], where X_4 is the minimal resolution of $(E_{\sqrt{-1}} \times E_{\sqrt{-1}})/\langle \mathrm{diag}(\sqrt{-1}, -\sqrt{-1}) \rangle$. However, according to the explicit description of $\mathrm{Aut}(X_4)$ by Vinberg [Vi], X_4 has no automorphism of order 7, a contradiction. Therefore, only the second case can happen and one has $(H^2) = 2n = 4$ (and $T_X = \mathrm{diag}(14, 14)$). $\square$

Now the following Claim will complete the proof of the main Theorem.

Claim (2.11). *$(X, \tilde{G})$ is isomorphic to $(X_{168}, L_2(7) \times \mu_4)$ defined in the Introduction.*

Proof. Since $S_X^G = \mathbf{Z}H$, $|H|$ has no fixed components. Indeed, the fixed part of $|H|$ must be also G-stable but is of negative definite [SD]. Therefore, the ample linear system $|H|$ is free [ibid.]. Note that $\dim|H| = 3$ by the Riemann-Roch formula and the fact $(H^2) = 4$. Then $|H|$ defines a morphism $\Phi := \Phi_{|H|} : X \to \mathbf{P}^3$. This Φ is either an embedding to a quartic surface S or a finite double cover of an integral quadratic surface Q. Note that $\tilde{G}$ acts on the image as a projectively linear transformation. Moreover, the action of G on the image is faithful even in the second case, because G is simple. Recall that the degrees of the projectively linear irreducible representations of G are $1, 3, 4, 6, 7, 8$ [ATLAS] and that the two 3-dimensional irreducible representations are transformed by the outer automorphism of G. Then the action of G on the image is induced by the irreducible decomposition $\mathbf{C}^4 = V_1 \oplus V_3$ or $\mathbf{C}^4 = V_4$. (Note that the action of G on $\mathbf{P}^3$ is linearized in the first case but not in the second case. More precisely, in the second case, only the action of $\mathrm{SL}(2, \mathbf{F}_7)$, i.e. the central extension of G corresponding to the Schur multiplier 2, is linearized.)

Let us first consider the second case. By [Ed, pp. 198–200 and p. 166], G has no invariant hypersurface of degree 2 but only one invariant hypersurface of degree 4:

$$f := x_0^4 + 6\sqrt{2}x_0x_1x_2x_3 + x_1x_2^3 + x_2x_3^3 + x_3x_1^3 = 0 \,,$$

where the homogeneous coordinates $[x_0 : x_1 : x_2 : x_3]$ are chosen in such a way that an order 7-element of G, say α, is represented by the following diagonal matrix:

$$A = \begin{pmatrix} 1 & 0 & 0 & 0 \\ 0 & \zeta_7^6 & 0 & 0 \\ 0 & 0 & \zeta_7^3 & 0 \\ 0 & 0 & 0 & \zeta_7^5 \end{pmatrix} \,.$$

The Simple Group of Order 168 and K3 Surfaces   179

Thus, Φ is an embedding and one has $S = (f = 0)$. Recall that $\tilde{G}$ fits in with the exact sequence

$$1 \to G \to \tilde{G} \to \mu_4 \to 1 \,,$$

where the last map is the representation of $\tilde{G}$ on $H^0(S, \Omega_S^2) = \mathbf{C}\omega_S$. Let $g \in \tilde{G}$ be a lift of $\zeta_4 \in \mu_4$. Then $g^*\omega_S = \zeta_4\omega_S$. Since G is a normal subgroup of $\tilde{G}$, one can define an element $c_g \in \mathrm{Aut}_{\mathrm{group}}(G)$ by $G \ni x \mapsto g^{-1}xg \in G$ for all $x \in G$. Since $\mathrm{Out}(G) \simeq C_2$ [ATLAS], $(c_g)^2$ is then an inner automorphism of G, i.e. there exists an element $y \in G$ such that $g^{-2}xg^2 = y^{-1}xy$ for all $x \in G$. Set $k = g^2 y^{-1}$. Then one has $k^{-1}xk = x$ for all $x \in G$, $k^*\omega_S = -\omega_S$ and $2|\mathrm{ord}(k)$. Therefore, replacing k by k^{2l+1} if necessary, one obtains an element $h \in \tilde{G}$ such that $h^{-1}xh = x$ for all $x \in G$, $h^*\omega_S = -\omega_S$ and $\mathrm{ord}(h) = 2^n$. Choose a representative (h_{ij}) of h in $\mathrm{GL}(4, \mathbf{C})$. Then for A above one has $(h_{ij})A(h_{ij})^{-1} = cA$ $(c \in \mathbf{C})$ in $\mathrm{GL}(4, \mathbf{C})$. This implies $c^{2^n} = 1$ and $\{1, \zeta_7^6, \zeta_7^3, \zeta_7^5\} = \{c, c\zeta_7^6, c\zeta_7^3, c\zeta_7^5\}$. Thus $c = 1$ and one has $(h_{ij})A(h_{ij})^{-1} = A$, i.e. $(h_{ij})A = A(h_{ij})$ in $\mathrm{GL}(4, \mathbf{C})$. This readily implies that (h_{ij}) is also a diagonal matrix, and one may write that

$$(h_{ij}) = \begin{pmatrix} 1 & 0 & 0 & 0 \\ 0 & h_{11} & 0 & 0 \\ 0 & 0 & h_{22} & 0 \\ 0 & 0 & 0 & h_{33} \end{pmatrix}.$$

Then, by the shape of f and by the fact that $h^*f = c'f$ for some $c' \in \mathbf{C}$, one has $c' = 1$ and $h_{11}h_{22}h_{33} = 1$. However, this would yield $(h_{ij})^*f = f$ and $\det(h_{ij}) = 1$, thereby $h^*\omega_S = \omega_S$, a contradiction to the previous equality $h^*\omega_S = -\omega_S$. Hence the second case cannot happen.

Let us next consider the first case. Let us choose the homogeneous coordinates $[x_0 : x_1 : x_2 : x_3]$ such that x_0 is the coordinate of V_1 and that x_i $(1 \le i \le 3)$ are the coordinates of V_3 described as in the Introduction.

Let us first consider the case where Φ is a double covering. Write an equation of Q as

$$ax_0^2 + x_0 f_1(x_1, x_2, x_3) + f_2(x_1, x_2, x_3) = 0 \,.$$

Since G is simple and acts on x_0 as an identity, we have $g^*(f_1) = f_1$ and $g^*(f_2) = f_2$ for all $g \in G$. Since there are no non-trivial G-invariant linear and quadratic forms in three variables [Bu, §267], one has $f_1 = f_2 = 0$. However, then Q is not integral, a contradiction. Hence Φ is an embedding. Let us write an equation of S as

$$\begin{aligned} F = {} & ax_0^4 + x_0^3 f_1(x_1, x_2, x_3) \\ & + x_0^2 f_2(x_1, x_2, x_3) + x_0 f_3(x_1, x_2, x_3) + f_4(x_1, x_2, x_3) = 0 \,. \end{aligned}$$

Then $g^*(f_i) = f_i$ for all $g \in G$ and all $1 \le i \le 4$. Thus by [ibid.], we have $f_i = 0$ for all $1 \le i \le 3$, $f_4(x_1, x_2, x_3) = b(x_1 x_2^3 + x_2 x_3^3 + x_3 x_1^3)$ and

$F(x_0, x_1, x_2, x_3) = ax_0^4 + b(x_1x_2^3 + x_2x_3^3 + x_3x_1^3)$. Here $a \neq 0$ and $b \neq 0$, because S is non-singular. Therefore, by multiplying coordinates suitably, one may adjust the equation of S as

$$x_0^4 + x_1x_2^3 + x_2x_3^3 + x_3x_1^3 = 0 \ .$$

Hence $S \simeq X_{168}$ and $L_2(7) \times \mu_4$ acts on S as described in the Introduction, where by the construction, the action $L_2(7)$ also coincides with the given action of G on X. Since S is a K3 surface and has no non-zero global holomorphic vector fields, the projectively linear automorphism group G'' of $S \subset \mathbf{P}^3$ is finite. This G'' satisfies $G'' \supset L_2(7) \times \mu_4$ and $G'' \supset \tilde{G}$. Thus $4 \cdot 168 | |G''|$ and one has $|G''| = |L_2(7) \times \mu_4| = |\tilde{G}| = 4 \cdot 168$. Hence $\tilde{G} = G'' = L_2(7) \times \mu_4$ as projectively linear automorphism groups of S. Now we are done. $\square$

Remark (2.12).

(1) By the proof of (2.10), we have that $T_X = \mathrm{diag}(14, 14)$ if $|\tilde{G}/G| = 4$. In particular, $T_{X_{168}} = \mathrm{diag}(14, 14)$.
(2) Now one can easily check that the two K3 surfaces X_{168} and X'_{168} in the Introduction are not isomorphic to each other. Note that X'_{168} has a G-stable ample class H of degree 2. Therefore, if $T_{X'_{168}}$ is isomorphic to $T_{X_{168}} = \mathrm{diag}\ (14, 14)$, then $[L^G : T_{X'_{168}} \oplus \mathbf{Z}H]^2 = 2 \cdot 14^2/196 = 2$, a contradiction.

References

[ATLAS] J. H. Conway et al., *ATLAS of Finite Groups*, Clarendon Press, Oxford, 1985.

[Bu] W. Burnside, *Theory of Groups of Finite Order*, Dover Publications, Inc., New York, 1955.

[CS] J. H. Conway and N. J. A. Sloane, Sphere packings, lattices and groups, *Grundlehren der Math. Wissenschaften* (1999), Springer-Verlag, New York.

[Ed] W. L. Edge, The Klein group in three dimensions, *Acta Math.* **79** (1947), 153–222.

[EDM] *Encyclopedic Dictionary of Mathematics*, Vol. II (1947) (Appendix B, Table 5.I), Translated from the second Japanese edition, MIT Press, Cambridge, Mass.-London, 1977, 885–1750.

[Ko1] S. Kondo, Niemeier Lattices, Mathieu groups, and finite groups of symplectic automorphisms of $K3$ surfaces, *Duke Math. J.* **92** (1998), 593–598.

[Ko2] S. Kondo, The maximum order of finite groups of automorphisms of $K3$ surfaces, *Amer. J. Math.* **121** (1999), 1245–1252.

[MO] N. Machida and K. Oguiso, On $K3$ surfaces admitting finite non-symplectic group actions, *J. Math. Sci. Univ. Tokyo* **5** (1998), 273–297.

[Ma] D. Markushevich, Resolution of $\mathbf{C}^3/H_{168}$, *Math. Ann.* **308** (1997), 279–289.

The Simple Group of Order 168 and K3 Surfaces   181

[Mu1]   S. Mukai, Finite groups of automorphisms of $K3$ surfaces and the Mathieu group, *Invent. Math.* **94** (1988), 183–221.

[Mu2]   S. Mukai, Lattice theoretic construction of symplectic actions on $K3$ surfaces (an Appendix to [Ko1]), *Duke Math. J.* **92** (1998), 599–603.

[Ni]   V. V. Nikulin, Finite automorphism groups of Kahler $K3$ surfaces, *Trans. Moscow Math. Soc.* **38** (1980), 71–135.

[Og]   K. Oguiso, On the complete classification of Calabi-Yau threefolds of Type III_0, in: *Higher Dimensional Complex Varieties*, de Gruyter, 329–340, 1996.

[OZ]   K. Oguiso and D. -Q. Zhang, On the most algebraic $K3$ surfaces and the most extremal log Enriques surfaces, *Amer. J. Math.* **118** (1996), 1277–1297.

[SI]   T. Shioda and H. Inose, On singular $K3$ surfaces, in: *Complex Analysis and Algebraic Geometry*, 119–136, Iwanami Shoten, Tokyo, 1977.

[SD]   B. Saint-Donat, Projective models of $K3$ surfaces, *Amer. J. Math.* **96** (1974), 602–639.

[Vi]   E. B. Vinberg, The two most algebraic $K3$ surfaces, *Math. Ann.* **265** (1983), 1–21.

Added in Proof

In this added in proof, we continue to employ the same notation, eg. $G = L_2(7)$. After submitting our note, we noticed the following Propositions. These answer questions asked by I. Dolgachev around January 2001:

Proposition 1. *In the main Theorem (1), one has $|\tilde{G}/G| \neq 3$.*

Proposition 2 *There are infinitely many non-isomorphic algebraic K3 surfaces X such that $G \subset \mathrm{Aut}(X)$.*

In Proposition 2, we already know that there are at most countably many such algebraic K3 surfaces X (Remark (2.2)).

Proof of Proposition 1

Assuming to the contrary, we shall derive a contradiction. Recall that $\mathrm{rk}S_X = 20$ and $\mathrm{rk}T_X = 2$ if X is an algebraic K3 surface admitting an $L_2(7)$-action.

Claim 1. $\tilde{G} \simeq G \times \mu_3$.

Proof. Let $\tilde{g} \in \tilde{G}$ be an element such that $\tilde{g}^*\omega_X = \zeta_3\omega_X$. Replacing $\tilde{g}$ by its power $\tilde{g}^n$ such that $(n, 3) = 1$, one may assume that $\mathrm{ord}(\tilde{g}) = 3^m$ for some positive integer m. Since G is a normal subgroup of $\tilde{G}$, one has $c_{\tilde{g}}(x) := \tilde{g}^{-1}x\tilde{g} \in G$ if $x \in G$, thereby $c_{\tilde{g}} \in \mathrm{Aut}(G)$. Since $\mathrm{Out}(G) = C_2$ and $\mathrm{ord}(\tilde{g}) = 3^m$, one has then $c_{\tilde{g}} \in \mathrm{Inn}(G)$, i.e. there is $y \in G$ such that $\tilde{g}^{-1}x\tilde{g} = y^{-1}xy$ for all $x \in G$. Now, replacing $\tilde{g}$ by $\tilde{g}y^{-1}$, one has $\tilde{g}^{-1}x\tilde{g} = x$ for all $x \in G$ and $\tilde{g}^*\omega_X = \zeta_3\omega_X$. Then $\tilde{g}^3 \in G$ and is also in the center of G. Note that the center of G is $\{\mathrm{id.}\}$ for G being simple, non-commutative. Then $\tilde{g}^3 = \mathrm{id.}$ and $\tilde{g}$ gives a desired splitting of the exact sequence $1 \to G \to \tilde{G} \to \mu_3 \to 1$. □

Claim 2. T_X *is isomorphic to* $\begin{pmatrix} 14 & -7 \\ -7 & 14 \end{pmatrix}$ *(and the degree of the primitive invariant polarization is* 12*).*

Proof. The argument is the same as in (2.10) but is based on the following three facts instead: Lemma (2.8)(1) (instead of (2)); $[L^G : T_X \oplus \mathbf{Z}H] = 3$ (In the course of proof of (2.9)); and the fact that X admits no automorphisms of order 7 if T_X is isomorphic to $\begin{pmatrix} 2 & -1 \\ -1 & 2 \end{pmatrix}$ [Vi]. Further details are left to the readers. $\qquad\square$

Let us next consider the irreducible decomposition of the natural linear action of G on $S_X \otimes \mathbf{C}$. We adopt the notation in [ATLAS]. We denote the irreducible representation of the character χ_i in [ibid.] by V_i. Then, dimV_i is 1, 3, 3, 6, 7, 8 if $i = 1, 2, 3, 4, 5, 6$.

Claim 3. *The irreducible decomposition of the natural action of G on $S_X \otimes \mathbf{C}$ is:*

$$S_X \otimes \mathbf{C} = V_1 \oplus V_4 \oplus V_4' \oplus V_5 ,$$

where V_4' is a copy of V_4.

Proof. Set $S_X \otimes \mathbf{C} = \oplus_{i=1}^6 V_i^{\oplus n_i}$. Here we have $n_1 = 1$ by (2.1)(1) and also $n_2 = n_3$ as V_2 and V_3 are (complex) conjugate to each other. Then by counting the dimension, one has

$$20 = n_1 + 6n_2 + 6n_4 + 7n_5 + 8n_6 .$$

Recall that for $g \in G$, one has

- $\chi_{\text{top}}(X^g) = 2 + \text{tr}(g^*|S_X) + \text{tr}(g^*|T_X) = 4 + \text{tr}(g^*|S_X)$
- $\chi_{\text{top}}(X^g)$ is 8, 6, 4, 3 if $\text{ord}(g) = 2, 3, 4, 7$.

The first equality is nothing but the Lefschetz fixed point formula and the second one is the result of Nikulin [Ni]. By applying these two formula and the character table in [ATLAS] for elements of G of order 2, 3, 4, 7 respectively, one obtains:

$$n_1 - 2n_2 + 2n_4 - n_5 = 4,$$

$$n_1 + n_5 - n_6 = 2,$$

$$n_1 + 2n_2 - n_5 = 0,$$

$$n_1 - n_2 - n_4 + n_6 = -1 .$$

Combining all of these together, we find that $n_1 = 1$, $n_2 = n_3 = 0$, $n_4 = 2$, $n_5 = 1$, $n_6 = 0$. This gives the result. $\qquad\square$

Let τ be a generator of μ_3. We regard $\tau \in \tilde{G}$ through the isomorphism found in Claim 1. Since $\tau^{-1}g\tau = g$ for all $g \in G$ and τ is of order 3, one has $\tau(V_i) = V_i$ for each i and $\tau(V_4') = V_4'$. Then by Schur's Lemma, $\tau|V_i$, $\tau|V_i'$ are all scalar multiplications. Note that $\tau|V_1 = \text{id.}$ for $S_X^{\tilde{G}} \neq \{0\}$. Set $\tau|V_4 = \zeta_3^a$, $\tau|V_4' = \zeta_3^b$, and $\tau|V_5 = \zeta_3^c$, where $a, b, c \in \mathbf{Z}/3$. Since $\tau|S_X \otimes \mathbf{C}$ is defined over S_X, the multiplicities of eigenvalues ζ_3 and ζ_3^2 of $\tau|S_X \otimes \mathbf{C}$ are the same. Therefore, $c = 0$ and $a + b = 0$, i.e. (a, b, c) is either $(0, 0, 0)$ or $(1, 2, 0)$.

Consider first the case $(a, b, c) = (0, 0, 0)$. Then $\tau^*|S_X = \text{id.}$ (and $\tau^*\omega_X = \zeta_3\omega_X$). However T_X would then be a 3-elementary lattice by [OZ2, Lemma (1.3)], a contradiction to Claim 2.

Let us consider next the case $(a, b, c) = (1, 2, 0)$. Let g be an order 2 element of G. Set $h := \tau g \in \tilde{G}$. Then h is of order 6 and satisfies $h^3 = g$. In particular, one has $X^h \subset X^g$. Here X^g is an 8-point set by [Ni]. Thus, X^h also consists of finitely many points (possibly empty), thereby $\chi_{\text{top}}(X^h) \geq 0$. On the other hand, by using the Lefschetz fixed point formula, the fact $(a, b, c) = (1, 2, 0)$, Claim 3, and the character table [ATLAS], one calculates

$$\begin{aligned}
\chi_{\text{top}}(X^h) &= 2 + \text{tr}(h^*|S_X) + \text{tr}(h^*|T_X) \\
&= 2 + \{1 + \text{tr}(g|V_4)(\zeta_3 + \zeta_3^2) + \text{tr}(g|V_5) \cdot 1\} + (\zeta_3 + \zeta_3^2) \\
&= 2 + 1 + 2 \cdot (-1) + (-1) \cdot 1 + (-1) = -1 < 0 ,
\end{aligned}$$

a contradiction to the previous inequality $\chi_{\text{top}}(X^h) \geq 0$. Now we are done. $\qquad\square$

Proof of Proposition 2

Let Λ be the K3 lattice, i.e. the lattice $U^{\oplus 3} \oplus E_8(-1)^{\oplus 2}$. Choosing a marking $\tau : H^2(X_{168}, \mathbf{Z}) \simeq \Lambda$, we set $\Lambda_0 := \tau(H^2(X_{168}, \mathbf{Z})^G)$. This Λ_0 is a positive definite even lattice of rank 3 whose $\mathbf{R}$ linear extension is spanned by the image of the classes of the invariant ample class η, $\text{Re}(\omega_{X_{168}})$ and $\text{Im}(\overline{\omega}_{X_{168}})$. Fixing the Ricci flat Kähler metric g on X_{168} such that the cohomology class of the associated $(1, 1)$-form is η and regarding $\Lambda_0 \otimes \mathbf{R}$ as a HK 3-space, one obtains the twister family $f : \mathcal{X} \to \mathbf{P}^1$ with $\mathcal{X}_0 = X_{168}$ (See for instance [Be, Exposé X]). This f is a smooth non-isotrivial family of (not necessarily algebraic) K3 surfaces $\mathcal{X}_t$. Denote by ω_t a nowhere vanishing holomorphic two form on $\mathcal{X}_t$ and by η_t the Kähler class on $\mathcal{X}_t$ associated with g. Let us fix a marking $\tilde{\tau} : R^2 f_* \mathbf{Z} \simeq \Lambda$ such that $\tilde{\tau}_0 = \tau$. (Here we used the fact that $\mathbf{P}^1$ is simply-connected.) By the construction, for each $t \in \mathbf{P}^1$, the HK 3-space $\Lambda_0 \otimes \mathbf{R}$ is spanned by the three vectors $\tilde{\tau}_t(\eta_t)$, $\tilde{\tau}_t(\omega_t)$ and $\tilde{\tau}_t(\overline{\omega}_t)$. In particular, we have $\rho(\mathcal{X}_t) \geq 19$ for all $t \in \mathbf{P}^1$. There are then infinitely many t such that $\rho(\mathcal{X}_t) = 20$ by [Og2]. Such $\mathcal{X}_t$ is necessarily algebraic by [SI] and $\tilde{\tau}_t(T_{\mathcal{X}_t})$ is a primitive sublattice of rank 2 of Λ_0. Using the marking $\tilde{\tau}$, let us define the

184   Keiji Oguiso and De-Qi Zhang

(real) period map:

$$\iota \circ p : \mathbf{P}^1 \to \{[\omega] \in \mathbf{P}(\Lambda \otimes \mathbf{C}) | (\omega, \omega) = 0, (\omega, \overline{\omega}) > 0\}$$
$$\simeq \{T \in \mathrm{Gr}^+(2, \Lambda \otimes \mathbf{R}) | T \text{ is positive definite}\} .$$

Since p is a complex analytic map and $\mathbf{P}^1$ is compact, $\iota \circ p$ is finite. Therefore, for each rank two sublattice T (of Λ_0), there are at most finitely many $t \in \mathbf{P}^1$ such that $\tilde{\tau}_t(T_{\mathcal{X}_t}) = T$. Hence, by the global Torelli Theorem for K3 surfaces with the maximum Picard number 20 ([SI]), the family f contains infinitely many non-isomorphic algebraic K3 surfaces. Now the following Claim completes the proof of Proposition 2:

Claim $\mathcal{X}_t$ satisfies $G \subset \mathrm{Aut}(\mathcal{X}_t)$ for all $t \in \mathbf{P}^1$.

This Claim also shows that there are uncountably many (non-algebraic) K3 surfaces admitting $L_2(7)$-actions.

Proof. Since $G|(\tilde{\tau}_0)^{-1}(\Lambda_0) = \{\mathrm{id}.\}$ and $\eta_t, \mathrm{Re}(\omega_t), \mathrm{Im}(\omega_t) \in (\tilde{\tau}_t)^{-1}(\Lambda_0 \otimes \mathbf{R})$, we see that $(\tilde{\tau}_t)^{-1} \circ (\tilde{\tau}_0) \circ G \circ (\tilde{\tau}_0)^{-1} \circ (\tilde{\tau}_t)$ is an effective Hodge isometry of $H^2(\mathcal{X}_t, \mathbf{Z})$. This action is also faithful, beacuse G is simple and $G|(\tilde{\tau}_0)^{-1}(\Lambda) \neq \{\mathrm{id}.\}$. Hence $G \subset \mathrm{Aut}(\mathcal{X}_t)$ for each $t \in \mathbf{P}^1$ by the global Torelli Theorem for K3 surfaces. $\square$

References added

[Be] A. Beauville, Geométrie des surfaces K3, *Astérisque* **126** (1985).

[Og2] K. Oguiso, Picard numbers in a fmily of hyperkähler manifolds; a supplement to the article of R. Borcherds, L. Katzarkov, T. Pantev, N. I. Shepherd-Barron, (2000) preprint submitted.

[OZ2] K. Oguiso and D. -Q. Zhang, On Vorontsov's Theorem on $K3$ surfaces with non-symplectic group actions, *Proc. AMS* **128** (2000), 1571–1580.

[Ya] S. T. Yau, On the Ricci curvature of a compact Kähler manifold and the complex Monge-Ampére equations I, *Comm. Pure Appl. Math.* **31** (1978), 339 –411.

A Precise L^2 Division Theorem

Takeo Ohsawa

Department of Mathematics, Nagoya University, Chikusa-ku, Nagoya 464-8602, Japan
E-mail address: ohsawa@math.nagoya-u.ac.jp

Table of Contents

0 Introduction...185
1 L^2 Extension Theorem on Complex Manifolds..............186
2 Extension and Division......................................188
3 Proof of Theorem...189
    References...191

0 Introduction

The purpose of this paper is to prove the following.

Theorem 1. *Let D be a bounded pseudoconvex domain in $\mathbb{C}^n$ and let $z = (z_1, \ldots, z_n)$ be the coordinate of $\mathbb{C}^n$. Then there exists a constant C depending only on the diameter of D such that, for any plurisubharmonic function φ on D and for any holomorphic function f on D satisfying*

$$\int_D |f(z)|^2 e^{-\varphi(z) - 2n \log |z|} d\lambda < \infty$$

there exists a vector valued holomorphic function $g = (g_1, \ldots, g_n)$ on D satisfying

$$f(z) = \sum_{i=1}^n z_i g_i(z)$$

and

$$\int_D |g(z)|^2 e^{-\varphi(z) - 2(n-1) \log |z|} d\lambda \leq C \int_D |f(z)|^2 e^{-\varphi(z) - 2n \log |z|} d\lambda .$$

Here $d\lambda$ denotes the Lebesgue measure.

This must be compared to the following theorem of Skoda [S-1].

2000 *Mathematics Subject Classification*: 32W05, 32T05.

Theorem 2. *Let Ω be a pseudoconvex domain in $\mathbb{C}^n$, let φ be a plurisubharmonic function on Ω and let $h = (h_1, \ldots, h_p)$ be a vector of holomorphic functions on Ω. If f is a holomorphic function on Ω such that*

$$\int_{\Omega \backslash h^{-1}(0)} |f|^2 |h|^{-2k-2} (1 + \Delta \log |h|) e^{-\varphi} d\lambda < \infty,$$

where $k = \inf(n, p-1)$, there exists a vector of holomorphic functions $g = (g_1, \ldots, g_p)$ satisfying

$$f = \sum_{i=1}^p g_i h_i$$

and

$$\int_\Omega |g|^2 |h|^{-2k} (1 + |z|^2)^{-2} e^{-\varphi} d\lambda < \infty.$$

We note that the following is essentially contained in [S-3].

Theorem 3. *Let D be a bounded pseudoconvex domain in $\mathbb{C}^n$. Then, for any $\varepsilon > 0$ there exists a positive number C_ε such that, for any plurisubharmonic function φ on D and for any holomorphic function f on D satisfying*

$$\int_D |f(z)|^2 e^{-\varphi(z) - (2n+\varepsilon) \log |z|} d\lambda < \infty$$

there exists a vector valued holomorphic function $g = (g_1, \ldots, g_n)$ on D satisfying

$$f = \sum_{i=1}^n z_i g_i(z)$$

and

$$\int_D |g(z)|^2 e^{-\varphi(z) - (2n-2+\varepsilon) \log |z|} d\lambda \le C \int_D |f(z)|^2 e^{-\varphi(z) - (2n+\varepsilon) \log |z|} d\lambda.$$

The proof of Theorem 1 is completely independent of Skoda's theory and relies on an L^2 extension theorem on complex manifolds in [O]. It is very likely that one can extend this new method to recover and refine all the known L^2 division theorems in [S-1]–[S-4], [D].

1 L^2 Extension Theorem on Complex Manifolds

Let M be a complex manifold of dimension n and let $E \longrightarrow M$ be a holomorphic vector bundle. Given a continuous volume form dV on M and a continuous fiber metric h on E, we denote by $A^2(M, E, h, dV)$ the Hilbert space of square integrable (or L^2) holomorphic section of E with respect to h and dV. The L^2 extension theorem will be stated with respect to the space of E-valued L^2 holomorphic n-forms, which is naturally identified with the space $A^2(M, E \otimes K_M, h \otimes (dV)^{-1}, dV)$. here K_M denotes the canonical line bundle of M.

Let S be a closed complex submanifold of M and let Ψ be a continuous function on $M \backslash S$ such that, for any compact subset $K \subset M$ the map $\Psi :$ $K \backslash S \longrightarrow (-\infty, \sup_{K \backslash S} \Psi]$ is proper. For any continuous function f on M with compact support and for any $t > 0$ we put

$$\mathrm{Av}^k(t, f, dV, \Psi) = \frac{2k}{\sigma_{2k-1}} \int_{\Psi^{-1}((-t-1, -t))} f e^{-\Psi} dV .$$

Here σ_{2k-1} denotes the volume of the $(2k-1)$-dimensional unit sphere.

The residue of dV with respect to Ψ, denoted by $dV[\Psi]$, is defined as the minimal element of the set of positive measures dV' on S such that for any k

$$\int_S f dV' \geq \varliminf_{t \to \infty} \mathrm{Av}^k(t, f, dV, \Psi)$$

holds for any nonnegative continuous function f whose support is compact and intersects with S only along the k-codimensional components. Clearly the residue exists whenever the set of such measures dV' is nonempty.

For any smooth fiber metric of E, say h, its curvature form is denoted by Θ_h and is naturally regarded as a Hermitian quadratic form on the fibers of $E \otimes T_M^{1,0}$. Here $T_M^{1,0}$ denotes the holomorphic tangent bundle of M. The inequality $\Theta_h \geq 0$ will mean that the quadratic form Θ_h is everywhere semi-positive.

In what follows we shall assume that Ψ is negative, smooth and satisfies the following condition.

(#) If S is k-codimensional around a point x, there exists a local coordinate $(z_1, \ldots, z_n)$ on a neighborhood U of x such that $z_1 = \cdots = z_k = 0$ on $S \cap U$ and

$$\sup_{U \backslash S} \left| \Psi(z) - k \log \sum_{i=1}^{k} |z_i|^2 \right| < \infty .$$

In this situation the following extension Theorem holds.

Theorem 1.1 (cf. [O]). *Let M, E, h, S, dV and Ψ be as above. Assume that there exists a closed subset $X \subset M$ such that*

a) X *is locally negligible with respect to L^2 holomorphic functions*

and

b) $M \backslash X$ *is a Stein manifold which intersects with every component of S.*

If moreover $\Theta_{he^{-\Psi}} \geq 0$ and $\Theta_{he^{-(1+\delta)\Psi}} \geq 0$ on $M \backslash S$ for some $\delta > 0$, then there exists a bounded linear operator I from $A^2(S, E \otimes K_M, h \otimes (dV)^{-1}, dV[\Psi])$ to $A^2(M, E \otimes K_M, h \otimes (dV)^{-1}, dV)$ whose norm does not exceed a constant depending only on δ such that $I(f)|_S = f$ for any f.

For the proof of our division theorem, an extension theorem under a more restrictive assumption suffices :

Theorem 1.2 (Corollary of Theorem 1.1). *Let M, E, h, S and dV be as above, and assume that a) and b) hold. If moreover S is everywhere of codimension one and there exists a fiber metric b of $[S]^*$ such that $\Theta_h + \mathrm{Id}_E \otimes \Theta_b$ and $\Theta_h + (1+\delta)\mathrm{Id}_E \otimes \Theta_b$ are both semipositive for some $\delta > 0$, then there exists, for any canonical section s of $[S]$ and for any relatively compact locally pseudoconvex open subset Ω of M, a bounded linear operator I from $A^2(S \cap \Omega, E \otimes K_M, h \otimes (dV)^{-1}, dV[\log|s|^2])$ to $A^2(\Omega, E \otimes K_M, h \otimes (dV)^{-1}, dV)$ such that $I(f)|_S = f$. Here the norm of I does not exceed a constant depending only on δ and $\sup_\Omega |s|$.*

2 Extension and Division

Let N be a complex manifold and let $E \longrightarrow N$ be a holomorphic vector bundle of rank r. We put

$$P(E) = (E \backslash \text{zero section})/\mathbb{C}^* .$$

Then $P(E)$ is a holomorphic fiber bundle over N whose typical fiber is isomorphic to $\mathbb{P}^{r-1}$. Let $L(E)$ be the tautological line bundle over $P(E)$, i.e.

$$L(E) = \coprod_{\ell \in P(E)} \ell$$

where the points of $P(E)$ are identified with complex linear subspaces of dimension one in the fibers of E.

Let $\mathcal{O}(E)$ denote the sheaf of germs of holomorphic section of E. Then we have a natural isomorphism

$$H^0(N, \mathcal{O}(E)) \simeq H^0(P(E^*), \mathcal{O}(L(E^*)^*))$$

which arises from the commutative diagram

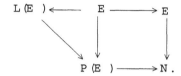

Let $\gamma : E \longrightarrow Q$ be a surjective homomorphism from E to another holomorphic vector bundle Q. Then we have a natural injective holomorphic map

$$P(Q^*) \hookrightarrow P(E^*)$$

and a commutative diagram:

$$L(E^*)^*|_{P(Q^*)} \longleftarrow \pi^* E|_{P(Q^*)}$$

$$\downarrow_{\wr} \qquad\qquad \downarrow$$

$$L(Q^*)^* \longleftarrow \pi^* Q|_{P(Q^*)}.$$

Identifying $L(E^*)^*|_{P(Q^*)}$ with $L(Q^*)^*$ by this, we obtain a commutative diagram:

$$H^0(N, E \otimes L) \xrightarrow{\sim} H^0(P(E^*), \pi^* L \otimes L(E^*)^*)$$

$$\downarrow_{\gamma_*} \qquad\qquad\qquad \downarrow_{\rho_\gamma}$$

$$H^0(N, Q \otimes L) \xrightarrow{\sim} H^0(P(Q^*), \pi^* L \otimes L(Q^*)^*)$$

for any holomorphic line bundle L over N. Here ρ_γ denotes the natural restriction map. We note that $P((E \otimes L)^*)$ is naturally identified with $P(E^*)$.

In this way the problem of lifting the sections (the division problem) is transferred to an extension problem.

If a volume form dV on N and fiber metrics of E and L are given, the above diagram induces its L^2 counterpart:

$$A^2(N, E \otimes L) \xrightarrow{\sim} A^2(P(E^*), \pi^* L \otimes L(E^*)^*)$$

$$\downarrow_{\gamma_*} \qquad\qquad\qquad \downarrow_{\rho_\gamma} \qquad\qquad (\star)$$

$$A^2(N, Q \otimes L) \xrightarrow{\sim} A^2(P(Q^*), \pi^* L \otimes L(Q^*)^*)$$

where the horizontal arrows are isometries and the fiber metric of Q is the induced metric.

3   Proof of Theorem

First of all we note that, in Theorem 1.1 (resp. in Theorem 1.2), for any plurisubharmonic function φ on M (resp. on Ω), the conclusion is also true for the spaces $A^2(S, E \otimes K_M, he^{-\varphi} \otimes (dV)^{-1}, dV[\Psi])$ and $A^2(M, E \otimes K_M, he^{-\varphi} \otimes (dV)^{-1}, dV)$ (resp. for the spaces $A^2(S \cap \Omega, E \otimes K_M, he^{-\varphi} \otimes (dV)^{-1}, dV[\log|s|^2])$ and $A^2(\Omega, E \otimes K_M, he^{-\varphi} \otimes (dV)^{-1}, dV)$), because on every compact subset of $M \backslash X$ (resp. $\Omega \backslash X$), φ is the limit of a decreasing sequence of C^∞ plurisubharmonic functions. We shall apply Theorem 1.2 in this more general formulation.

Letting L be the trivial line bundle, the problem of finding a right inverse for the morphism

$$A^2(M, E) \longrightarrow A^2(M, Q)$$

190   Takeo Ohsawa

is equivalent to finding one for

$$A^2(P(E^*), L(E^*)^*) \longrightarrow A^2(P(Q^*), L(Q^*)^*) .$$

In order to apply Theorem 1.2, let $\sigma : \widetilde{P}(E^*) \longrightarrow P(E^*)$ be the monoidal transform along $P(Q^*)$ and consider the restriction map

$$A^2(\widetilde{P}(E^*), \sigma^*L(E^*)^*) \longrightarrow A^2(\sigma^{-1}(P(Q^*)), \sigma^*L(Q^*)^*)$$

or equivalently the map

$$A^2(\widetilde{P}(E^*), (\sigma^*L(E^*)^* \otimes K^*_{\widetilde{P}(E^*)}) \otimes K_{\widetilde{P}(E^*)})$$
$$\longrightarrow A^2(\sigma^{-1}(P(Q^*)), (\sigma^*L(Q^*)^* \otimes K^*_{\widetilde{P}(E^*)}) \otimes K_{\widetilde{P}(E^*)}) .$$

Here the volume form on $\widetilde{P}(E^*)$ is induced from

$$dV_h = \bigwedge^{r-1} \sqrt{-1}\partial\bar{\partial} \log h(\zeta, \zeta) \wedge dV$$

and a fiber metric b of the bundle $[\sigma^{-1}(P(Q^*))]$ via the isomorphism

$$K_{\widetilde{P}(E^*)} \simeq \sigma^* K_{P(E^*)} \otimes [\sigma^{-1}(P(Q^*))]^{\otimes(k-1)} .$$

Here k is the codimension of $P(Q^*)$ in $P(E^*)$. As the (singular) fiber metric of $\sigma^*L(E^*)^* \otimes K^*_{\widetilde{P}(E)}$ we take

$$\sigma^*(h(\zeta, \zeta)e^{-\varphi}dV_h) \cdot b^{k-1} .$$

To prove Theorem 1 we put

$$N = D\backslash\{0\}$$
$$dV = \bigwedge^{n} \sqrt{-1}\partial\bar{\partial}(|z|^2 + \log |z|^2)$$
$$E = \mathbb{C}^n \times N, \ h = \mathrm{Id}$$
$$Q = \mathbb{C} \times N$$

and

$$\gamma((\zeta_1, \ldots, \zeta_n), z) = \left(\sum_{i=1}^{n} \zeta_i z_i, z\right) .$$

To apply Theorem 1.2 we consider the extensions

$$\widehat{E} = \mathbb{C}^n \times \mathbb{C}^n \xrightarrow{\widehat{\pi}} \mathbb{C}^n, \ \widehat{Q} = \mathbb{C} \times (\mathbb{C}^n\backslash 0)$$

and note that the closure of the image of $P(\widehat{Q}^*)$ in $P(\widehat{E}^*) (\supset P(E^*))$, say $\overline{P(\widehat{Q}^*)}$, is nothing but the monoidal transform of $\mathbb{C}^n$ centered at 0.

Let $\sigma : \widetilde{P}(\widehat{E}^*) \longrightarrow P(\widehat{E}^*)$ be the monoidal transform along $\overline{P(\widehat{Q}^*)}$. We observe that for any complex line ℓ in the projective space associated to $\widehat{\pi}^{-1}(0)$, the normal bundle $N_{\overline{P(\widehat{Q}^*)}/P(\widehat{E}^*)}$ satisfies

$$N_{\overline{P(\widehat{Q}^*)}/P(\widehat{E}^*)}|_\ell \simeq \mathcal{O}^{n-2} \oplus \mathcal{O}(1) \ .$$

Therefore $[\sigma^{-1}(\overline{P(\widehat{Q}^*)})]^*$ admits a fiber metric b such that, with respect to the induced fiber metric $\widehat{h}$ of $\sigma^* L(E^*)^*$,

$$\Theta_b + (1+\varepsilon)\Theta_{\widehat{h}} > 0 \qquad\qquad (\star\star)$$

for any $\varepsilon > 0$.

On the other hand

$$K_{\widetilde{P}(\widehat{E}^*)} \simeq \sigma^* K_{P(\widehat{E}^*)} \otimes [\sigma^{-1}(\overline{P(\widehat{Q}^*)})]^{\otimes(n-2)} \ ,$$

so that

$$\sigma^* L(\widehat{E}^*)^* \otimes K_{\widetilde{P}(\widehat{E}^*)}^* \simeq \sigma^* L(\widehat{E}^*)^* \otimes \sigma^* K_{P(\widehat{E}^*)}^* \otimes ([\sigma^{-1}(\overline{P(\widehat{Q}^*)})]^*)^{\otimes(n-2)} \ .$$

Since $\Theta_{\sigma^* dV_h} \geq n\Theta_{\widehat{h}}$ from the definition of dV_h, we have

$$\Theta_{\widehat{h} \otimes \sigma^* dV_h \otimes b^{n-2}} + (1+\delta)\Theta_b \geq (1+n)\Theta_{\widehat{h}} + (n-1+\delta)\Theta_b$$
$$\text{on } \widetilde{P}(\widehat{E}^*) \backslash \sigma^{-1}(\widehat{\pi}^{-1}(0)) \ .$$

By (†) the right hand side of the inequality is positive if $1 - n < \delta < 2$.

Hence, if $n \geq 2$ for any plurisubharmonic function φ on D one may apply Theorem 1.2 and obtain the desired conclusion. (Note that the difference of the weights in the L^2 estimate of the solution is caused by the singularity at 0 of the induced metric on the fibers of $\mathbb{C} \times N$.) If $n = 1$, the assertion is trivially true.

References

[D]   Demailly, J. -P., Scindage holomorphe d'un morphisme de fibrés vectoriels semi-positifs av ec estimation L^2, Séminaire P. Lelong-H. Skoda (Analyse) années 1980/81 et Colloque de Wimereux, *LNM* **919** (1982), 77–107.

[O]   Ohsawa, T., On the extension of L^2 holomorphic functions V – effects of generalization, *Nagoya Math. J.* **161** (2001), 1–21; (Erratum, to appear).

[S-1] Skoda, H., Applications des techniques L^2 à la théorie des idéaux d'une algèbre de fonctions holomorphes avec poids, *Ann. Sci. Ec. Norm. Sup.* **5** (1972), 545–579.

[S-2] Skoda, H., Morphismes surjectifs et fibrés linéaires semi-positifs, Séminaire P. Lelong-H. Skoda (Analyse), 17e année, 1976–1977, *LNM* **694**, Springer, Berlin, Heidelberg, New York.

[S-3] Skoda, H., Morphismes surjectifs de fibrés vectoriels semi-positifs, *Ann. Sci. Ec. Norm. Sup.* **11** (1978), 577–611.

[S-4] Skoda, H., Relèvement des sections globales dans les fibrés semi-positifs. Séminaire P. Lelong-H. Skoda (Analyse) 19e année, 1978–1979, *LNM* **822**.

Irreducible Degenerations of Primary Kodaira Surfaces

Stefan Schröer[1] and Bernd Siebert[2]

[1] Mathematische Fakultät, Ruhr-Universität, 44780 Bochum, Germany
E-mail address: s.schroeer@ruhr-uni-bochum.de
[2] Mathematische Fakultät, Ruhr-Universität, 44780 Bochum, Germany
E-mail address: bernd.siebert@ruhr-uni-bochum.de

Abstract We classify irreducible d-semistable degenerations of primary Kodaira surfaces. As an application we construct a canonical partial completion for the moduli space of primary Kodaira surfaces.

Table of Contents

0 Introduction..193
1 Smooth Kodaira Surfaces..................................195
2 D-semistable Surfaces with Trivial Canonical Class.........197
3 Hopf Surfaces..199
4 Ruled Surfaces over Elliptic Curves......................207
5 Rational Surfaces and Honeycomb Degenerations...........211
6 The Completed Moduli Space and its Boundary............217
References..221

0   Introduction

A smooth compact complex surface with trivial canonical bundle is a K3 surface, a 2-dimensional complex torus, or a primary Kodaira surface. Normal crossing degenerations of such surfaces have attracted much attention. For example, Kulikov analyzed projective degenerations of K3 surfaces [13]. His results were generalized by Persson and Pinkham to degenerations of surfaces with trivial canonical bundle whose central fiber has algebraic components [21]. Conversely, Friedman characterized the singular K3 surfaces from Kulikov's list that deform to smooth K3 surfaces by his notion of d-semistability [6].

For non-Kähler degenerations no general results seem to be known. We chose primary Kodaira surfaces as our object of study because they lack some complications that the more interesting classes of tori and K3 surfaces have. On the other hand, Kodaira surfaces have a lot in common with tori and,

---

2000 *Mathematics Subject Classification*: 14D15, 14D22, 14J15, 32G05, 32G13, 32J05, 32J15.

194 Stefan Schröer and Bernd Siebert

via the Kummer construction, with K3 surfaces. The phenomena one sees in degenerations of Kodaira surfaces should therefore also be observable in the other classes of surfaces with trivial canonical bundle.

By a result of Borcea the moduli space $\mathfrak{K}$ of smooth Kodaira surfaces is isomorphic to a countable union of copies of $\mathbb{C} \times \Delta^*$ [4]. The parameters correspond to the j-invariant of the elliptic base and a refined j-invariant of the fiber. As Δ^* can not be completed to a closed Riemann surface by adding finitely many points, it is clear that by studying degenerations we can at most hope for a partial completion of $\mathfrak{K}$. The explicit form of the moduli space also suggests the existence of families leading to a degeneration of the base or the fiber to a nodal elliptic curve. We will see that this is indeed the case. This leads to a partial completion $\mathbb{C} \times \Delta^* \subset \mathbb{P}^1 \times \Delta \setminus (\infty, 0)$ of each component of $\mathfrak{K}$. However, as elliptic curves can degenerate to any k-cycle of rational curves, this is not the full story. Rather we obtain a whole hierarchy of such completions, that are linked by non-Hausdorff phenomena at the boundary. Moreover, we will see that at the most interesting point $(\infty, 0)$ the picture becomes complicated. We were not able to fully clarify what happens there. If one restricts to normal crossing surfaces one should certainly blow up this point. On the other hand, we will also make the presumably not so surprising observation that, as in the case of abelian varieties [1], it does not suffice to restrict to normal crossing varieties. We did however find families of generalized Kodaira surfaces mapping properly to $\Delta \times \mathbb{P}^1$, whose singularities are at most products of normal crossing singularities.

The bulk of the paper is concerned with a classification of irreducible, d-semistable, locally normal crossing surfaces X with $K_X = 0$. Three different types occur. Our main result is a description of the resulting completion $\mathfrak{K} \subset \overline{\mathfrak{K}}$, derived by deformation theory. The three different types correspond to three different parts in the boundary $\mathfrak{B} = \overline{\mathfrak{K}} \setminus \mathfrak{K}$. Each part is a countable union of copies of Δ^* or $\mathbb{C}$. Locally along the boundary divisor the completion $\mathfrak{K} \subset \overline{\mathfrak{K}}$ looks like the blowing-up of $(\infty, 0) \in \mathbb{P}^1 \times \Delta$, with two points on the exceptional divisor removed.

Some examples for degenerations of Kodaira surfaces have previously been given by Friedman and Shepherd-Barron in [7].

This article is divided into six sections. The first section contains general facts about smooth Kodaira surfaces. In the second section we describe their potential degenerations and show that three types are possible. Sections 3–5 contain an analysis of each type. In the final section we assemble our results in terms of moduli spaces, complemented by some examples of smoothable surfaces with singularities that are products of normal crossings.

We thank the referee for suggestions concerning Theorem 3.10 and the interpretation of the completed moduli space after Proposition 6.1.

This paper is dedicated to Hans Grauert on the occasion of his 70th birthday. The second author wants to take this opportunity to express his

Degenerations of Kodaira Surfaces 195

gratitude for the support and mathematical stimulus he received from him as one of his last students. It was a great pleasure to learn from him.

1 Smooth Kodaira Surfaces

In this section we collect some facts on primary Kodaira surfaces. Suppose B, E are two elliptic curves, and endow E with a group structure. Let $f : X \to B$ be a holomorphic principal E-bundle. The canonical bundle formula gives $K_X = 0$, so the Kodaira dimension is $\kappa(X) = 0$. As a topological space, X is the product of the 1-sphere with a 1-sphere-bundle $g : M \to B$. Let $e(g) \in H^2(B, \mathbb{Z})$ be its Euler class. The homological Gysin sequence

$$H_2(B, \mathbb{Z}) \xrightarrow{e(g)} H_0(B, \mathbb{Z}) \longrightarrow H_1(M, \mathbb{Z}) \longrightarrow H_1(B, \mathbb{Z}) \longrightarrow 0$$

and the Künneth formula yield $H_1(X, \mathbb{Z}) = \mathbb{Z}^3 \oplus \mathbb{Z}/d\mathbb{Z}$ with $d = e(g) \cap [B]$. We call the integer $d \geq 0$ the *degree* of X. Bundles of degree $d = 0$ are 2-dimensional complex tori. A smooth compact complex surface X with an elliptic bundle structure of degree $d > 0$ is called a *primary Kodaira surface*. For simplicity, we refer to such surfaces as Kodaira surfaces.

Kodaira surfaces have three invariants. The first invariant is the degree $d > 0$. It determines the underlying topological space. The Universal Coefficient Theorem gives

$$H^1(X, \mathbb{Z}) = \mathbb{Z}^3 \quad \text{and} \quad H^2(X, \mathbb{Z}) = \mathbb{Z}^4 \oplus \mathbb{Z}/d\mathbb{Z} .$$

Hence the degree d is also the order of the torsion subgroup of the Néron-Severi group $\mathrm{NS}(X) = \mathrm{Pic}(X)/\mathrm{Pic}^0(X)$.

The second invariant is the j-invariant of B. It depends only on X: Since $b_1(X) = 3$ is odd, X is nonalgebraic. Moreover, since $f : X \to B$ has connected fibers it is the algebraic reduction and so the fibration structure does not depend on choices.

The third invariant is an element $\alpha \in \Delta^*$. Here $\Delta^* = \{z \in \mathbb{C} \mid 0 < |z| < 1\}$ is the punctured unit disk. The Jacobian Pic^0_X has dimension $h^1(\mathcal{O}_X) = 2$, and the quotient $\mathrm{Pic}^0_X / \mathrm{Pic}^0_B$ is isomorphic to $\mathbb{C}^*$. We have $h^2(\mathcal{O}_X(-X_b)) = h^0(\mathcal{O}_X(X_b)) = 1$ for each $b \in B$. So the map on the left in the exact sequence

$$H^1(X, \mathcal{O}_X) \longrightarrow H^1(X_b, \mathcal{O}_{X_b}) \longrightarrow H^2(X, \mathcal{O}_X(-X_b)) \longrightarrow H^2(X, \mathcal{O}_X) \longrightarrow 0$$

is surjective. Thus $\mathrm{Pic}^0_X / \mathrm{Pic}^0_B \to \mathrm{Pic}^0_{X_b}$ is an epimorphism. By semicontinuity, the kernel equals $\mathrm{NS}(B) = \mathbb{Z}$ and is generated by a well-defined element $\alpha \in \Delta^*$.

According to [3], p. 145, the principal E-bundle X is an associated fibre bundle $X = (L \setminus 0) \times_{\mathbb{C}^*} \mathbb{C}^* / \langle \alpha \rangle$ for some line bundle $L \to B$ of degree d. One can check that this α agrees with the invariant α above. It follows that the isomorphism classes of Kodaira surfaces correspond bijectively to the triples

$(d, j, \alpha) \in \mathbb{Z}_{>0} \times \mathbb{C} \times \Delta^*$. Moreover, each Kodaira surface with invariant (d, j, α) is the quotient of a properly discontinuous free $\mathbb{Z}^2$-action on $\mathbb{C}^* \times \mathbb{C}^*$ defined by

$$\Phi(z_1, z_2) = (\beta z_1, z_1^d z_2) \quad \text{and} \quad \Psi(z_1, z_2) = (z_1, \alpha z_2).$$

Here $\beta \in \Delta^*$ is defined as follows: Consider the j-invariant as a function $j : H \to \mathbb{C}$ on the upper half plane. It factors over $\exp : H \to \Delta^*$, $\tau \mapsto \exp(2\pi\sqrt{-1}\tau)$. Now $\beta = \exp(\tau)$ with $j(\tau) = j$. This uniformization illustrates Borcea's result on the moduli space of Kodaira surfaces [4]. Moreover, it suggests the application of toric geometry for the construction of degenerations.

The following observation will be useful in the sequel:

Lemma 1.1. *Suppose $X \to B$ is a principal E-bundle of degree $d \geq 0$. Let G be a finite group of order w acting on X. If the induced action on B is free, then w divides d, and the quotient X/G is a principal E-bundle of degree d/w.*

Proof. Set $X' = X/G$ and $B' = B/G$. Then B' is an elliptic curve, $B \to B'$ is Galois and the induced fibration $X' \to B'$ is a principal E-bundle. To calculate its degree d', we use a characterization of d in terms of $\pi_1(X)$. Since the universal covering of E is contractible the higher homotopy groups of E vanish. The beginning of the homotopy sequence of the fibrations $X \to B$, $X' \to B'$ thus leads to the following diagram of central extensions:

$$
\begin{array}{ccccccccc}
0 & \longrightarrow & \pi_1(E) & \longrightarrow & \pi_1(X) & \longrightarrow & \pi_1(B) & \longrightarrow & 0 \\
& & {\scriptstyle\mathrm{id}}\downarrow & & \downarrow & & \downarrow & & \\
0 & \longrightarrow & \pi_1(E) & \longrightarrow & \pi_1(X') & \longrightarrow & \pi_1(B') & \longrightarrow & 0\,.
\end{array}
$$

Let $h_1, h_2 \in \pi_1(X')$ map to generators of $\pi_1(B')$. Since $[\pi_1(B') : \pi_1(B)] = w$ there exists an integral matrix $\left(\begin{smallmatrix} a & b \\ c & d \end{smallmatrix}\right)$ with determinant w such that

$$g_1 = ah_1 + bh_2, \quad g_2 = ch_1 + dh_2$$

map to generators of $\pi_1(B)$. Then $g_1, g_2 \in \pi_1(X) \subset \pi_1(X')$ and $[g_1, g_2]$, $[h_1, h_2]$ are generators for the commutator subgroups $[\pi_1(X), \pi_1(X)]$ and $[\pi_1(X'), \pi_1(X')]$ respectively. Now the degrees d, d' being the orders of the torsion subgroups of $H_1(X, \mathbb{Z}) = \pi_1(X)/[\pi_1(X), \pi_1(X)]$ and of $H_1(X', \mathbb{Z}) = \pi_1(X')/[\pi_1(X'), \pi_1(X')]$ they can be expressed as divisibility of a generator of the commutator subgroup. Write $[h_1, h_2] = d' \cdot x$ for some primitive $x \in \pi_1(E)$. Then

$$[g_1, g_2] = [ah_1 + bh_2, ch_1 + dh_2] = (ad - bc)[h_1, h_2] = wd' \cdot x$$

shows $d = wd'$ as claimed. $\qquad\square$

2 D-semistable Surfaces with Trivial Canonical Class

Our objective is the study of degenerations of smooth Kodaira surfaces. We consider the following class of singular surfaces:

Definition 2.1. A reduced compact complex surface X is called *admissible* if it is irreducible, has locally normal crossing singularities, is d-semistable, and satisfies $K_X = 0$.

The sheaf of first order deformations $\mathcal{T}_X^1 = \mathcal{E}xt^1(\Omega_X^1, \mathcal{O}_X)$ is supported on $D = \text{Sing}(X)$. Following Friedman [6], we call a locally normal crossing surface X *d-semistable* if $\mathcal{T}_X^1 \simeq \mathcal{O}_D$. This is a necessary condition for the existence of a global smoothing with smooth total space.

Smooth admissible surfaces are either K3 surfaces, 2-dimensional complex tori, or Kodaira surfaces. Throughout this section, X will be a *singular* admissible surface. We will see that three types of such surfaces are possible. A finer classification is deferred to subsequent sections.

Let $\nu : S \to X$ be the normalization, $C \subset S$ its reduced ramification locus, $D \subset X$ the reduced singular locus, and $\varphi : C \to D$ the induced morphism. Then S is smooth. The surface X can be recovered from the commutative diagram

$$
\begin{array}{ccc}
C & \longrightarrow & S \\
\varphi \downarrow & & \downarrow \nu \\
D & \longrightarrow & X \,,
\end{array}
$$

which is cartesian and cocartesian. It gives rise to a long exact Mayer-Vietoris sequence

$$
\cdots \longrightarrow H^p(X, \mathcal{O}_X) \longrightarrow H^p(S, \mathcal{O}_S) \oplus H^p(D, \mathcal{O}_D) \longrightarrow H^p(C, \mathcal{O}_C) \longrightarrow \cdots .
$$

The ideals $\mathcal{O}_S(-C) \subset \mathcal{O}_S$ and $\mathcal{I}_D = \nu_*(\mathcal{O}_S(-C)) \subset \mathcal{O}_X$ are the conductor ideals of the inclusion $\mathcal{O}_X \subset \nu_*(\mathcal{O}_S)$. They coincide with the relative dualizing sheaf

$$
\nu_*(\omega_{S/X}) = \text{Hom}(\nu_*(\mathcal{O}_S), \mathcal{O}_X) \,.
$$

Hence $K_S = -C$; in particular, the Kodaira dimension is $\kappa(S) = -\infty$. The Enriques-Kodaira classification of surfaces ([3, Chap. VI]) tells us that S is either ruled or has $b_1(S) = 1$. In the latter case, one says that S is a surface of *class VII*.

Following Deligne and Rapoport [5] we call the seminormal curve obtained from $\mathbb{P}^1 \times \mathbb{Z}/n\mathbb{Z}$ by the relations $(0, i) \sim (\infty, i+1)$, a *Néron polygon*.

Lemma 2.2. *Each connected component of $C \subset S$ is an elliptic curve or a Néron polygon. For each singular connected component $D' \subset D$, the number of irreducible components in $\nu^{-1}(D') \subset C$ is a multiple of 6.*

198   Stefan Schröer and Bernd Siebert

Proof. The adjunction formula gives $\omega_C = \mathcal{O}_C$. Obviously, C has ordinary nodes. By [5, Lemma 1.3], each component of C is an elliptic curve or a Néron polygon. Let s be the number of singularities and c the number of irreducible components in D'. The preimage $\nu^{-1}(D') \subset S$ is a disjoint union of Néron polygons. It has $2c$ irreducible components and $3s$ singularities. For Néron polygons, these numbers coincide. □

We will use below the following *triple point formula* for d-semistable normal crossing surfaces:

Lemma 2.3 ([20, Cor. 2.4.2]). *Let $D' \subset D$ be an irreducible component and $C_1' \cup C_2' \subset C$ its preimage in S. Then $-(C_1')^2 - (C_2')^2$ is the number of triple points of D'.*

Let $g : S \to S'$ be a minimal model and $C' = g(C)$. Then $K_{S'} = -C'$, so $C' \subset S'$ has ordinary nodes, and $g : S \to S'$ is a sequence of blowing-ups with centers over $\text{Sing}(C')$. Recall that S is a *Hopf surface* if its universal covering space is isomorphic to $\mathbb{C}^2 \setminus 0$.

Proposition 2.4. *Suppose S is nonalgebraic. Then S is a Hopf surface. The curve $C = C_1 \cup C_2$ is a disjoint union of elliptic curves, and X is obtained by gluing them together.*

Proof. Since S has algebraic dimension $a(S) < 2$, no curve on S has positive selfintersection. Suppose C contains a Néron polygon $C' = C_1 \cup \ldots \cup C_m$. As X is d-semistable, the triple point formula Lemma 2.3 implies $C_i^2 = -1$. So $(C')^2 = \sum C_i^2 + \sum_{i \neq j} C_i \cdot C_j = -m + 2m > 0$, contradiction. Hence C is the disjoint union of elliptic curves. Consider the exact sequence

$$H^0(S, \mathcal{O}_S) \longrightarrow H^0(C, \mathcal{O}_C) \longrightarrow H^1(S, \omega_S) .$$

By the Kodaira classification, $b_1(S) = 1$, consequently $h^1(\mathcal{O}_S) = 1$, thus C has at most two components. Suppose $\varphi : C \to D$ induces a bijection of irreducible components. Then each component of C double-covers its image in D. So the map on the left in the exact sequence

$$H^1(S, \mathcal{O}_S) \oplus H^1(D, \mathcal{O}_D) \longrightarrow H^1(C, \mathcal{O}_C) \longrightarrow H^2(X, \mathcal{O}_X) \longrightarrow 0$$

is surjective, contradicting $K_X = 0$. Consequently $C = C_1 \cup C_2$ consists of two elliptic curves which are glued together in X. Let $n \geq 0$ be the number of blowing-ups in $S \to S'$. The normal bundle of C has degree $K_S^2 = K_{S'}^2 - n = -n$. Again by the triple point formula it follows $n = 0$. Hence S is minimal. By [18, Prop. 3.1], S is a Hopf or Inoue surface. The latter case is impossible here because Inoue surfaces contain at most one elliptic curve ([18, Prop. 1.5]). □

Proposition 2.5. *Suppose S is nonrational algebraic. Then S is a $\mathbb{P}^1$-bundle over an elliptic curve B. The curve $C = C_1 \cup C_2$ is the disjoint union of two sections, and X is obtained by gluing them together.*

Proof. By the Enriques classification, the minimal model S' is a $\mathbb{P}^1$-bundle over a nonrational curve B. By Lüroth's Theorem, C does not contain Néron polygons, hence is the disjoint union of elliptic curves. The Hurwitz formula implies that B is also an elliptic curve. Since K_S has degree -2 on the ruling, we infer $C = -K_S$ has at most two irreducible components. As in the preceding proof, we conclude that C consists of two components, which are identified in X. Let $n \geq 0$ be the number of blowing-ups in $S \to S'$. The normal bundle of C has degree $K_S^2 = K_{S'}^2 - n = -n$. Since X is d-semistable, $n = 0$ follows. Hence S is a $\mathbb{P}^1$-bundle. $\square$

Proposition 2.6. *Suppose S is a rational surface. Then it is the blowing-up of a Hirzebruch surface in two points P_1, P_2 in disjoint fibers F_1, F_2. The curve C is a Néron 6-gon, consisting of the two exceptional divisors, the strict transforms of F_1, F_2, and the strict transforms of two disjoint sections whose union contains P_1 and P_2. The surface X is obtained by identifying pairs of irreducible components in C.*

Proof. The minimal model S' is either $\mathbb{P}^2$ or a Hirzebruch surface of degree $e \neq 1$. Since each representative of $-K_{S'}$ is connected the curve C must be connected. Suppose C is an elliptic curve. Then the map on the left in the exact sequence

$$H^1(S, \mathcal{O}_S) \oplus H^1(D, \mathcal{O}_D) \longrightarrow H^1(C, \mathcal{O}_C) \longrightarrow H^2(X, \mathcal{O}_X) \longrightarrow 0$$

is surjective, contradicting $K_X = 0$. Consequently, $C = C_1 \cup \ldots \cup C_m$ is a Néron polygon. The triple point formula gives $\sum C_i^2 = -m$, hence

$$K_S^2 = \sum C_i^2 + \sum_{i \neq j} C_i \cdot C_j = -m + 2m = m \,.$$

Let $n \geq 0$ be the number of blowing-ups in $S \to S'$, so $K_S^2 = K_{S'}^2 - n$. For $S' = \mathbb{P}^2$ this gives $m = 9 - n$. If S' is a Hirzebruch surface, $m = 8 - n$ holds instead. According to Lemma 2.2, the natural number m is a multiple of 6. The only possibilities left are $S' = \mathbb{P}^2$ and C' a Néron 3-gon, or S' a Hirzebruch surface and C' a Néron 4-gon. From this it easily follows that X is obtained from a blowing-up of a Hirzebruch surface as stated. $\square$

Remark 2.7. For the results of this section one can weaken the hypothesis that X is d-semistable. It suffices to assume that the invertible $\mathcal{O}_D$-module $\mathcal{T}_X^1$ is numerically trivial. One might call such surfaces *numerically d-semistable*. They will occur as locally trivial deformations of d-semistable surfaces.

3 Hopf Surfaces

In this section we analyze the geometry and the deformations of admissible surfaces X whose normalization S is *nonalgebraic*. By Proposition 2.4, such S

200 Stefan Schröer and Bernd Siebert

is a Hopf surface containing a union of two disjoint isomorphic elliptic curves $C = C_1 \cup C_2$.

As a topological space, Hopf surfaces are certain fibre bundles over the 1-sphere, whose fibres are quotients of the 3-sphere (see [10, Thm. 9], and [11]). By Hartog's Theorem, the action of $\pi_1(S)$ on the universal covering space $\mathbb{C}^2 \setminus 0$ extends to $\mathbb{C}^2$, fixing the origin. We call a Hopf surface *diagonizable* if $\pi_1(S)$ is contained in the maximal torus $\mathbb{C}^* \times \mathbb{C}^* \subset \mathrm{GL}_2(\mathbb{C})$ up to conjugacy inside the group of all biholomorphic automorphisms of $\mathbb{C}^2$ fixing the origin. This is precisely the class of Hopf surfaces we are interested in:

Proposition 3.1. *A Hopf surface S is diagonizable if and only if it contains two disjoint elliptic curves $C_1, C_2 \subset S$ with $-K_S = C_1 + C_2$. Moreover, a diagonizable Hopf surface has fundamental group $\pi_1(S) = \mathbb{Z} \oplus \mathbb{Z}/n\mathbb{Z}$.*

Proof. Suppose $\pi_1(S) \subset \mathrm{GL}_2(\mathbb{C})$ consists of diagonal matrices. Let C_1, $C_2 \subset S$ be the images of $\mathbb{C}^* \times 0$ and $0 \times \mathbb{C}^*$, respectively. The invariant meromorphic 2-form $dz_1/z_1 \wedge dz_2/z_2$ on $\mathbb{C}^2$ yields $-K_S = C_1 + C_2$.

Conversely, assume the condition $-K_S = C_1 + C_2$. An element $\alpha \in \pi_1(S)$ is called a *contraction* if $\lim_{n\to\infty} \alpha^n(U) = \{0\}$ holds for the unit ball U in $\mathbb{C}^2$. According to a classical result [22, eq. 44], in suitable coordinates a contraction takes the form

$$\alpha(z_1, z_2) = (\alpha_1 z_1 + \lambda z_2^m, \alpha_2 z_2)$$

with $0 < |\alpha_1| \le |\alpha_2| < 1$ and $\lambda(\alpha_1 - \alpha_2^m) = 0$. We claim that each contraction $\alpha \in \pi_1(S)$ has $\lambda = 0$. Suppose not. The quotient of $\mathbb{C}^2 \setminus 0$ by $\langle \alpha \rangle$ is a primary Hopf surface and defines a finite étale covering $f : S' \to S$. Let $C' \subset S'$ be the image of $\mathbb{C}^* \times 0$. Then [12, p. 696], gives $-K_{S'} = (m+1)C'$. On the other hand,

$$-K_{S'} = -f^*(K_S) = f^{-1}(C_1) + f^{-1}(C_2)$$

holds. Consequently, a nonempty subcurve of $(m+1)C'$ is base point free, so S' is elliptic. Now according to [12, Thm. 31], ellipticity of $S' = (\mathbb{C}^2 \setminus 0)/\langle \alpha \rangle$ implies $\lambda = 0$, contradiction.

Next we claim that $\pi_1(S)$ is contained in $\mathrm{GL}_2(\mathbb{C})$, at least up to conjugacy. Suppose not. A nonlinear fundamental group $\pi_1(S)$ is necessarily abelian ([12, Thm. 47]). By [10, p. 231], there is a contraction $\alpha \in \pi_1(S)$ with $\lambda \ne 0$, contradiction.

Now we can assume $\pi_1(S) \subset \mathrm{GL}_2(\mathbb{C})$. Write $\pi_1(S) = G \cdot Z$ as a semidirect product, where $G = \{\gamma \in \pi_1(S) \mid |\det(\gamma)| = 1\}$, and $Z \simeq \mathbb{Z}$ is generated by a contraction $\alpha = (\alpha_1, \alpha_2)$, acting diagonally. If $\alpha_1 \ne \alpha_2$ then [23, p. 24], ensures that S is diagonizable. It remains to treat the case $\alpha_1 = \alpha_2$. Then Z is central and we only have to show that G is abelian. Let S' be the quotient of $\mathbb{C}^2 \setminus 0$ by $\langle \alpha \rangle$. The canonical projection $\mathbb{C}^2 \setminus 0 \to \mathbb{P}^1$ induces an elliptic bundle $S' \to \mathbb{P}^1$. This defines an elliptic structure $g : S = S'/G \longrightarrow \mathbb{P}^1/G$. Suppose $\mathbb{P}^1 \to \mathbb{P}^1/G$ has more than 2 ramification points. Let F_b be the reduced fiber over $g \in \mathbb{P}^1/G$ with multiplicity m_b and $C_i = F_{b_i}$; the canonical bundle formula yields

$$
\begin{aligned}
0 = C_1 + C_2 + K_S &= F_{b_1} + F_{b_2} - 2F + \sum_{b \in \mathbb{P}^1/G} (m_b - 1)F_b \\
&= (1 - m_{b_1})F_{b_1} + (1 - m_{b_2})F_{b_2} + \sum_{b \in \mathbb{P}^1/G} (m_b - 1)F_b > 0,
\end{aligned}
$$

contradiction. Consequently there are only 2 ramification points. Therefore G is a subgroup of $\mathbb{C}^*$, hence abelian. This finishes the proof of the equivalence.

As for the fundamental group, it must have rank 1 for $b_1 = 1$. If there were more than one generator needed for the torsion part the action could not be free on $(\mathbb{C}^* \times 0) \cup (0 \times \mathbb{C}^*)$. $\qquad\square$

Suppose that S is a diagonizable Hopf surface with

$$
\pi_1(S) \simeq \mathbb{Z} \oplus \mathbb{Z}/n\mathbb{Z} \subset \mathrm{GL}_2(\mathbb{C})
$$

consisting of diagonal matrices. The free part of $\pi_1(S)$ is generated by a contraction $\alpha = (\alpha_1, \alpha_2)$ with $\alpha_1, \alpha_2 \in \Delta^*$. The torsion part is generated by a pair $\zeta = (\zeta_1, \zeta_2)$ of primitive n-th roots of unity. Note that they must be primitive, again since the action is free on the coordinate axes minus the origin. Choose $\tau_i \in \mathbb{C}$ and $n_i \in \mathbb{Z}$ with $\alpha_i = \exp(2\pi\sqrt{-1}\tau_i)$ and $\zeta_i = \exp(2\pi\sqrt{-1}n_i/n)$. Consider the lattice

$$
\Lambda_i = \langle \tau_i, n_i/n, 1 \rangle = \mathbb{Z} \cdot \tau_i + \mathbb{Z} \cdot 1/n \subset \mathbb{C}.
$$

The images $C_1, C_2 \subset S$ of $\mathbb{C}^* \times 0$ and $0 \times \mathbb{C}^*$ take the form

$$
C_i = \mathbb{C}^* / \langle \alpha_i, \zeta_i \rangle = \mathbb{C}/\Lambda_i .
$$

Without loss of generality we can assume $C = C_1 \cup C_2$. In fact, if S is not elliptic then C_1, C_2 are the only curves on S at all. Otherwise the elliptic fibration comes from a rational function of the form z_1^a/z_2^b on $\mathbb{C}^2 \setminus 0$. Then $|ab| \neq 1$ iff the elliptic fibration $S \to \mathbb{P}^1$ has multiple fibers. In this case the canonical bundle formula implies $C = C_1 \cup C_2$. Otherwise we can apply a linear coordinate change on $\mathbb{C}^2$ to achieve that the preimages of the components of C are $\mathbb{C}^* \times 0$ and $0 \times \mathbb{C}^*$.

Now assume there is an isomorphism $\psi : C_2 \to C_1$. Set $D = C_1$, $C = C_1 \cup C_2$, and $\varphi = \mathrm{id} \cup \psi : C \to D$. The cocartesian diagram

$$
\begin{array}{ccc}
C & \longrightarrow & S \\
\varphi \downarrow & & \downarrow \nu \\
D & \longrightarrow & X
\end{array}
$$

defines a normal crossing surface X. Since each translation of the elliptic curve C_2 extends to an automorphism of S fixing C_1, we can assume that

202   Stefan Schröer and Bernd Siebert

$\psi : C_2 \to C_1$ respects the origin. Hence ψ is defined by a homothety $\mu : \mathbb{C} \to \mathbb{C}$ with $\mu \Lambda_2 = \Lambda_1$. In other words, there is a matrix

$$\begin{pmatrix} a & c \\ b & d \end{pmatrix} \in \mathrm{SL}_2(\mathbb{Z}) \qquad \text{with} \qquad \begin{aligned} \mu \tau_2 &= a\tau_1 + b/n \\ \mu/n &= c\tau_1 + d/n \, . \end{aligned} \tag{3.1}$$

The following two propositions describe the properties of the surface X in terms of $\pi_1(S) \subset \mathrm{GL}_2(\mathbb{C})$ and $\mu \in \mathbb{C}$.

Proposition 3.2. *The condition $K_X = 0$ holds if and only if $a = d = 1$ and $c = 0$.*

Proof. As $\nu^*(K_X) = 0$ in any case K_X is numerically trivial. The condition $K_X = 0$ is thus equivalent to $H^2(X, \mathcal{O}_X) \neq 0$. Assume $K_X = 0$. Then the map on the left in the exact sequence

$$H^1(S, \mathcal{O}_S) \oplus H^1(D, \mathcal{O}_D) \longrightarrow H^1(C, \mathcal{O}_C) \longrightarrow H^2(X, \mathcal{O}_X) \longrightarrow 0$$

is not surjective. Hence $H^1(\mathcal{O}_D)$ and $H^1(\mathcal{O}_S)$ have the same image in $H^1(\mathcal{O}_C)$. According to the commutative diagram

$$\begin{array}{ccccc} H^1(S, \mathcal{O}_S) & \longrightarrow & H^1(C, \mathcal{O}_C) & \overset{\varphi^*}{\longleftarrow} & H^1(D, \mathcal{O}_D) \\ {\scriptstyle \text{bij}} \downarrow & & {\scriptstyle \text{inj}} \downarrow & & {\scriptstyle \text{inj}} \downarrow \\ H^1(S, \mathbb{C}) & \longrightarrow & H^1(C, \mathbb{C}) & \overset{\varphi^*}{\longleftarrow} & H^1(D, \mathbb{C}) \end{array}$$

from Hodge theory, the image of $H^1(D, \mathbb{C})$ in $H^1(C, \mathbb{C})$ contains the image of $H^1(S, \mathbb{C})$. So the composition

$$H^1(S, \mathbb{C}) \longrightarrow H^1(C, \mathbb{C}) \overset{(\mathrm{id}, \psi^*)^{-1}}{\longrightarrow} H^1(D, \mathbb{C}) \oplus H^1(D, \mathbb{C})$$

factors over the diagonal $H^1(D, \mathbb{C}) \subset H^1(D, \mathbb{C})^{\oplus 2}$. Dually, the composition

$$H_1(D, \mathbb{C}) \oplus H_1(D, \mathbb{C}) \overset{(\mathrm{id}, \psi_*)^{-1}}{\longrightarrow} H_1(C, \mathbb{C}) \longrightarrow H_1(S, \mathbb{C})$$

factors over the addition map $H_1(D, \mathbb{C})^{\oplus 2} \to H_1(D, \mathbb{C})$. In other words, the composition

$$\Lambda_1 \overset{(\mathrm{id}, -\mathrm{id})}{\longrightarrow} \Lambda_1 \oplus \Lambda_1 \overset{\mathrm{id} \,\oplus\, \left(\begin{smallmatrix} d & -c \\ -b & a \end{smallmatrix} \right)}{\longrightarrow} \Lambda_1 \oplus \Lambda_2 \overset{(1,0,1,0)}{\longrightarrow} \mathbb{Z}$$

is zero. It is described by $(1 - d, c)$; thus $d = 1$, $c = 0$, and in turn $a = 1$. Hence the condition is necessary. The converse is shown in a similar way. $\square$

Remark 3.3. Inserting the above conditions into (3.1), one sees that the homothety $\mu : \mathbb{C} \to \mathbb{C}$ must be the identity, and then $\tau_2 = \tau_1 + b/n$. In particular S is elliptically fibered.

Proposition 3.4. *Suppose $K_X = 0$. Then X is d-semistable if and only if the congruences $(n_1 - n_2)^2 \equiv 0$ and $b(n_1 - n_2) \equiv 0$ modulo n hold.*

Proof. Let N_i be the normal bundle of $C_i \subset S$. We can identify $\mathcal{T}_X^1$ with $N_2 \otimes \psi^*(N_1)$. The bundle N_1 can be obtained as quotient of the normal bundle of $\mathbb{C}^* \times 0$ in $\mathbb{C}^* \times \mathbb{C}^*$, which is trivial, by the induced action of $\pi_1(S) = \mathbb{Z} \oplus \mathbb{Z}/n\mathbb{Z}$: $N_1 = \mathbb{C}^* \times \mathbb{C}^*/(\alpha, \zeta)$ with

$$\alpha \cdot (z_1, v) = (\alpha_1 z_1, \alpha_2 v) \quad \text{and} \quad \zeta \cdot (z_1, v) = (\zeta_1 z_1, \zeta_2 v) .$$

After pull back along $\mathbb{C} \to \mathbb{C}^*$, $z \mapsto \exp(2\pi\sqrt{-1}z)$, the bundle N_1 can also be seen as the quotient of the trivial bundle $\mathbb{C} \times \mathbb{C}$ by $\pi_1(C_1) = \Lambda_1$ acting via

$$\tau_1 \cdot (w, v) = (w + \tau_1, \alpha_2 v), \quad n_1/n \cdot (w, v) = (w + n_1/n, \zeta_2 v), \quad 1 \cdot (w, v) = (w + 1, v) .$$

Choose m_1 with $m_1 n_1 \equiv 1$ modulo n. Then $1/n \in \Lambda_1$ acts via $1/n \cdot (w, v) = (w + 1/n, \zeta_2^{m_1} v)$. An analogous situation holds for N_2. Inserting $\Psi = \left(\begin{smallmatrix} 1 & 0 \\ b & 1 \end{smallmatrix}\right)$ we obtain that $\mathcal{T}_X^1 = N_2 \otimes \psi^*(N_1)$ is the quotient of $\mathbb{C} \times \mathbb{C}$ by the action

$$\tau_2 \cdot (w, v) = (\tau_2 + w, \alpha_1 \alpha_2 \zeta_2^{bm_1} v) \quad \text{and} \quad 1/n \cdot (w, v) = (1/n + w, \zeta_1^{m_2} \zeta_2^{m_1} v) .$$

The isomorphism class of $\mathcal{T}_X^1 = N_2 \otimes \psi^*(N_1)$ depends only on the element

$$\tau_2 \longmapsto \alpha_1 \alpha_2 \zeta_2^{bm_1}, \quad 1/n \longmapsto \zeta_1^{m_2} \zeta_2^{m_1}$$

in $\operatorname{Hom}_{\mathbb{Z}}(\Lambda_2, \mathbb{C}^*)$. To check for triviality of $\mathcal{T}_X^1$, consider the commutative diagram

$$
\begin{array}{ccc}
\operatorname{Hom}_{\mathbb{Z}}(\Lambda_2, \mathbb{C}) & \xrightarrow{\ \text{pr}\ } \operatorname{Hom}_{\mathbb{C}}(\mathbb{C}, \overline{\mathbb{C}}) \xrightarrow{\ \cong\ } H^1(C_2, \mathcal{O}_{C_2}) \\
\ \downarrow{\scriptstyle \exp} & \ \downarrow{\scriptstyle \exp} \\
\operatorname{Hom}_{\mathbb{Z}}(\Lambda_2, \mathbb{C}^*) & \xrightarrow{\qquad} \operatorname{Pic}(C_2) \xleftarrow[\cong]{\quad} H^1(C_2, \mathcal{O}_{C_2}^*) ,
\end{array}
$$

explained in [15], Sect. I.2. Here pr is the projection of $\operatorname{Hom}_{\mathbb{Z}}(\Lambda_2, \mathbb{C}) \cong \operatorname{Hom}_{\mathbb{R}}(\mathbb{C}, \mathbb{C})$ onto the $\mathbb{C}$-antilinear homomorphisms $\operatorname{Hom}_{\mathbb{C}}(\mathbb{C}, \overline{\mathbb{C}})$. Lifting the homomorphism defining $N_2 \otimes \varphi^*(N_1)$ from $\operatorname{Hom}_{\mathbb{Z}}(\Lambda_2, \mathbb{C}^*)$ to $\operatorname{Hom}_{\mathbb{Z}}(\Lambda_2, \mathbb{C})$, we obtain

$$\tau_2 \longmapsto \tau_1 + \tau_2 + \frac{bm_1 n_2}{n}, \quad 1/n \longmapsto \frac{m_2 n_1 + m_1 n_2}{n} .$$

Using the diagram, we infer that X is d-semistable if and only if this homomorphism is $\mathbb{C}$-linear up to an integral homomorphism. The latter condition is equivalent to the equation

$$\left(\tau_1 + \tau_2 + \frac{bm_1 n_2}{n} + e\right) - \tau_2 n\left(\frac{m_2 n_1 + m_1 n_2}{n} + f\right) = 0$$

for certain integers e, f. Now proceed as follows: Substitute $\tau_1 = \tau_2 - b/n$; the coefficients of τ_1 and 1 give two equations; since e and f are variable this leads to two congruences modulo n; finally use $n_1 m_1 \equiv 1$, $n_2 m_2 \equiv 1$ modulo n to deduce the stated congruences. $\qquad\square$

204   Stefan Schröer and Bernd Siebert

We seek a coordinate free description of X. We call an automorphism of a Néron 1-gon a *rotation* if the induced action on the normalization $\mathbb{P}^1$ fixes the branch points (compare [5, §3.6]).

Proposition 3.5. *Let X be a complex surface. The following are equivalent:*

(i) *X is admissible with nonalgebraic normalization S.*

(ii) *There is an elliptic principal bundle $X' \to B'$ of degree $e > 0$ over the Néron 1-gon B' and a (cyclic) Galois covering $g : X' \to X$ such that the Galois group G acts effectively on B' via rotations.*

Proof. Suppose (ii) holds. The normalization S' of X' is an elliptic principal bundle of degree $e > 0$ over $\mathbb{P}^1$, hence nonalgebraic. Since $g : X' \to X$ induces a finite morphism $g' : S' \to S$, the surface S is nonalgebraic as well. It is easy to check $K_X = 0$ using the assumption that G acts via rotations.

Conversely, assume (i). As we have seen, S is a diagonizable Hopf surface. The exact sequence

$$0 \longrightarrow \pi_1(S') \longrightarrow \pi_1(S) \longrightarrow \mathrm{PGL}_2(\mathbb{C})$$

determines another diagonizable Hopf surface S' and a Galois covering $S' \to S$. The Galois group $G = \pi_1(S)/\pi_1(S')$ is cyclic of order $m = n/\gcd(n, n_1 - n_2, b)$, and the order of the torsion subgroup in $\pi_1(S')$ is $n' = \gcd(n, n_1 - n_2)$, as simple computations show.

The projection $\mathbb{C}^2 \setminus 0 \to \mathbb{P}^1$ induces an elliptic principal bundle $f' : S' \to \mathbb{P}^1$. The Galois group G acts effectively on $\mathbb{P}^1$ and fixes $0, \infty$. Let $C_1', C_2' \subset S'$ be the fibres over the fixed points. The quotient $\mathbb{P}^1/G$ is a projective line, and we obtain an induced elliptic structure $f : S \to \mathbb{P}^1/G$. Note that f has two multiple fibres of order m, whose reductions are C_1, C_2. Next, the canonical isomorphism $\psi' : C_2' \to C_1'$ gives a normal crossing surface X'. Clearly, it is an elliptic principal bundle of degree $e = n' > 0$ over the Néron 1-gon B' obtained from $\mathbb{P}^1$ by the relation $0 \sim \infty$. The congruences $(n_1 - n_2)^2 \equiv 0 \equiv b(n_1 - n_2)$ modulo n are equivalent to $n'(n_1 - n_2) \equiv 0 \equiv n'b$. A straightforward calculation shows that this ensures that the free G-action on S' descends to a (free) action on X'. The quotient is $X = X'/G$. $\qquad\square$

Remark 3.6. At this point we are in position to classify the admissible surfaces with nonalgebraic normalization. The elliptic principal bundles $X' \to B'$ are in one-to-one correspondence with pairs $(\alpha, e) \in \Delta^* \times \mathbb{N}$, while up to isomorphism the action of G on B' only depends on $|G|$. A further discrete invariant belongs to possibly non-isomorphic lifts of the G-action to X'.

We call the invariants e and $w = |G|$ of X the *degree* and the *warp* of X respectively. The congruences in Proposition 3.4 imply that w divides e. This is important for the construction of smoothings:

Theorem 3.7. *Suppose X is an admissible surface (Definition 2.1) with nonalgebraic normalization, of degree e and warp w. Then X deforms to a smooth Kodaira surface of degree e/w.*

Proof. Let $X' \to X$ be the Galois covering from Proposition 3.5. It suffices to construct a G-equivariant deformation of X'.

The deformation of X' will be obtained as infinite quotient of a toric variety belonging to an infinite fan. Our construction fibers over a similar construction of a smoothing of the Néron 1-gon due to Mumford ([2, Ch.I, 4]) that we now recall for the reader's convenience. Mumford considered the fan generated by the cones

$$\bar{\sigma}_m = \big\langle (m,1), (m+1,1) \big\rangle, \quad m \in \mathbb{Z}.$$

There is a $\mathbb{Z}$-action on the associated toric variety W belonging to the linear (right-) action by $\begin{pmatrix} 1 & 0 \\ 1 & 1 \end{pmatrix}$ on the fan. Let $q : W \to \mathbb{C}$ be the morphism coming from the projection $\mathrm{pr}_2 : \mathbb{Z}^2 \to \mathbb{Z}$. Note that the central fiber is an infinite chain of rational curves (a Néron ∞-gon), while the general fiber is isomorphic to $\mathbb{C}^*$. Now $\mathbb{Z}$ acts fiberwise and the action is proper and free over the preimage of $\Delta \subset \mathbb{C}$. Moreover, $\mathbb{Z}$ acts transitively on the set of irreducible components of the central fiber. The quotient is the desired deformation of the Néron 1-gon.

For our purpose, consider the infinite fan in $\mathbb{Z}^3$ generated by the two-dimensional cones

$$\sigma_m = \Big\langle \big(m, e \cdot \tbinom{m}{2}, 1\big), \big(m+1, e \cdot \tbinom{m+1}{2}, 1\big) \Big\rangle, \quad m \in \mathbb{Z}.$$

Let V be the corresponding 3-dimensional smooth toric variety and $V \to \mathbb{C}$ the toric morphism belonging to the projection $\mathrm{pr}_3 : \mathbb{Z}^3 \to \mathbb{Z}$. Now the special fibre V_0 is isomorphic to a $\mathbb{C}^*$-bundle over the Néron ∞-gon. As neighbouring cones σ_m, σ_{m+1} are not coplanar, its degree over each irreducible component of the Néron ∞-gon is nonzero. A simple computation shows that its degree is e. This time the $\mathbb{Z}$-action on the fan is given by the automorphism

$$\Phi = \begin{pmatrix} 1 & e & 0 \\ 0 & 1 & 0 \\ 1 & 0 & 1 \end{pmatrix} \in \mathrm{SL}_3(\mathbb{Z}).$$

Again the induced $\mathbb{Z}$-action on the preimage $U \subset V$ of $\Delta \subset \mathbb{C}$ is proper and free. Choose $\alpha \in \Delta^*$ such that the fibres of $X' \to B'$ are isomorphic to $\mathbb{C}^*/\langle \alpha^w \rangle$. Let $\Psi : V \to V$ be the automorphism extending the action of $(1, \alpha, 1) \in (\mathbb{C}^*)^3$. Then $\mathfrak{X}' = U/\langle \Phi, \Psi^w \rangle$ is a smooth 3-fold, endowed with a projection $\mathfrak{X}' \to \mathbb{C}$. The general fibres $\mathfrak{X}_t$, $t \in \Delta^*$, are smooth Kodaira surfaces of degree e with fibres $\mathbb{C}^*/\langle \alpha^w \rangle$ and basis $\mathbb{C}^*/\langle t \rangle$. The special fibre $\mathfrak{X}'_0$ is isomorphic to X'.

Finally we extend the G-action on X' to $\mathfrak{X}'$. The automorphism $\Psi : V \to V$ descends to a free G-action on $\mathfrak{X}'$. Replacing α by another primitive w-th root of α^w if necessary, the induced action on $\mathfrak{X}'_0$ coincides with the given action on X'. Consequently, $\mathfrak{X} = \mathfrak{X}'/G$ is the desired smoothing of $X = \mathfrak{X}_0$. The action of Ψ on B' is free. Therefore, according to Lemma 1.1, the general fibres $\mathfrak{X}_t$ are smooth Kodaira surfaces of degree e/w. $\qquad\square$

206   Stefan Schröer and Bernd Siebert

The next task is to determine the versal deformation of X. We have to calculate the relevant cohomology groups:

Lemma 3.8. *For a locally normal crossing surface X with $K_X = 0$ and non-algebraic normalization it holds $h^0(\Theta_X) = 1$, $h^1(\Theta_X) = 1$ and $h^2(\Theta_X) = 0$.*

Proof. The Ext spectral sequence together with a local computation gives

$$H^p(X, \Theta_X) = \mathrm{Ext}^p(\Omega^1_X/\tau_X, \mathcal{O}_X),$$

see [6, p. 88ff]. Here $\tau_X \subset \Omega^1_X$ denotes the torsion subsheaf. In view of Serre duality

$$\mathrm{Ext}^p(\Omega^1_X/\tau_X, \mathcal{O}_X)^\vee \simeq H^{2-p}(X, \Omega^1_X/\tau_X)$$

it thus suffices to compute the cohomology of Ω^1_X/τ_X. Consider the exact sequence

$$0 \longrightarrow \Omega^1_X/\tau_X \longrightarrow \Omega^1_S \longrightarrow \Omega^1_D \longrightarrow 0$$

of $\mathcal{O}_X$-modules ([6, Prop. 1.5]). Note that the map on the right is an *alternating sum*. The inclusion $H^0(X, \Omega^1_X/\tau_X) \subset H^0(S, \Omega^1_S) = 0$ yields $h^2(\Theta_X) = 0$. Moreover,

$$0 \longrightarrow H^0(D, \Omega^1_D) \longrightarrow H^1(X, \Omega^1_X/\tau_X) \longrightarrow H^1(S, \Omega^1_S)$$

and $h^{1,1}(S) = 0$ gives $h^1(\Theta_X) = 1$. With $b_3(S) = 1$, $h^1(\omega_S) = h^1(\mathcal{O}_S) = 1$ and degeneracy of the Fröhlicher spectral sequence for smooth compact surfaces ([3, IV, Thm. 2.7]) we have $h^{1,2}(S) = 0$. Now the exact sequence

$$0 \longrightarrow H^1(D, \Omega^1_D) \longrightarrow H^2(X, \Omega^1_X/\tau_X) \longrightarrow H^2(S, \Omega^1_S)$$

gives $h^0(\Theta_X) = 1$.   □

Proposition 3.9. *Let X be as in Theorem 3.7. Then $\dim T^0_X = 1$, $\dim T^1_X = 2$ and $\dim T^2_X = 1$.*

Proof. Since X is locally a complete intersection the E_2 term of the spectral sequence $E_2^{p,q} = H^p(\mathcal{T}^q_X) \Rightarrow T^{p+q}_X$ has at most one non-trivial differential, which is $H^0(\mathcal{T}^1_X) \to H^2(\Theta_X)$. The previous proposition shows first degeneracy at E_2 level and in turn gives the stated values for $\dim T^i_X$.   □

Theorem 3.10. *X is an admissible surface with nonalgebraic normalization, of degree e and warp w. Then the semiuniversal deformation $p : \mathfrak{X} \to V$ of X has a smooth, 2-dimensional base V. The locally trivial deformations are parameterized by a smooth curve $V' \subset V$, and $V \setminus V'$ corresponds to smooth Kodaira surfaces of degree e/w.*

Proof. Since $h^1(\Theta_X) = 1$ and $h^2(\Theta_X) = 0$, the locally trivial deformations are unobstructed, and V' is a smooth curve. Since $\dim T^1_X = 2$ and since X deforms to smooth Kodaira surfaces of degree e/w, which move in a

2-dimensional family, the base V is a smooth surface. We saw in the proof of the previous proposition that the restriction map $T^1 \to H^0(X, \mathcal{T}_X^1)$ is surjective. According to [6, Prop. 2.5], the total space $\mathfrak{X}$ is smooth. Now Sard's Lemma implies that the projection $\mathfrak{X} \to V$ is smooth over $V \setminus V'$, at least after shrinking V. $\qquad\square$

Remark 3.11. The referee pointed out that V' should have an interpretation as versal deformation space of the elliptic curve $D \subset X$. This is indeed the case: Since the restriction of the Kodaira-Spencer map to $T_{V'}$ generates $H^1(\Theta_X)$ it suffices to show that the composition

$$H^1(\Theta_X) \longrightarrow H^1(\Theta_X \otimes \mathcal{O}_D) \longrightarrow H^1(\Theta_D) \qquad (3.2)$$

is surjective. This statement is stable under étale covers, so we may assume that there is an elliptic fibration $p : X \to B$ over the Néron 1-gon (Proposition 3.5). The double curve D is the fiber over the node of B, and by relative duality $R^1 p_*(\Theta_{X/B}) = \mathcal{O}_B$. The base change theorem implies that the restriction map $R^1 p_*(\Theta_{X/B}) \to H^1(\Theta_D)$ is surjective. On global sections this map is nothing but the composition of

$$\mathbb{C} = H^0(B, R^1 p_*(\Theta_{X/B})) \longrightarrow H^1(X, \Theta_X)$$

with (3.2). Therefore the latter map is surjective too.

4 Ruled Surfaces over Elliptic Curves

In this section we study admissible surfaces X whose normalization S is *algebraic and nonrational*, as in Proposition 2.5.

Let B' be an elliptic curve and $f : S \to B'$ a $\mathbb{P}^1$-bundle with two disjoint sections $C_1, C_2 \subset S$. Put $C = C_1 \cup C_2$ and $D = C_1$. Let $\psi : C_2 \to C_1$ be an isomorphism and $\varphi = \mathrm{id} \cup \psi : C \to D$ the induced double covering. The cocartesian diagram

$$\begin{array}{ccc} C & \longrightarrow & S \\ \varphi \downarrow & & \downarrow \nu \\ D & \longrightarrow & X \end{array}$$

defines a normal crossing surface X. By abuse of notation we write ψ also for the induced automorphism $(f|_{C_1}) \circ \psi \circ (f|_{C_2})^{-1}$ of B'.

Proposition 4.1. *We have $K_X = 0$ if and only if $\psi : B' \to B'$ is a translation.*

Proof. As in the proof of Proposition 3.2, K_X is trivial iff the images of $H^1(S, \mathcal{O}_S)$ and $H^1(D, \mathcal{O}_D)$ in $H^1(C, \mathcal{O}_C)$ coincide. Notice that the map $\phi^* : H^1(\mathcal{O}_D) \to H^1(\mathcal{O}_C) \simeq H^1(\mathcal{O}_D) \oplus H^1(\mathcal{O}_D)$ is the diagonal embedding. It does not depend on ϕ.

208   Stefan Schröer and Bernd Siebert

Suppose ψ is a translation. Then the images of $H^1(\mathcal{O}_S)$ and $H^1(\mathcal{O}_D)$ both agree with $(f|_C)^* H^1(\mathcal{O}_{B'})$.

Conversely, assume that ψ is not a translation. Then it acts on $H^1(\mathcal{O}_E)$ as a nontrivial root of unity. Consequently, $H^1(\mathcal{O}_S)$ and $H^1(\mathcal{O}_C)$ have different images in $H^1(\mathcal{O}_D)$. $\qquad\square$

The next task is to calculate the sheaf of first order deformations $\mathcal{T}_X^1$. We call the degree $e = -\min\{A^2 \mid A \subset S \text{ a section}\}$ of S also the *degree* of X. In our case $e \geq 0$, since $S \to B'$ has two disjoint sections. Let w be the order, possibly 0, of the automorphism $\psi : B' \to B'$. Call it the *warp* of X.

Proposition 4.2. *Suppose $K_X = 0$. Then X is d-semistable if and only if the warp w divides the degree e.*

Proof. We identify C_1, C_2 with B' via $f : S \to B'$. Set $\mathcal{L}_i = f_*(\mathcal{I}_{C_i}/\mathcal{I}_{C_i}^2)$. Then $\deg(\mathcal{L}_i) = -C_i^2$ and $\mathcal{L}_1 \otimes \mathcal{L}_2 = \mathcal{O}_{B'}$. On the other hand, $\mathcal{T}_X^1$ is the dual of $\mathcal{L}_1 \otimes \psi^*(\mathcal{L}_2)$. The kernel of the homomorphism

$$\phi_{\mathcal{L}} : B' \longrightarrow \operatorname{Pic}^0(B'), \quad b \longmapsto T_b^*(\mathcal{L}) \otimes \mathcal{L}^{-1}$$

is the subgroup $B'_e \subset B'$ of e-torsion points ([14, Lem. 4.7]). Choose an origin $0 \in B'$ and some $b \in B'$ such that ψ is the translation $T_b : B' \to B'$. So $\mathcal{T}_X^1$ is trivial if and only if $b \in B'_e$, which means $w \mid e$. $\qquad\square$

Remark 4.3. Here we see that a complete set of invariants of admissible X with nonrational algebraic normalization and positive degree is: The j-invariant of B', the degree $e > 0$, and a translation ψ of B' of finite order. In fact, $\mathbb{P}^1$ bundles over B' of degree $e > 0$ with two disjoint sections are of the form $\mathbb{P}(\mathcal{O}_{B'} \oplus L^\vee)$ with L a line bundle of degree e. Up to pull-back by translations the latter are all isomorphic.

We come to the construction of smoothings:

Theorem 4.4. *Suppose X is an admissible surface (Definition 2.1) with algebraic, nonrational normalization, of degree e (possibly 0) and warp w. Then X deforms to an elliptic principal bundle of degree e/w over an elliptic curve.*

Proof. In order to construct the desired smoothing, we first pass to a Galois covering of X. The ruling $f : S \to B'$ yields a bundle $X \to B$ over the isogenous elliptic curve $B = B'/\langle\psi\rangle$ whose fibres are Néron w-gons. Let $G \subset \operatorname{Pic}^0(B') \subset \operatorname{Aut}(B')$ be the group of order w generated by ψ. Consider the surface $S' = S \times G$ and the isomorphisms

$$\psi_j : C_2 \times \{\psi^j\} \longrightarrow C_1 \times \{\psi^{j+1}\}, \quad (s, \psi^j) \longmapsto (\psi(s), \psi^{j+1})$$

for $j \in \mathbb{Z}/w\mathbb{Z}$. The corresponding relation defines a normal crossing surface X' with w irreducible components. Clearly, X' is d-semistable with $K_{X'} = 0$.

Degenerations of Kodaira Surfaces   209

The surface S' is endowed with the free G-action $\psi : (s, \psi^j) \mapsto (s, \psi^{j+1})$. It descends to a free G-action on X' with quotient $X = X'/G$.

We proceed similarly as in Theorem 3.7. Let V be the smooth toric variety corresponding to the infinite fan in $\mathbb{Z}^3$ generated by the cones

$$\sigma_n = \big\langle (0, n, 1), (0, n+1, 1) \big\rangle, \quad n \in \mathbb{Z}.$$

The projection $\mathrm{pr}_3 : \mathbb{Z}^3 \to \mathbb{Z}$ defines a toric morphism $V \to \mathbb{C}$. The special fibre V_0 is the product of a Néron ∞-gon with $\mathbb{C}^*$. The automorphism

$$\Psi = \begin{pmatrix} 1 & 0 & 0 \\ 0 & 1 & 0 \\ 0 & 1 & 1 \end{pmatrix} \in \mathrm{SL}_3(\mathbb{Z}).$$

acting from the right on row vectors maps σ_n to σ_{n+1}. By abuse of notation we also write Ψ for the induced automorphism of V. Choose $\alpha \in \Delta^*$ with $B' = \mathbb{C}^*/\langle \alpha^w \rangle$ and $B = \mathbb{C}^*/\langle \alpha \rangle$. The automorphism $(t_1, t_2, t_3) \mapsto (\alpha t_1, t_1^{e/w} t_2, t_3)$ of $(\mathbb{C}^*)^3$ extends to another automorphism Φ of V. The action of $\langle \Phi, \Psi \rangle$ on the preimage $U \subset V$ of $\Delta \subset \mathbb{C}$ is proper and free. Hence $\mathcal{X}' = U/\langle \Phi^w, \Psi^w \rangle$ is a smooth 3-fold, endowed with a projection $\mathcal{X}' \to \Delta$. The general fibres $\mathcal{X}'_t$, $t \in \Delta^*$, are smooth Kodaira surfaces of degree e. The special fibre is $\mathcal{X}'_0 = X'$. The action of $(\Phi\Psi)^w$ descends to a free G-action on $\mathcal{X}'$, which coincides with the given action on $\mathcal{X}'_0 = X'$. Hence $\mathcal{X} = \mathcal{X}'/G$ is the desired smoothing. $\square$

We head for the calculation of the versal deformation of X.

Proposition 4.5. *Suppose $K_X = 0$. Then the following holds:*

(i) *If $e > 0$, then $h^0(\Theta_X) = 1$, $h^1(\Theta_X) = 2$, and $h^2(\Theta_X) = 1$.*
(ii) *If $e = 0$, then $h^0(\Theta_X) = 2$, $h^1(\Theta_X) = 3$, and $h^2(\Theta_X) = 1$.*

Proof. As in the proof of Lemma 3.8 we use $H^p(\Theta_X) \simeq H^{2-p}(\Omega^1_X/\tau_X)$ and the exact sequence

$$0 \longrightarrow \Omega^1_X/\tau_X \longrightarrow \Omega^1_S \longrightarrow \Omega^1_D \longrightarrow 0.$$

Recall $h^{1,0}(S) = h^{1,0}(D) = 1$. The map on the right in

$$0 \longrightarrow H^0(X, \Omega^1_X/\tau_X) \longrightarrow H^0(S, \Omega^1_S) \longrightarrow H^0(D, \Omega^1_D)$$

is zero, since it is an alternating sum, so $h^2(\Theta_X) = 1$. Next we consider

$$0 \longrightarrow H^0(D, \Omega^1_D) \longrightarrow H^1(X, \Omega^1_X/\tau_X) \longrightarrow H^1(S, \Omega^1_S) \longrightarrow H^1(D, \Omega^1_D).$$

We have $h^{1,1}(S) = 2$ and $h^{1,1}(D) = 1$. The class of a fibre $F \subset S$ in $H^{1,1}(S)$ maps to zero in $H^{1,1}(D)$. For $e = 0$, this also holds for the class of the section

210   Stefan Schröer and Bernd Siebert

$C_1 \subset S$, and $h^1(\Theta_X) = 3$ follows. For $e > 0$, the image of the section does not vanish, and $h^1(\Theta_X) = 2$ holds instead. Finally, the sequence

$$H^1(S, \Omega_S^1) \longrightarrow H^1(D, \Omega_D^1) \longrightarrow H^2(X, \Omega_X^1/\tau_X) \longrightarrow H^2(S, \Omega_S^1) \longrightarrow 0$$

is exact. Now $h^{1,2}(S) = 1$ yields $h^0(\Theta_X) = 2$ for $e = 0$, and $h^0(\Theta_X) = 1$ for $e > 0$. $\square$

Corollary 4.6. *Suppose $K_X = 0$ and $e = 0$. Then X does not deform to a smooth Kodaira surface.*

Proof. Write $S = \mathbb{P}(\mathcal{O}_{B'} \oplus \mathcal{L})$ for some invertible $\mathcal{O}_{B'}$-module $\mathcal{L}$ of degree 0. Moving the gluing parameter ψ and the isomorphism class of B' and $\mathcal{L}$ gives a 3-dimensional locally trivial deformation of X. Since $h^1(\Theta_X) = 3$, the space parameterizing the locally trivial deformations in the semiuniversal deformation $p : \mathcal{X} \to V$ of X is a smooth 3-fold. Now each fibre $\mathcal{X}_t$ deforms to an elliptic principal bundle of degree 0 (Thm. 4.4), and the embedding dimension of V is $\dim T_X^1 = 4$. This shows that V is a smooth 4-fold with an open dense set parameterizing complex tori. Hence no fibre $\mathcal{X}_t$ is isomorphic to a smooth Kodaira surface. $\square$

Proposition 4.7. *Let X be as in Theorem 4.4 with $e > 0$. Then $\dim T_X^0 = 1$, $\dim T_X^1 = 3$ and $\dim T_X^2 = 2$.*

Proof. In view of Proposition 4.5 we only have to show degeneracy of the spectral sequence of tangent cohomology $E_2^{p,q} = H^p(\mathcal{T}_X^q) \Rightarrow T_X^{p+q}$ at E_2 level, cf. Proposition 3.9. This is the case iff $T_X^1 \to H^0(\mathcal{T}_X^1)$ is surjective. Now by d-semistability $H^0(\mathcal{T}_X^1)$ is one-dimensional and any generator has no zeros. Geometrically degeneracy of the spectral sequence therefore means the existence of a deformation of X over $\Delta_\varepsilon := \operatorname{Spec} \mathbb{C}[\varepsilon]/\varepsilon^2$ that is not locally trivial. This is what we established by explicit construction in Theorem 4.4. More precisely, let $\mathcal{X} \to \Delta$ be a deformation of X. For $P \in D$ the image of the Kodaira-Spencer class $\kappa \in T_X^1$ of this deformation in $\mathcal{T}_{X,P}^1 \simeq \mathcal{E}xt^1(\Omega_{X,P}^1, \mathcal{O}_{X,P})$ is the Kodaira-Spencer class of the induced deformation of the germ of X along D. In appropriate local coordinates such deformations have the form $xy - \varepsilon f(z) = 0$ with f inducing the section of $\mathcal{E}xt^1(\Omega_{X,P}^1, \mathcal{O}_{X,P})$. Therefore $f \not\equiv 0$ iff the total space $\mathcal{X}$ is smooth at P. $\square$

Theorem 4.8. *Suppose X is admissible with algebraic, nonrational normalization, of degree $e > 0$ and warp w. Let $p : \mathcal{X} \to V$ be the semiuniversal deformation of X. Then $V = V_1 \cup V_2$ has two irreducible components, and the following holds:*

(i) *V_2 is a smooth surface and parameterizes the locally trivial deformations.*
(ii) *$V_1 \cap V_2$ is a smooth curve and parameterizes the d-semistable locally trivial deformations.*

(iiii) V_1 *is a smooth surface, and* $V_1 \setminus V_1 \cap V_2$ *parameterizes smooth Kodaira surfaces of degree* e/w.

Proof. A similar situation has been found by Friedman in his study of deformations of d-semistable K3 surfaces. We follow the proof of [6, Thm. 5.10] closely.

Deformation theory provides a holomorphic map $h : T_X^1 \to T_X^2$ with $V = h^{-1}(0)$, whose linear term is zero, and whose quadratic term is given by the Lie bracket $1/2[v,v]$ (cf. e.g. [19]). Moving the two parameters $j(E) \in \mathbb{C}$, $\psi \in B'$ and using $h^1(\Theta_X) = 2$, we see that the locally trivial deformations are parameterized by a smooth surface $V_2 \subset T_X^1$. Let $h_2 : T_X^1 \to \mathbb{C}$ be a holomorphic map with $V_2 = h_2^{-1}(0)$, and $h_1 : T_X^1 \to T_X^2$ a holomorphic map with $h = h_1 h_2$. Let $L_1 : T_X^1 \to T_X^2$ and $L_2 : T_X^1 \to \mathbb{C}$ be the corresponding tangential maps and set $V_1 = h_1^{-1}(0)$.

We proceed by showing that $V_1 \cap V_2$ is a smooth curve, or equivalently that the intersection of $\mathrm{kern}(L_1)$ with $H^1(\Theta_X) = \mathrm{kern}(L_2)$ is 1-dimensional. The smoothing of X constructed in Theorem 4.4 obviously has a smooth total space. As in the proof of Proposition 3.9 we see that its Kodaira-Spencer class $k \in T_X^1$ generates $\mathcal{T}_X^1$.

Let $L : H^1(\Theta_X) \to H^1(\mathcal{T}_X^1)$ be the linear map $v \mapsto [v,k]$. By [6, Prop. 5], its kernel are the first order deformations which remain d-semistable. Since we can destroy d-semistability by moving the gluing $\psi \in \mathrm{Pic}^0(E)$, this kernel is 1-dimensional. As in [6, p. 109], one shows that for $v \in H^1(\Theta_X)$:

$$L(v) = [v,k] = 1/2[v+k, v+k] = L_1(v+k) \cdot L_2(v+k) = L_1(v) \cdot L_2(k) .$$

Using $L_2(k) \neq 0$ we infer $\mathrm{kern}\, L_1 \cap H^1(\Theta_X) = \mathrm{kern}\, L \simeq \mathbb{C}$. It follows that $V_1 \cap V_2$ is a smooth curve, which parameterizes the d-semistable locally trivial deformations, and V_1 must be a smooth surface. Our local computation together with the interpretation of $j(E) \in \mathbb{C}$ as coordinate on $V_1 \cap V_2$ shows that V_1 is nothing but the base of the smoothing of X constructed in Theorem 4.4. In particular, $V_1 \setminus V_1 \cap V_2$ parameterizes smooth Kodaira surfaces. $\square$

5 Rational Surfaces and Honeycomb Degenerations

The goal of this section is to analyze rational admissible surfaces. According to Proposition 2.6, the ramification curve $C \subset S$ of the normalization $\nu : S \to X$ is a Néron 6-gon $C = C_0 \cup \ldots \cup C_5$ with $K_S = -C$. We suppose $C_i \cap C_{i+1} \neq \emptyset$, regarding the indices as elements in $\mathbb{Z}/6\mathbb{Z}$. The singular locus $D \subset X$ is the seminormal curve with normalization $\mathbb{P}^1 \times \mathbb{Z}/3\mathbb{Z}$ and imposed relations $(0,i) \sim (0,j)$ and $(\infty,i) \sim (\infty,j)$ for $i,j \in \mathbb{Z}/3\mathbb{Z}$. The gluing $\varphi : C \to D$ identifies pairs of irreducible components in C. Since the resulting surface should be normal crossing the fibers of $S \to X$ have cardinality at most 3. There are two possibilities left. The first alternative is

$\varphi(C_i) = \varphi(C_{i+3})$ and $\varphi(C_i \cap C_{i+1}) = \varphi(C_{i+2} \cap C_{i+3})$ for all $i \in \mathbb{Z}/6\mathbb{Z}$. We call it *untwisted* gluing. The second alternative is $\varphi(C_i) = \varphi(C_{i+3})$, $\varphi(C_{i+1}) = \varphi(C_{i-1})$, $\varphi(C_{i+2}) = \varphi(C_{i-2})$ and $\varphi(C_i \cap C_{i+1}) = \varphi(C_{i-1} \cap C_{i-2})$ for some $i \in \mathbb{Z}/6\mathbb{Z}$. Call it *twisted* gluing. The distinction is important for the canonical class:

Proposition 5.1. *The condition $K_X = 0$ holds if and only if the gluing $\varphi : C \to D$ is untwisted.*

Proof. As in Propositions 3.2 and 4.1 the condition $K_X = 0$ is equivalent to $H^2(\mathcal{O}_X) \neq 0$. By the exact sequence

$$H^1(S, \mathcal{O}_S) \oplus H^1(D, \mathcal{O}_D) \longrightarrow H^1(C, \mathcal{O}_C) \longrightarrow H^2(X, \mathcal{O}_X) \longrightarrow 0$$

and with $h^{0,1}(S) = 0$ this holds precisely if $\varphi^* : H^1(\mathcal{O}_D) \to H^1(\mathcal{O}_C)$ vanishes. The bicoloured graphs attached to the curves C, D ([5], section 3.5) are

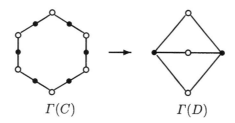

$\Gamma(C)$ $\qquad\qquad$ $\Gamma(D)$

Here the white vertices are the irreducible components, the black vertices are the singularities, and the edges are the ramification points on the normalization. Since the curves in question are seminormal, the map $\varphi^* : H^1(\mathcal{O}_D) \to H^1(\mathcal{O}_C)$ coincides with $\varphi^* : H^1(\Gamma(D), \mathbb{C}) \to H^1(\Gamma(C), \mathbb{C})$. A direct calculation left to the reader shows that it vanishes if and only if φ is untwisted. $\square$

From now on we assume that the gluing $\varphi : C \to D$ is untwisted. According to Proposition 2.6, the surface S is obtained from a Hirzebruch surface $f' : S' \to \mathbb{P}^1$ by blowing-up twice. Our conventions are that $C_0, C_3 \subset S$ are sections of the induced fibration $f : S \to \mathbb{P}^1$, that $C_1 \cup C_2$, $C_4 \cup C_5$ are the fibres over $0, \infty \in \mathbb{P}^1$, and that C_1, C_4 are exceptional for the contraction $r : S \to S'$. Let us also assume that the image of C_0 in S' is a minimal section. If $e \geq 0$ is the degree of the Hirzebruch surface S', this means $C_0^2 = -e - 1$ and $C_3^2 = e - 1$.

Set $D_i = \varphi(C_i)$ for $i \in \mathbb{Z}/3\mathbb{Z}$. The space of all untwisted gluings $\varphi : C \to D$ is a torsor under

$$\operatorname{Aut}^0(D) = \prod \operatorname{Aut}^0(D_i) \simeq (\mathbb{C}^*)^3. \tag{5.1}$$

The action of $\phi \in \operatorname{Aut}^0(D)$ is given by composing $\varphi \mid C_0 \cup C_1 \cup C_2$ with ϕ. We call $\varphi \mid C_0 \cup C_3$ the *vertical gluing*. The ruling $f : S \to \mathbb{P}^1$ yields a

Degenerations of Kodaira Surfaces 213

preferred vertical gluing, which identifies points of the same fibre. Every other vertical gluing differs from the preferred one by a *vertical gluing parameter* $\zeta \in \mathbb{C}^*$. We call its order $w \geq 0$ the *warp* of X. The warp is important for the calculation of $\mathcal{T}_X^1$:

Proposition 5.2. *The surface X is d-semistable if and only if the warp w divides the degree e.*

Proof. The inclusions $D_i \cup D_j \subset D$ give an injection

$$\operatorname{Pic}(D) \subset \prod_{i \neq j} \operatorname{Pic}(D_i \cup D_j) \, .$$

We proceed by calculating the class of $\mathcal{T}_X^1 \mid D_0 \cup D_1$. Consider the normal crossing surface $\bar{S}$ defined by the cocartesian diagram

$$
\begin{array}{ccc}
C_2 \cup C_5 & \longrightarrow & S \\
\downarrow & & \downarrow \\
D_2 & \longrightarrow & \bar{S} \, .
\end{array}
$$

Let $\bar{C}_i \subset \bar{S}$ be the images of $C_i \subset S$. The ideals $\mathcal{I}_{01}, \mathcal{I}_{34} \subset \mathcal{O}_{\bar{S}}$ of the Weil divisors $\bar{C}_{01} = \bar{C}_0 \cup \bar{C}_1$ and $\bar{C}_{34} = \bar{C}_3 \cup \bar{C}_4$ are invertible. A local computation shows that

$$(\mathcal{T}_X^1 \mid D_0 \cup D_1)^\vee \simeq \mathcal{I}_{01}/\mathcal{I}_{01}^2 \otimes \psi^*(\mathcal{I}_{34}/\mathcal{I}_{34}^2) \, ,$$

where $\psi : \bar{C}_{01} \longrightarrow \bar{C}_{34}$ is the isomorphism obtained from the induced gluing $\bar{\varphi} : \bar{C}_{01} \cup \bar{C}_{34} \to D_0 \cup D_1$. Another local computation gives

$$\mathcal{I}_{01}/\mathcal{I}_{01}^2 \mid \bar{C}_0 \simeq \mathcal{I}_{C_0}/\mathcal{I}_{C_0}^2(-C_0 \cap C_{-1}) \, , \quad \mathcal{I}_{01}/\mathcal{I}_{01}^2 \mid \bar{C}_1 \simeq \mathcal{I}_{C_1}/\mathcal{I}_{C_1}^2(-C_0 \cap C_1) \, .$$

It follows that $\mathcal{T}_X^1 \mid D_0 \cup D_1$ is trivial if and only if $\zeta^e = 1$. A similar argument applies to $D_0 \cup D_2$ and $D_1 \cup D_2$. $\qquad \square$

Remark 5.3. The previous two propositions yield the following classification of admissible X with rational normalization and positive degree e: According to (5.1), the isomorphism class of X is determined by $e > 0$ and 3 gluing parameters. By Proposition 5.2 the vertical one is an e-th root of unity ζ. For (e, ζ) fixed the automorphisms of S act on the remaining two horizontal gluing parameters with quotient isomorphic to $\mathbb{C}^*$.

We come to the existence of smoothings:

Theorem 5.4. *Suppose X is an admissible surface (Definition 2.1) with rational normalization, of degree e (possibly 0) and warp w. Then X deforms to an elliptic principal bundle of degree e/w over an elliptic curve.*

Proof. It should not be too surprising that the construction is a modification of both the constructions in Theorems 3.7 and 4.4.

First, we simplify matters by passing to a Galois covering. Let $G \subset \mathbb{C}^*$ be the group of order w generated by the vertical gluing parameter $\zeta \in \mathbb{C}^*$. Since $w|e$, the G-action $z \mapsto \zeta z$ on $\mathbb{P}^1$ lifts to a G-action on the Hirzebruch surface S', and hence to the blowing-up S. The diagonal action on $\tilde{S} = S \times G$ is free. Let $\psi_i : C_i \to C_{i+3}$ be the isomorphisms induced by the gluing $\varphi : C \to D$. This gives G-equivariant isomorphisms

$$\psi_{ij} : C_i \times \{\zeta^j\} \longrightarrow C_{i+3} \times \{\zeta^{j+1}\}, \quad (s, \zeta^j) \longmapsto (\psi_i(s), \zeta^{j+1}).$$

Let X' be the normal crossing surface obtained from $\tilde{S}$ modulo the relation imposed by the ψ_{ij}. Then G acts freely on X' with quotient $X = X'/G$.

The next task is to construct a G-equivariant smoothing of X'. Again toric geometry enters the scene. Consider the infinite fan in $\mathbb{Z}^4$ generated by the cones

$$\sigma_{m,n} = \left\langle (m, e \cdot \binom{m}{2}, 1, 0), (m+1, e \cdot \binom{m+1}{2}, 1, 0), (0, n, 0, 1), (0, n+1, 0, 1) \right\rangle$$

for $m, n \in \mathbb{Z}$. Let V be the corresponding 4-dimensional smooth toric variety. The projection $\mathrm{pr}_{34} : \mathbb{Z}^4 \to \mathbb{Z}^2$ defines a toric morphism $\mathrm{pr}_{34} : V \to \mathbb{C}^2$. The special fibre V_0 is isomorphic to a bundle of Néron ∞-gons over the Néron ∞-gon. Let $f : \hat{\mathbb{C}}^2 \to \mathbb{C}^2$ be the blowing-up of the origin $0 \in \mathbb{C}^2$ with the closed toric orbits removed and $E \subset \hat{\mathbb{C}}^2$ the exceptional set. It is isomorphic to $\mathbb{C}^*$. The cartesian diagram

defines a smooth toric 4-fold $\hat{V}$, which is an open subset of the blowing-up of V with center $V_0 \subset V$. The exceptional divisor $\hat{V}_E = \hat{V}_0$ of $\hat{V} \to V$ is isomorphic to $E \times V_0$. The exceptional divisor E will be a parameter space of the whole construction via the $\mathbb{C}^*$-bundle $\hat{\mathbb{C}} \to E$ induced by the canonical map $\mathbb{C}^2 \setminus 0 \to \mathbb{P}^1$.

Let $Z \subset \hat{V}$ be the set of 1-dimensional toric orbits. Each fibre Z_t, $t \in E$, consists of the discrete set of non-normal-crossing singularities in $\hat{V}_t \simeq V_0$, as illustrated by Fig. 1. Let $\bar{V} \to \hat{V}$ be the blowing-up with center $Z \subset \hat{V}$. The exceptional divisor $\bar{Z} \subset \bar{V}$ is easy to determine. Each fibre $\hat{V}_t$, $t \in E$, is smooth, except for the points in Z_t. At such points a local equation for $\hat{V}_t$ is $T_1 T_2 = t T_3 T_4$, which is an affine cone over a smooth quadric in $\mathbb{P}^2$. Hence $\bar{Z}$ is a disjoint union of smooth quadrics in $\mathbb{P}^3$, each isomorphic to $\mathbb{P}^1 \times \mathbb{P}^1$. The whole exceptional divisor is $\bar{Z} \simeq E \times Z \times \mathbb{P}^1 \times \mathbb{P}^1$. The picture over a fixed $t \in E$ is depicted in Fig. 2. Here the octagons lie on the strict transform of V_t, and the squares are contained in the exceptional divisor $\bar{Z}_t$.

Degenerations of Kodaira Surfaces   215

Fig. 1

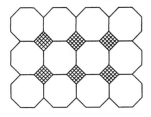
Fig. 2

We seek to contract $\bar{Z} \subset \bar{V}$ along one of the two rulings of $\mathbb{P}^1 \times \mathbb{P}^1$. Let $F \subset \bar{Z}$ be a $\mathbb{P}^1$-fibre. A local calculation shows that $\mathcal{O}_F(\bar{Z})$ has degree -1, and that the Cartier divisors $K_{\bar{V}}$ and $\mathcal{O}_{\bar{V}}(\bar{Z})$ coincide in a neighborhood of $F \subset \bar{V}$. So the Nakano contraction criterion [17] applies, and there exists a contraction $\bar{V} \to \hat{V}$ which restricts to the projection

$$\mathrm{pr}_{123} : \bar{Z} = E \times Z \times \mathbb{P}^1 \times \mathbb{P}^1 \longrightarrow E \times Z \times \mathbb{P}^1 \ .$$

The special fibre $\tilde{V}_0$ resembles an infinite system of honeycombs:

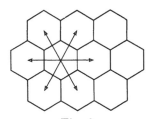
Fig. 3

Finally, we want to arrive at compact surfaces, so we seek to divide out a cocompact group action. Consider the two commuting automorphisms

$$\Phi = \begin{pmatrix} 1 & e & 0 & 0 & 0 \\ 0 & 1 & 0 & 0 & 0 \\ 1 & 0 & 1 & 0 & 0 \\ 0 & 0 & 0 & 1 & 0 \\ 0 & 1 & 0 & 0 & 1 \end{pmatrix} \quad \text{and} \quad \Psi = \begin{pmatrix} 1 & 0 & 0 & 0 & 0 \\ 0 & 1 & 0 & 0 & 0 \\ 0 & 0 & 1 & 0 & 0 \\ 0 & 1 & 0 & 1 & 0 \\ 0 & 0 & 0 & 0 & 1 \end{pmatrix}$$

of $\mathbb{Z}^5$ acting from the right on row vectors. We regard our fan in $\mathbb{Z}^4$ as a fan in $\mathbb{Z}^5 = \mathbb{Z}^4 \oplus \mathbb{Z}$. Torically, the trivial $\mathbb{Z}$-factor amounts to going over to $V \times \mathbb{C}^*$, which will give a horizontal gluing parameter. Since $\Phi(\sigma_{m,n}) = \sigma_{m+1,n}$ and $\Psi(\sigma_{m,n}) = \sigma_{m,n+1}$, we get an induced action of the discrete group $\mathbb{Z}^2 = \langle \Phi, \Psi \rangle$ on $V \times \mathbb{C}^*$. Let $U \subset V$ be the preimage of $\Delta^2 \subset \mathbb{C}^2$. The action is proper and free on $U \times \mathbb{C}^*$. Let $\tilde{U} \subset \tilde{V}$ be the corresponding preimage. It is easy to see that the action on $V \times \mathbb{C}$ induces an action on $\tilde{V} \times \mathbb{C}^*$ which is proper and

216   Stefan Schröer and Bernd Siebert

free on $\tilde{U} \times \mathbb{C}^*$. So the quotient $\mathfrak{X}' = (\tilde{U} \times \mathbb{C}^*)/\langle \Phi, \Psi^w \rangle$ is a smooth complex 5-fold. The action on the special fibre is indicated in Fig. 3 by the arrows. The general fibres $\mathfrak{X}'_t$, $(t, \lambda) \in f^{-1}(\Delta^* \times \Delta^*) \times \mathbb{C}^*$ are elliptic bundles of degree e over the elliptic curve $\mathbb{C}^*/\langle t_1 \rangle$ with fibre $\mathbb{C}^*/\langle t_2 \rangle$, where $f(t) = (t_1, t_2)$. Some special fibre $\mathfrak{X}'_t$, $t \in E \times \mathbb{C}^*$, is isomorphic to X', since t_5 moves through all horizontal gluings.

It remains to extend the G-action on X' to a G-action on $\mathfrak{X}'$. The automorphism

$$(\mathbb{C}^*)^5 \longrightarrow (\mathbb{C}^*)^5, \quad (t_1, t_2, t_3, t_4, t_5) \longmapsto (\zeta t_1, t_2 t_4, t_3, t_4, t_5)$$

of the torus extends to an automorphism of the torus embedding V. It induces a free G-action on $\mathfrak{X}'$, which is the desired extension. Set $\mathfrak{X} = \mathfrak{X}'/G$. General fibres $\mathfrak{X}_t$, $t \in (\mathbb{C}^2 \setminus 0) \times \mathbb{C}^*$ are elliptic bundles of degree e/w over an elliptic curve. Some special fibre $\mathfrak{X}_t$, $t \in E \times \mathbb{C}^*$ is isomorphic to X. $\qquad \square$

We turn our attention to the versal deformation of X and calculate the relevant cohomology groups:

Proposition 5.5. *Suppose $K_X = 0$. Then the following holds:*

(i) *If $e > 0$, then $h^0(\Theta_X) = 1$, $h^1(\Theta_X) = 2$, and $h^2(\Theta_X) = 0$.*
(ii) *If $e = 0$, then $h^0(\Theta_X) = 2$, $h^1(\Theta_X) = 3$, and $h^2(\Theta_X) = 0$.*

Proof. As in Lemmas 3.8 and 4.5 we use $H^p(X, \Theta_X) \simeq H^{2-p}(X, \Omega^1_X/\tau_X)$ and the exact sequence

$$0 \longrightarrow \Omega^1_X/\tau_X \longrightarrow \Omega^1_S \longrightarrow \Omega^1_{\tilde{D}} \longrightarrow 0 \,.$$

Here $\tilde{D}$ is the normalization of D. The inclusion $H^0(\Omega^1_X/\tau_X) \subset H^0(\Omega^1_S)$ and $b_1(S) = 0$ gives $h^2(\Theta_X) = 0$. Next consider the exact sequence

$$0 \longrightarrow H^1\left(X, \Omega^1_X/\tau_X\right) \longrightarrow H^1\left(S, \Omega^1_S\right) \longrightarrow H^1\left(\tilde{D}, \Omega^1_{\tilde{D}}\right).$$

The task is to determine the rank of the map on the right. This is done as in [6], p. 91: One has to compute the images in $H^{1,1}(\tilde{D})$ of the classes of $C_i \subset S$ using intersection numbers on S. For $e > 0$ this gives $h^1(\Theta_X) = 2$, whereas for $e = 0$ the result is $h^1(\Theta_X) = 3$. We leave the actual computation to the reader. Finally, the exact sequence

$$H^1\left(S, \Omega^1_S\right) \longrightarrow H^1\left(\tilde{D}, \Omega^1_{\tilde{D}}\right) \longrightarrow H^2\left(X, \Omega^1_X/\tau_X\right) \longrightarrow H^2\left(S, \Omega^1_S\right),$$

together with the preceding observations and $b_3(S) = 0$ gives the stated values for $h^0(\Theta_X)$. $\qquad \square$

Corollary 5.6. *Suppose $K_X = 0$ and $e = 0$. Then X does not deform to a smooth Kodaira surface.*

Proof. The locally trivial deformations of X are unobstructed since $h^2(\Theta_X) = 0$. They define a smooth 3-fold $V' \subset V$ in the base of the semi-universal deformation $\mathfrak{X} \to V$. It is easy to see that it is given by the three gluing parameters in $\varphi : C \to D$. Since each fibre $\mathfrak{X}_t$ deforms to a complex torus (Thm. 5.4), V is smooth of dimension 4 and no fibre can be isomorphic to a smooth Kodaira surface. $\square$

Proposition 5.7. *Let X be as in Theorem 5.4 with $e > 0$. Then $\dim T_X^0 = 1$, $\dim T_X^1 = 3$ and $\dim T_X^2 = 2$.*

Proof. This is shown as the similar statement of Proposition 4.7. $\square$

Theorem 5.8. *Suppose X is admissible with rational normalization, of degree $e > 0$ and warp w. Let $p : \mathfrak{X} \to V$ be the semiuniversal deformation of X. Then $V = V_1 \cup V_2$ consists of two irreducible components, and the following holds:*

(i) *V_2 is a smooth surface parameterizing the locally trivial deformations.*
(ii) *$V_1 \cap V_2$ is a smooth curve parameterizing the locally trivial deformations which remain d-semistable.*
(iii) *$V_1 \setminus V_1 \cap V_2$ parameterizes smooth Kodaira surfaces of degree e/w.*

Proof. The argument is as in the proof of Theorem 4.8. $\square$

6 The Completed Moduli Space and its Boundary

In this final section, we take up a global point of view and analyze degenerations of smooth Kodaira surfaces in terms of moduli spaces. Here we use the word "moduli space" in the broadest sense, namely as a topological space whose points correspond to Kodaira surfaces. We do not discuss wether it underlies a coarse moduli space or even an analytic stack or analytic orbispace.

Let $\mathfrak{K}_d = \mathbb{C} \times \Delta^*$ be the moduli space of smooth Kodaira surfaces of degree $d > 0$, and $\mathfrak{K} = \cup_{d>0}\mathfrak{K}_d$ their union. The points of $\mathfrak{K}$ correspond to the isomorphism classes $[X]$ of smooth Kodaira surfaces. The topology on $\mathfrak{K}$ is induced by what we suggest to call the *versal topology* on the set $\mathfrak{M}$ of isomorphism classes of all compact complex spaces. The versal topology is the finest topology on $\mathfrak{M}$ rendering continuous all maps $V \to \mathfrak{M}$ defined by flat families $\mathfrak{X} \to V$ which are versal for all fibres. The complex structure on $\mathfrak{K}$ comes from Hodge theory: One can view $\mathfrak{K}_d$ as the period domain of polarized pure Hodge structures of weight 2 on $H^2(X, \mathbb{Z})$/torsion divided by the automorphism group of this lattice. According to [4], the induced structure of a ringed space on $\mathfrak{K}_d$ is the usual complex structure on $\mathbb{C} \times \Delta^*$.

Let $\mathfrak{K} \subset \overline{\mathfrak{K}}$ be the space obtained by adding all admissible surfaces in the closure of $\mathfrak{K} \subset \mathfrak{M}$. The surfaces parameterized by the boundary $\mathfrak{B} = \overline{\mathfrak{K}} \setminus \mathfrak{K}$ are called *d-semistable* Kodaira surfaces. These are nothing but the admissible

218   Stefan Schröer and Bernd Siebert

surfaces deforming to smooth Kodaira surfaces. The boundary decomposes into three parts

$$\mathfrak{B} = \mathfrak{B}^h \cup \mathfrak{B}^r \cup \mathfrak{B}^e$$

according to the three types of admissible surfaces. Here $\mathfrak{B}^h$ refers to the surfaces whose normalization are Hopf surfaces, $\mathfrak{B}^r$ to the surfaces with rational normalization, and $\mathfrak{B}^e$ to surfaces whose normalization is ruled over an elliptic base. We refer to these parts of $\overline{\mathfrak{K}}$ as Hopf, rational and elliptic ruled stratum respectively.

Proposition 6.1. *The irreducible components of the boundary $\mathfrak{B}$ are smooth complex curves. The components in $\mathfrak{B}^h$ are isomorphic to $\Delta^* \simeq \{\infty\} \times \Delta^*$, the components in $\mathfrak{B}^r$ are isomorphic to $\mathbb{C}^*$, and the components in $\mathfrak{B}^e$ are isomorphic to $\mathbb{C} \simeq \mathbb{C} \times \{0\}$.*

Proof. This follows from Remarks 3.6, 4.3 and 5.3. ☐

Locally, the completion $\mathfrak{K} \subset \overline{\mathfrak{K}}$ is isomorphic to the blowing-up of $(\infty, 0) \in \mathbb{P}^1 \times \Delta$, but with the points $0, \infty \in \mathbb{P}^1$ on the exceptional divisor removed. This follows in particular from the construction in Theorem 5.4 of a family over the blow up of $\mathbb{C}^2$ with 2 points removed, with fiber over $(t_1, t_2) \in \mathbb{C}^* \times \mathbb{C}^*$ a smooth Kodaira surface with invariants $(j, \alpha) = (\exp(t_1), t_2)$. To complete further we need to enlarge the class of generalized Kodaira surfaces. The least one would hope for is that any family $\mathfrak{X}^* \to \Delta^*$ of generalized Kodaira surfaces, that can be completed to a proper family over Δ, can be completed by a generalized Kodaira surface.

We first show that this is impossible if we consider only normal crossing surfaces.

Theorem 6.2. *Let $\mathfrak{X} \to \Delta$ be a degeneration of d-semistable Kodaira surfaces with elliptically ruled normalization. Assume that $\mathfrak{X}$ is bimeromorphic to a Kähler manifold, and that the j-invariant of the base of $\mathfrak{X}_t$ tends to 0 with $t \in \Delta$. Then $\mathfrak{X}_0$ is not of normal crossing type.*

Proof. Let $\tilde{\mathfrak{X}} \to \mathfrak{X}$ be the normalization. Let $\mathfrak{B}$ be the component of the Douady space of holomorphic curves in $\tilde{\mathfrak{X}}$ that contain a fiber of $\tilde{\mathfrak{X}}_t \to B_t$ for general $t \in \Delta$. Since these curves are contained in fibers of $\tilde{\mathfrak{X}} \to \Delta$ there is a map $\mathfrak{B} \to \Delta$. By the Kähler assumption this map is proper [8]. Let $\tilde{\mathfrak{X}}' \to \mathfrak{B}$ be the universal family. The universal map $\tilde{\mathfrak{X}}' \to \tilde{\mathfrak{X}}$ is an isomorphism over Δ^*, hence bimeromorphic. By desingularisation we may dominate this map by successively blowing up $\tilde{\mathfrak{X}}$ in smooth points and curves. This can be arranged to keep the property "normal crossing". In other words, we can assume that $\tilde{\mathfrak{X}}' = \tilde{\mathfrak{X}}$ is a fibration by rational curves over a degenerating family $\mathfrak{B} \to \Delta$ of elliptic curves.

Let $\mathfrak{C}', \mathfrak{C}'' \subset \tilde{\mathfrak{X}}$ be the two components of the preimage of the singular locus of $\mathfrak{X}$ mapping onto Δ. For any $b \in \mathfrak{B}$ the corresponding rational curve

Degenerations of Kodaira Surfaces   219

F_b intersects $\mathfrak{C}', \mathfrak{C}''$ in one point each. Let $b_0 \in \mathfrak{B}_0$ be a node of the nodal elliptic curve over $0 \in \Delta$. Then each intersection $\mathfrak{C}' \cap F_{b_0}$ and $\mathfrak{C}'' \cap F_{b_0}$ gives a point of multiplicity at least 2 on $\tilde{\mathfrak{X}}_0$. As $\mathfrak{C}', \mathfrak{C}''$ are identified under $\tilde{\mathfrak{X}} \to \mathfrak{X}$ this leads to a point of multiplicity at least 4 on $\mathfrak{X}_0$. $\square$

There are however various completions if we admit *products of normal crossing singularities*. In dimension 2 the only such singularity that is not normal crossing is a point (X, x) of multiplicity 4. It has as completed local ring

$$\mathcal{O}^\wedge_{X,x} = \mathbb{C}[[T_1, \ldots, T_4]]/(T_1 T_2, T_3 T_4) \ .$$

In particular, it is a complete intersection and hence still Gorenstein. We will call (X, x) a *quadrupel point*. The singular locus $C \subset X$ consists of the four coordinate lines $C_1, \ldots, C_4$, and C_i, C_j lie on the same irreducible component iff $i - j \not\equiv 2$ modulo 4. The embedding dimension of (X, x) is 4. It is therefore not embeddable into a smooth 3-fold. Its appearance here is perhaps not so surprising, as it occurs also in certain compactifications of moduli of polarized abelian surfaces [16].

We were nevertheless unable to select a natural class of generalized Kodaira surfaces that would satisfy the mentioned completeness property. Therefore we content ourselves to define two surfaces with quadrupel points, connecting the elliptic ruled stratum to the Hopf stratum and to the rational stratum respectively. A surface connecting the Hopf stratum with the rational stratum could not be found, although such surfaces should probably exist.

Fix an integer $e > 0$. Let S be the Hirzebruch surface of degree e. Choose two sections $C_0, C_2 \subset S$ with $C_0^2 = -e$, $C_2^2 = e$, and let C_1, C_3 be the fibers over $0, \infty \in \mathbb{P}^1$. We define the surface X_1 by gluing C_1, C_3 by any isomorphism (all choices give isomorphic results), and C_0, C_2 by an isomorphism of finite order w over $\mathbb{P}^1$. Note that X_1 is normal crossing except at one quadrupel point. As in the proof of Proposition 5.1 one checks $K_{X_1} = 0$.

Proposition 6.3. *Suppose w divides e. There is a family $\mathfrak{X} \to \mathbb{C}^2$ with the following properties:*

1. $\mathfrak{X}_0 \simeq X_1$
2. *the fibers over $\mathbb{C}^* \times \{0\}$ are d-semistable Kodaira surfaces of warp w and degree e with elliptic ruled normalization*
3. *the fibers over $\{0\} \times \mathbb{C}^*$ are d-semistable Kodaira surfaces of warp w and degree e with nonalgebraic normalization*

Proof. In the proof of Theorem 5.4 we constructed a toric morphism $V \to \mathbb{C}^2$ with central fiber a bundle of Néron ∞-gons over the Néron ∞-gon. The restriction of $\Phi, \Psi \in \mathrm{GL}(\mathbb{Z}^5)$ defined there to $\mathbb{Z}^4 \simeq \mathbb{Z}^4 \oplus \{0\}$ defines an action of $\mathbb{Z}^2$ on the fan defining V. The induced action on V is proper and free, and commutes with the map to $\mathbb{C}^2$. We obtain a proper family $\mathfrak{X}' = V/\mathbb{Z}^2 \to \mathbb{C}^2$. By construction of V there is also an action of $G \simeq \mathbb{Z}/w\mathbb{Z}$ on $\mathfrak{X}'/\mathbb{C}^2$. The family $\mathfrak{X}'/G \to \mathbb{C}^2$ has the desired properties. $\square$

Finally we study a surface connecting the rational and the elliptic stratum, under the expense of changing the warp. Again we fix an integer $e > 0$. Let $S = S' \cup S''$ be the disjoint union of two Hirzebruch surfaces of degrees $e+1$ and $e-1$ respectively. Choose two sections $C'_0, C'_2 \subset S'$ with $(C'_0)^2 = -(e+1)$ and $(C'_2)^2 = e+1$, and let C'_1, C'_3 be the fibres over $0, \infty \in \mathbb{P}^1$. Make the analogous choices C''_0, C''_2 etc. for S''. We glue C'_i to C''_{i+2} in a way that preserves the intersection points indicated in the following figure by circles and crosses.

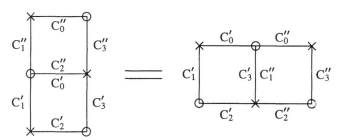

Fig. 4

For odd i the isomorphism $C'_i \to C''_i$ can be chosen arbitrarily; for even i we glue compatibly with the projections $S' \to \mathbb{P}^1$, $S'' \to \mathbb{P}^1$. The result is a reduced surface X_2 that is normal crossing except at two points corresponding to the circles and the crosses in the figure. It is a bundle of Néron 1-gons over $\mathbb{P}^1$. Again as in the proof of Proposition 5.1, one verifies that $K_{X_2} = 0$ holds.

Proposition 6.4. 1. X_2 deforms to d-semistable Kodaira surfaces with rational normalization of degree e and warp 1.
2. X_2 deforms to d-semistable Kodaira surfaces with elliptic ruled normalization of degree $2e$ and warp 2.

Proof. (i) The idea is to modify a fiberwise degeneration of Hirzebruch surfaces. Let $\widetilde{S}'$, $\widetilde{S}''$ be Hirzebruch surfaces of degree e. Define curves $\widetilde{C}'_i \subset \widetilde{S}'$ and $\widetilde{C}''_i \subset \widetilde{S}''$ as above.

Let $p : \widetilde{\mathfrak{Z}} \to \Delta$ be a flat family whose general fibres $\widetilde{\mathfrak{Z}}_t$, $t \neq 0$, are Hirzebruch surfaces of degree e; the closed fiber $\widetilde{\mathfrak{Z}}_0$ is the union of $\widetilde{S}'$ and $\widetilde{S}''$ with $\widetilde{C}'_2$ and $\widetilde{C}''_0$ identified, analogously to the left-half of Fig. 4. Such a family can be constructed either by toric methods or as follows. Let $L \to \mathbb{P}^1$ be the $\mathbb{C}^*$-bundle of degree e. Let $\mathfrak{C} \to \Delta$ be a versal deformation of the nodal rational curve $xy = 0$ in $\mathbb{P}^2$. The action $t \cdot (z, w) = (tz, t^{-1}w)$ of $\mathbb{C}^*$ on $\mathfrak{C}_0$ extends to an action on $\mathfrak{C}$ over Δ. We may take $\widetilde{\mathfrak{Z}} = (L \times \mathfrak{C})/\mathbb{C}^*$ with $\mathbb{C}^*$ acting diagonally.

The points $\widetilde{C}'_0 \cap \widetilde{C}'_1$ and $\widetilde{C}''_2 \cap \widetilde{C}''_3$ on $\widetilde{\mathfrak{Z}}_0$ can be lifted to sections of $\widetilde{\mathfrak{Z}} \to \Delta$. Let $\mathfrak{Z} \to \widetilde{\mathfrak{Z}}$ be the blowing-up of these sections. The strict transform of $\widetilde{C}'_0 \cup \widetilde{C}''_2$ is a Cartier divisor on $\mathfrak{Z}_0$. It can be extended to an effective Cartier divisor on $\mathfrak{Z}$, whose associated line bundle we denote by $\mathcal{L}$. Consider the

Degenerations of Kodaira Surfaces   221

factorization $3 \to \Delta \times \mathbb{P}^1$ of $3 \to \Delta$. The relative base locus of $\mathcal{L}$ over $\Delta \times \mathbb{P}^1$ is obviously finite. According to the Fujita-Zariski Theorem ([9, Thm. 1.10]) the corresponding invertible $\mathcal{O}_3$-module $\mathcal{L}$ is relatively semiample over $\mathbb{P}^1 \times \Delta$. Moreover, $\mathcal{L}$ is relatively ample over $\mathbb{P}^1 \times \Delta^*$. Let $3 \to \mathfrak{S}$ be the corresponding contraction. The exceptional locus of this contraction is the strict transform of $\widetilde{C}_1' \cup \widetilde{C}_3''$.

Now $\mathfrak{S} \to \Delta$ has as general fiber the blowing up of a Hirzebruch surface of degree e in two points as needed for the construction of d-semistable Kodaira surfaces with rational normalization. The central fiber consists of a union of two Hirzebruch surfaces of degrees $e+1$ and $e-1$ respectively, with a section of degree $e+1$ glued to a section of degree $-(e-1)$. So this is a partial normalization of X_2. The gluing morphism $\varphi : C \to D$ on the special fibre $\mathfrak{S}_0 = S$ extends to a gluing morphism over Δ. The corresponding cocartesian diagram gives the desired flat family $\mathfrak{X} \to \Delta$ with $\mathfrak{X}_0 = X_2$.

(ii) For the elliptic case consider the partial normalization Y of X_2 obtained from S by identifying C_1' with C_3'' and C_3' with C_1'', see the right half of Fig. 4. It is the projective closure of a line bundle over the Néron 2-gon. In this picture the two disjoint sections $C_0 = C_0' \cup C_0''$, $C_2 = C_2' \cup C_2''$ are the zero section and the section at infinity. Let $\mathfrak{B} \to \Delta$ be a smoothing of B. Since the line bundle defining Y extends to $\mathfrak{B}$ there exists an extension $\mathfrak{Y} \to \mathfrak{B}$ of $Y \to B$ with disjoint sections $\mathfrak{C}_0, \mathfrak{C}_2 \subset \mathfrak{Y}$. The gluing $C_0 \to C_2$ is given by an automorphism of B of order 2. This automorphism can be extended to an automorphism of $\mathfrak{B}/\Delta$ of the same order. An appropriate identification of the sections now yields the desired deformation of X_2. $\square$

References

[1] V. Alexeev, Complete moduli in the presence of semiabelian group action, preprint, 1999.

[2] A. Ash, D. Mumford, M. Rapoport, Y. Tai, Smooth compactifications of locally symmetric varieties, Math Sci Press, Brookline, 1975.

[3] W. Barth, C. Peters, A. Van de Ven, Compact complex surfaces, *Ergeb. Math. Grenzgebiete* **4**(3) (1984), Springer, Berlin etc.

[4] C. Borcea, Moduli for Kodaira surfaces, *Compos. Math.* **52** (1984), 373–380.

[5] P. Deligne, M. Rapoport, Les schémas de modules de courbes elliptiques, in: P. Deligne, W. Kuyk (eds.), *Modular Functions of one Variable II*, pp. 143–316. *Lect. Notes Math.* **349** (1973) Springer, Berlin etc.

[6] R. Friedman, Global smoothings of varieties with normal crossings, *Ann. Math.* **118**(2) (1983), 75–114.

[7] R. Friedman, N. Shepherd-Barron, Degenerations of Kodaira surfaces, in: R. Friedman, D. Morrison (eds.), *The Birational Geometry of Degenerations*, pp. 261–275. *Prog. Math.* **29** (1983), Birkhäuser, Boston etc.

[8] A. Fujiki, On the Douady space of a compact complex space in the category $\mathcal{C}$. II, *Publ. Res. Inst. Math. Sci.* **20** (1984), 461–489.

[9] T. Fujita, Semipositive line bundles, *J. Fac. Sci. Univ. Tokyo* **30** (1983), 353–378.

222   Stefan Schröer and Bernd Siebert

[10] M. Kato, Topology of Hopf surfaces, *J. Math. Soc. Japan* **27** (1975), 222–238.

[11] M. Kato, Erratum to "Topology of Hopf surfaces," *J. Math. Soc. Japan* **41** (1989), 173–174.

[12] K. Kodaira, On the structure of compact complex analytic surfaces II, III, *Am. J. Math.* **88** (1966), 682–721; **90** (1968), 55–83.

[13] V. Kulikov, Degenerations of K_3 surfaces and Enriques surfaces, *Math. USSR, Izv.* **11** (1977), 957–989.

[14] H. Lange, C. Birkenhake, Complex abelian varieties, *Grundlehren Math. Wiss.* **302** (1992), Springer, Berlin etc.

[15] D. Mumford, Abelian varieties, *Tata Institute of Fundamental Research Studies in Mathematics* **5** (1970), Oxford University Press, London

[16] I. Nakamura, On moduli of stable quasi abelian varieties, *Nagoya Math. J.* **58** (1975), 149–214.

[17] S. Nakano, On the inverse of monoidal transformation, *Publ. Res. Inst. Math. Sci., Kyoto Univ.* **6** (1971), 483–502.

[18] K. Nishiguchi, Canonical bundles of analytic surfaces of class VII$_0$, in: H. Hijikata, H. Hironaka et al. (eds.), *Algebraic Geometry and Commutative Algebra II*, pp. 433–452, Academic Press, Tokyo, 1988.

[19] V. Palamodov, Deformations of complex spaces, *Russ. Math. Surveys* **31**(3) (1976), 129–197.

[20] U. Persson, On degenerations of algebraic surfaces, *Mem. Am. Math. Soc.* **189** (1977).

[21] U. Persson, H. Pinkham, Degeneration of surfaces with trivial canonical bundle, *Ann. Math.* **113**(2) (1981), 45–66.

[22] S. Sternberg, Local contractions and a theorem of Poincaré, *Amer. J. Math.* **79** (1957), 809–824.

[23] J. Wehler, Versal deformation of Hopf surfaces. *J. Reine Angew. Math.* **328** (1981), 22–32.

Extension of Twisted Pluricanonical Sections with Plurisubharmonic Weight and Invariance of Semipositively Twisted Plurigenera for Manifolds Not Necessarily of General Type

Yum-Tong Siu[*]

Department of Mathematics, Harvard University, Cambridge, MA, USA
E-mail address: siu@math.harvard.edu

Abstract Let X be a holomorphic family of compact complex projective algebraic manifolds with fibers X_t over the open unit 1-disk Δ. Let K_{X_t} and K_X be respectively the canonical line bundles of X_t and X. We prove that, if L is a holomorphic line bundle over X with a (possibly singular) metric $e^{-\varphi}$ of semipositive curvature current on X such that $e^{-\varphi}|_{X_0}$ is locally integrable on X_0, then for any positive integer m, any $s \in \Gamma\left(m K_{X_0} + L\right)$ with $|s|^2 e^{-\varphi}$ locally bounded on X_0 can be extended to an element of $\Gamma\left(X, m K_X + L\right)$. In particular, $\dim \Gamma\left(X_t, m K_{X_t} + L\right)$ is independent of t for φ smooth. The case of trivial L gives the deformational invariance of the plurigenera. The method of proof uses an appropriately formulated effective version, with estimates, of the argument in the author's earlier paper on the invariance of plurigenera for general type. A delicate point of the estimates involves the use of metrics as singular as possible for $p K_{X_0} + a_p L$ on X_0 to make the dimension of the space of L^2 holomorphic sections over X_0 bounded independently of p, where a_p is the smallest integer $\geq \frac{p-1}{m}$. These metrics are constructed from s. More conventional metrics, independent of s, such as generalized Bergman kernels are not singular enough for the estimates.

Table of Contents

Introduction. 224

1 Review of Existing Argument for Invariance of Plurigenera. . 228

2 Global Generation of Multiplier Ideal Sheaves with Estimates.234

3 Extension Theorems of Ohsawa-Takegoshi Type from Usual Basic Estimates with Two Weight Functions. 241

4 Induction Argument with Estimates. 248

5 Effective Version of the Process of Taking Powers and Roots of Sections. 256

6 Remarks on the Approach of Generalized Bergman Kernels. 264

References. 276

---

[*] Partially supported by a grant from the National Science Foundation.
2000 *Mathematics Subject Classification*: 32G05, 32C35.

224   Yum-Tong Siu

0 Introduction

For a holomorphic family of compact complex projective algebraic manifolds, the plurigenera of a fiber are conjectured to be independent of the fiber. The case when the fibers are of general type was proved in [Siu98]. Generalizations were made by Kawamata [Kaw99] and Nakayama [Nak98] and, in addition, they recast the transcendentally formulated methods in [Siu98] into a completely algebraic geometric setting. Recently Tsuji put on the web a preprint on the deformational invariance of the plurigenera for manifolds not necessarily of general type [Tsu01], in which, in addition to the techniques of [Siu98], he uses his theory of analytic Zariski decomposition and generalized Bergman kernels. Tsuji's approach of generalized Bergman kernels naturally reduces the problem of the deformational invariance of the plurigenera to a growth estimate on the generalized Bergman kernels. This crucial estimate is still lacking. We will explain it briefly in (1.5.2) and discuss it in more details in §6 at the end of this paper.

In this paper we use an appropriately formulated effective version, with estimates, of the argument in [Siu98] to prove the following extension theorem (Theorem 0.1) which implies the invariance of semipositively twisted plurigenera (Corollary 0.2). The results in this paper can be regarded as generalizations of the deformational invariance of plurigenera.

Theorem 0.1. *Let $\pi : X \to \Delta$ be a holomorphic family of compact complex projective algebraic manifolds over the open unit 1-disk $\Delta = \{z \in \mathbf{C} \,|\, |z| < 1\}$. For $t \in \Delta$, let $X_t = \pi^{-1}(t)$ and K_t be the canonical line bundle of X_t. Let L be a holomorphic line bundle over X with a (possibly singular) metric $e^{-\varphi}$ whose curvature current $\frac{\sqrt{-1}}{2\pi}\partial\bar{\partial}\varphi$ is semi-positive on X such that $e^{-\varphi}|_{X_0}$ is locally integrable on X_0. Let m be any positive integer. Then any element $s \in \Gamma(X_0, m K_{X_0} + L)$ with $|s|^2 e^{-\varphi}$ locally bounded on X_0 can be extended to an element $\tilde{s} \in \Gamma(X, m K_X + L)$ in the sense that $\tilde{s}|_{X_0} = s \wedge \pi^*(dt)$, where K_X is the canonical line bundle of X.*

Corollary 0.2. *Let $\pi : X \to \Delta$ be a holomorphic family of compact complex projective algebraic manifolds over the open unit 1-disk Δ with fiber X_t. Let L be a holomorphic line bundle over X with a smooth metric $e^{-\varphi}$ whose curvature form $\frac{\sqrt{-1}}{2\pi}\partial\bar{\partial}\varphi$ is semi-positive on X. Let m be any positive integer. Then the complex dimension of $\Gamma(X_t, m K_{X_t} + L)$ is independent of t for $t \in \Delta$.*

Corollary 0.3. *Let $\pi : X \to \Delta$ be a holomorphic family of compact complex projective algebraic manifolds over the open unit 1-disk Δ with fiber X_t. Let m be any positive integer. Then the complex dimension of $\Gamma(X_t, m K_{X_t})$ is independent of t for $t \in \Delta$.*

So far as the logical framework is concerned, the method in this paper simply follows that of [Siu98], the only difference being the monitoring of estimates in this paper. Some of the estimates are quite delicate. The estimates depend on a choice, at the beginning, of singular metrics for the twisted pluri-canonical line bundles of the initial fiber. In contrast to the metrics chosen in [Siu98], for the effective argument in this paper metrics as singular as possible have to be chosen for the twisted pluricanonical bundles on X_0, as long as the relevant sections on X_0 remain L^2. The use of usual abstractly-defined general metrics for the twisted pluricanonical line bundle of the initial fiber, such as generalized Bergman kernels on X_0, would contribute an uncontrollable factor in the final estimate (see (1.5.2) below). If one uses generalized Bergman kernels on X as for example in [Tsu01], there are difficulties with norm changes similar to the norm-change problems encountered in the papers of Nash [Nas54], Moser [Mos61], and Grauert [Gra60]. The norm changes in our case mean the need to shrink the domain for supremum estimates in each of an infinite number of steps. Our situation here is different from those which could be handled by the norm-change techniques of Nash [Nas54], Moser [Mos61], and Grauert [Gra60] (see §6, in particular (6.4)). The difficulty here with the use of generalized Bergman kernels cannot be overcome.

The method of this paper should be applicable to give the deformational invariance of the plurigenera twisted by a numerically effective line bundle, i.e., Corollary 0.2 in which L is assumed to be numerically effective instead of having a smooth metric with semi-positive curvature. In this paper we give the easier semi-positive case to avoid one more layer of complication in our estimates. For the numerically effective case of Corollary 0.2 we have to use a sufficiently ample line bundle A so that $m\,L + A$ is very ample for any positive integer m and to keep track of the limiting behavior of the m-th root of some canonically defined smooth metric of $m\,L + A$. In order not to be distracted from the main arguments of this paper by another lengthy peripheral limiting process, we will leave the numerically effective case of Corollary 0.2 to another occasion.

Unlike the case of general type (see (1.5.1)), for the proof in this paper a genuine limiting process is being used. It is not clear whether this proof can be translated into a completely algebraic geometric setting. Of course, instead of L being semi-positive, the algebraic geometric formulation of Corollary 0.2 would have to assume that L is generated by global holomorphic sections on X. The difficulty of translating the limiting process into an algebraic geometric setting occurs already for the case of trivial L (Corollary 0.3).

A more general formulation of the deformational invariance of plurigenera is for the setting of a holomorphic family of compact Kähler manifolds. Such a setting is completely beyond the reach of the methods of [Siu98] and this paper. For that setting, the only known approach is that of Levine [Lev83] in which he uses Hodge theory to extend a pluricanonical section from the initial fiber to its finite neighborhood of second order over a double point of the base. The conjecture for the Kähler case is the following.

Conjecture 0.4. (Conjecture on Deformational Invariance of Plurigenera for the Kähler Case) Let $\pi : X \to \Delta$ be a holomorphic family of compact Kähler

226   Yum-Tong Siu

manifolds over the open unit 1-disk Δ with fiber X_t. Then for any positive integer m the complex dimension of $\Gamma(X_t, m K_{X_t})$ is independent of t for $t \in \Delta$.

The case of twisting by a numerically effective line bundle can be formulated for the Kähler case in the form of a conjecture as follows.

Conjecture 0.5. (Conjecture on Deformational Invariance of Plurigenera for the Kaehler Case with Twisting by Numerically Effective Line Bundles) Let $\pi : X \to \Delta$ be a holomorphic family of compact Kähler manifolds over the open unit 1-disk Δ with fiber X_t. Let L be a holomorphic line bundle on X which is numerically effective in the sense that, for any strictly negative $(1,1)$-form ω on X and any compact subset W of X, there exists a smooth metric for L whose curvature form is $\geq \omega$ on W. Then for any positive integer m the complex dimension of $\Gamma(X_t, (L|X_t) + m K_{X_t})$ is independent of t for $t \in \Delta$.

Since the complex dimension of $\Gamma(X_t, (L|X_t) + m K_{X_t})$ is always upper semi-continuous as a function of t, to prove its independence on t it suffices to show that every element s of $\Gamma(X_{t_0}, (L|X_{t_0}) + m K_{X_{t_0}})$ can be extended to an element $\tilde{s}$ of $\Gamma(X, L + m K_X)$ in the sense that $s \wedge \pi^*(dt) = \tilde{s}|X_{t_0}$. Thus Corollaries 0.2 and 0.3 follow readily from Theorem 0.1. Most of the rest of this paper is devoted to the proof of Theorem 0.1. For notational simplicity, in referring to the extension of a twisted pluricanonical section on X_0 to X, we identify K_{X_0} with $K_X|X_0$ by the map which is defined by the wedge product with $\pi^*(dt)$.

In the proof of the deformational invariance of the plurigenera in [Siu98], there are the following four major ingredients.

(i) Global generation of the multiplier ideal sheaf after twisting by a sufficiently ample line bundle (by the method of Skoda [Sko72]).

(ii) Extension of sections from the initial fiber which are L^2 with respect to a singular metric on the family with semi-positive curvature current (by the method of Ohsawa-Takegoshi [OT87]).

(iii) An induction argument on m, which uses the two preceding ingredients and regards a section of the m-canonical bundle as a top-degree form with values in the $(m-1)$-canonical bundle, in order to construct singular metrics on the $(m-1)$-canonical bundle on the family and extend sections of the m-canonical bundle from the initial fiber.

(iv) The process of raising a section on X_0 to a high power ℓ and later taking the ℓ-th root after extending to X its product with the canonical section s_D of some fixed effective line bundle D on X. For the case of general type, when $m K_X$ is written as the sum of D and a sufficiently ample line bundle A for some large m, multiplication by s_D provides us with the twisting by A which is needed for the global generation of the multiplier ideal sheaf in the first ingredient. This technique of taking powers, extending the product with s_D, and taking roots is to eliminate the effect of multiplication by s_D, or equivalently to eliminate the effect of multiplication of a section of A.

The rough and naive motivation underlying the idea of the proof in [Siu98] is that, if one could write an element $s^{(m)}$ of $\Gamma(X_0, m K_{X_0})$ as a sum of terms, each of which is the product of an element $s^{(1)}$ of $\Gamma(X_0, K_{X_0})$ and an element $s^{(m-1)}$ of $\Gamma(X_0, (m-1) K_{X_0})$, then one can extend $s^{(m)}$ to an element of $\Gamma(X, m K_X)$ by induction on m. Of course, in general it is clearly impossible to so express $s^{(m)}$ as a sum of such products. However, one could successfully implement a very much modified form of this rough and naive motivation by using the above four ingredients in the case of general type. The actual proof in [Siu98] by induction on m, which uses the modified form of the argument, appears very different from this rough and naive motivation and is not recognizable as related in any way to it, but it was in fact from such a rough and naive motivation that the actual proof in [Siu98] evolved. The modification is to require only $s^{(1)}$ to be just a local holomorphic function and twist each $m K_{X_0}$ by the same sufficiently ample line bundle A on X so that locally the absolute value of $s^{(m)}$ can be estimated by the sum of absolute values of elements of $\Gamma(X_0, (m-1) K_{X_0} + A)$. Such an estimate enables us to inductively get the extension and then to use the technique of taking powers and roots to get rid of the twisting by A.

The assumption of general type is used in the fourth ingredient listed above. For manifolds not necessarily of general type, in the fourth ingredient multiplication by s_D is replaced by multiplication by a section s_A of A and one has to pass to limit as the integer, which is both the power and the root order, goes to infinity, in order to eliminate the effect of multiplication by s_A. To carry out the limiting argument, one needs to have a good control of the estimates, which necessitates the use of an effective version of the argument of [Siu98].

Besides proving Theorem 0.1, the paper includes in §3 a simple approach to extension theorems which uses the usual basic estimates and two weight functions (see §3) and explains in §6 the delicate point of the estimates, especially why there are difficulties with the simple natural approach of using generalized Bergman kernels.

Every metric of holomorphic line bundles in this paper is allowed to be singular, but, so far as regularity is concerned, its curvature current is assumed to be no worse than the sum of a smooth form and a semi-positive current. For a metric $e^{-\varphi}$ of a holomorphic line bundle, we use the notation $\mathcal{I}_\varphi$ to denote the *multiplier ideal sheaf* which consists of all holomorphic function germs f such that $|f|^2 e^{-\varphi}$ is locally integrable. Multiplier ideal sheaves were introduced by Nadel [Nad89] for his vanishing theorem. Nadel's vanishing theorem, in the algebraic setting and in the special case of algebraically-definable

228   Yum-Tong Siu

singular metrics, is reducible to the vanishing theorem of Kawamata-Viehweg [Kaw82], [Vie82].

The structure of this paper is as follows. In §1 we review the argument of [Siu98] for the deformational invariance of plurigenera for the case of general type. The purpose of the review is to first present the logical framework of the argument without the estimates so that it can serve as a guide for following the later complicated details of the estimates in the effective version. Sections 2, 3, 4, and 5 with detailed estimates correspond respectively to the first, second, third, and fourth ingredients listed above. Section 6 contains some remarks on the difficulties of using generalized Bergman kernels for the effective arguments for the problem of the deformational invariance of plurigenera for manifolds not necessarily of general type.

1 Review of Existing Argument for Invariance of Plurigenera

The first two ingredients of the argument of [Siu98] for the deformational invariance of plurigenera for general type uses the global generation of multiplier ideal sheaves (by the method of Skoda [Sko72]) and an extension theorem of Ohsawa-Takegoshi type [OT87]. Let us first recall the precise statements of these two results.

1.1 *Global Generation of Multiplier Ideal Sheaves* [Siu98, p.664, Prop. 1]

Let L be a holomorphic line bundle over an n-dimensional compact complex manifold Y with a Hermitian metric which is locally of the form $e^{-\xi}$ with ξ plurisubharmonic. Let $\mathcal{I}_\xi$ be the multiplier ideal sheaf of the Hermitian metric $e^{-\xi}$. Let E be an ample holomorphic line bundle over Y such that for every point P of Y there are a finite number of elements of $\Gamma(Y, E)$ which all vanish to order at least $n + 1$ at P and which do not simultaneously vanish outside P. Then $\Gamma(Y, \mathcal{I}_\xi \otimes (L + E + K_Y))$ generates $\mathcal{I}_\xi \otimes (L + E + K_Y)$ at every point of Y.

Theorem 1.2. *(Extension Theorem of Ohsawa-Takegoshi Type)* [Siu98, p.666, Prop. 2]. *Let $\gamma : Y \to \Delta$ be a projective algebraic family of compact complex manifolds parametrized by the open unit 1-disk Δ. Let $Y_0 = \gamma^{-1}(0)$ and let n be the complex dimension of Y_0. Let L be a holomorphic line bundle with a Hermitian metric which locally is represented by $e^{-\chi}$ such that $\sqrt{-1}\partial\bar\partial\chi \geq \omega$ in the sense of currents for some smooth positive $(1,1)$-form ω on Y. Let $0 < r < 1$ and $\Delta_r = \{\, t \in \Delta \mid |t| < r \,\}$. Then there exists a positive constant A_r with the following property. For any holomorphic L-valued n-form f on Y_0 with*

$$\int_{Y_0} |f|^2 e^{-\chi} < \infty \,,$$

there exists a holomorphic L-valued $(n+1)$-form $\tilde{f}$ on $\gamma^{-1}(\Delta_r)$ such that $\tilde{f}|_{Y_0} = f \wedge \gamma^*(dt)$ at points of Y_0 and

$$\int_Y |\tilde{f}|^2 e^{-\chi} \leq A_r \int_{Y_0} |f|^2 e^{-\chi} .$$

Note that Theorem 2.1 (respectively Theorem 3.1) below is an effective version of Theorem 1.1 (respectively Theorem 1.2).

1.3 *Induction Argument in Axiomatic Formulation*

To prepare for later adaptation to the effective version, we now formulate axiomatically the induction argument of [Siu98]. The induction argument is the precise formulation of the rough and naive motivation explained in the Introduction. The main idea of the induction argument is to start with suitable metrics $e^{-\varphi_p}$ for $p K_{X_0} + D$ on X_0 with semi-positive curvature current on X_0. The induction step is to use a sufficiently ample line bundle A to extend L^2 sections of $m K_{X_0} + D + A$ with respect to $e^{-\varphi_{m-1}}$ from X_0 to X and to produce a metric $e^{-\chi_m}$ of $m K_X + D + A$ on X with semi-positive curvature current so that L^2 sections of $(m+1) K_{X_0} + D + A$ on X_0 with respect to $e^{-\varphi_m}$ are L^2 with respect to $e^{-\chi_m}|_{X_0}$.

Fix a positive integer m_0 and a holomorphic line bundle D over X. Assume that A is a sufficiently ample line bundle on X so that for any point P of X_0 there are a finite number of elements of $\Gamma(X_0, A)$ which all vanish to order at least $n+1$ at P and which do not simultaneously vanish outside P. Thus the theorem 1.1 of global generation of multiplier ideal sheaf on X_0 holds with twisting by $A|X_0$.

Suppose we have

(i) a metric $e^{-\varphi_p}$ for $p K_{X_0} + D$ over X_0 with semi-positive curvature current on X_0 for $0 \leq p < m_0$, and

(ii) a metric $e^{-\varphi_D}$ of D over X, with semi-positive curvature current on X,

such that

(a) $\mathcal{I}_{\varphi_p} \subset \mathcal{I}_{\varphi_{p-1}}$ on X_0 for $0 < p < m_0$,
(b) $\mathcal{I}_{\varphi_D|_{X_0}}$ agrees with $\mathcal{I}_{\varphi_0}$ on X_0.

Proposition 1.4. *If a holomorphic family $\pi : X \to \Delta$ of compact complex projective algebraic manifolds satisfies Assumptions (i), (ii) and Conditions (a), (b) of (1.3), then any element $f \in \Gamma(X_0, m_0 K_{X_0} + D + A)$ which locally belongs to $\mathcal{I}_{\varphi_{m_0-1}}$ can be extended to an element $\tilde{f} \in \Gamma(X, m_0 K_X + D + A)$.*

Proof. By (1.1) and Condition (a) of (1.3), we can cover X_0 by a finite number of open subsets U_λ $(1 \leq \lambda \leq \Lambda)$ so that, for some nowhere zero element $\xi_\lambda \in \Gamma(U_\lambda, -K_{X_0})$, we can write

$$\xi_\lambda f|_{U_\lambda} = \sum_{k=1}^{N_{m_0}-1} b_k^{(m_0-1,\lambda)} s_k^{(m_0-1)} , \tag{1.4.1}$$

230   Yum-Tong Siu

where $b_k^{(m_0-1,\lambda)}$ is a holomorphic function on U_λ and

$$s_k^{(m_0-1)} \in \Gamma\left(X_0, \mathcal{I}_{\varphi_{m_0}-2}\left((m_0-1)K_{X_0}+D+A\right)\right) .$$

Moreover, we can assume that the open subsets U_λ are chosen small enough so that, for $1 \le p \le m_0 - 2$, inductively by (1.1) we can write

$$\xi_\lambda\, s_j^{(p+1)}\big|U_\lambda = \sum_{k=1}^{N_p} b_{j,k}^{(p,\lambda)} s_k^{(p)} , \tag{1.4.2}$$

where $b_{j,k}^{(p,\lambda)}$ is a holomorphic function on U_λ and

$$s_k^{(p)} \in \Gamma\left(X_0, \mathcal{I}_{\varphi_{p-1}}\left(p\,K_{X_0}+D+A\right)\right) .$$

For this we have used already Condition (a) and have also used Assumption (i) of (1.3) in order to apply (1.1) on the global generation of multiplier ideal sheaves.

We are going to show, by induction on $1 \le p < m_0$, the following claim.

(1.4.3) *Claim*

$$s_j^{(p)} \in \Gamma\left(X_0, p\,K_{X_0}+D+A\right)$$

can be extended to

$$\tilde{s}_j^{(p)} \in \Gamma\left(X, p\,K_X+D+A\right)$$

for $1 \le j \le N_p$ and $1 \le p < m_0$.

The case $p = 1$ of Claim (1.4.3) clearly follows from Assumption (ii) and Condition (b) of (1.3) and the extension theorem (Theorem 1.2). Suppose Claim (1.4.3) has been proved for Step p and we are going to prove Step $p+1$, where p is replaced by $p+1$.

Let

$$\chi_p = \log \sum_{k=1}^{N_p} \left|\tilde{s}_k^{(p)}\right|^2$$

so that $e^{-\chi_p}$ is a metric of $p\,K_X+D+A$. Since $\tilde{s}_j^{(p)}|X_0 = s_j^{(p)}$ for $1 \le p \le N_p$ by Step p of Claim 1.4.3, it follows from (1.4.2) that the germ of $s_j^{(p+1)}$ at any point of X_0 belongs to $\mathcal{I}_{\chi_p|X_0}$ for $1 \le j \le N_{p+1}$. By the extension theorem 1.2,

$$s_j^{(p+1)} \in \Gamma\left(X_0, (p+1)\,K_{X_0}+D+A\right)$$

can be extended to

$$\tilde{s}_j^{(p+1)} \in \Gamma\left(X, (p+1)\,K_X+D+A\right) ,$$

finishing the verification of Step $p+1$ of Claim (1.4.3).

Invariance of Twisted Plurigenera for General Manifolds 231

Since $\tilde{s}_j^{(m_0-1)}|X_0 = s_j^{(m_0-1)}$ for $1 \leq p \leq N_{m_0-1}$ from Claim (1.4.1), by (1.4.3) the germ of f at any point of X_0 belongs to $\mathcal{I}_{X_{m_0-1}|x_0}$. By the extension theorem 1.2

$$f \in \Gamma(X_0, m_0 K_{X_0} + D + A)$$

can be extended to

$$\tilde{f} \in \Gamma(X, (m_0 K_X + D + A)) .$$

$\square$

1.5 Metrics for the Induction Argument for the Case of General Type

We now discuss how the metrics $e^{-\varphi_p}$ in the axiomatic formulation of the induction argument can be chosen for the case of general type.

In [Siu98], a sufficiently large positive number a is chosen so that $aK_X = D + A$ for some effective divisor D. Let

$$s_1^{(m)}, \cdots, s_{q_m}^{(m)} \in \Gamma(X_0, m K_{X_0})$$

be a basis over $\mathbf{C}$. Let s_D be the canonical section of D whose divisor is D and h_A be a smooth metric of A with positive curvature form.

For the case of general type, one can simply use

$$\varphi_p = \log \sum_{j=1}^{q_{m_0}} \left| s_j^{(m_0)} \right|^{\frac{2p}{m_0}} + \log |s_D|^2$$

for $0 \leq p \leq m_0 - 1$.

1.5.1 In the proof of the invariance of plurigenera in [Siu98] the infinite series

$$\varphi = \log \sum_{m=1}^{\infty} \varepsilon_m \sum_{j=1}^{q_m} \left| s_j^{(m)} \right|^{\frac{2}{m}}$$

is used to define

$$\varphi_p = p\,\varphi + \log |s_D|^2 ,$$

where a sequence of positive numbers ε_m ($1 \leq m < \infty$) is chosen which decreases sufficiently rapidly, as m increases, to guarantee local convergence of the infinite series φ. The infinite series φ was introduced in [Siu98] more for notational expediency than for absolute necessity. In effective arguments where estimates have to be controlled, clearly we should avoid the complications of unnecessary infinite processes. So the metric given in (1.5) is more to our advantage than the one given in [Siu98] in the form of an infinite series.

232   Yum-Tong Siu

1.5.2 *Reason for Metrics for Pluricanonical Bundles on Initial Fiber as Singular as Possible.* As a matter of fact, in order to guarantee convergence in the effective arguments where D is set to be 0, the more singular the metric $e^{-\varphi_p}$ of $p K_{X_0}$ is, the easier it is to control the estimates. For the extension $s^{(m_0)} \in \Gamma(X_0, m_0 K_{X_0})$ to X in the effective argument, we will use the metric

$$e^{-\varphi_p} = \frac{1}{\left|s^{(m_0)}\right|^{\frac{2p}{m_0}}} .$$

It is introduced in (4.1) as $e^{-p\psi}$. (The metric $e^{-p\psi}$ in (4.1) contains also the contribution from the semi-positive line bundle L and would be equal to $e^{-\varphi_p}$ when L is the trivial line bundle. Here for simplicity we assume $L = 0$ in our discussion here in $(1.5.2)$.)

The reason why we want more singularity for $e^{-\varphi_p} = e^{-p\psi}$ is that, for the final estimates, a uniform bound, independent of p, is needed for the complex dimension of $\Gamma\left(X_0, \mathcal{I}_{\varphi_p}(p K_{X_0} + A)\right)$. If some other less singular metrics are used for $p K_{X_0}$, one may run into the difficulty that the complex dimension of $\Gamma\left(X_0, \mathcal{I}_{\varphi_p}(p K_{X_0} + A)\right)$ grows as a positive power of p. When we take the ℓ-th root of the extension of $\left(s^{(m_0)}\right)^{\ell} s_A$ from X_0 to X, where $s_A \in \Gamma(X, A)$, and let $\ell \to \infty$, in the estimate there is a contribution of the factor which is the ℓ-th root of the product of the complex dimension of $\Gamma\left(X_0, \mathcal{I}_{\varphi_p}(p K_{X_0} + A)\right)$ for $0 \leq p < \ell m_0$. If the complex dimension of $\Gamma\left(X_0, \mathcal{I}_{\varphi_p}(p K_{X_0} + A)\right)$ grows at least like a positive power of p, a positive power of the factor $(\ell!)^{\frac{1}{\ell}}$ occurs in the final estimate and becomes unbounded as ℓ goes to infinity. The occurrence of such a factor is due to the estimate in $(5.3.4)$. So the use of usual abstractly-defined general metrics for $p K_{X_0}$, such as generalized Bergman kernels, would not work in the effective argument.

1.6 *Taking Powers and Roots of Sections for the Case of General Type.*

We take $s \in \Gamma(X_0, m_0 K_{X_0})$. To finish the last step of the proof of the deformational invariance of the plurigenera for general type, we need to extend s to an element of $\Gamma(X, m_0 K_{X_0})$.

Take a positive integer ℓ and let $m = m_0 \ell$. We use the formula (1.5) to define φ_p for the larger range $0 \leq p \leq m - 1$. Then the germ of $s^{\ell} s_D$ at any point of X_0 belongs to $\mathcal{I}_{\varphi_{m-1}}$. By Theorem 1.1 on the global generation of multiplier ideal sheaves, we can write

$$s^{\ell} s_D = \sum_{k=1}^{q_m} b_k s_k^{(m)} \tag{1.6.1}$$

locally on X_0 for some local holomorphic functions b_k on X_0 and for

$$s_k^{(m)} \in \Gamma\left(X_0, \mathcal{I}_{\varphi_{m-1}}(m K_{X_0} + D + A)\right) .$$

Then, by Proposition 1.4 with m_0 replaced by m, $s_k^{(m)}$ can be extended to

$$\tilde{s}_k^{(m)} \in \Gamma\left(X, m\,K_X + D + A\right) = \Gamma\left(X_0, (m+a)\,K_X\right)$$

for $1 \le k \le q_m$. Let

$$\chi_m = \log \sum_{k=1}^{q_m} \left|\tilde{s}_k^{(m)}\right|^2 .$$

From (1.6.1) the germ of of $s^\ell s_D$ at any point of X_0 belongs to the multiplier ideal sheaf $\mathcal{I}_{\chi_m|X_0}$.

We now apply Hölder's inequality. Let h_D be a smooth metric for D without any curvature condition. Let $h_{K_{X_0}}$ be a smooth metric for K_{X_0} without any curvature condition. Then $dV_{X_0} := (K_{X_0})^{-1}$ is a smooth volume form of X_0. Let $\frac{1}{\ell} + \frac{1}{\ell'} = 1$ for ℓ sufficiently large. Since $\frac{1}{\ell'} = 1 - \frac{1}{\ell}$ and $\ell' = \frac{\ell}{\ell-1}$, it follows that Hölder's inequality gives

$$\int_{X_0} |s|^2 e^{-\left(\frac{m_0}{m+a}\right)\chi_m}\, dV_{X_0}$$

$$= \int_{X_0} \left(\left|s\,(s_D)^{\frac{1}{\ell}}\right|^2 h_D\, e^{-\left(\frac{m_0}{m+a}\right)\chi_m}\right) \left(\frac{1}{\left|(s_D)^{\frac{1}{\ell}}\right|^2 h_D}\right) dV_{X_0}$$

$$\le \left(\int_{X_0} |s^\ell s_D|^2\, h_D\, e^{-\ell\left(\frac{m_0}{m+a}\right)\chi_m} dV_{X_0}\right)^{\frac{1}{\ell}} \left(\int_{X_0} \frac{1}{\left(h_D\,|s_D|^2\right)^{\frac{1}{\ell-1}}} dV_{X_0}\right)^{1-\frac{1}{\ell}}$$

$$\le \left(\sup_{X_0} h_{K_X}\, e^{\left(\frac{1}{m+a}\right)\chi_m}\right)^{\frac{a}{\ell}} \left(\int_{X_0} \frac{|s^\ell s_D|^2\, h_D\, e^{-\chi_m} dV_{X_0}}{\left(h_{K_{X_0}}\right)^a}\right)^{\frac{1}{\ell}} .$$

$$\cdot \left(\int_{X_0} \frac{dV_{X_0}}{\left(h_D\,|s_D|^2\right)^{\frac{1}{\ell-1}}}\right)^{1-\frac{1}{\ell}}$$

which is finite for ℓ sufficiently large. Hence

$$\int_{X_0} |s|^2 e^{-\left(\frac{m_0-1}{m+a}\right)\chi_m} \le \left(\sup_{X_0} h_{K_X}\, e^{\left(\frac{1}{m+a}\right)\chi_m}\right) \int_{X_0} |s|^2 e^{-\left(\frac{m_0}{m+a}\right)\chi_m}\, dV_{X_0} < \infty .$$

By the extension theorem 1.2, s can be extended to an element

$$\tilde{s} \in \Gamma\left(X, m_0\,K_X\right) .$$

This finishes the review of the argument of [Siu98].

2 Global Generation of Multiplier Ideal Sheaves with Estimates

Now we give the effective version, with estimates, of the global generation of multiplier ideal sheaves.

Theorem 2.1. *(Effective Version of Global Generation of Multiplier Ideal Sheaves). Assume that for every point P_0 of X_0 one has a coordinate chart $\tilde{U}_{P_0} = \{|z^{(P_0)}| < 2\}$ of X_0 with coordinates*

$$z^{(P_0)} = \left(z_1^{(P_0)}, \cdots, z_n^{(P_0)} \right)$$

centered at P_0 such that the set U_{P_0} of points of $\tilde{U}_{P_0}$ where $|z^{(P_0)}| < 1$ is relatively compact in $\tilde{U}_{P_0}$. Let ω_0 be a Kähler form of X_0. Let C_{X_0} be a positive number such that the supremum norm of $\bar{\partial} z_j^{(P_0)}$ with respect to ω_0 is $\leq C_{X_0}$ on U_{P_0} for $1 \leq j \leq n$. Let $0 < r_1 < r_2 \leq 1$. Let A be an ample line bundle over X_0 with a smooth metric h_A of positive curvature. Assume that, for every point P_0 of X_0, there exists a singular metric h_{A,P_0} of A, whose curvature current dominates $c_A \omega_0$ for some positive constant c_A, such that

$$\frac{h_A}{|z^{(P_0)}|^{2(n+1)}} \leq h_{A,P_0}$$

on U_{P_0} and

$$\sup_{r_1 \leq |z^{(P_0)}| \leq r_2} \frac{h_{A,P_0}(z^{(P_0)})}{h_A(z^{(P_0)})} \leq C_{r_1, r_2}$$

and

$$\sup_{X_0} \frac{h_A}{h_{A,P_0}} \leq C^{\sharp}$$

for some constants C_{r_1, r_2} and $C^{\sharp} \geq 1$ independent of P_0. Let m be a positive integer. Assume that there is a (possibly singular) metric $e^{-\varphi_{m-1}}$ for $(m-1) K_{X_0}$ with semi-positive curvature current. Let

$$C^{\flat} = 2n \left(\frac{1}{r_1^{2(n+1)}} + 1 + C^{\sharp} \frac{1}{c_A} C_{r_1, r_2} \left(\frac{2 r_2 C_{X_0}}{r_2^2 - r_1^2} \right)^2 \right).$$

Let $0 < r < 1$ and let

$$\hat{U}_{P_0, r} = U_{P_0} \cap \left\{ |z^{(P_0)}| < \frac{r}{n \sqrt{C^{\flat}}} \right\}.$$

Let N_m be the complex dimension of the subspace of all elements

$$s \in \Gamma \left(X_0, m K_{X_0} + A \right)$$

Invariance of Twisted Plurigenera for General Manifolds   235

such that

$$\int_{X_0} |s|^2 \, e^{-\varphi_{m-1}} \, h_A < \infty \, .$$

Then there exist

$$\sigma_1^{(m)}, \cdots, \sigma_{N_m}^{(m)} \in \Gamma\left(X_0, m\,K_{X_0} + A\right)$$

with

$$\int_{X_0} \left|\sigma_k^{(m)}\right|^2 e^{-\varphi_{m-1}} h_A \leq 1$$

($1 \leq k \leq N_m$) such that, for any $P_0 \in X_0$ and for any holomorphic section s of $m\,K_{X_0} + A$ over U_{P_0} with

$$\int_{U_{P_0}} |s|^2 e^{-\varphi_{m-1}} h_A = C_s < \infty \, ,$$

one can find holomorphic functions $b_{P_0,m,k}$ on $\hat{U}_{P_0,r}$ such that

$$s = \sum_{k=1}^{N_m} b_{P_0,m,k} \, \sigma_k^{(m)}$$

on $\hat{U}_{P_0,r}$ and

$$\sup_{\hat{U}_{P_0,r}} \sum_{k=1}^{N_m} |b_{P_0,m,k}|^2 \leq \frac{1}{(1-r)^2} \, C^{\flat} \, C_s \, .$$

The proof of Theorem 2.1 will depend on the following lemma.

Lemma 2.2. *Under the assumption of Theorem 2.1, given any holomorphic section s of $m\,K_{X_0} + A$ over U_{P_0} with*

$$\int_{U_{P_0}} |s|^2 e^{-\varphi_{m-1}} h_A = C_s < \infty$$

there exist $\sigma \in \Gamma\left(X_0, m\,K_{X_0} + A\right)$ and $v_j \in \Gamma\left(U_{P_0}, m\,K_{X_0} + A\right)$ for $1 \leq j \leq n$ such that $s - \sigma = \sum_{j=1}^{n} z_j^{(P_0)} v_j$ with

$$\int_{X_0} |\sigma|^2 e^{-\varphi_{m-1}} h_A \leq C^{\flat} \, C_s$$

and

$$\int_{U_{P_0}} |v_j|^2 \, e^{-\varphi_{m-1}} h_A \leq C^{\flat} \, C_s \, .$$

236   Yum-Tong Siu

Before we prove Lemma 2.2, we recall the following result of Skoda [Sko72, Th. 1, pp.555-556].

Theorem 2.3. *(Skoda). Let Ω be a pseudoconvex domain in $\mathbf{C}^n$ and ψ be a plurisubharmonic function on Ω. Let $g_1, \ldots, g_p$ be holomorphic functions on Ω. Let $\alpha > 1$ and $q = \inf(n, p - 1)$. Then for every holomorphic function f on Ω such that*

$$\int_\Omega |f|^2 |g|^{-2\alpha q - 2} e^{-\psi} d\lambda < \infty ,$$

there exist holomorphic functions $h_1, \ldots, h_p$ on Ω such that

$$f = \sum_{j=1}^{p} g_j h_j$$

and

$$\int_\Omega |h|^2 |g|^{-2\alpha q} e^{-\psi} d\lambda \leq \frac{\alpha}{\alpha - 1} \int_\Omega |f|^2 |g|^{-2\alpha q - 2} e^{-\psi} d\lambda ,$$

where

$$|g| = \left(\sum_{j=1}^{p} |g_j|^2 \right)^{\frac{1}{2}} , \qquad |h| = \left(\sum_{j=1}^{p} |h_j|^2 \right)^{\frac{1}{2}} ,$$

and $d\lambda$ is the Euclidean volume element of $\mathbf{C}^n$.

2.4 *Proof of Lemma 2.2.*

We take a smooth cut-off function $0 \leq \varrho(x) \leq 1$ of a single real variable x so that $\varrho(x) = 1$ for $x \leq r_1^2$ and $\varrho(x) = 0$ for $x > r_2^2$ and

$$\left| \frac{d\varrho}{dx}(x) \right| \leq \frac{2}{r_2^2 - r_1^2} .$$

Let $\varrho_{P_0} = \varrho \left(\left| z^{(P_0)} \right|^2 \right)$. Then the supremum norm of $\bar{\partial} \varrho_{P_0}$ with respect to ω_0 is no more than

$$\frac{2 \, r_2 \, C_{X_0}}{r_2^2 - r_1^2} .$$

Consider $s \bar{\partial} \varrho_{P_0}$. Then

$$\int_{X_0} \left| s \, \bar{\partial} \varrho_{P_0} \right|^2 e^{-\varphi_{m-1}} h_{A,P_0}$$

$$= \int_{X_0} \left| s \, \bar{\partial} \varrho_{P_0} \right|^2 e^{-\varphi_{m-1}} h_A \left(\frac{h_{A,P_0}}{h_A} \right)$$

$$\leq \left(\sup_{r_1 \leq |z^{(P_0)}| \leq r_2} \frac{h_{A,P_0}}{h_A} \right) \int_{X_0} \left| s \, \bar{\partial} \varrho_{P_0} \right|^2 e^{-\varphi_{m-1}} h_A \leq C_{r_1, r_2} \left(\frac{2 \, r_2 C_{X_0}}{r_2^2 - r_1^2} \right)^2 C_s .$$

We now solve the equation $\bar{\partial}u = s\,\bar{\partial}\varrho_{P_0}$ with the estimate

$$\int_{X_0} |u|^2\, e^{-\varphi_m}\, h_{A,P_0} \le \frac{1}{c_A}\, C_{r_1,r_2} \left(\frac{2\,r_2\,C_{X_0}}{r_2^2 - r_1^2}\right)^2 C_s\,.$$

Let $\sigma = s\varrho_{P_0} - u$. Then

$$\begin{aligned}
\int_{X_0} |\sigma|^2 e^{-\varphi_{m-1}} h_A &= \int_{X_0} |s\varrho_{P_0} - u|^2 e^{-\varphi_{m-1}} h_A \\
&\le 2\int_{X_0} \left(|s|^2 + |u|^2\right) e^{-\varphi_{m-1}} h_A \\
&\le 2\,C_s + 2\int_{X_0} |u|^2\, e^{-\varphi_{m-1}} h_{A,P_0} \left(\frac{h_A}{h_{A,P_0}}\right) \\
&\le 2\,C_s + 2\left(\sup_{X_0} \frac{h_A}{h_{A,P_0}}\right) \int_{X_0} |u|^2\, e^{-\varphi_{m-1}} h_{A,P_0} \\
&\le 2\left(1 + C^{\sharp}\frac{1}{c_A}\, C_{r_1,r_2} \left(\frac{2\,r_2\,C_{X_0}}{r_2^2 - r_1^2}\right)^2\right) C_s
\end{aligned}$$

which is $\le C^{\flat}\, C_s$. Since $s - \sigma = (1 - \varrho_{P_0})\,s + u$, it follows that

$$\begin{aligned}
\int_{U_{P_0}} \frac{|s-\sigma|^2}{|z^{(P_0)}|^{2(n+1)}}\, e^{-\varphi_m} h_A &\le 2\int_{U_{P_0}} \frac{|(1-\varrho_{P_0})\,s|^2 + |u|^2}{|z^{(P_0)}|^{2(n+1)}}\, e^{-\varphi_m} h_A \\
&\le 2\left(\frac{1}{r_1^{2(n+1)}} + \frac{1}{c_A}\, C_{r_1,r_2} \left(\frac{2\,r_2\,C_{X_0}}{r_2^2 - r_1^2}\right)^2\right) C_s\,,
\end{aligned}$$

because $\varrho_{P_0} = 1$ on $\left\{\left|z^{(P_0)}\right| < r_1\right\}$ and

$$\frac{h_A}{\left|z^{(P_0)}\right|^{2(n+1)}} \le h_{A,P_0}$$

on U_{P_0}. By Skoda's theorem 2.3 with $g_j = z_j$, $p = n$, and $\alpha = \frac{n}{n-1}$, we can write

$$s - \sigma = \sum_{j=1}^{n} z_j^{(P_0)}\, v_j\,,$$

where v_j is a holomorphic section of $m\,K_{X_0} + A$ on U_{P_0} with

$$\begin{aligned}
\int_{U_{P_0}} \frac{|v_j|^2}{|z^{(P_0)}|^{2n}}\, e^{-\varphi_{m-1}} h_A &\le n\int_{U_{P_0}} \frac{|s-\sigma|^2}{|z^{(P_0)}|^{2(n+1)}}\, e^{-\varphi_{m-1}} h_A \\
&\le 2n\left(\frac{1}{r_1^{2(n+2)}} + \frac{1}{c_A}\, C_{r_1,r_2} \left(\frac{2\,r_2\,C_{X_0}}{r_2^2 - r_1^2}\right)^2\right) C_s \le C^{\flat}\, C_s
\end{aligned}$$

for $1 \le j \le n$. Since $U_{P_0} = \{\left|z^{(P_0)}\right| < 1\}$, it follows that

238   Yum-Tong Siu

$$\int_{U_{P_0}} |v_j|^2 \, e^{-\varphi_{m-1}} h_A \leq C^\flat \, C_s \, . \qquad\qquad \square$$

2.5 *Proof of Theorem 2.1.*

Let

$$\sigma_1^{(m)}, \cdots, \sigma_{N_m}^{(m)} \in \Gamma\left(X_0, m\, K_{X_0} + A\right)$$

be an orthonormal basis, with respect to $e^{-\varphi_{m-1}} h_A$ on X_0, of the subspace of all elements $\sigma^{(m)} \in \Gamma\left(X_0, m\, K_{X_0} + A\right)$ such that

$$\int_{X_0} \left|\sigma^{(m)}\right|^2 e^{-\varphi_{m-1}} \, h_A < \infty \, .$$

By Lemma 2.2 there exist

$$\sigma \in \Gamma\left(X_0, m\, K_{X_0} + A\right)$$

and

$$v_j \in \Gamma\left(U_{P_0}, m\, K_{X_0} + A\right)$$

for $1 \leq j \leq n$ such that $s - \sigma = \sum_{j=1}^n z_j^{(P_0)} v_j$ with

$$\int_{X_0} |\sigma|^2 e^{-\varphi_{m-1}} h_A \leq C^\flat \, C_s$$

and

$$\int_{U_{P_0}} |v_j|^2 \, e^{-\varphi_{m-1}} h_A \leq C^\flat \, C_s \, .$$

We can uniquely write $\sigma = \sum_{k=1}^{N_m} c_k \, \sigma_k^{(m)}$ with $c_k \in \mathbf{C}$ for $1 \leq k \leq N_m$. Then $s = \sum_{k=1}^{N_m} c_k \, \sigma_k^{(m)} + \sum_{j=1}^n z_j^{(P_0)} v_j$ on U_{P_0} and

$$\sum_{k=1}^{N_m} |c_k|^2 \leq C^\flat \, C_s \, .$$

We are going to iterate this process. We do it with s replaced by v_j and use induction. This iteration process is simply an effective version, with estimates, of Nakayama's lemma in algebraic geometry. For notational convenience in the iteration and induction process, we rewrite the index $1 \leq j \leq n$ as $1 \leq j_1 \leq n$. We now replace s by v_{j_1} in the preceding argument and get

$$v_{j_1} = \sum_{k=1}^{N_m} c_{k,j_1} \sigma_k^{(m)} + \sum_{j_2=1}^n z_{j_2}^{(P_0)} \, v_{j_1,j_2}$$

with $c_{k,j_1} \in \mathbf{C}$ and $v_{j_1,j_2} \in \Gamma\left(U_{P_0}, m\, K_{X_0} + A\right)$ such that

$$\sum_{k=1}^{N_m} |c_{k,j_1}|^2 \leq C^\flat \int_{U_{P_0}} |v_{j_1}|^2 e^{-\varphi_{m-1}} h_A \leq \left(C^\flat\right)^2 C_s$$

and

$$\int_{U_{P_0}} |v_{j_1,j_2}|^2 \, e^{-\varphi_{m-1}} h_A \le C^{\flat} \int_{U_{P_0}} |v_{j_1}|^2 e^{-\varphi_{m-1}} h_A \le \left(C^{\flat}\right)^2 C_s \, .$$

By induction on ℓ and applying Lemma 2.2 with s replaced by $v_{j_1,\cdots,j_{\ell-1}}$, we get

$$v_{j_1,\cdots,j_{\ell-1}} = \sum_{k=1}^{N_m} c_{k,j_1,\cdots,j_{\ell-1}} \sigma_k^{(m)} + \sum_{j_\ell=1}^{n} z_{j_2}^{(P_0)} v_{j_1,\cdots,j_{\ell-1},j_\ell}$$

with $c_{k,j_1,\cdots,j_{\ell-1}} \in \mathbf{C}$ and $v_{j_1,\cdots,j_\ell} \in \Gamma\left(U_{P_0}, m\,K_{X_0} + A\right)$ such that

$$\sum_{k=1}^{N_m} |c_{k,j_1,\cdots,j_\nu}|^2 \le \left(C^{\flat}\right)^{\nu+1} C_s$$

and

$$\int_{U_{P_0}} |v_{j_1,\cdots,j_\ell}|^2 \, e^{-\varphi_{m-1}} h_A \le \left(C^{\flat}\right)^{\ell} C_s \, .$$

Thus

$$s = \sum_{j_1,\cdots,j_\ell=1}^{n} v_{j_1,\cdots,j_\ell} z_{j_1} \cdots z_{j_\ell}$$

$$+ \sum_{k=1}^{N_m} \left(c_k + \sum_{\nu=1}^{\ell-1} \sum_{j_1,\cdots,j_\nu=1}^{n} c_{k,j_1,\cdots,j_\nu} z_{j_1}^{(P_0)} \cdots z_{j_\nu}^{(P_0)} \right) \sigma_k^{(m)} \, .$$

Let

$$b_{P_0,m,k} = c_k + \sum_{\nu=1}^{\infty} \sum_{j_1,\cdots,j_\nu=1}^{n} c_{k,j_1,\cdots,j_\nu} z_{j_1}^{(P_0)} \cdots z_{j_\nu}^{(P_0)} \, .$$

We are going to verify that the series defining $b_{P_0,m,k}$ converges to a holomorphic function on $\hat{U}_{P_0,r}$ and

$$\sup_{\hat{U}_{P_0,r}} \sum_{k=1}^{N_m} |b_{P_0,m,k}|^2 \le \frac{1}{(1-r)^2} C^{\flat} C_s \, .$$

Since

$$\sup_{\hat{U}_{P_0,r}} \left| z_{j_1}^{(P_0)} \cdots z_{j_\nu}^{(P_0)} \right|^2 \le \left(\frac{r^2}{n^2 \, C^{\flat}} \right)^{\nu}$$

it follows that on $\hat{U}_{P_0,r}$ and for $1 \le \nu < \infty$ one has the estimate

$$\sum_{k=1}^{N_m} \left| \sum_{j_1,\cdots,j_\nu=1}^{n} c_{k,j_1,\cdots,j_\nu} z_{j_1}^{(P_0)} \cdots z_{j_\nu}^{(P_0)} \right|^2$$

$$\le \left(\frac{r^2}{n^2\,C^\flat}\right)^\nu \sum_{k=1}^{N_m} \left| \sum_{j_1,\cdots,j_\nu=1}^{n} c_{k,j_1,\cdots,j_\nu} \right|^2$$

$$\le \left(\frac{r^2}{n^2\,C^\flat}\right)^\nu n^\nu \sum_{k=1}^{N_m} \sum_{j_1,\cdots,j_\nu=1}^{n} |c_{k,j_1,\cdots,j_\nu}|^2$$

$$= \left(\frac{r^2}{n^2\,C^\flat}\right)^\nu n^\nu \sum_{j_1,\cdots,j_\nu=1}^{n} \sum_{k=1}^{N_m} |c_{k,j_1,\cdots,j_\nu}|^2$$

$$\le \left(\frac{r^2}{n^2\,C^\flat}\right)^\nu n^\nu \sum_{j_1,\cdots,j_\nu=1}^{n} \left(C^\flat\right)^{\nu+1} C_s = r^{2\nu}\,C^\flat\,C_s. \qquad (2.5.1)$$

Let F_ν denotes the N_m-tuple

$$F_\nu = (F_{\nu,1},\cdots,F_{\nu,N_m})$$

of holomorphic functions on $\hat{U}_{P_0,r}$ with

$$F_{\nu,k} = \sum_{j_1,\cdots,j_\nu=1}^{n} c_{k,j_1,\cdots,j_\nu} z_{j_1}^{(P_0)} \cdots z_{j_\nu}^{(P_0)}$$

for $\nu \ge 1$ and $F_{0,k} = c_k$. Then the pointwise L^2 norm $|F_\nu|$ of F_ν is the function

$$\left(|F_{\nu,1}|^2 + \cdots + |F_{\nu,N_m}|^2\right)^{\frac{1}{2}}.$$

By (2.5.1),

$$|F_\nu| \le r^\nu \sqrt{C^\flat\,C_s}$$

on $\hat{U}_{P_0,r}$. Hence

$$\left| \sum_{\nu=0}^{\infty} F_\nu \right| \le \left(\sum_{\nu=0}^{\infty} r^\nu \right) \sqrt{C^\flat\,C_s} = \frac{1}{1-r}\sqrt{C^\flat\,C_s}.$$

Since

$$b_k = \sum_{\nu=0}^{\infty} F_{\nu,k},$$

it follows that b_k is holomorphic on $\hat{U}_{P_0,r}$ and

$$\sup_{\hat{U}_{P_0,r}} \sum_{k=1}^{N_m} |b_{P_0,m,k}|^2 \le \frac{1}{(1-r)^2}\,C^\flat\,C_s. \qquad \square$$

3 Extension Theorems of Ohsawa-Takegoshi Type from Usual Basic Estimates with Two Weight Functions

In this section we are going to state and derive the extension theorem of Ohsawa-Takegoshi type with estimates which we need for the effective version of the arguments of [Siu98]. Such extension theorems originated in a paper of Ohsawa-Takegoshi [OT87] and generalizations were made by Manivel [Man93] and a series of papers of Ohsawa [Ohs88], [Ohs94], [Ohs95], [Ohs01]. Ohsawa's series of papers [Ohs88], [Ohs94], [Ohs95], [Ohs01] contain more general results, which were proved from identities in Kähler geometry and specially constructed complete metrics. Here we use the simple approach of the usual basic estimates with two weight functions. We choose to derive here the extension theorem we need instead of just quoting from more general results, because the simple approach given here gives a clearer picture what and why additional techniques of solving the $\bar{\partial}$ equation other than the standard ones are required for the proof of the extension result. The derivation given here is essentially the same as the one given in [Siu96] with the modifications needed for the present case of no strictly positive lower bound for the curvature current. The only modifications consist of the use of $|\langle u, dw \rangle|$ instead of $|u|$ in some inequalities between (3.5.2) and (3.6.1). The modification simply replaces the strictly positive lower bound of the curvature current in all directions by the strictly positive lower bound of the curvature just for the direction normal to the hypersuface from which the holomorphic section is extended. The precise statement which we need is the following.

Theorem 3.1. *Let Y be a complex manifold of complex dimension n. Let w be a bounded holomorphic function on Y with nonsingular zero-set Z so that dw is nonzero at any point of Z. Let L be a holomorphic line bundle over Y with a (possibly singular) metric $e^{-\kappa}$ whose curvature current is semipositive. Assume that there exists a hypersurface V in Y such that $V \cap Z$ is a subvariety of codimension at least 1 in Z and $Y - V$ is the union of a sequence of Stein subdomains Ω_ν of smooth boundary and Ω_ν is relatively compact in $\Omega_{\nu+1}$. If f is an L-valued holomorphic $(n-1)$-form on Z with*

$$\int_Z |f|^2 \, e^{-\kappa} < \infty \,,$$

then $f dw$ can be extended to an L-valued holomorphic n-form F on Y such that

$$\int_Y |F|^2 \, e^{-\kappa} \leq 8\pi \, e \, \sqrt{2 + \frac{1}{e}} \left(\sup_Y |w|^2 \right) \int_Z |f|^2 \, e^{-\kappa} \,.$$

We will devote most of this section to the proof of Theorem 3.1. We will fix ν and solve the problem on Ω_ν, instead of on Y, and we will do it with the estimate on L^2 norms which is independent of ν and then we will take limit as $\nu \to \infty$. For notational simplicity, in the presentation of our argument,

242   Yum-Tong Siu

we will drop the index ν in Ω_ν and simply denote Ω_ν by Ω. After dividing w by the supremum of $|w|$ on Y, we can assume without loss of generality that the supremum norm of w on Y is no more than 1. Moreover, since $\Omega_{\nu+1}$ is Stein there exists a holomorphic L-valued holomorphic $(n-1)$-form $\tilde{f}$ on $\Omega_{\nu+1}$ such that $(f \wedge dw)|(\Omega_{\nu+1} \cap Z)$ is the restriction of $\tilde{f}$ to $\Omega_{\nu+1} \cap Z$. Of course, when such an extension $\tilde{f}$ is obtained simply by the Stein property of $\Omega_{\nu+1}$, we do not have any L^2 norm estimate on $\tilde{f}$ which is independent of ν.

3.2 *Functional Analysis Preliminaries.*

We recall the standard technique of using functional analysis and Hilbert spaces to solve the $\bar{\partial}$ equation. Consider an operator T which later will be an operator modified from $\bar{\partial}$. Let S be an operator such that $S\,T = 0$. The operator S later will be an operator modified from the $\bar{\partial}$ operator of the next step in the Dolbeault complex. Given g with $Sg = 0$ we would like to solve the equation $Tu = g$. The equation $Tu = g$ is equivalent to $(v, Tu) = (v, g)$ for all $v \in \operatorname{Ker} S \cap \operatorname{Dom} T^*$, which means $(T^*v, u) = (v, g)$ for all $v \in \operatorname{Ker} S \cap \operatorname{Dom} T^*$. To get a solution u it suffices to prove that the map $T^*v \to (v, g)$ can be extended to a bounded linear functional, which means that there exists a positive constant C such that $|(v, g)| \leq C\|T^*v\|$ for all $v \in \operatorname{Ker} S \cap \operatorname{Dom} T^*$. In that case we can solve the equation $Tu = g$ with $\|u\| \leq C$. We could also use the equivalent inequality

$$|(v, g)|^2 \leq C^2 \left(\|T^*v\|^2 + \|Sv\|^2 \right)$$

for all $v \in \operatorname{Dom} S \cap \operatorname{Dom} T^*$.

3.3 *Bochner-Kodaira Formula with Two Weights.*

The crucial point of the argument is the use of two different weights. One weight is for the norm and the other is for the definition of the adjoint of $\bar{\partial}$. We now derive the formula for the Bochner-Kodaira formula in which the weight for the norm is different from the weight used to define the adjoint of $\bar{\partial}$. Formulas, of such a kind, for different weight functions were already given in the literature in the nineteen sixties by authors such as Hörmander (for example, [Hör66]). There is nothing particular new here, except that we need the statement in the form precisely stated below for our case at hand.

We start with a weight $e^{-\varphi}$ and use the usual Bochner-Kodaira formula for this particular weight. Let η be a positive-valued function and let $e^{-\psi} = \frac{e^{-\varphi}}{\eta}$. We will use the weight $e^{-\psi}$ for the definition of the adjoint of $\bar{\partial}$. We use $\bar{\partial}^*_\varphi$ (respectively $\bar{\partial}^*_\psi$) to denote the formal adjoint of $\bar{\partial}$ with respect to the weight function $e^{-\varphi}$ (respectively $e^{-\psi}$). We agree to use the summation convention that, when a lower-case Greek index appears twice in a term, once with a bar and once without a bar, we mean the contraction of the two indices by the Kähler metric tensor. An index without (respectively with) a bar inside

Invariance of Twisted Plurigenera for General Manifolds 243

the complex conjugation of a factor is counted as an index with (respectively without) a bar. We use $\langle \cdot, \cdot \rangle$ to denote the pointwise inner product. Let $\bar{\nabla}$ be the covariant differentiation in the $(0,1)$-direction. The formula we seek is the following.

Proposition 3.4. Let Ω be defined by $r < 0$ so that $|dr|$ with respect to the Kähler metric is identically 1 on the boundary $\partial \Omega$ of Ω. Let u be an $(n,1)$-form in the domain of the actual adjoint of $\bar{\partial}$ on Ω. Then

$$
\int_\Omega \left\langle \bar{\partial}_\psi^* u, \bar{\partial}_\psi^* u \right\rangle e^{-\varphi} + \int_\Omega \left\langle \bar{\partial} u, \bar{\partial} u \right\rangle e^{-\varphi}
$$
$$
= \int_{\partial \Omega} \overline{u_{\bar{\beta}}} u_{\bar{\alpha}} (\partial_{\bar{\beta}} \partial_\alpha r) e^{-\varphi} + \int_\Omega \left\langle \nabla u, \nabla u \right\rangle e^{-\varphi}
$$
$$
+ \int_\Omega \overline{u_{\bar{\beta}}} u_{\bar{\alpha}} \left(\partial_{\bar{\beta}} \partial_\alpha \psi \right) e^{-\varphi} - \int_\Omega \left(\frac{\partial_{\bar{\alpha}} \partial_\beta \eta}{\eta} \right) u_\alpha \overline{u_\beta} e^{-\varphi}
$$
$$
+ 2 \operatorname{Re} \int_\Omega \left(\frac{\partial_\alpha \eta}{\eta} \right) u_{\bar{\alpha}} \overline{\left(\bar{\partial}_\psi^* u \right)} e^{-\varphi}.
$$

Proof. The usual Bochner-Kodaira formula for a domain with smooth boundary for the same weight (also known as the basic estimate) gives

$$
\int_\Omega \left\langle \bar{\partial}_\varphi^* u, \bar{\partial}_\varphi^* u \right\rangle e^{-\varphi} + \int_\Omega \left\langle \bar{\partial} u, \bar{\partial} u \right\rangle e^{-\varphi}
$$
$$
= \int_{\partial \Omega} \overline{u_{\bar{\beta}}} u_{\bar{\alpha}} (\partial_{\bar{\beta}} \partial_\alpha r) e^{-\varphi} + \int_\Omega \left\langle \nabla u, \nabla u \right\rangle e^{-\varphi}
$$
$$
+ \int_\Omega \overline{u_{\bar{\beta}}} u_{\bar{\alpha}} \left(\partial_{\bar{\beta}} \partial_\alpha \varphi \right) e^{-\varphi}. \tag{3.4.1}
$$

The relation between the formal adjoints of $\bar{\partial}$ for different weights is as follows:

$$
\bar{\partial}_\varphi^* u = -e^\varphi \partial_\alpha \left(e^{-\varphi} u_{\bar{\alpha}} \right) = -\frac{e^\psi}{\eta} \partial_\alpha \left(\eta \, e^{-\psi} u_{\bar{\alpha}} \right) = -\frac{\partial_\alpha \eta}{\eta} u_{\bar{\alpha}} + \bar{\partial}_\psi^* u.
$$

Thus

$$
\left| \bar{\partial}_\varphi^* u \right|^2 e^{-\varphi} = \left| -\frac{\partial_\alpha \eta}{\eta} u_{\bar{\alpha}} + \bar{\partial}_\psi^* u \right|^2 e^{-\varphi}
$$
$$
= \left| \frac{\partial_\alpha \eta}{\eta} u_{\bar{\alpha}} \right|^2 e^{-\varphi} - 2 \operatorname{Re} \left(\frac{\partial_\alpha \eta}{\eta} u_{\bar{\alpha}} \overline{\left(\bar{\partial}_\psi^* u \right)} \right) e^{-\varphi} + \left| \bar{\partial}_\psi^* u \right|^2 e^{-\varphi}.
$$

We now rewrite (3.4.1) as

$$
\int_\Omega \left\langle \bar{\partial}_\psi^* u, \bar{\partial}_\psi^* u \right\rangle e^{-\varphi} + \int_\Omega \left\langle \bar{\partial} u, \bar{\partial} u \right\rangle e^{-\varphi}
$$
$$
= \int_{\partial \Omega} \overline{u_{\bar{\beta}}} u_{\bar{\alpha}} (\partial_{\bar{\beta}} \partial_\alpha r) e^{-\varphi} + \int_\Omega \left\langle \nabla u, \nabla u \right\rangle e^{-\varphi}
$$
$$
+ \int_\Omega \overline{u_{\bar{\beta}}} u_{\bar{\alpha}} \left(\partial_{\bar{\beta}} \partial_\alpha \varphi \right) e^{-\varphi} - \int_\Omega \left| \frac{\partial_\alpha \eta}{\eta} u_{\bar{\alpha}} \right|^2 e^{-\varphi}
$$

$$+ 2\,\mathrm{Re}\int_\Omega \frac{\partial_\alpha \eta}{\eta}\, u_{\bar\alpha}\, \overline{\left(\bar\partial_\psi^* u\right)} e^{-\varphi}\,. \tag{3.4.2}$$

From $\varphi = \psi - \log \eta$ it follows that

$$\partial\bar\partial\varphi = \partial\bar\partial\psi - \frac{\partial\bar\partial\eta}{\eta} + \frac{\partial\eta \wedge \bar\partial\eta}{\eta^2}\,.$$

Hence we can rewrite (3.4.2) as

$$\int_\Omega \left\langle \bar\partial_\psi^* u, \bar\partial_\psi^* u \right\rangle e^{-\varphi} + \int_\Omega \left\langle \bar\partial u, \bar\partial u \right\rangle e^{-\varphi}$$
$$= \int_{\partial\Omega} \overline{u_{\bar\beta}}u_{\bar\alpha}(\partial_{\bar\beta}\partial_\alpha r)e^{-\varphi} + \int_\Omega \left\langle \nabla u, \nabla u \right\rangle e^{-\varphi}$$
$$+ \int_\Omega \overline{u_{\bar\beta}}u_{\bar\alpha}\left(\partial_{\bar\beta}\partial_\alpha \psi\right) e^{-\varphi} - \int_\Omega \left(\frac{\partial_{\bar\alpha}\partial_\beta \eta}{\eta}\right) u_\alpha \overline{u_\beta}e^{-\varphi}$$
$$+ 2\,\mathrm{Re}\int_\Omega \left(\frac{\partial_\alpha \eta}{\eta}\right) u_{\bar\alpha}\, \overline{\left(\bar\partial_\psi^* u\right)} e^{-\varphi}\,. \qquad \square$$

3.5 *Choice of Two Different Weights.*

Since Ω is weakly pseudoconvex, the Levi form of r is semi-positive at every point of the boundary $\partial\Omega$ of Ω. The inequality in Proposition 3.4 becomes

$$\int_\Omega \left\langle \bar\partial_\psi^* u, \bar\partial_\psi^* u \right\rangle e^{-\varphi} + \int_\Omega \left\langle \bar\partial u, \bar\partial u \right\rangle e^{-\varphi} \geq \int_\Omega \overline{u_{\bar\beta}}u_{\bar\alpha}\left(\partial_{\bar\beta}\partial_\alpha \psi\right) e^{-\varphi}$$
$$- \int_\Omega \left(\frac{\partial_{\bar\alpha}\partial_\beta \eta}{\eta}\right) u_\alpha \overline{u_\beta}e^{-\varphi} + 2\,\mathrm{Re}\int_\Omega \left(\frac{\partial_\alpha \eta}{\eta}\right) u_{\bar\alpha}\, \overline{\left(\bar\partial_\psi^* u\right)} e^{-\varphi}\,. \tag{3.5.1}$$

Take any positive number $A > e$, where e is the base of the natural logarithm. Let

$$\varepsilon_0 = \sqrt{\frac{A}{e} - 1}\,.$$

For any positive $\varepsilon < \varepsilon_0$, we let

$$\eta = \log \frac{A}{|w|^2 + \varepsilon^2}\,,$$

$$\gamma = \frac{1}{|w|^2 + \varepsilon^2}\,.$$

Then $\eta > 1$ on Ω, because the supremum norm of w is no more than 1 on Ω.

$$-\partial_w \partial_{\bar w} \eta = \frac{\varepsilon^2}{(|w|^2 + \varepsilon^2)^2}\,,$$

$$\partial_w \eta = -\frac{\bar w}{|w|^2 + \varepsilon^2}\,,$$

$$\partial_{\bar w} \eta = -\frac{w}{|w|^2 + \varepsilon^2}\,.$$

Invariance of Twisted Plurigenera for General Manifolds   245

We have the estimate

$$\left| 2\mathrm{Re} \int_\Omega \frac{\partial_\alpha \eta}{\eta} u_{\bar\alpha}(\bar\partial^*_\psi u) e^{-\varphi} \right|$$

$$\leq 2 \int_\Omega \frac{|w|}{|w|^2 + \varepsilon^2} |\langle u, dw \rangle| \, |\bar\partial^*_\psi u| \, e^{-\psi}$$

$$\leq \int_\Omega \frac{|w|^2}{|w|^2 + \varepsilon^2} |\langle u, dw \rangle|^2 e^{-\psi} + \int_\Omega \frac{1}{|w|^2 + \varepsilon^2} |\bar\partial^*_\psi u|^2 e^{-\psi}$$

$$= \int_\Omega \frac{|w|^2}{|w|^2 + \varepsilon^2} |\langle u, dw \rangle|^2 e^{-\psi} + \int_\Omega \gamma |\bar\partial^*_\psi u|^2 e^{-\psi} . \qquad (3.5.2)$$

Choose $\psi = |w|^2 + \kappa$. Since

$$\eta(\partial_\alpha \partial_{\bar\beta} \psi) u_{\bar\alpha} \overline{u_{\bar\beta}} \geq |\langle u, dw \rangle|^2 \geq \frac{|w|^2}{|w|^2 + \varepsilon^2} |\langle u, dw \rangle|^2 ,$$

it follows that

$$\int_\Omega \overline{u_{\bar\beta}} u_{\bar\alpha} \left(\partial_{\bar\beta} \partial_\alpha \psi \right) e^{-\varphi} + \int_\Omega \gamma |\bar\partial^*_\psi u|^2 e^{-\psi}$$

$$= \int_\Omega \eta(\partial_\alpha \partial_{\bar\beta} \psi) u_{\bar\alpha} \overline{u_{\bar\beta}} e^{-\psi} + \int_\Omega \gamma |\bar\partial^*_\psi u|^2 e^{-\psi}$$

$$\geq \int_\Omega \frac{|w|^2}{|w|^2 + \varepsilon^2} |\langle u, dw \rangle|^2 e^{-\psi} + \int_\Omega \gamma |\bar\partial^*_\psi u|^2 e^{-\psi}$$

$$\geq \left| 2\mathrm{Re} \int_\Omega \frac{\partial_\alpha \eta}{\eta} u_{\bar\alpha}(\bar\partial^*_\psi u) e^{-\varphi} \right| ,$$

where the last inequality is from (3.5.2). Adding $\int_\Omega \gamma |\bar\partial^*_\psi u|^2 e^{-\psi}$ to both sides of (3.5.1), we obtain

$$\int_\Omega \left\langle (\eta + \gamma) \bar\partial^*_\psi u, \bar\partial^*_\psi u \right\rangle e^{-\psi} + \int_\Omega \left\langle \eta \, \bar\partial u, \bar\partial u \right\rangle e^{-\psi}$$

$$\geq \int_\Omega \overline{u_{\bar\beta}} u_{\bar\alpha} \left(\partial_{\bar\beta} \partial_\alpha \psi \right) e^{-\varphi} + \int_\Omega \gamma |\bar\partial^*_\psi u|^2 e^{-\psi}$$

$$- \int_\Omega \left(\frac{\partial_{\bar\alpha} \partial_\beta \eta}{\eta} \right) u_\alpha \overline{u_\beta} e^{-\varphi} + 2\mathrm{Re} \int_\Omega \left(\frac{\partial_\alpha \eta}{\eta} \right) u_{\bar\alpha} \overline{\left(\bar\partial^*_\psi u \right)} e^{-\varphi}$$

$$\geq - \int_\Omega \left(\frac{\partial_{\bar\alpha} \partial_\beta \eta}{\eta} \right) u_\alpha \overline{u_\beta} e^{-\varphi}$$

$$= \int_\Omega \frac{\varepsilon^2}{(|w|^2 + \varepsilon^2)^2} |\langle u, dw \rangle|^2 e^{-\psi} . \qquad (3.5.3)$$

246   Yum-Tong Siu

We now consider the operator T defined by $Tu = \bar{\partial}(\sqrt{\eta + \gamma}\, u)$ and the operator S defined by $Su = \sqrt{\eta}\, \bar{\partial}u$. Then $ST = 0$ and we can rewrite (3.5.3) as

$$\|T^*u\|^2_{\Omega,\psi} + \|Su\|^2_{\Omega,\psi} \geq \int_\Omega \frac{\varepsilon^2}{(|w|^2 + \varepsilon^2)^2}\, |\langle u, dw\rangle|^2\, e^{-\psi}\,. \qquad (3.5.4)$$

Here $\|\cdot\|_{\Omega,\psi}$ means the L^2 norm over Ω with respect to the weight function $e^{-\psi}$.

3.6 *Choice of Cut-Off Function.*

Choose any positive number $\delta < 1$. Choose a C^∞ function $0 \leq \varrho(x) \leq 1$ of a single real variable x on $[0, \infty)$ so that the support of ϱ is in $[0, 1]$ and $\varrho(x)$ is identically 1 on $[0, \frac{\delta}{2}]$ and the supremum norm of $\frac{\partial}{\partial x}\varrho(x)$ on $[0, 1]$ is no more than $1 + \delta$.

Let $\varrho_\varepsilon(w) = \varrho\left(\frac{|w|^2}{\varepsilon^2}\right)$ and let

$$g_\varepsilon = \frac{\left(\tilde{f} \wedge dw\right) \bar{\partial}\varrho_\varepsilon}{w} = \frac{\tilde{f} \wedge dw}{\varepsilon^2}\, \varrho'\left(\frac{|w|^2}{\varepsilon^2}\right)\, d\bar{w}\,.$$

We would like to solve the equation $T h_\varepsilon = g_\varepsilon$ for some $(n, 0)$-form h_ε on Ω. For that we would like to verify the inequality

$$|(u, g_\varepsilon)_{\Omega,\psi}|^2 \leq C^2 \left(\|T^* u\|^2_{\Omega,\psi} + \|S u\|^2_{\Omega,\psi}\right)$$

for some positive constant C and for all $u \in \operatorname{Dom} S \cap \operatorname{Dom} T^*$. Here $(\cdot, \cdot)_{\Omega,\psi}$ means the inner product on Ω with respect to the weight function $e^{-\psi}$. We have

$$|(u, g_\varepsilon)_{\Omega,\psi}|^2 = \int_\Omega |\langle u, g_\varepsilon\rangle|\, e^{-\psi}$$

$$= \int_\Omega \left|\left\langle u, \frac{\tilde{f} \wedge dw}{\varepsilon^2}\, \varrho'\left(\frac{|w|^2}{\varepsilon^2}\right)\, d\bar{w}\right\rangle\right| e^{-\psi}$$

$$\leq \left(\int_\Omega \left|\frac{\tilde{f} \wedge dw}{\varepsilon^2}\, \varrho'\left(\frac{|w|^2}{\varepsilon^2}\right)\right|^2 \frac{(|w|^2 + \varepsilon^2)^2}{\varepsilon^2}\, e^{-\psi}\right)$$

$$\cdot \left(\int_\Omega |\langle u, d\bar{w}\rangle|^2\, \frac{\varepsilon^2}{(|w|^2 + \varepsilon^2)^2}\, e^{-\psi}\right)$$

$$\leq C_\varepsilon(\|T^* u\|^2_{\Omega,\psi} + \|S u\|^2_{\Omega,\psi})\,,$$

where

$$C_\varepsilon = \int_\Omega \left|\frac{\tilde{f} \wedge dw}{\varepsilon^2}\, \varrho'\left(\frac{|w|^2}{\varepsilon^2}\right)\right|^2 \frac{(|w|^2 + \varepsilon^2)^2}{\varepsilon^2}\, e^{-\psi}$$

Invariance of Twisted Plurigenera for General Manifolds   247

and the last inequality is from (3.5.4). We can solve $\bar{\partial}(\sqrt{\eta + \gamma}\, h_\varepsilon) = g_\varepsilon$ with

$$\int_\Omega |h_\varepsilon|^2\, e^{-\psi} \leq C_\varepsilon \ . \tag{3.6.1}$$

As $\varepsilon \to 0$, we have the following bound for the limit of C_ε.

$$\limsup_{\varepsilon \to 0} C_\varepsilon \leq \left(\int_{\Omega \cap Y} |f|^2 e^{-\kappa} \right) \left(\limsup_{\varepsilon \to 0} (1 + \delta)^2 \int_{|w| \leq \varepsilon} \frac{(|w|^2 + \varepsilon^2)^2}{\varepsilon^6} |dw|^2 \right)$$

$$\leq 8\pi\, (1 + \delta)^2 \int_{\Omega \cap Y} |f|^2 e^{-\kappa} \ . \tag{3.6.2}$$

3.7 *Final Step in the Proof of Theorem 3.1.*

We now set
$$F_\varepsilon = \varrho_\varepsilon \tilde{f} \wedge dw - w\sqrt{\eta + \gamma}\, h_\varepsilon \ .$$

Then
$$\lim_{\varepsilon \to 0} \int_\Omega \left| \varrho_\varepsilon \tilde{f} \wedge dw \right|^2 e^{-\kappa} = 0 \ ,$$

because $\tilde{f} \wedge dw$ is smooth in a relatively compact open neighborhood of $\bar{\Omega}$ in $Y - V$ and the support of $\varrho_\varepsilon \tilde{f} \wedge dw$ approaches a set of measure zero in Ω as $\varepsilon \to 0$. The supremum norm of $w\sqrt{\eta + \gamma}$ on $\Omega \subset \{|w| < 1\}$ is no more than the square root of

$$\sup_{0 < x \leq 1} x^2 \left(\log A + \log \frac{1}{x^2 + \varepsilon^2} + \frac{1}{x^2 + \varepsilon^2} \right) \leq \log A + \frac{1}{e} + 1 \ ,$$

because the maximum of $y \log \frac{1}{y}$ on $0 < y \leq 1$ occurs at $y = \frac{1}{e}$ where its value is $\frac{1}{e}$ as one can easily verify by checking the critical points of $y \log \frac{1}{y}$. Since A is any number greater than the base of natural logarithm e and δ is any positive number, when we take limit as $A \to e$ and $\delta \to 0$ and $\nu \to \infty$ and we use (3.6.2) and

$$\int_\Omega |h_\varepsilon|^2\, e^{-\kappa} \leq e\, C_\varepsilon$$

from (3.6.1) and $\sup_\Omega |w| \leq 1$, it follows that the limit of F is an L-valued holomorphic n-form on Y whose restriction to Z is $f \wedge dw$ with the following estimate on its norm.

$$\int_Y |F|^2 e^{-\kappa} \leq 8\pi\, e\, \sqrt{2 + \frac{1}{e}} \int_Z |f|^2 e^{-\kappa} \ .$$

This finishes the proof of Theorem 3.1. The following version of extension from submanifolds of higher codimension follows from successive applications of Theorem 3.1.

248   Yum-Tong Siu

Theorem 3.8. *Let Y be a complex manifold of complex dimension n. Let $1 \leq k \leq n$ be an integer and $w_1, \cdots, w_k$ be bounded holomorphic functions on Y whose common zero-set is a complex submanifold Z of complex codimension k in Y (with multiplicity 1 at every point of it). Let L be a holomorphic line bundle over Y with a (possibly singular) metric $e^{-\kappa}$ whose curvature current is semipositive. Assume that there exists a hypersurface V in Y such that $V \cap Z$ is a subvariety of dimension $\leq n - k - 1$ in Z and $Y - V$ is the union of a sequence of Stein subdomains Ω_ν of smooth boundary and Ω_ν is relatively compact in $\Omega_{\nu+1}$. If f is an L-valued holomorphic $(n-k)$-form on Z with*

$$\int_Z |f|^2 \, e^{-\kappa} < \infty \, ,$$

then $f \, dw_1 \wedge \cdots \wedge dw_k$ can be extended to an L-valued holomorphic n-form F on Y such that

$$\int_Y |F|^2 \, e^{-\kappa} \leq \left(8\pi \, e \, \sqrt{2 + \frac{1}{e}} \right)^k \left(\sup_Y |w_1 \cdots w_k|^2 \right) \int_Z |f|^2 \, e^{-\kappa} \, .$$

Proof. We can find a hypersurface V_ν in Ω_ν such that $V_\nu \cap Z$ is of complex dimension $\leq n - k - 1$ in Z and $dw_1 \wedge \cdots \wedge dw_k$ is nowhere zero on $\Omega_\nu - V_\nu$. We can now apply Theorem 3.1 to the case with Y replaced

$$(\Omega_\nu - V_\nu) \cap \{w_\ell = \cdots = w_k = 0\}$$

and Z replaced by

$$(\Omega_\nu - V_\nu) \cap \{w_{\ell+1} = \cdots = w_k = 0\}$$

and w replaced by $w_{\ell+1}$ for $0 \leq \ell < k$ and use descending induction on ℓ. The theorem now follows by removable singularity for L^2 holomorphic functions and the independence of the constants on $\Omega_\nu - V_\nu$ so that one can pass to limit as $\nu \to \infty$. $\square$

4   Induction Argument with Estimates

In this section we carefully keep track of the estimates in the induction argument of [Siu98]. We follow the logical framework set forth in the axiomatic formulation of the induction argument in (1.3). The effective version of the induction argument estimates the L^2 norm of the quotient of the absolute value of a twisted $(p+1)$-canonical section by the maximum of the absolute value of twisted p-canonical sections. Again, this is the implementation of the rough and naive motivation explained in the Introduction. The effective versions of the global generation of multiplier ideal sheaves (2.1) and the extension theorem of Ohsawa-Takegoshi type (3.1) will be used in the process. As explained in the Introduction, metrics for the relevant bundles on the

Invariance of Twisted Plurigenera for General Manifolds   249

initial fibers have to be as singular as possible (see (5.2)). Given an element of $s^{(m_0)} \in \Gamma\left(X_0, m_0\, K_{X_0} + L\right)$, we use the $\left|s^{(m_0)}\right|^{\frac{-2p}{m_0}}$ for the construction of metrics for a properly twisted $p\, K_{X_0}$. To avoid fractional multiples of L, for coefficients of L in (4.1) below we have to introduce integers closest to the quotient $\frac{p}{m_0}$.

Proposition 4.1. *Let A, U_{P_0}, and $C^\flat$ be as in Proposition 2.1. Let U_λ $(1 \leq \lambda \leq \Lambda)$ be a covering of X_0 such that each U_λ is of the form*

$$U_{P_\lambda} \cap \left\{ \left| z^{(P_\lambda)} \right| < \frac{1}{2n\sqrt{C^\flat}} \right\}$$

for some point P_λ of X_0 (i.e., $U_\lambda = \hat{U}_{P_\lambda, r}$ with $r = \frac{1}{2}$). Let $C^\heartsuit$ be a positive constant such that, for any holomorphic line bundle E over X with a (possibly singular) metric h_E of semi-positive curvature current and for any element

$$s \in \Gamma\left(X_0, E + K_{X_0}\right)$$

with

$$\int_{X_0} |s|^2\, h_E < \infty\,,$$

there exists an extension

$$\tilde{s} \in \Gamma\left(X, E + K_X\right)$$

such that

$$\int_X |\tilde{s}|^2\, h_E \leq C^\heartsuit \int_{X_0} |s|^2\, h_E\,.$$

(According to Theorem 3.1 the constant $C^\heartsuit$ can be taken to be $8\pi e\sqrt{2 + \frac{1}{e}}$.) Let $\{\varrho_\lambda\}_{1 \leq \lambda \leq \Lambda}$ be a partition of unity subordinate to the covering $\{U_\lambda\}_{1 \leq \lambda \leq \Lambda}$ of X_0. For $1 \leq \lambda \leq \Lambda$, let

$$\begin{aligned}
\tau_{\lambda, A} &\in \Gamma\left(U_\lambda, A\right)\,, \\
\tau_{\lambda, L} &\in \Gamma\left(U_\lambda, L\right)\,, \\
\xi_\lambda &\in \Gamma\left(U_\lambda, -K_{X_0}\right)
\end{aligned}$$

be all nowhere zero. Let L be a holomorphic line bundle over X with a (possibly singular) metric $e^{-\varphi}$ whose curvature current is semi-positive on X. Let m_0 be an integer ≥ 2 and let

$$s^{(m_0)} \in \Gamma\left(X_0, m_0\, K_{X_0} + L\right)$$

be non identically zero such that $\left|s^{(m_0)}\right|^2 e^{-\varphi}$ is locally bounded on X_0. Let

$$C^{(\%)} = \int_{X_0} \left|s^{(m_0)}\right|^{\frac{2}{m_0}} e^{-\frac{1}{m_0}\varphi}\,.$$

250 Yum-Tong Siu

Let $\psi = \frac{1}{m_0}\log\left|s^{(m_0)}\right|^2$ so that $e^{-m_0\psi}$ is a metric for $m_0 K_{X_0} + L$ and $e^{\psi - \frac{1}{m_0}\varphi}$ is locally bounded on X_0. Let ℓ be a positive number ≥ 2 and let $m_1 = \ell\, m_0$. For $1 \leq p \leq m_1$, let a_p denote the smallest integer which is no less than $\frac{p-1}{m_0}$. Let $d_p = a_p - \frac{p-1}{m_0}$ for $1 \leq p \leq m_1$ and $c_p = a_p - a_{p-1}$ for $2 \leq p \leq m_1$. Let $C^\natural$ be the maximum of

$$
\begin{cases}
4\,C^\flat C^{(\%)}\sup_{U_{P_\lambda}}\left(\left|\xi_\lambda \tau_{\lambda,L}^{-c_p}\right|^2 e^{\psi + \left(c_p - \frac{1}{m_0}\right)\varphi}\right) & \text{for } 1 \leq \lambda \leq \Lambda \text{ and } 2 \leq p < m_1, \\[2ex]
4\,C^\flat C^{(\%)}\sup_{U_{P_\lambda}}\left(\left|\xi_\lambda \tau_{\lambda,L}^{-c_{m_1}} \tau_{\lambda,A}\right|^2 e^{\psi + \left(c_{m_1} - \frac{1}{m_0}\right)\varphi}h_A\right) & \text{for } 1 \leq \lambda \leq \Lambda.
\end{cases}
$$

For $1 \leq p < m_1$ let N_p be the complex dimension of the subspace of all elements $s \in \Gamma\left(X_0, p\,K_{X_0} + a_p L + A\right)$ such that

$$
\int_{X_0} |s|^2\, e^{-(p-1)\psi - d_p\varphi}\, h_A < \infty\,.
$$

Then there exist

$$
s_1^{(p)}, \cdots, s_{N_p}^{(p)} \in \Gamma\left(X_0, p\,K_{X_0} + a_p L + A\right)
$$

for $1 \leq p < m_1$ with

$$
\int_{X_0} \left|s_j^{(p)}\right|^2 e^{-(p-1)\psi - d_p\varphi}\, h_A \leq 1
$$

for $1 \leq j \leq N_p$ such that

(i) $s_j^{(p)}$ can be extended to

$$
\tilde{s}_j^{(p)} \in \Gamma\left(X, p\,K_X + a_p L + A\right)
$$

for $1 \leq p < m_1$ and $1 \leq j \leq N_p$,

(ii)

$$
\int_X \frac{\left|\tilde{s}_j^{(p+1)}\right|^2 e^{-c_{p+1}\varphi}}{\max_{1 \leq k \leq N_p}\left|\tilde{s}_k^{(p)}\right|^2} \leq C^\heartsuit\, C^\natural \int_{X_0} \sum_{\lambda=1}^{\Lambda} \frac{\varrho_\lambda\left|\tau_{\lambda,L}^{c_{p+1}}\right|^2 e^{-c_{p+1}\varphi}}{|\xi_\lambda|^2}
$$

for $1 \leq p \leq m_1 - 2$ and $1 \leq j \leq N_{p+1}$,

(iii)

$$
\int_X \left|\tilde{s}_j^{(1)}\right|^2 h_A \leq C^\heartsuit
$$

for $1 \leq j \leq N_1$, and

(iv)

$$\int_{X_0} \frac{\left|\left(s^{(m_0)}\right)^\ell s_A\right|^2 e^{-c_{m_1}\varphi}}{\max_{1\leq k\leq N_{m_1-1}}\left|\tilde{s}_k^{(m_1-1)}\right|^2} \leq C^\natural \int_{X_0} \sum_{\lambda=1}^\Lambda \frac{\varrho_\lambda \left|s_A \tau_{\lambda,L}^{c_{m_1}}\right|^2 e^{-c_{m_1}\varphi}}{\left|\xi_\lambda \tau_{\lambda,A}\right|^2},$$

where s_A is any holomorphic section of A over X.

In particular, $\left(s^{(m_0)}\right)^\ell s_A$ can be extended to

$$\hat{s}^{(m_1)} \in \Gamma\left(X, m_1 K_X + \ell L + A\right)$$

such that

$$\int_X \frac{\left|\hat{s}^{(m_1)}\right|^2 e^{-c_{m_1}\varphi}}{\max_{1\leq k\leq N_{m_0-1}}\left|\tilde{s}_k^{(m_1-1)}\right|^2} \leq C^\heartsuit C^\natural \int_{X_0} \sum_{\lambda=1}^\Lambda \frac{\varrho_\lambda \left|s_A \tau_{\lambda,L}^{c_{m_1}}\right|^2 e^{-c_{m_1}\varphi}}{\left|\xi_\lambda \tau_{\lambda,A}\right|^2}.$$

Proof of Proposition 4.1. The nonnegative integers a_p and the numbers $0 \leq d_p < 1$ are introduced so that $e^{-(p-1)\psi-d_p\varphi}$ is a metric for the line bundle $(p-1)K_{X_0} + a_p L$ on X_0. The integers c_p are introduced so that multiplication by $\xi_\lambda \tau_{\lambda,L}^{-c_p}$ defines a map from the sheaf of germs of holomorphic sections of $p K_{X_0} + a_p L$ on U_λ to the sheaf of germs of holomorphic sections of $(p-1)K_{X_0} + a_{p-1}L$ on U_λ. We note that

$$a_1 = d_1 = 0, \quad a_{m_1} = \ell, \quad d_{m_1} = \frac{1}{m_0}, \quad 0 \leq c_p \leq 1 \text{ for } 2 \leq p \leq m_1. \quad (4.1.1)$$

For $1 \leq p \leq m_1$ the local function

$$(p-1)\psi + d_p\varphi = \frac{p-1}{m_0}\log\left|s^{(m_0)}\right|^2 + d_p\varphi \text{ is plurisubharmonic.} \quad (4.1.2)$$

By the definition of $C^{(\%)}$ and the value of d_{m_1} from (4.1.1) we have

$$\int_{X_0}\left|\left(s^{(m_0)}\right)^\ell\right|^2 e^{-(m_1-1)\psi-d_{m_1}\varphi} = C^{(\%)}. \quad (4.1.3)$$

Now we are going to apply Theorem 2.1 on the global generation of multiplier ideal sheaves with estimates to prove the following claim by descending induction on $1 \leq p \leq m_1 - 1$. This claim involving estimates corresponds to the step containing (1.4.1) and (1.4.2) in the proof of Proposition 1.4.

252   Yum-Tong Siu

(4.1.4) *Claim.* There exist

(i)
$$s_1^{(p)}, \cdots, s_{N_p}^{(p)} \in \Gamma\left(X_0, p\,K_{X_0} + a_p\,L + A\right)$$
for $1 \le p \le m_1 - 1$,

(ii) holomorphic functions $b_{j,k}^{(p,\lambda)}$ on U_λ, for $1 \le p \le m_1 - 2$, $1 \le j \le N_{p+1}$, $1 \le k \le N_p$, and $1 \le \lambda \le \Lambda$, and

(iii) holomorphic functions $b_k^{(m_1-1,\lambda)}$ on U_λ for $1 \le k \le N_{m_1-1}$ and $1 \le \lambda \le \Lambda$ such that, for $1 \le \lambda \le \Lambda$,

(a)
$$\xi_\lambda \,\tau_{\lambda,A}\, \tau_{\lambda,L}^{-c_{m_1}} \left(s^{(m_0)}\right)^\ell = \sum_{k=1}^{N_{m_1-1}} b_k^{(m_1-1,\lambda)}\, s_k^{(m_1-1)}$$
on U_λ,

(b)
$$\xi_\lambda \,\tau_{\lambda,L}^{-c_{p+1}}\, s_j^{(p+1)} = \sum_{k=1}^{N_p} b_{j,k}^{(p,\lambda)}\, s_k^{(p)}$$
on U_λ for $1 \le p \le m_1 - 2$ and $1 \le j \le N_{p+1}$,

(c)
$$\sup_{U_\lambda} \sum_{1 \le k \le N_{m_1-1}} \left| b_k^{(m_1-1,\lambda)} \right|^2 \le C^\natural,$$

(d)
$$\sup_{U_\lambda} \sum_{k=1}^{N_p} \left| b_{j,k}^{(p,\lambda)} \right|^2 \le C^\natural$$
for $1 \le j \le N_{p+1}$ and $1 \le k \le N_p$ and $1 \le p \le m_1 - 2$, and

(e)
$$\int_{X_0} \left| s_k^{(p)} \right|^2 e^{-(p-1)\psi - d_p\,\varphi}\, h_A \le 1$$
for $1 \le k \le N_p$ and $1 \le p \le m_1 - 1$.

To verify Claim (4.1.4) by descending induction on $1 \le p \le m_1 - 1$, we do first the step $p = m_1 - 1$ and start out with

$$\xi_\lambda \,\tau_{\lambda,A}\, \tau_{\lambda,L}^{-c_{m_1}} \left(s^{(m_0)}\right)^\ell \in \Gamma\left(U_\lambda, (m_1 - 1)\, K_{X_0} + a_{m_1-1}L + A\right).$$

By (4.1.3) we have

$$\int_{U_{P_\lambda}} \left| \xi_\lambda \,\tau_{\lambda,A}\, \tau_{\lambda,L}^{-c_{m_1}} \left(s^{(m_0)}\right)^\ell \right|^2 e^{-(m_1-2)\psi - d_{m_1-1}\varphi}\, h_A$$

$$\le C^{(\%)} \sup_{U_{P_\lambda}} \left(\left| \xi_\lambda \,\tau_{\lambda,A}\, \tau_{\lambda,L}^{-c_{m_1}} \right|^2 e^{\psi + \left(c_{m_1} - \frac{1}{m_0}\right)\varphi}\, h_A \right).$$

By applying Theorem 2.1 with $r = \frac{1}{2}$ to the line bundle $(m_1 - 2) K_{X_0} + a_{m_1-1} L$ with the metric $e^{-(m_1-2)\psi - d_{m_1-1}\varphi}$, from (4.1.2) we obtain

$$s_1^{(m_1-1)}, \cdots, s_{N_{m_1-1}}^{(m_1-1)} \in \Gamma\left(X_0, (m_1-1) K_{X_0} + a_{m_1-1} L + A\right)$$

and holomorphic functions $b_k^{(m_1-1,\lambda)}$ on U_λ, for $1 \le k \le N_{m_1-1}$, such that

$$\xi_\lambda \, \tau_{\lambda,A} \, \tau_{\lambda,L}^{-c_{m_1}} \left(s^{(m_0)}\right)^\ell = \sum_{k=1}^{N_{m_1-1}} b_k^{(m_1-1,\lambda)} \, s_k^{(m_1-1)}$$

and

$$\sup_{U_\lambda} \sum_{k=1}^{N_{m_1-1}} \left| b_{j,k}^{(m_1-1,\lambda)} \right|^2$$

$$\le 4 \, C^\flat \int_{U_{P_\lambda}} \left| \xi_\lambda \, \tau_{\lambda,L}^{-c_{m_1}} \, \tau_{\lambda,A} \left(s^{(m_0)}\right)^\ell \right|^2 e^{-(m_1-2)\psi - d_{m_1-1}\,\varphi}$$

$$\le 4 \, C^\flat \sup_{U_{P_\lambda}} \left(\left| \xi_\lambda \, \tau_{\lambda,L}^{-c_{m_1}} \, \tau_{\lambda,A} \right|^2 , e^{\psi + \left(c_{m_1} - \frac{1}{m_0}\right)\varphi} \, h_A \right)$$

$$\cdot \left(\int_{U_{P_\lambda}} \left| \left(s^{(m_0)}\right)^\ell \right|^2 e^{-(m_1-1)\psi - d_{m_1}\varphi} \right)$$

$$\le 4 \, C^\flat \, C^{(\%)} \sup_{U_{P_\lambda}} \left(\left| \xi_\lambda \, \tau_{\lambda,L}^{-c_{m_1}} \, \tau_{\lambda,A} \right|^2 e^{\psi + \left(c_{m_1} - \frac{1}{m_0}\right)\varphi} \, h_A \right) \le C^\natural$$

and

$$\int_{X_0} \left| s_k^{(m_1-1)} \right|^2 e^{-(m_1-2)\psi - d_{m_1-1}\,\varphi} \, h_A \le 1 \,.$$

Thus we have (4.1.4)(a) and (4.1.4)(c) and the case $p = m_1 - 1$ of (4.1.4)(e). This finishes the initial step $p = m_1 - 1$.

Suppose we have done this for Step p and we would like to do it for the next step which is Step $p - 1$. Since

$$\xi_\lambda \, \tau_{\lambda,L}^{-c_p} \, s_k^{(p)} \in \Gamma\left(U_\lambda, (p-1) K_{X_0} + a_{p-1} L + A\right)$$

with

$$\int_{U_{P_\lambda}} \left| \xi_\lambda \, \tau_{\lambda,L}^{-c_p} \, s_k^{(p)} \right|^2 e^{-(p-2)\psi - d_{p-1}\,\varphi} \, h_A$$

$$\le \sup_{U_{P_\lambda}} \left(\left| \xi_\lambda \, \tau_{\lambda,L}^{-c_p} \right|^2 e^{\psi + \left(c_p - \frac{1}{m_0}\right)\varphi} \right) \int_{U_{P_\lambda}} \left| s_k^{(p)} \right|^2 e^{-(p-1)\psi - d_p\,\varphi} \, h_A$$

$$\le \sup_{U_{P_\lambda}} \left(\left| \xi_\lambda \, \tau_{\lambda,L}^{-c_p} \right|^2 e^{\psi + \left(c_p - \frac{1}{m_0}\right)\varphi} \right)$$

254   Yum-Tong Siu

from (4.1.4)(e) of Step p, it follows from Theorem 2.1 with $r = \frac{1}{2}$ applied to the line bundle $(p-1)K_{X_0} + a_{p-1}L$ with the metric $e^{-(p-2)\psi - d_{p-1}\varphi}$ and from (4.1.2) that we obtain

$$s_k^{(p-1)} \in \Gamma\left(X_0, (p-1)K_{X_0} + a_{p-1}L + A\right)$$

for $1 \le k \le N_{p-1}$ and holomorphic functions $b_{j,k}^{(p-1,\lambda)}$ on U_λ, for $1 \le j \le N_p$ and $1 \le k \le N_{p-1}$ such that

$$\xi_\lambda \, \tau_{\lambda,L}^{-c_p} \, s_j^{(p)} = \sum_{k=1}^{N_{p-1}} b_{j,k}^{(p-1,\lambda)} \, s_k^{(p-1)}$$

and

$$\sup_{U_\lambda} \sum_{k=1}^{N_{p-1}} \left| b_{j,k}^{(p-1,\lambda)} \right|^2 \le 4\, C^\flat \, \sup_{U_{P_\lambda}} \left(\left| \xi_\lambda \, \tau_{\lambda,L}^{-c_p} \right|^2 e^{\psi + \left(c_p - \frac{1}{m_0}\right)\varphi} \right) \le C^\natural .$$

Thus we have (4.1.4)(b) and (4.1.4)(d) and (4.1.4)(e). This finishes the the verification of Claim (4.1.4) by induction.

Next we use induction on $1 \le p < m_1$ to verify the following claim. This claim involving estimates corresponds to Claim (1.4.3) in the proof of Proposition 1.4.

(4.1.5) *Claim.* One can extend $s_j^{(p)}$ to

$$\tilde{s}_j^{(p)} \in \Gamma\left(X, p\,K_X + a_p\,L + A\right)$$

for $1 \le j \le N_p$ and $1 \le p < m_1$ such that the extension

$$\tilde{s}_j^{(p)} \in \Gamma\left(X, p\,K_X + a_p\,L + A\right)$$

satisfies the inequalities (ii) and (iii) in the statement of Proposition 4.1.

To start the induction process when $p = 1$, we simply observe that, since the curvature form of the metric h_A of A is positive on X and since

$$\int_{X_0} \left| s_k^{(1)} \right|^2 h_A \le 1,$$

it follows from the assumption on $C^\heartsuit$ that $s_j^{(1)}$ can be extended to

$$\tilde{s}_j^{(1)} \in \Gamma\left(X, K_X + A\right)$$

with

$$\int_X \left| \tilde{s}_j^{(1)} \right|^2 h_A \le C^\heartsuit,$$

Invariance of Twisted Plurigenera for General Manifolds   255

which is inequality (iii) in the statement of Proposition 4.1.

Suppose Step p has been proved and we go to the next step which is Step $p + 1 < m_1$. Since

$$\xi_\lambda \, \tau_{\lambda,L}^{-c_{p+1}} s_j^{(p+1)} = \sum_{k=1}^{N_p} b_{j,k}^{(p,\lambda)} \, s_k^{(p)}$$

by $(4.1.4)(b)$ and

$$\sup_{U_\lambda} \sum_{k=1}^{N_p} \left| b_{j,k}^{(p,\lambda)} \right|^2 \leq C^\natural$$

by $(4.1.4)(d)$, it follows that

$$\int_{X_0} \frac{\left| s_j^{(p+1)} \right|^2 e^{-c_{p+1}\,\varphi}}{\max_{1 \leq k \leq N_p} \left| \tilde{s}_k^{(p)} \right|^2} = \int_{X_0} \frac{\sum_{\lambda=1}^\Lambda \varrho_\lambda \left| \left(\frac{\tau_{\lambda,L}^{c_{p+1}}}{\xi_\lambda} \right) \tau_{\lambda,L}^{-c_{p+1}} \xi_\lambda \, s_j^{(p+1)} \right|^2 e^{-c_{p+1}\,\varphi}}{\max_{1 \leq k \leq N_p} \left| s_k^{(p)} \right|^2}$$

$$\leq \left(\sup_{U_\lambda} \sum_{k=1}^{N_p} \left| b_{j,k}^{(p,\lambda)} \right|^2 \right) \int_{X_0} \sum_{\lambda=1}^\Lambda \frac{\varrho_\lambda \left| \tau_{\lambda,L}^{c_{p+1}} \right|^2 e^{-c_{p+1}\,\varphi}}{\left| \xi_\lambda \right|^2}$$

$$\leq C^\natural \int_{X_0} \sum_{\lambda=1}^\Lambda \frac{\varrho_\lambda \left| \tau_{\lambda,L}^{c_{p+1}} \right|^2 e^{-c_{p+1}\,\varphi}}{\left| \xi_\lambda \right|^2} \, .$$

By the assumption on $C^\heartsuit$, one can extend $s_j^{(p+1)}$ to

$$\tilde{s}_j^{(p+1)} \in \Gamma\left(X, (p+1)K_X + a_{p+1}\,L + A \right)$$

with

$$\int_X \frac{\left| \tilde{s}_j^{(p+1)} \right|^2 e^{-c_{p+1}\,\varphi}}{\max_{j=1}^{N_p} \left| \tilde{s}_j^{(p)} \right|^2} \leq C^\heartsuit \, C^\natural \int_{X_0} \sum_{\lambda=1}^\Lambda \frac{\varrho_\lambda \left| \tau_{\lambda,L}^{c_{p+1}} \right|^2 e^{-c_{p+1}\,\varphi}}{\left| \xi_\lambda \right|^2} \, ,$$

which is inequality (ii) in the statement of Proposition 4.1. This finishes the verification of Claim $(4.1.5)$ by induction.

Finally, since

$$\xi_\lambda \, \tau_{\lambda,A} \, \tau_{\lambda,L}^{-c_{m_1}} \left(s^{(m_0)} \right)^\ell = \sum_{k=1}^{N_{m_1}-1} b_k^{(m_1-1,\lambda)} \, s_k^{(m_1-1)}$$

by $(4.1.4)(a)$ and

$$\sup_{U_\lambda} \sum_{k=1}^{N_{m_1}-1} \left| b_k^{(m_1-1,\lambda)} \right|^2 \leq C^\natural$$

by (4.1.4)(c), it follows that

$$
\int_{X_0} \frac{\left|\left(s^{(m_0)}\right)^\ell s_A\right|^2 e^{-c_{m_1}\varphi}}{\max_{1\le k\le N_p}\left|\tilde{s}_k^{(m_1-1)}\right|^2}
$$

$$
= \int_{X_0} \frac{\sum_{\lambda=1}^\Lambda \varrho_\lambda \left|\left(\frac{s_A \tau_{\lambda,L}^{c_{m_1}}}{\xi_\lambda}\right)\tau_{\lambda,L}^{-c_{m_1}}\xi_\lambda\left(s^{(m_0)}\right)^\ell\right|^2 e^{-c_{m_1}\varphi}}{\max_{1\le k\le N_p}\left|s_k^{(p)}\right|^2}
$$

$$
\le \left(\sum_{k=1}^{N_{m_1}-1}\sup_{U_\lambda}\left|b_k^{(m_1-1,\lambda)}\right|^2\right)\int_{X_0}\sum_{\lambda=1}^\Lambda \frac{\varrho_\lambda\left|s_A\tau_{\lambda,L}^{c_{r_1}}\right|^2 e^{-c_{m_1}\varphi}}{\left|\xi_\lambda\tau_{\lambda,A}\right|^2}
$$

$$
\le C^\natural \int_{X_0}\sum_{\lambda=1}^\Lambda \frac{\varrho_\lambda\left|s_A\tau_{\lambda,L}^{c_{m_1}}\right|^2 e^{-c_{m_1}\varphi}}{\left|\xi_\lambda\tau_{\lambda,A}\right|^2},
$$

which is inequality (iv) in the statement of Proposition 4.1. By the assumption on $C^\heartsuit$, $\left(s^{(m_0)}\right)^\ell s_A$ can be extended to

$$
\hat{s}^{(m_1)} \in \Gamma\left(X, m_1 K_X + \ell L + A\right)
$$

such that

$$
\int_X \frac{\left|\hat{s}^{(m_1)}\right|^2 e^{-c_{m_1}\varphi}}{\max_{1\le k\le N_{m_1-1}}\left|\tilde{s}_k^{(m_1-1)}\right|^2} \le C^\heartsuit C^\natural \int_{X_0}\sum_{\lambda=1}^\Lambda \frac{\varrho_\lambda\left|s_A\tau_{\lambda,L}^{c_{m_1}}\right|^2 e^{-c_{m_1}\varphi}}{\left|\xi_\lambda\tau_{\lambda,A}\right|^2}.
$$

$\square$

5 Effective Version of the Process of Taking Powers and Roots of Sections

This section corresponds to the effective version of the fourth ingredient listed in the Introduction, which is the process of raising a section on the initial fiber to a high power and later taking the root of the same order after extending its product with a holomorphic section of some fixed line bundle on X.

5.1 To apply Theorem 2.1 on the global generation of the multiplier ideal sheaf with estimates, we have to introduce an auxiliary sufficiently ample line bundle A. To eliminate the undesirable effects from A at the end, we take the ℓ-th power of an m_0-canonical section on X_0 to be extended and then we multiply the power by some section of A on X to make the extension of the product to X possible by Proposition 4.1 and then we take the ℓ-th root of the extension and let $\ell \to \infty$.

The goal of the limiting process is to produce a metric for $(m_0 - 1) K_X$ on X so that the m_0-canonical section on X_0 has finite L^2 norm on X_0 with respect to it. For the limiting process, we have to control the estimates to guarantee the convergence of the limit. We do this by using the concavity of the logarithmic function and the sub-mean-value property of the logarithm of the absolute value of a holomorphic function. A delicate point is that we have to get the bound on the dimension of certain spaces of sections independent of ℓ. It is because of this delicate point that, as explained in the Introduction and (1.5.2), the metric chosen for the pluricanonical line bundles of the initial fiber at the beginning of the effective argument has to be as singular as possible. For that purpose, one cannot use usual abstractly-defined general metrics for the pluricanonical line bundle of the initial fiber such as generalized Bergman kernels. The bound for the dimension in question will be given below in (5.2).

Fix a positive number m_0 and take a non identically zero

$$s^{(m_0)} \in \Gamma\left(X_0, m_0 K_{X_0} + L\right) .$$

Our goal is to extend it to an element of $\Gamma\left(X, m_0 K_X + L\right)$. Since the case of $m_0 = 1$ is an immediate consequence of the extension theorem 3.1, we can assume without loss of generality that $m_0 \geq 2$.

Fix an element $s_A \in \Gamma\left(X, A\right)$ such that its zero-set in X_0 is of complex codimension at least 1 in X_0. We are going to apply Proposition 4.1 and let $\ell \to \infty$. We need the following lemma concerning the bound on the dimension of certain spaces of sections independent of ℓ.

Lemma 5.2. *In the notations of the statement of Proposition 4.1, let $\tilde{N}_{m_0}$ be the maximum of the complex dimension of $\Gamma\left(X_0, k K_{X_0} + a L + A\right)$ for integers $1 \leq k \leq m_0$ and $a = 0, 1$. Then $N_p \leq \tilde{N}_{m_0}$ for $1 \leq p < m_0 \ell$ for any integer $\ell \geq 2$.*

Proof. We follow the notations of Proposition 4.1. For $1 \leq p < m_1$ let Γ_p be the subspace of all elements $s \in \Gamma\left(X_0, p K_{X_0} + a_p L + A\right)$ such that

$$\int_{X_0} |s|^2 \, e^{-(p-1)\psi - d_p \varphi} \, h_A < \infty .$$

Let b_p be the largest integer not exceeding $\frac{p-1}{m_0}$. For any $s \in \Gamma_p$, the definition of Γ_p implies that

$$\frac{s}{\left(s^{(m_0)}\right)^{b_p}}$$

is a holomorphic section of $(p - b_p m_0) K_{X_0} + (a_p - b_p) L + A$ over X_0. Thus the map

$$\Xi_p : \Gamma_p \to \Gamma\left(X_0, (p - b_p m_0) K_{X_0} + (a_p - b_p) L + A\right)$$

defined by

$$\Xi_p(s) = \frac{s}{\left(s^{(m_0)}\right)^{b_p}}$$

258 Yum-Tong Siu

is injective. We conclude that

$$\dim_{\mathbf{C}} \Gamma_p \leq \dim_{\mathbf{C}} \Gamma \left(X_0, (p - b_p m_0) K_{X_0} + (a_p - b_p) L + A \right)$$

Since

$$b_p \leq \frac{p-1}{m_0}, \quad b_p + 1 > \frac{p-1}{m_0}, \quad a_p \geq \frac{p-1}{m_0}, \quad a_p - 1 < \frac{p-1}{m_0},$$

it follows that $1 \leq p - b_p m_0 \leq m_0$ and $0 \leq a_p - b_p \leq 1$. The definitions of $\tilde{N}_{m_0}$ and N_p imply that $N_p \leq \tilde{N}_{m_0}$ for $1 \leq p < \ell m_0$ for any integer $\ell \geq 2$.

The reason for imposing the condition $1 \leq p < \ell m_0$ in the conclusion, instead of just getting the conclusion for all positive integers p, is that technically in the statement of Proposition 4.1 the number N_p is defined only for $p < m_1$ and $m_1 = \ell m_0$. $\qquad\square$

5.3 *Application of Induction Argument.*

Take an arbitrary integer $\ell \geq 2$. We now apply Proposition 4.1 to

$$s^{(m_0)} \in \Gamma \left(X_0, m_0 K_{X_0} + L \right)$$

so that we can extend

$$\left(s^{(m_0)} \right)^{\ell} s_A \in \Gamma \left(X_0, \ell m_0 K_{X_0} + \ell L + A \right)$$

to

$$\hat{s}^{(\ell m_0, A)} \in \Gamma \left(X, \ell m_0 K_X + \ell L + A \right).$$

This extension from Proposition 4.1 comes with estimates. We use the notations $N_{p,A}$, ϱ_λ, ξ_λ, $\tau_{\lambda,A}$, $C^{\natural}$, and $C^{\heartsuit}$ of Proposition 4.1. By Lemma 5.2, $N_{p,A} \leq N_{m_0}$ for all $1 \leq p \leq \ell m_0 - 1$.

Let $C^{\diamond}$ be the maximum of the following positive numbers

$$C^{\heartsuit}, \quad C^{\heartsuit} C^{\natural} \int_{X_0} \sum_{\lambda=1}^{\Lambda} \frac{\varrho_\lambda \left| s_A \, \tau_{\lambda,L}^{c_{m_1}} \right|^2 e^{-c_{m_1} \varphi}}{\left| \xi_\lambda \, \tau_{\lambda,A} \right|^2},$$

$$C^{\heartsuit} C^{\natural} \int_{X_0} \sum_{\lambda=1}^{\Lambda} \frac{\varrho_\lambda \left| \tau_{\lambda,L}^{c_{p+1}} \right|^2 e^{-c_{p+1} \varphi}}{\left| \xi_\lambda \right|^2} \quad \text{for } 1 \leq p \leq m_1 - 2.$$

The finiteness of $C^{\diamond}$ follows from $0 \leq c_p \leq 1$ in 4.1.1 and from the local integrability of $e^{-\varphi}|_{X_0}$ on X_0. According to Proposition 4.1 we have

$$s_1^{(p)}, \cdots, s_{N_p}^{(p)} \in \Gamma \left(X_0, p \, K_{X_0} + a_p \, L + A \right)$$

for $1 \leq p < m_1$ such that

(i) the section $s_j^{(p)}$ can be extended to

$$\tilde{s}_j^{(p)} \in \Gamma\left(X, p K_X + a_p L + A\right)$$

for $1 \le p < m_1$ and $1 \le p \le N_p$

(ii)

$$\int_X \frac{\left|\tilde{s}_j^{(p+1)}\right|^2 e^{-c_{p+1}\varphi}}{\max_{1 \le k \le N_p}\left|\tilde{s}_k^{(p)}\right|^2} \le C^\diamond \qquad (5.3.1)$$

for $1 \le p \le m_1 - 2$ and $1 \le j \le N_{p+1}$,

(iii)

$$\int_X \left|\tilde{s}_j^{(1)}\right|^2 h_A \le C^\diamond \qquad (5.3.2)$$

for $1 \le j \le N_1$, and

(iv)

$$\int_X \frac{\left|\hat{s}^{(m_1)}\right|^2 e^{-c_{m_1}\varphi}}{\max_{1 \le k \le N_{m_1-1}}\left|\tilde{s}_k^{(m_1-1)}\right|^2} \le C^\diamond . \qquad (5.3.3)$$

From (5.3.1) we have

$$\int_X \frac{\max_{1 \le k \le N_{p+1}}\left|\tilde{s}_j^{(p+1)}\right|^2 e^{-c_{p+1}\varphi}}{\max_{1 \le k \le N_p}\left|\tilde{s}_k^{(p)}\right|^2} \le \sum_{j=1}^{N_{p+1}} \int_X \frac{\left|\tilde{s}_j^{(p+1)}\right|^2 e^{-c_{p+1}\varphi}}{\max_{1 \le k \le N_p}\left|\tilde{s}_k^{(p)}\right|^2}$$

$$\le N_{p+1} C^\diamond \le \tilde{N}_{m_0} C^\diamond \qquad (5.3.4)$$

for $1 \le p \le m_1 - 2$ and $1 \le j \le N_{p+1}$. It is for this step that we need the bound on N_{p+1} given in Lemma 5.2 which is independent of ℓ.

5.4 *Supremum Estimates by the Concavity of Logarithm and the Sub-Mean-Value Property of the Logarithm of the Absolute Value of a Holomorphic Function.*

To continue with our estimates, we have to switch at this point to the use of supremum norms. The reason is that we are estimating the ℓ-th root of the product of $\ell\, m_0$ factors, for each of which we have only an L^2 estimate and we cannot continue with L^2 estimates by using the Hölder inequality. We switch to supremum estimates by using the concavity of logarithm and the sub-mean-value property of the logarithm of the absolute value of a holomorphic function. The use of the sub-mean-value property for switching to

260   Yum-Tong Siu

supremum estimates necessitates the shrinking of the domain on which estimates are made. For notational convenience we assume that the family X can be extended to a larger one over a larger disk in $\mathbf{C}$.

More precisely, without loss of generality we can assume the following. The given family $\pi : X \to \Delta$ can be extended to a holomorphic family $\tilde{\pi} : \tilde{X} \to \tilde{\Delta}$ of compact complex projective algebraic manifolds so that $\tilde{\Delta}$ is an open disk of radius > 1 in $\mathbf{C}$ centered at 0. We can assume that the coordinate charts U_λ $(1 \leq \lambda \leq \Lambda)$ from Proposition 4.1 are restrictions of coordinate charts in $\tilde{X}$ in the following way. We have coordinate charts W_λ in $\tilde{X}$ $(1 \leq \lambda \leq \Lambda)$ with coordinates

$$\left(z_1^{(\lambda)}, \cdots, z_n^{(\lambda)}, t \right)$$

so that

$$\bigcup_{\lambda=1}^{\Lambda} W_\lambda = \tilde{X} .$$

Moreover, we assume that for each $1 \leq \lambda \leq \Lambda$ we have a relative compact open subset $\tilde{U}_\lambda$ of W_λ such that

$$\tilde{U}_\lambda = W_\lambda \bigcap \left\{ \left| z_1^{(\lambda)} \right| < 1, \cdots, \left| z_n^{(\lambda)} \right| < 1, |t| < 1 \right\}$$

and

$$\bigcup_{\lambda=1}^{\Lambda} \tilde{U}_\lambda = X$$

and

$$U_\lambda = X_0 \cap \tilde{U}_\lambda .$$

We can also assume that there exists nowhere zero elements

$$\hat{\xi}_\lambda \in \Gamma\left(W_\lambda, -K_{\tilde{X}} \right), \quad \hat{\tau}_{\lambda,A} \in \Gamma\left(W_\lambda, A \right), \quad \hat{\tau}_{\lambda,L} \in \Gamma\left(W_\lambda, L \right)$$

for $1 \leq \lambda \leq \Lambda$ such that

$$\xi_\lambda = \hat{\xi}_\lambda \,|\, U_\lambda \,,$$
$$\tau_{\lambda,A} = \hat{\tau}_{\lambda,A} \,|\, U_\lambda \,,$$
$$\tau_{\lambda,L} = \hat{\tau}_{\lambda,L} \,|\, U_\lambda$$

for $1 \leq \lambda \leq \Lambda$.

There exists some $0 < r_0 < 1$ such that, if we let $X' = \pi^{-1}\left(\Delta_{r_0} \right)$ and

$$U'_\lambda = \tilde{U}_\lambda \bigcap \left\{ \left| z_1^{(\lambda)} \right| < r_0, \cdots, \left| z_n^{(\lambda)} \right| < r_0, |t| < r_0 \right\},$$

for $1 \leq \lambda \leq \Lambda$, then $X' = \bigcup_{\lambda=1}^{\Lambda} U'_\lambda$. Let $dV_{z^{(\lambda)},t}$ be the Euclidean volume form in the coordinates system

$$\left(z^{(\lambda)}, t \right) = \left(z_1^{(\lambda)}, \cdots, z_n^{(\lambda)}, t \right).$$

From (5.3.4) we have

$$\frac{1}{\pi^{n+1}} \int_{\tilde{U}_\lambda} \left(\log \max_{1 \le j \le N_{p+1}} \left| \tau_{\lambda,A}^{-1} \tau_{\lambda,L}^{-a_{p+1}} \xi_\lambda^{p+1} \tilde{s}_j^{(p+1)} \right|^2 \right) dV_{z(\lambda),t}$$

$$- \frac{1}{\pi^{n+1}} \int_{\tilde{U}_\lambda} \left(\log \max_{1 \le k \le N_p} \left| \tau_{\lambda,A}^{-1} \tau_{\lambda,L}^{-a_p} \xi_\lambda^p \tilde{s}_k^{(p)} \right|^2 \right) dV_{z(\lambda),t}$$

$$= \frac{1}{\pi^{n+1}} \int_{\tilde{U}_\lambda} \left(\log \frac{\max_{1 \le j \le N_{p+1}} \left| \tau_{\lambda,A}^{-1} \tau_{\lambda,L}^{-a_{p+1}} \xi_\lambda^{p+1} \tilde{s}_j^{(p+1)} \right|^2}{\max_{1 \le k \le N_p} \left| \tau_{\lambda,A}^{-1} \tau_{\lambda,L}^{-a_p} \xi_\lambda^p \tilde{s}_k^{(p)} \right|^2} \right) dV_{z(\lambda),t}$$

$$\le \log \left(\frac{1}{\pi^{n+1}} \int_{\tilde{U}_\lambda} \frac{\max_{1 \le j \le N_{p+1}} \left| \tau_{\lambda,A}^{-1} \tau_{\lambda,L}^{-a_{p+1}} \xi_\lambda^{p+1} \tilde{s}_j^{(p+1)} \right|^2}{\max_{1 \le k \le N_p} \left| \tau_{\lambda,A}^{-1} \tau_{\lambda,L}^{-a_p} \xi_\lambda^p \tilde{s}_k^{(p)} \right|^2} dV_{z(\lambda),t} \right)$$

$$\le \log \left(\sup_{\tilde{U}_\lambda} \left(\frac{1}{\pi^{n+1}} \left| \tau_{\lambda,L}^{-c_{p+1}} \xi_\lambda \right|^2 e^{c_{p+1}\varphi} dV_{z(\lambda),t} \right) \right.$$

$$\left. \cdot \int_{\tilde{U}_\lambda} \frac{\max_{1 \le j \le N_{p+1}} \left| \tilde{s}_j^{(p+1)} \right|^2 e^{-c_{p+1}\varphi}}{\max_{1 \le k \le N_p} \left| \tilde{s}_k^{(p)} \right|^2} \right)$$

$$\le \log \left(\tilde{N}_{m_0} C^\diamond \sup_{\tilde{U}_\lambda} \left(\frac{1}{\pi^{n+1}} \left| \tau_{\lambda,L}^{-c_{p+1}} \xi_\lambda \right|^2 e^{c_{p+1}\varphi} dV_{z(\lambda),t} \right) \right) \quad (5.4.1)$$

for $1 \le p \le m_1 - 2$. Here for the first inequality we have used the concavity of the logarithm. Summing (5.4.1) from $p = 1$ to $p = m_1 - 2$, we get

$$\frac{1}{\pi^{n+1}} \int_{\tilde{U}_\lambda} \left(\log \max_{1 \le j \le N_{m_1-1}} \left| \tau_{\lambda,A}^{-1} \tau_{\lambda,L}^{-a_{m_1-1}} \xi_\lambda^{m_1-1}, \tilde{s}_j^{(m_1-1)} \right|^2 \right) dV_{z(\lambda),t}$$

$$\le \frac{1}{\pi^{n+1}} \int_{\tilde{U}_\lambda} \left(\log \max_{1 \le k \le N_1} \left| \tau_{\lambda,A}^{-1} \xi_\lambda \tilde{s}_k^{(1)} \right|^2 \right) dV_{z(\lambda),t}$$

$$+ \sum_{p=1}^{m_1-2} \log \left(\tilde{N}_{m_0} C^\diamond \sup_{\tilde{U}_\lambda} \left(\frac{1}{\pi^{n+1}} \left| \tau_{\lambda,L}^{-c_{p+1}} \xi_\lambda \right|^2 e^{c_{p+1}\varphi} dV_{z(\lambda),t} \right) \right). \quad (5.4.2)$$

Likewise, from (5.3.3)

$$\frac{1}{\pi^{n+1}} \int_{\tilde{U}_\lambda} \left(\log \left| \tau_{\lambda,A}^{-1} \tau_{\lambda,L}^{-a_{m_1}} \xi_\lambda^{m_1} \hat{s}^{(m_1)} \right|^2 \right) dV_{z(\lambda),t}$$

$$- \frac{1}{\pi^{n+1}} \int_{\tilde{U}_\lambda} \left(\log \max_{1 \le k \le N_{m_1-1}} \left| \tau_{\lambda,A}^{-1} \tau_{\lambda,L}^{-a_{m_1}-1} \xi_\lambda^{m_1-1} \tilde{s}_k^{(m_1-1)} \right|^2 \right) dV_{z(\lambda),t}$$

$$= \frac{1}{\pi^{n+1}} \int_{\tilde{U}_\lambda} \left(\log \frac{\left| \tau_{\lambda,A}^{-1} \tau_{\lambda,L}^{-a_{m_1}} \xi_\lambda^{m_1} \hat{s}^{(m_1)} \right|^2}{\max_{1 \le k \le N_{m_1-1}} \left| \tau_{\lambda,A}^{-1} \tau_{\lambda,L}^{-a_{m_1}-1} \xi_\lambda^{m_1-1} \tilde{s}_k^{(m_1-1)} \right|^2} \right) dV_{z(\lambda),t}$$

$$\le \log \left(\frac{1}{\pi^{n+1}} \int_{\tilde{U}_\lambda} \frac{\left| \tau_{\lambda,A}^{-1} \tau_{\lambda,L}^{-a_{m_1}} \xi_\lambda^{m_1} \hat{s}^{(m_1)} \right|^2}{\max_{1 \le k \le N_{m_1-1}} \left| \tau_{\lambda,A}^{-1} \tau_{\lambda,L}^{-a_{m_1}-1} \xi_\lambda^{m_1-1} \tilde{s}_k^{(m_1-1)} \right|^2} dV_{z(\lambda),t} \right)$$

$$\le \log \left(\sup_{\tilde{U}_\lambda} \left(\frac{1}{\pi^{n+1}} \left| \tau_{\lambda,L}^{-c_{m_1}} \xi_\lambda \right|^2 e^{c_{m_1} \varphi} dV_{z(\lambda),t} \right) \right.$$

$$\left. \cdot \int_{\tilde{U}_\lambda} \frac{\left| \hat{s}^{(m_1)} \right|^2 e^{-c_{m_1} \varphi}}{\max_{1 \le k \le N_{m_1-1}} \left| \tilde{s}_k^{(m_1-1)} \right|^2} \right)$$

$$\le \log \left(C^\diamond \sup_{\tilde{U}_\lambda} \left(\frac{1}{\pi^{n+1}} \left| \tau_{\lambda,L}^{-c_{m_1}} \xi_\lambda \right|^2 e^{c_{m_1} \varphi} dV_{z(\lambda),t} \right) \right). \tag{5.4.3}$$

Again here for the first inequality we have used the concavity of the logarithm. Adding up (5.4.2) and (5.4.3), we get

$$\frac{1}{\pi^{n+1}} \int_{\tilde{U}_\lambda} \left(\log \left| \tau_{\lambda,L}^{-a_{m_1}} \tau_{\lambda,A}^{-1} \xi_\lambda^{m_1} \hat{s}^{(m_1)} \right|^2 \right) dV_{z(\lambda),t}$$

$$\le \frac{1}{\pi^{n+1}} \int_{\tilde{U}_\lambda} \left(\log \max_{1 \le k \le N_1} \left| \tau_{\lambda,A}^{-1} \xi_\lambda \tilde{s}_k^{(1)} \right|^2 \right) dV_{z(\lambda),t}$$

$$+ \sum_{p=1}^{m_1-1} \log \left(\tilde{N}_{m_0} C^\diamond \sup_{\tilde{U}_\lambda} \left(\frac{1}{\pi^{n+1}} \left| \tau_{\lambda,L}^{-c_{p+1}} \xi_\lambda \right|^2 e^{c_{p+1} \varphi} dV_{z(\lambda),t} \right) \right). \tag{5.4.4}$$

By the sub-mean-value property of plurisubharmonic functions

$$\sup_{U_\lambda'} \log \left| \tau_{\lambda,A}^{-1} \tau_{\lambda,L}^{-a_{m_1}} \xi_\lambda^{m_1} \hat{s}^{(m_1)} \right|^2$$

$$\leq \frac{1}{\left(\pi \left(1 - r_0\right)^2\right)^{n+1}} \int_{\tilde{U}_\lambda} \left(\log \left|\tau_{\lambda,A}^{-1} \tau_{\lambda,L}^{-a_{m_1}} \xi_\lambda^{m_1} \hat{s}^{(m_1)}\right|^2\right) dV_{z(\lambda),t}. \quad (5.4.5)$$

Let $C^{\clubsuit}$ be defined as the number such that $\frac{1}{m_0} \log C^{\clubsuit}$ is the maximum of

$$\frac{1}{(1 - r_0)^{2(n+1)}} \log \left(\tilde{N}_{m_0} C^{\diamond} \sup_{\tilde{U}_\lambda} \left(\frac{1}{\pi^{n+1}} \left|\tau_{\lambda,L}^{-c_p+1} \xi_\lambda\right|^2 e^{c_p+1 \varphi} dV_{z(\lambda),t}\right)\right)$$

for $1 \leq \lambda \leq \Lambda$ and $1 \leq p \leq m_1 - 1$. Since c_{p+1} only takes the value 0 and 1 according to (4.1.1), the number $C^{\clubsuit}$ is independent of ℓ.

Let $\hat{C}$ be defined by

$$\log \hat{C} = \frac{1}{\left(\pi \left(1 - r_0\right)^2\right)^{n+1}} \sup_{1 \leq \lambda \leq \Lambda} \int_{\tilde{U}_\lambda} \left(\log \max_{1 \leq k \leq N_1} \left|\tau_{\lambda,A}^{-1} \xi_\lambda \hat{s}_k^{(1)}\right|^2\right) dV_{z(\lambda),t}.$$

Since $m_1 = \ell m_0$ and $a_{m_1} = \ell$, from (5.4.4) and (5.4.5) we obtain

$$\sup_{1 \leq \lambda \leq \Lambda} \sup_{U_\lambda'} \left|\tau_{\lambda,A}^{-1} \tau_{\lambda,L}^{-\ell} \xi_\lambda^{\ell m_0} \hat{s}^{(\ell m_0)}\right|^2 \leq \hat{C} \left(C^{\clubsuit}\right)^{\ell m_0 - 1}. \quad (5.4.6)$$

5.5 Construction of Metric as Limit.

For $1 \leq \lambda \leq \Lambda$ let χ_λ be the function on U_λ' which is the upper semi-continuous envelope of

$$\limsup_{\ell \to \infty} \log \left|\tau_{\lambda,L}^{-\ell} \xi_\lambda^{\ell m_0} \hat{s}^{(\ell m_0)}\right|^{\frac{2}{\ell}}.$$

From the definition of χ_λ, we have

$$\sup_{X_0 \cap U_\lambda'} \left(\left|\tau_{\lambda,L}^{-1} \xi_\lambda^{m_0} s^{(m_0)}\right|^2 e^{-\chi_\lambda}\right) \leq 1 \quad (5.5.1)$$

for $1 \leq \lambda \leq \Lambda$. By (5.4.6), we have

$$\sup_{U_\lambda'} \chi_\lambda \leq m_0 \log C^{\clubsuit} \quad (5.5.2)$$

for $1 \leq \lambda \leq \Lambda$. Moreover, from the definition of χ_λ we have the transformation rule

$$e^{-\chi_\lambda} = \left|\frac{\tau_{\lambda,L}^{-1} \xi_\lambda^{m_0}}{\tau_{\mu,L}^{-1} \xi_\mu^{m_0}}\right|^2 e^{-\chi_\mu}$$

on $U_\lambda' \cap U_\mu'$ for $1 \leq \lambda, \mu \leq \Lambda$. Hence the collection $\{e^{-\chi_\mu}\}_{1 \leq \lambda \leq \Lambda}$ defines a metric $e^{-\chi}$ for $(m_0 K_X + L)\big|X'$ with respect to the local trivializations

$$m_0 K_X + L \to \mathcal{O}_X$$

on U'_λ defined by multiplication by $\tau_{\lambda,L}^{-1}\xi_\lambda^{m_0}$ on U'_λ for $1 \leq \lambda \leq \Lambda$. More precisely,

$$e^{-\chi} = \frac{|\xi_\lambda^{m_0}|^2 \, e^{-\chi_\lambda}}{|\tau_{\lambda,L}|^2}$$

on U'_λ for $1 \leq \lambda \leq \Lambda$. The curvature current of $e^{-\chi}$ is semi-positive and

$$\int_{X_0} \left|s^{(m_0)}\right|^2 e^{-\left(\frac{m_0-1}{m_0}\right)\chi - \left(\frac{1}{m_0}\right)\varphi}$$

$$= \sum_{1 \leq \lambda \leq \Lambda} \int_{X_0 \cap U'_\lambda} \left(\left|\tau_{\lambda,L}^{-1}\xi_\lambda^{m_0} s^{(m_0)}\right|^2 e^{-\chi_\lambda}\right) e^{\frac{1}{m_0}\chi_\lambda} \left(|\tau_{\lambda,L}|^2 e^{-\varphi}\right)^{\frac{1}{m_0}} \left(\frac{\varrho_\lambda}{|\xi_\lambda|^2}\right)$$

$$\leq C^{\spadesuit} \sum_{1 \leq \lambda \leq \Lambda} \int_{X_0 \cap U'_\lambda} \left(|\tau_{\lambda,L}|^2 e^{-\varphi}\right)^{\frac{1}{m_0}} \frac{\varrho_\lambda}{|\xi_\lambda|^2} < \infty,$$

where for the last inequality (5.5.1) and (5.5.2) are used. The finiteness of

$$\int_{X_0 \cap U'_\lambda} \left(|\tau_{\lambda,L}|^2 e^{-\varphi}\right)^{\frac{1}{m_0}} \frac{\varrho_\lambda}{|\xi_\lambda|^2}$$

follows from the local integrability of $e^{-\varphi}|_{X_0}$ on X_0 and $m_0 \geq 1$.

By Theorem 3.1, we can extend $s^{(m_0)}$ to an element of $\Gamma(X', m_0 K_X + L)$. Extension to $\Gamma(X, m_0 K_X + L)$ follows from Stein theory and Grauert's coherence of the proper direct image of coherent sheaves [Gra60] as follows.

Let $\mathcal{F}$ be the coherent sheaf on Δ which is the zeroth direct image $R^0 \pi_* (\mathcal{O}_X (m_0 K_X + L))$ of the zeroth direct image, under π, of the sheaf $\mathcal{O}_X (m_0 K_X + L)$ of germs of holomorphic sections of $m_0 K_X + L$ on X. The Stein property of Δ and the coherence of $\mathcal{F}$ imply the surjectivity of the map

$$\Gamma(\Delta, \mathcal{F}) \to \mathcal{F}_0 / \mathbf{m}_{\Delta,0} \mathcal{F}_0,$$

where $\mathcal{F}_0$ is the stalk of $\mathcal{F}$ at 0 and $\mathbf{m}_{\Delta,0}$ is the maximum ideal of Δ at 0. This finishes the proof of Theorem 0.1.

6 Remarks on the Approach of Generalized Bergman Kernels

As discussed in (1.5.2), (5.1), (5.2), and (5.3.4), the metric for the twisted pluricanonical bundles on X_0 should be chosen as singular as possible to avoid the undesirable unbounded growth of the dimension of the space of L^2 holomorphic sections. Usual naturally-defined metrics such as generalized Bergman kernels would not be singular enough to give the uniform bound of the dimension of the space of L^2 holomorphic sections. The uniform bound of the dimensions of section spaces is needed when we use the concavity of the logarithm for the estimates which leads to (5.3.4). There are a number of other natural alternative ways of estimation which do not need such a uniform bound of dimensions of section spaces. Here we make some remarks on the difficulties of such natural alternatives in the case of generalized Bergman kernels on X as used, for example, in [Tsu01].

Generalized Bergman kernels use square integrable (possibly twisted) pluricanonical sections for definition, instead of just square integrable canonical sections used for the usual Bergman kernels. To use square integrable (possibly twisted) m-canonical sections, we need a metric for the (possibly twisted) $(m-1)$-canonical bundle. There are more than one way of constructing a metric for the (possibly twisted) $(m-1)$-canonical bundle in order to define the generalized Bergman kernel. In order to make the definition applicable to the problem of the deformational invariance of the plurigenera, one has to use the inductive definition, which inductively constructs a metric for the (possibly twisted) $(m-1)$-canonical bundle. Such an inductive definition is defined as follows.

6.1 *Inductively Defined Generalized Bergman Kernels.*

Let Y be a complex manifold of complex dimension n and E be a holomorphic line bundle over Y. Let $e^{-\kappa}$ be a (possibly singular) metric for $K_Y + E$. We inductively define a generalized Bergman kernel Φ_m as follows, so that $\frac{1}{\Phi_m}$ is a metric for $m\,K_Y + E$. Let $\Phi_1 = e^\kappa$ and

$$\Phi_m(P) = \sup\left\{ \left|\sigma^{(m)}(P)\right|^2 \,\middle|\, \sigma^{(m)} \in \Gamma\left(Y, m\,K_Y + E\right), \int_Y \frac{\left|\sigma^{(m)}\right|^2}{\Phi_{m-1}} \leq 1 \right\}$$

for $m \geq 2$. Equivalently, we can define

$$\Phi_m = \sum_j \left|\sigma_j^{(m)}\right|^2$$

for an orthonormal basis $\left\{\sigma_j^{(m)}\right\}_j$ of the Hilbert space Γ_m of all $\sigma^{(m)} \in \Gamma\left(Y, m\,K_Y + E\right)$ with

$$\int_Y \frac{\left|\sigma^{(m)}\right|^2}{\Phi_{m-1}} < \infty.$$

The reason for the equivalence is that any $\sigma^{(m)} \in \Gamma_m$ with

$$\int_Y \frac{\left|\sigma^{(m)}\right|^2}{\Phi_{m-1}} = 1$$

can be written as $\sum_j c_j \sigma_j^{(m)}$ with $\sum_j |c_j|^2 = 1$ and that for any $P \in Y$ one can first choose $\left\{\sigma_j^{(m)}\right\}_{j \geq 2}$ as the orthonormal basis for the subspace of Γ_m consisting of all elements vanishing at P and then add $\sigma_1^{(m)}$.

266   Yum-Tong Siu

6.2 *How Generalized Bergman Kernels Are Used.*

The use of generalized Bergman kernels is meant for the fourth ingredient listed in the Introduction. Consider the case of L being trivial. Use X as Y. Use as E a sufficiently ample line bundle A over X so that we have a metric $e^{-\kappa}$ for $K_X + E$ with positive curvature on X. Let $s^{m_0} \in \Gamma(X_0, m_0 K_{X_0})$ be given which is to be extended to an element of $\Gamma(X, m_0 K_X)$. Let s_A be an element of $\Gamma(X, A)$ whose restriction to X_0 is not identically zero. The setting is the following situation after (5.3). For any integer $\ell \geq 2$, $\left(s^{(m_0)}\right)^\ell s_A$ can be extended to $\hat{s}^{(\ell m_0)} \in \Gamma(X, \ell m_0 K_X + A)$ such that

$$\int_X \frac{\left|\hat{s}^{(\ell m_0)}\right|^2}{\Phi_{\ell m_0 - 1}} \leq \left(C^\diamond\right)^{\ell m_0}$$

for some positive constant $C^\diamond$ independent of ℓ. Hence

$$\left|\hat{s}^{(\ell m_0)}\right|^2 \leq \left(C^\diamond\right)^{\ell m_0} \Phi_{\ell m_0} .$$

Suppose on X_0 one were able to prove that $(\Phi_m)^{\frac{1}{m}}$ is locally bounded on X_0 as $m \to \infty$. If we denote by Θ the upper semi-continuous envelope of the limit superior of $(\Phi_m)^{\frac{1}{m}}$ as $m \to \infty$, then $\frac{1}{\Theta}$ is a metric of K_X with semi-positive curvature current and

$$\int_{X_0} \frac{\left|s^{(m_0)}\right|^2}{\Theta^{m_0}} \leq \left(C^\diamond\right)^\ell .$$

Thus $s^{(m_0)}$ can be extended to an element of $\Gamma(X, m_0 K_X)$. In the rest of §6 we will discuss the difficulties of various approaches to get the local bound of $(\Phi_m)^{\frac{1}{m}}$ as $m \to \infty$.

6.3 *Bounds of Roots of Generalized Bergman Kernels.*

We now return to the general setting of Y in (6.2). Assume that on Y we have a finite number of coordinate charts $\tilde{G}_\lambda$ with coordinates

$$\left(z_1^{(\lambda)}, \cdots, z_n^{(\lambda)}\right) .$$

We assume that there exists nowhere zero elements

$$\xi_\lambda \in \Gamma\left(\tilde{G}_\lambda, -K_Y\right),$$
$$\tau_{\lambda, E} \in \Gamma\left(\tilde{G}_\lambda, E\right).$$

Let dV_λ be the Euclidean volume of the coordinate chart $\tilde{G}_\lambda$. For $0 < r \leq 1$ let

$$G_{\lambda, r} = \tilde{G}_\lambda \cap \left\{\left|z_1^{(\lambda)}\right| < r, \cdots, \left|z_n^{(\lambda)}\right| < r\right\} .$$

Assume that $G_{\lambda,r}$ is relatively compact in $\tilde{G}_\lambda$ for $1 \leq \lambda \leq \Lambda$ and $0 < r \leq 1$. Let $G_\lambda = G_{\lambda,1}$. Let

$$\Theta_{m,r} = \sup_{1 \leq \lambda \leq \Lambda} \sup_{G_{\lambda,r}} \Phi_m \left| \xi_\lambda^m \tau_{\lambda,E}^{-1} \right|^2,$$

$$C_1^\clubsuit = \sup_{1 \leq \lambda \leq \Lambda} \sup_{G_\lambda} \left| \xi_\lambda \tau_{\lambda,E}^{-1} \right|^2 e^\kappa,$$

$$C_2^\clubsuit = \sup_{1 \leq \lambda,\mu \leq \Lambda} \sup_{G_\lambda \cap G_\mu} \left| \frac{\xi_\lambda}{\xi_\mu} \right|^2,$$

$$C_3^\clubsuit = \sup_{1 \leq \lambda,\mu \leq \Lambda} \sup_{G_\lambda \cap G_\mu} \left| \frac{\tau_{\lambda,E}}{\tau_{\mu,E}} \right|^2.$$

As explained in (6.2), the kind of estimates required for the invariance of plurigenera is a bound for $(\Theta_{m,r})^{\frac{1}{m}}$ as $m \to \infty$. We are going to inductively estimate $\Theta_{m,r}$ in terms of $C_1^\clubsuit$ and $C_2^\clubsuit$ and show where the difficulty lies in getting a finite local bound for $\limsup_{m \to \infty} (\Theta_{m,r})^{\frac{1}{m}}$. For the estimates we will first use the plurisubharmonicity of the absolute value squares of holomorphic sections. Later we will discuss the use of the concavity of the logarithm.

6.4 Difficulty of Shrinking Domain from Sub-Mean-Value Property.

Take $0 < r_0 < 1$. Fix an arbitrary point $P_0 \in G_{\lambda,r_0}$ for some $1 \leq \lambda \leq \Lambda$. For $0 < r < 1 - r_0$ let

$$\Delta_r^{(P_0,\lambda)} = G_\lambda \cap \left\{ \left| z_1^{(\lambda)} - z_1^{(\lambda)}(P_0) \right| < r, \cdots, \left| z_n^{(\lambda)} - z_n^{(\lambda)}(P_0) \right| < r \right\}.$$

By the definition of Φ_m there exists some

$$\sigma^{(m)} \in \Gamma(Y, m K_Y + E)$$

such that

$$\int_Y \frac{\left| \sigma^{(m)} \right|^2}{\Phi_{m-1}} \leq 1$$

and

$$\Phi_m(P_0) = \left| \sigma^{(m)}(P_0) \right|^2.$$

268 Yum-Tong Siu

For $m \geq 2$ and $0 < r \leq 1 - r_0$, by the sub-mean-value property of the absolute value square of a holomorphic function, we have

$$\left(\Phi_m \left|\xi_\lambda^m \tau_{\lambda,E}^{-1}\right|^2\right)(P_0) = \left|\sigma^{(m)} \xi_\lambda^m \tau_{\lambda,E}^{-1}\right|^2 (P_0)$$

$$\leq \frac{1}{(\pi r^2)^n} \int_{\Delta_r^{(P_0,\lambda)}} \left|\sigma^{(m)} \xi_\lambda^m \tau_{\lambda,E}^{-1}\right|^2 dV_\lambda$$

$$= \frac{1}{(\pi r^2)^n} \int_{\Delta_r^{(P_0,\lambda)}} \frac{\left|\sigma^{(m)}\right|^2}{\Phi_{m-1}} \left(\Phi_{m-1} \left|\xi_\lambda^{m-1} \tau_{\lambda,E}^{-1}\right|^2\right) |\xi_\lambda|^2 dV_\lambda$$

$$\leq \left(\sup_{\Delta_r^{(P_0,\lambda)}} \Phi_{m-1} \left|\xi_\lambda^{m-1} \tau_{\lambda,E}^{-1}\right|^2\right) \left(\sup_{\Delta_r^{(P_0,\lambda)}} |\xi_\lambda|^2 dV_\lambda\right) \frac{1}{(\pi r^2)^n} \int_{\Delta_r^{(P_0,\lambda)}} \frac{\left|\sigma^{(m)}\right|^2}{\Phi_{m-1}}$$

$$\leq \left(\sup_{\Delta_r^{(P_0,\lambda)}} \Phi_{m-1} \left|\xi_\lambda^{m-1} \tau_{\lambda,E}^{-1}\right|^2\right) \left(\sup_{\Delta_r^{(P_0,\lambda)}} |\xi_\lambda|^2 dV_\lambda\right) \frac{1}{(\pi r^2)^n}. \quad (6.4.1)$$

Choose positive numbers $r_1, \cdots, r_{m-1}$ such that $\sum_{j=1}^{m-1} r_j \leq 1 - r_0$. Let $\delta_j = 1 - (r_1 + \cdots + r_{j-1})$. From (6.4.1) and by induction on m we have

$$\left(\Phi_m \left|\xi_\lambda^m \tau_{\lambda,E}^{-1}\right|^2\right)(P_0) \leq \left(\sup_{\Delta_{\delta_1}^{(P_0,\lambda)}} \left|\xi_\lambda \tau_{\lambda,E}^{-1}\right|^2 e^\kappa\right) \prod_{j=1}^{m-1} \sup_{\Delta_{\delta_j}^{(P_0,\lambda)}} \frac{|\xi_\lambda|^2 dV_\lambda}{(\pi r_j^2)^n} \quad (6.4.2)$$

for $m \geq 2$. The estimate is for $P_0 \in G_{\lambda,r_0}$ when $r_0 < 1 - \sum_{j=1}^{m-1} r_j$. When we increase m, we have to add more and more r_j $(1 \leq j < m)$. Their sum have to remain less than 1 and the m-th root of their product have to remain bounded in m.

We have to decide between two choices. The first choice is to use smaller and smaller r_j to maintain a fixed positive lower bound for r_0. This choice results in the lack of a finite bound for the m-th root of $\prod_{j=1}^{m-1} \frac{1}{r_j^2}$ as m goes to infinity. The second choice is to keep the m-th root of $\prod_{j=1}^{m-1} \frac{1}{r_j^2}$ bounded in m. This choice forces r_0 to become eventually negative so that the domain of supremum estimates becomes empty.

This kind of difficulties with norm change, involving supremum norms on different domains, is similar to the norm-change problems encountered in the papers of Nash [Nas54], Moser [Mos61], and Grauert [Gra60]. However, the situation here is different from those which could be handled by the norm-change techniques of Nash [Nas54], Moser [Mos61], and Grauert [Gra60].

Invariance of Twisted Plurigenera for General Manifolds 269

6.5 *The Compact Case and Difficulty from Transition Functions.*

For the resolution of the difficulty explained in (6.4), let us consider the case of a compact Y. When Y is compact, a shrunken cover of Y is still a cover of Y and the difficulty of the shrinking domain for the supremum estimate seems not to exist. However, if we use this approach to resolve the problem in the compact case, the transition functions of the m-canonical bundle would give us trouble.

When we consider a point of an unshrunken member of the cover as being in another shrunken member, for the estimation of the bound in question we have to use the transition function of the m-canonical bundle whose bound grows as the m-th power of a positive constant. Let us more precisely describe the contribution from the transition functions of the m-canonical bundle.

From (6.4.1) we obtain

$$\sup_{1 \leq \lambda \leq \Lambda} \sup_{G_{\lambda,r_0}} \Phi_m \left| \xi_\lambda^m \tau_{\lambda,E}^{-1} \right|^2 \leq \frac{C_1^\clubsuit}{\left(\pi \left(1 - r_0\right)^2 \right)^n} \Theta_{m-1,1} \tag{6.5.1}$$

for $m \geq 2$. Since Y is compact, we can assume $Y = \bigcup_{1 \leq \lambda \leq \Lambda} G_{\lambda,r_0}$. Any $P_0 \in G_\lambda$ belongs to some G_{μ,r_0} and

$$\left(\Phi_m \left| \xi_\lambda^m \tau_{\lambda,E}^{-1} \right|^2 \right)(P_0) = \frac{\left| \xi_\lambda^m \tau_{\lambda,E}^{-1} \right|^2}{\left| \xi_\nu^m \tau_{\lambda,E}^{-1} \right|^2} \left(\Phi_m \left| \xi_\mu^m \tau_{\lambda,E}^{-1} \right|^2 \right)(P_0)$$

$$\leq \left(C_2^\clubsuit \right)^m C_3^\clubsuit \left(\Phi_m \left| \xi_\mu^m \tau_{\mu,E}^{-1} \right|^2 \right)(P_0)$$

$$\leq \left(C_2^\clubsuit \right)^m \frac{C_1^\clubsuit C_3^\clubsuit}{\left(\pi \left(1 - r_0\right)^2 \right)^n} \Theta_{m-1,1}$$

for $m \geq 2$ by (6.5.1). Hence

$$\Theta_{m,1} \leq \left(C_2^\clubsuit \right)^m \frac{C_1^\clubsuit C_3^\clubsuit}{\left(\pi \left(1 - r_0\right)^2 \right)^n} \Theta_{m-1,1} \tag{6.5.2}$$

for $m \geq 2$. By induction on $m \geq 2$, we conclude from (6.5.2) that

$$\Theta_{m,1} \leq \left(\frac{C_1^\clubsuit C_3^\clubsuit}{\left(\pi \left(1 - r_0\right)^2 \right)^n} \right)^{m-1} \left(C_2^\clubsuit \right)^{\frac{m(m+1)}{2} - 1} \Theta_{1,1} \tag{6.5.3}$$

for $m \geq 2$. When we take the m-th root of both sides of (6.5.3), we get

$$\left(\Theta_{m,1} \right)^{\frac{1}{m}} \leq \left(\frac{C_1^\clubsuit C_3^\clubsuit}{\left(\pi \left(1 - r_0\right)^2 \right)^n} \right)^{1 - \frac{1}{m}} \left(C_2^\clubsuit \right)^{\frac{m+1}{2} - \frac{1}{m}} \left(\Theta_{1,1} \right)^{\frac{1}{m}}$$

270   Yum-Tong Siu

for $m \geq 2$ and we would have trouble bounding the factor $\left(C_2^{\clubsuit}\right)^{\frac{m+1}{2}}$ as $m \to \infty$. So even in the compact case, the difficulty cannot be completely circumvented, because of the undesirable contribution from the transition functions of the m-canonical bundle.

6.6 Difficulty with Bounds for Nonempty Limit of Shrinking Domain.

Let us return to the general case in which Y is noncompact. We look more closely at the difficulty of the shrinking domain. In our case of $\pi : X \to \Delta$, we use the notations of (5.4) and let

$$\tilde{U}_{\lambda,r} = W_\lambda \cap \left\{ \left| z_1^{(\lambda)} \right| < r, \cdots, \left| z_n^{(\lambda)} \right| < r, |t| < r \right\}.$$

We apply the above argument to $Y = X$ and $G_{\lambda,r} = \tilde{U}_{\lambda,r}$ and with n replaced by $n+1$. Let $r_1, \cdots, r_{m-1}$ be positive numbers such that $r_1 + \cdots + r_{m-1} < 1$. With $\delta_j = 1 - (r_1 + \cdots + r_{j-1})$, from (6.4.2) we have

$$\sup_{\tilde{U}_{\lambda,\delta_m}} \Phi_m \left| \xi_\lambda^m \tau_{\lambda,E}^{-1} \right|^2 \leq \left(\sup_{\tilde{U}_{\lambda,\delta_1}} \left| \xi_\lambda \tau_{\lambda,E}^{-1} \right|^2 e^\kappa \right) \prod_{j=1}^{m-1} \sup_{\tilde{U}_{\lambda,\delta_j}} \frac{|\xi_\lambda|^2 \, dV_\lambda}{(\pi \, r_j^2)^n}$$

$$= \frac{1}{\pi^{n(m-1)} \prod_{j=1}^{m-1} r_j^2} \left(\sup_{\tilde{U}_{\lambda,\delta_1}} \left| \xi_\lambda \tau_{\lambda,E}^{-1} \right|^2 e^\kappa \right) \prod_{j=1}^{m-1} \sup_{\tilde{U}_{\lambda,\delta_j}} \left(|\xi_\lambda|^2 \, dV_\lambda \right)$$

for $m \geq 2$. After we take the m-th of both sides, we end up with a factor $\frac{1}{\left(\prod_{j=1}^m r_j^2\right)^{\frac{1}{m}}}$ which is not bounded in m if $\sum_{j=1}^\infty r_j < 1$. For example, if the sequence r_j is non-increasing and if there exists some positive number C such that $\frac{1}{\left(\prod_{j=1}^m r_j^2\right)^{\frac{1}{m}}} \leq C$ for all $m \geq 1$, then

$$\frac{1}{r_\ell^2} \leq \left(\prod_{j=\ell}^{2\ell} \frac{1}{r_j^2} \right)^{\frac{1}{\ell}} \leq C^2 \quad \text{for } \ell \geq 1$$

$$\implies \quad r_\ell \geq \frac{1}{C} \quad \text{for } \ell \geq 1$$

$$\implies \quad \sum_{\ell=1}^m r_\ell \geq \frac{m}{C} \to \infty \quad \text{as } m \to \infty.$$

It shows that the difficulty of the shrinking domain for the supremum estimate is an essential one when the generalized Bergman kernels are inductively defined.

6.7 *Non-Inductively Defined Generalized Bergman Kernels.*

We would like to point out that, if the metric for the $(m-1)$-canonical bundle is not constructed inductively in order to define the generalized Bergman kernel, then we would be able to avoid the difficulty with the bound of

$$\limsup_{m \to \infty} (\Theta_{m,r})^{\frac{1}{m}}.$$

This non-inductive definition works only when E is the trivial line bundle, which we assume only for the discussion here in (6.7). Let $e^{-\kappa_0}$ be a metric for K_Y. The non-inductive definition defines $\tilde{\Phi}_1 = e^{\kappa_0}$ and

$$\tilde{\Phi}_m(P) = \sup \left\{ \left| \sigma^{(m)}(P) \right|^2 \ \middle| \ \sigma^{(m)} \in \Gamma(Y, m\,K_Y) \int_Y \left| \sigma^{(m)} \right|^2 e^{-(m-1)\kappa_0} \leq 1 \right\}$$

for $m \geq 2$. We let

$$\tilde{\Theta}_{m,r} = \sup_{1 \leq \lambda \leq \Lambda} \sup_{G_{\lambda,r}} \tilde{\Phi}_m \, |\xi_\lambda^m|^2.$$

Again take $0 < r_0 < 1$ and fix an arbitrary point $P_0 \in G_{\lambda,r_0}$ for some $1 \leq \lambda \leq \Lambda$. By the definition of $\tilde{\Phi}_m$ there exists some

$$\sigma^{(m)} \in \Gamma(Y, m\,K_Y)$$

such that

$$\int_Y \left| \sigma^{(m)} \right|^2 e^{-(m-1)\kappa_0} \leq 1 \tag{6.7.1}$$

and

$$\tilde{\Phi}_m(P_0) = \left| \sigma^{(m)}(P_0) \right|^2.$$

Let $0 < r \leq 1 - r_0$. For $m \geq 2$, by the sub-mean-value property of the absolute value square of a holomorphic function, from (6.7.1) we have

$$
\begin{aligned}
\left(\tilde{\Phi}_m \, |\xi_\lambda^m|^2 \right)(P_0) &= \left| \sigma^{(m)} \xi_\lambda^m \right|^2 (P_0) \\
&\leq \frac{1}{(\pi r^2)^n} \int_{\Delta_r^{(P_0,\lambda)}} \left| \sigma^{(m)} \xi_\lambda^m \right|^2 dV_\lambda \\
&\leq \frac{1}{(\pi r^2)^n} \left(\sup_{\Delta_r^{(P_0,\lambda)}} |\xi_\lambda^m|^2 \, e^{(m-1)\kappa_0} \, dV_\lambda \right).
\end{aligned}
$$

Hence

$$\left(\tilde{\Theta}_{m,r_0} \right)^{\frac{1}{m}} \leq \left(\frac{1}{\left(\pi (1 - r_0)^2 \right)^n} \left(\sup_{1 \leq \lambda \leq \Lambda} \sup_{G_\lambda} |\xi_\lambda^m|^2 \, e^{(m-1)\kappa_0} \, dV_\lambda \right) \right)^{\frac{1}{m}}$$

272   Yum-Tong Siu

is bounded as $m \to \infty$. However, this kind of non-inductively defined generalized Bergman kernel is not useful to the problem of the deformational invariance of plurigenera. The induction argument of [Siu98] for the problem of the deformational invariance of plurigenera requires the inductive definition if generalized Bergman kernels are used. Moreover, the non-inductive definition works only when E is trivial, which is not useful for the case of manifolds not necessarily of general type.

6.8 *Hypothetical Situation of Sub-Mean-Value Property of Quotients.*

Now we return to the general case where E may not be trivial and $e^{-\kappa}$ is a metric for $K_Y + E$. We would like to remark that, if

$$\frac{\left|\sigma^{(m)}\xi_\lambda\right|^2}{\Phi_{m-1}}$$

were to have the sub-mean-value property for $\sigma^{(m)} \in \Gamma\left(Y, m\, K_Y\right)$, the difficulty of the shrinking of the domain of supremum estimate for the inductively defined generalized Bergman kernel would disappear. The reason is as follows.

Again fix an arbitrary point $P_0 \in G_{\lambda,r_0}$ for some $1 \le \lambda \le \Lambda$. By the definition of Φ_m there exists some

$$\sigma^{(m)} \in \Gamma\left(Y, m\, K_Y + E\right)$$

such that

$$\int_Y \frac{\left|\sigma^{(m)}\right|^2}{\Phi_{m-1}} \le 1 \tag{6.8.1}$$

and

$$\Phi_m\left(P_0\right) = \left|\sigma^{(m)}\left(P_0\right)\right|^2.$$

Let $0 < r \le 1 - r_0$. For $m \ge 2$, by the hypothetical sub-mean-value property of

$$\frac{\left|\sigma^{(m)}\xi_\lambda\right|^2}{\Phi_{m-1}},$$

we would have

$$\left(\frac{\Phi_m\left|\xi_\lambda\right|^2}{\Phi_{m-1}}\right)(P_0) = \left(\frac{\left|\sigma^{(m)}\xi_\lambda\right|^2}{\Phi_{m-1}}\right)(P_0)$$

$$\le \frac{1}{\left(\pi\, r^2\right)^n} \int_{\Delta_r^{(P_0,\lambda)}} \frac{\left|\sigma^{(m)}\xi_\lambda\right|^2}{\Phi_{m-1}}\, dV_\lambda$$

$$\le \frac{1}{\left(\pi\, r^2\right)^n} \left(\sup_{\Delta_r^{(P_0,\lambda)}} \left|\xi_\lambda\right|^2 dV_\lambda\right) \tag{6.8.2}$$

by (6.8.1). Hence by (6.8.2) and by induction on m, we have

$$
\left(\frac{\Phi_m \left| \xi_\lambda^m \tau_{\lambda,E}^{-1} \right|^2}{\Phi_1 \left| \xi_\lambda \tau_{\lambda,E}^{-1} \right|^2} \right)(P_0) \leq \left(\frac{1}{(\pi \, r^2)^n} \left(\sup_{\Delta_r^{(P_0,\lambda)}} |\xi_\lambda|^2 \, dV_\lambda \right) \right)^{m-1}
$$

and

$$
(\Theta_{m,r_0})^{\frac{1}{m}} \leq \left(\frac{1}{\left(\pi \, (1 - r_0)^2 \right)^n} \left(\sup_{1 \leq \lambda \leq \Lambda} \sup_{\Delta_r^{(P_0,\lambda)}} |\xi_\lambda|^2 \, dV_\lambda \right) \right)^{1 - \frac{1}{m}} (\Theta_{1,r_0})^{\frac{1}{m}}
$$

which is bounded as $m \to \infty$. Of course, unfortunately in general

$$
\frac{\left| \sigma^{(m)} \xi_\lambda \right|^2}{\Phi_{m-1}}
$$

does not have the sub-mean-value property for $\sigma^{(m)} \in \Gamma \, (Y, m \, K_Y + E)$.

6.9 *Growth of Dimension of Section Space When Concavity of Logarithm is Used.*

In the case of a compact Y, for $0 < r_0 < 1$ we can follow the method of (5.4) and use the concavity of the logarithm for the supremum estimate of $(\Theta_{m,r_0})^{\frac{1}{m}}$ as $m \to \infty$. Let N_m be the complex dimension of the space Γ_m of all $\sigma^{(m)} \in \Gamma \, (Y, m \, K_Y + E)$ such that

$$
\int_Y \frac{\left| \sigma^{(m)} \right|^2}{\Phi_{m-1}} < \infty .
$$

As in (5.4), for this method of estimation the bound of $(\Theta_{m,r_0})^{\frac{1}{m}}$ cannot be independent of m if the growth order of N_m is at least that of some positive power of m. More precisely, the estimation analogous to that of (5.4) is given as follows.

From the definition of m we have

$$
\int_Y \frac{\Phi_m}{\Phi_{m-1}} \leq N_m
$$

for $m \geq 2$. By the concavity of the logarithm,

$$
\frac{1}{\pi^n} \int_{G_\lambda} \left(\log \Phi_m \left| \xi_\lambda^m \tau_{\lambda,E}^{-1} \right|^2 \right) dV_\lambda - \frac{1}{\pi^n} \int_{G_\lambda} \left(\log \Phi_{m-1} \left| \xi_\lambda^{m-1} \tau_{\lambda,E}^{-1} \right|^2 \right) dV_\lambda
$$

$$
= \frac{1}{\pi^n} \int_{G_\lambda} \left(\log \frac{\Phi_m |\xi_\lambda|^2}{\Phi_{m-1}} \right) dV_\lambda
$$

$$
\leq \log \left(\frac{1}{\pi^n} \int_{G_\lambda} \frac{\Phi_m |\xi_\lambda|^2}{\Phi_{m-1}} dV_\lambda \right)
$$

274 Yum-Tong Siu

$$\leq \log\left(\frac{N_m}{\pi^n}\sup_{G_\lambda}|\xi_\lambda|^2\,dV_\lambda\right) \tag{6.9.1}$$

for $m \geq 2$. Adding up (6.9.1) from $m = 2$ to m, we obtain

$$\frac{1}{\pi^n}\int_{G_\lambda}\left(\log\Phi_m\left|\xi_\lambda^m\tau_{\lambda,E}^{-1}\right|^2\right)dV_\lambda$$

$$\leq \frac{1}{\pi^n}\int_{G_\lambda}\left(\log\left|\xi_\lambda\tau_{\lambda,E}^{-1}\right|^2 e^{-\kappa}\right)dV_\lambda + \sum_{j=2}^m\log\left(\frac{N_m}{\pi^n}\sup_{G_\lambda}|\xi_\lambda|^2\,dV_\lambda\right)$$

$$\leq \log\left(\frac{1}{\pi^n}\int_{G_\lambda}\left|\xi_\lambda\tau_{\lambda,E}^{-1}\right|^2 e^{-\kappa}dV_\lambda\right) + \sum_{j=2}^m\log\left(\frac{N_m}{\pi^n}\sup_{G_\lambda}|\xi_\lambda|^2\,dV_\lambda\right), \tag{6.9.2}$$

where for the last inequality the concavity of the logarithm is used. Using the sub-mean-value property of $\log\Phi_m\left|\xi_\lambda^m\tau_{\lambda,E}^{-1}\right|^2$, we obtain

$$\sup_{G_{\lambda,r_0}}\log\Phi_m\left|\xi_\lambda^m\tau_{\lambda,E}^{-1}\right|^2 \leq \frac{1}{\left(\pi\left(1-r_0\right)^2\right)^n}\int_{G_\lambda}\left(\log\Phi_m\left|\xi_\lambda^m\tau_{\lambda,E}^{-1}\right|^2\right)dV_\lambda \tag{6.9.3}$$

for $0 < r_0 < 1$. Let

$$\gamma = \frac{1}{\left(\pi\left(1-r_0\right)^2\right)^n},$$

$$C_1 = \frac{1}{\pi^n}\sup_{1\leq\lambda\leq\Lambda}\int_{G_\lambda}\left|\xi_\lambda\tau_{\lambda,E}^{-1}\right|^2 e^{-\kappa}dV_\lambda,$$

$$C_2 = \frac{1}{\pi^n}\sup_{G_\lambda}|\xi_\lambda|^2\,dV_\lambda.$$

By (6.9.2) and (6.9.3) we have

$$(\Theta_{m,r_0})^{\frac{1}{m}} \leq \left(C_1^{\frac{1}{m}}C_2^{1-\frac{1}{m}}\left(\prod_{j=2}^m N_m\right)^{\frac{1}{m}}\right)^\gamma$$

and whether $(\Theta_{m,r_0})^{\frac{1}{m}}$ is bounded in m depends on whether $\left(\prod_{j=2}^m N_m\right)^{\frac{1}{m}}$ is bounded in m.

If the order of growth of N_m is at least that of some positive power of m (i.e., $N_m \geq \alpha\,m^\beta - C$ for some positive α and β and C), then

$$\left(\prod_{j=2}^m N_m\right)^{\frac{1}{m}} \geq \hat{C}\,(m!)^{\frac{\beta}{m}}$$

Invariance of Twisted Plurigenera for General Manifolds   275

for some positive $\hat{C}$ and for m sufficiently large and is therefore unbounded in m.

Even if we start with some very singular $e^{-\kappa}$, the definition of Φ_m may make $\frac{1}{\Phi_m}$ gain some regularity in each step as m increases. As a result, the growth order of N_m may become comparable to that of the dimension of $\Gamma\left(Y, m K_Y + E\right)$ as $m \to \infty$.

Besides the difficulty of the growth order of N_m, this method of the concavity of the logarithm does not apply to the case of noncompact Y, which is what the use of generalized Bergman kernels for the deformation invariance of plurigenera would require.

6.10 *Modification of Generalized Bergman Kernels to Make the Method of Concavity of Logarithm Applicable to the Case of Open Manifolds.*

To make the method of the concavity of the logarithm applicable to the noncompact case of $\pi : X \to \Delta$ at hand, we can do the following. Let E be a holomorphic line bundle over X, $e^{-\kappa}$ be a metric of $K_X + E$, and $C^\heartsuit > 0$. Let Φ_m be the generalized Bergman metric for $m K_{X_0} + E|_{X_0}$ and $e^{-\kappa}$, as defined in (6.1) with Y equal to X_0. Let $\sigma_j^{(m)}$ $(1 \le j \le N_m)$ form an orthonormal basis of the space Γ_m of all $\sigma^{(m)} \in \Gamma\left(X_0, m K_{X_0} + E|_{X_0}\right)$ with

$$\int_{X_0} \frac{\left|\sigma^{(m)}\right|^2}{\Phi_{m-1}} < \infty.$$

Then inductively let $\hat{\sigma}_j^{(m)} \in \Gamma\left(X, m K_X + E\right)$ be an extension of $\sigma_j^{(m)}$ with

$$\int_X \frac{\left|\hat{\sigma}^{(m)}\right|^2}{\hat{\Phi}_{m-1}} < \infty$$

and

$$\hat{\Phi}_m = \max_{1 \le j \le N_m} \left|\hat{\sigma}_j^{(m)}\right|^2.$$

Define

$$\hat{\Theta}_{m,r} = \sup_{1 \le \lambda \le \Lambda} \sup_{G_{\lambda,r}} \hat{\Phi}_m \left|\xi_\lambda^m \tau_{\lambda,E}^{-1}\right|^2,$$

where $G_{\lambda,r}, \xi_\lambda^m, \tau_{\lambda,E}$ are as defined above when Y is replaced by X. We can then apply the method of the concavity of the logarithm to $\hat{\Phi}_m$ on X for the supremum estimate of $\left(\hat{\Theta}_{m,r}\right)^{\frac{1}{m}}$ as $m \to \infty$. Of course, $\hat{\Phi}_m$ is no longer the generalized Bergman kernel. We would still have the difficulty with the growth order of N_m when it is at least of the order of some positive power of m.

276   Yum-Tong Siu

6.11   The conclusion to these remarks in §6 is that, in the various approaches discussed above, the generalized Bergman kernels pose insurmountable difficulties when they are used for estimates in the problem of the deformational invariance of plurigenera for manifolds not necessarily of general type.

References

[Gra60] H. Grauert, Ein Theorem der analytischen Garbentheorie und die Modulräume komplexer Strukturen, *Inst. Hautes Études Sci. Publ. Math.* No. 5 (1960), 64 pp; Berichtigung zu der Arbeit "Ein Theorem der analytischen Garbentheorie und die Modulräume komplexer Strukturen", *Inst. Hautes Études Sci. Publ. Math.* No. 16 (1963), 35–36.

[Hör66] L. Hörmander, *An Introduction to Complex Analysis in Several Variables*, Van Nostrand, Princeton, New Jersey, 1966.

[Kaw82] Y. Kawamata, A generalization of Kodaira-Ramanujam's vanishing theorem, *Math. Ann.* **261** (1982), 43–46.

[Kaw99] Y. Kawamata, Deformations of canonical singularities, *J. Amer. Math. Soc.* **12** (1999), 85–92.

[Lev83] M. Levine, Pluri-canonical divisors on Kähler manifolds, *Invent. Math.* **74** (1983), 293–903.

[Man93] L. Manivel, Un thórème de prolongement L^2 de sections holomorphes d'un fibré hermitien, *Math. Zeitschr.* **212** (1993), 107–122.

[Mos61] J. Moser, On Harnack's theorem for elliptic differential equations, *Comm. Pure Appl. Math.* **14** (1961), 577–591.

[Nad89] A. Nadel, Multiplier ideal sheaves and the existence of Kähler-Einstein metrics of positive scalar curvature, *Proc. Natl. Acad. Sci. USA* **86** (1989), 7299–7300; *Ann. of Math.* **132** (1989), 549–596.

[Nak98] N. Nakayama, Invariance of the Plurigenera of Algebraic Varieties, preprint, *Research Inst. for Math. Sci.*, RIMS-1191, March 1998.

[Nas54] J. Nash, C^1 isometric imbeddings, *Ann. of Math.* **60** (1954), 383–396.

[Ohs88] T. Ohsawa, On the extension of L^2 holomorphic functions, II, *Publ. Res. Inst. Math. Sci.* **24** (1988), 265–275.

[Ohs94] T. Ohsawa, On the extension of L^2 holomorphic functions, IV, A new density concept, in *Geometry and Analysis on Complex Manifolds*, 157–170, World Sci. Publishing, River Edge, NJ, 1994.

[Ohs95] T. Ohsawa, On the extension of L^2 holomorphic functions, III. Negligible weights, *Math. Z.* **219** (1995), 215–225.

[Ohs01] T. Ohsawa, On the extension of L^2 holomorphic functions, V. Effects of generalization, *Nagoya Math. J.* **161** (2001), 1–21.

[OT87] T. Ohsawa and K. Takegoshi, On the extension of L^2 holomorphic functions, *Math. Z.* **195** (1987), 197–204.

[Siu96] Y.-T. Siu, The Fujita conjecture and the extension theorem of Ohsawa-Takegoshi, in *Geometric Complex Analysis* ed. Junjiro Noguchi *et al*, World Scientific, Singapore, New Jersey, London, Hong Kong, 1996, pp. 577–592.

[Siu98] Y.-T. Siu, Invariance of plurigenera, *Invent. Math.* **134** (1998), 661–673.

[Sko72] H. Skoda, Application des techniques L^2 à la théorie des ideaux d'un algèbre de fonctions holomorphes avec poids, *Ann. Sci. Ec. Norm. Sup.* **5** (1972), 548–580.

Invariance of Twisted Plurigenera for General Manifolds   277

[Tsu92]   H. Tsuji, Analytic Zariski decomposition, *Proc. Japan Acad.* **61** (1992), 161–163.

[Tsu99]   H. Tsuji, Existence and applications of analytic Zariski decompositions, in *Analysis and Geometry in Several Complex Variables*, ed. Komatsu and Kuranishi, *Trends in Math.*, Birkhäuser, 1999, pp. 253–272.

[Tsu01]   H. Tsuji, Deformational invariance of plurigenera, Preprint 2001, math. AG/0011257; (earlier version: Invariance of plurigenera of varieties with nonnegative Kodaira dimensions, math.AG/0012225).

[Vie82]   E. Viehweg, Vanishing theorems, *J. reine und angew. Math.* **335** (1982), 1–8.

Base Spaces of Non-Isotrivial Families of Smooth Minimal Models

Eckart Viehweg[1] and Kang Zuo[2] *

[1] Universität Essen, FB6 Mathematik, 45117 Essen, Germany
E-mail address: viehweg@uni-essen.de
[2] The Chinese University of Hong Kong, Department of Mathematics, Shatin, Hong Kong
E-mail address: kzuo@math.cuhk.edu.hk

Table of Contents

1 Differential Forms on Moduli Stacks 283
2 Mild Morphisms 287
3 Positivity and Ampleness 294
4 Higgs Bundles and the Proof of 1.4 301
5 Base Spaces of Families of Smooth Minimal Models 314
6 Subschemes of Moduli Stacks of Canonically Polarized
  Manifolds 316
7 A Vanishing Theorem for Sections of Symmetric Powers
  of Logarithmic One Forms 321
References 327

Given a polynomial h of degree n let $\mathcal{M}_h$ be the moduli functor of canonically polarized complex manifolds with Hilbert polynomial h. By [23] there exist a quasi-projective scheme M_h together with a natural transformation

$$\Psi : \mathcal{M}_h \to \mathrm{Hom}(\_, M_h)$$

such that M_h is a coarse moduli scheme for $\mathcal{M}_h$. For a complex quasi-projective manifold U we will say that a morphism $\varphi : U \to M_h$ factors through the moduli stack, or that φ is induced by a family, if φ lies in the image of $\Psi(U)$, hence if $\varphi = \Psi(f : V \to U)$.

Let Y be a projective non-singular compactification of U such that $S = Y \setminus U$ is a normal crossing divisor, and assume that the morphism $\varphi : U \to M_h$, induced by a family, is generically finite. For moduli of curves of genus $g \geq 2$, i.e. for $h(t) = (2t-1)(g-1)$, it is easy to show, that the existence of φ forces $\Omega^1_Y(\log S)$ to be big (see 1.1 for the definition), hence

---

* This work has been supported by the "DFG-Forschergruppe Arithmetik und Geometrie" and the "DFG-Schwerpunktprogramm Globale Methoden in der Komplexen Geometrie". The second named author is supported by a grant from the Research Grants Council of the Hong Kong Special Administrative Region, China (Project No. CUHK 4239/01P).
2000 *Mathematics Subject Classification*: 14D05, 14D22, 14F17, 14J60, 14E30.

280   Eckart Viehweg and Kang Zuo

that $S^m(\Omega^1_Y(\log S))$ contains an ample subsheaf of full rank for some $m > 0$. In particular, U cannot be an abelian variety or $\mathbb{C}^*$. By [15] the bigness of $\Omega^1_Y(\log S)$ implies even the Brody hyperbolicity of U. As we will see, there are other restrictions on U, as those formulated below in 0.1, 0.3, and 0.2.

In the higher dimensional case, i.e. if $\deg(h) > 1$, L. Migliorini, S. Kovács, E. Bedulev and the authors studied in [16], [12], [13], [14], [2], [24] [25] geometric properties of manifolds U mapping non-trivially to the moduli stack. Again, U cannot be $\mathbb{C}^*$, nor an abelian variety, and more generally it must be Brody hyperbolic, if φ is quasi-finite.

In general the sheaf $\Omega^1_Y(\log S)$ fails to be big (see Example 6.3). Nevertheless, building up on the methods introduced in [24] and [25] we will show that for m sufficiently large the sheaf $S^m(\Omega^1_Y(\log S))$ has enough global sections (see §1 for the precise statement), to exclude the existence of a generically finite morphism $\varphi : U \to M_h$, or even of a non-trivial morphism, for certain manifolds U.

Theorem 0.1 (see 5.2, 5.3 and 7.2). *Assume that U satisfies one of the following conditions*

a) *U has a smooth projective compactification Y with $S = Y \setminus U$ a normal crossing divisor and with $T_Y(-\log S)$ weakly positive.*

b) *Let $H_1 + \cdots + H_\ell$ be a reduced normal crossing divisor in $\mathbb{P}^N$, and $\ell < N/2$. For $0 \leq r \leq l$ define*

$$H = \bigcap_{j=r+1}^{\ell} H_j, \quad S_i = H_i|_H, \quad S = \sum_{i=1}^{r} S_i ,$$

and assume $U = H \setminus S$.

c) *$U = \mathbb{P}^N \setminus S$ for a reduced normal crossing divisor $S = S_1 + \cdots + S_\ell$ in $\mathbb{P}^N$, with $\ell < N$.*

Then a morphism $U \to M_h$, induced by a family, must be trivial.

In a) the sheaf $T_Y(-\log S)$ denotes the dual of the sheaf of one forms with logarithmic poles along S. The definition of "weakly positive" will be recalled in 1.1. Part a) of 0.1, for $S = \emptyset$, has been shown by S. Kovács in [14].

Considering $r = 0$ in 0.1, b), one finds that smooth complete intersections U in $\mathbb{P}^N$ of codimension $\ell < N/2$ do not allow a non-trivial morphism $U \to M_h$, induced by a family. In b) the intersection with an empty index set is supposed to be $H = \mathbb{P}^N$. So for $\ell < N/2$ part c) follows from b).

In general, 0.1, c), will follow from the slightly stronger statement in the second part of the next theorem. In fact, if one chooses general linear hyperplanes $D_0, \cdots, D_N$ then $D_0 + D_1 + \cdots + D_N + S$ remains a normal crossing divisor. and all morphism

$$\mathbb{P}^N \setminus (D_0 + D_1 + \cdots + D_N + S) \longrightarrow M_h ,$$

induced by a family, must be trivial.

Base Spaces of Non-Isotrivial Families   281

Theorem 0.2 (see 7.2).

a) *Assume that U is the complement of a normal crossing divisor S with strictly less than N components in an N-dimensional abelian variety. Then there exists no generically finite morphism $U \to M_h$, induced by a family.*

b) *For $Y = \mathbb{P}^{\nu_1} \times \cdots \times \mathbb{P}^{\nu_k}$ let*

$$D^{(\nu_i)} = D_0^{(\nu_i)} + \cdots + D_{\nu_i}^{(\nu_i)}$$

be coordinate axes in $\mathbb{P}^{\nu_i}$ and

$$D = \bigoplus_{i=1}^{k} D^{(\nu_i)} \, .$$

Assume that $S = S_1 + \cdots S_\ell$ is a divisor, such that $D + S$ is a reduced normal crossing divisor, and $\ell < \dim(Y)$. Then there exists no morphism $\varphi : U = Y \setminus (D + S) \to M_h$ with

$$\dim(\varphi(U)) > \mathrm{Max}\{\dim(Y) - \nu_i; \ i = 1, \ldots, k\} \, .$$

We do not know, whether the bound $\ell < \dim(Y)$ in 0.2 is really needed. If the infinitesimal Torelli theorem holds true for the general fibre, hence if the family $V \to U$ induces a generically finite map to a period domain, then the fundamental group of U should not be abelian. In particular U should not be the complement of a normal crossing divisor in $\mathbb{P}^N$.

In Sect. 6 we will prove different properties of U in case there exists a quasi-finite morphism $\varphi : U \to M_h$. Those properties will be related to the rigidity of generic curves in moduli stacks.

Theorem 0.3 (see 6.4 and 6.7). *Let U be a quasi-projective variety and let $\varphi : U \to M_h$ be a quasi-finite morphism, induced by a family. Then*

a) *U cannot be isomorphic to the product of more than $n = \deg(h)$ varieties of positive dimension.*

b) *$\mathrm{Aut}(U)$ is finite.*

Although we do not need it in its full strength, we could not resist to include a proof of the finiteness theorem 6.2, saying that for a projective curve C, for an open sub curve C_0, and for a projective compactification $\bar{U}$ of U, the morphisms $\pi : C \to \bar{U}$ with $\pi(C_0) \subset U$ are parameterized by a scheme of finite type.

We call $f : V \to U$ a (flat or smooth) family of projective varieties, if f is projective (flat or smooth) and all fibres connected. For a flat family, an invertible sheaf $\mathcal{L}$ on V will be called f-semi-ample, or relatively semi-ample over U, if for some $\nu > 0$ the evaluation of sections $f^* f_* \mathcal{L}^\nu \to \mathcal{L}^\nu$ is surjective.

282 Eckart Viehweg and Kang Zuo

The notion f-ampleness will be used if in addition for $\nu \gg 0$ the induced U-morphism $V \to \mathbb{P}(f_* \mathcal{L}^\nu)$ is an embedding, or equivalently, if the restriction of $\mathcal{L}$ to all the fibres is ample.

For families over a higher dimensional base $f : V \to U$, the non-isotriviality will be measured by an invariant, introduced in [20]. We define $\text{Var}(f)$ to be the smallest integer η for which there exists a finitely generated subfield K of $\overline{\mathbb{C}(U)}$ of transcendence degree η over $\mathbb{C}$, a variety F' defined over K, and a birational equivalence

$$V \times_U \text{Spec}(\overline{\mathbb{C}(U)}) \sim F' \times_{\text{Spec}(K)} \text{Spec}(\overline{\mathbb{C}(U)}) .$$

We will call f isotrivial, in case that $\text{Var}(f) = 0$. If $(f : V \to U) \in \mathcal{M}_h(U)$ induces the morphism $\varphi : U \to M_h$, then $\text{Var}(f) = \dim(\varphi(U))$.

Most of the results in this article carry over to families $V \to U$ with $\omega_{V/U}$ semi-ample. The first result without requiring local Torelli theorems, saying that there are no non-isotrivial families of elliptic surfaces over $\mathbb{C}^*$ or over elliptic curves, has been shown by K. Oguiso and the first named author [17]. It was later extended to all families of higher dimensional minimal models in [24].

Variant 0.4 *Let U be a quasi-projective manifold as in 0.1 or in 0.2. Then there exists no smooth family $f : V \to U$ with $\omega_{V/U}$ f-semi-ample and with $\text{Var}(f) = \dim(U)$.*

Variant 0.5 *For U a quasi-projective manifold let $f : V \to U$ be a smooth family with $\omega_{V/U}$ f-semi-ample and with $\text{Var}(f) = \dim(U)$. Then the conclusion a) and b) in 0.3 hold true.*

All the results mentioned will be corollaries of Theorem 1.4, formulated in the first section. It is closely related to some conjectures and open problems on differential forms on moduli stacks, explained in 1.5. The proof of 1.4, which covers sections 2, 3, and 4, turns out to be quite complicated, and we will try to give an outline at the end of the first section.

The methods are close in spirit to the ones used in [24] for Y a curve, replacing [24, Prop. 1.3], by [26], Theorem 0.1, and using some of the tools developed in [25]. So the first three and a half sections do hardly contain any new ideas. They are needed nevertheless to adapt methods and notations to the situation studied here, and hopefully they can serve as a reference for methods needed to study positivity problems over higher dimensional bases. The reader who just wants to get some idea on the geometry of moduli stacks should skip Sects. 2, 3 and 4 in a first reading and start with Sects. 1, 5, 6 and 7.

This article benefited from discussions between the first named author and S. Kovács. In particular we thank him for informing us about his results. We thank the referee for hints, how to improve the presentation of the results.

A first version of this paper was written during a visit of the first named author to the Institute of Mathematical Science and the Department of Mathematics at the Chinese University of Hong Kong. The final version, including Sect. 7, was finished during a visit of the second named author to the Department of Mathematics at the University of Essen. We both would like to all the members of the host institutes for their hospitality.

1 Differential Forms on Moduli Stacks

Our motivation and starting point are conjectures and questions on the sheaf of differential forms on moduli-stacks. Before formulating the technical main result and related conjectures and questions, let us recall some definitions.

Definition 1.1. Let $\mathcal{F}$ be a torsion free coherent sheaf on a quasi-projective normal variety Y and let $\mathcal{H}$ be an ample invertible sheaf.

a) $\mathcal{F}$ is generically generated if the natural morphism

$$H^0(Y, \mathcal{F}) \otimes \mathcal{O}_Y \longrightarrow \mathcal{F}$$

is surjective over some open dense subset U_0 of Y. If one wants to specify U_0 one says that $\mathcal{F}$ is globally generated over U_0.

b) $\mathcal{F}$ is weakly positive if there exists some dense open subset U_0 of Y with $\mathcal{F}|_{U_0}$ locally free, and if for all $\alpha > 0$ there exists some $\beta > 0$ such that

$$S^{\alpha \cdot \beta}(\mathcal{F}) \otimes \mathcal{H}^\beta$$

is globally generated over U_0. We will also say that $\mathcal{F}$ is weakly positive over U_0, in this case.

c) $\mathcal{F}$ is big if there exists some open dense subset U_0 in Y and some $\mu > 0$ such that

$$S^\mu(\mathcal{F}) \otimes \mathcal{H}^{-1}$$

is weakly positive over U_0. Underlining the role of U_0 we will also call $\mathcal{F}$ ample with respect to U_0.

Here, as in [20] and [25], we use the following convention: If $\mathcal{F}$ is a coherent torsion free sheaf on a quasi-projective normal variety Y, we consider the largest open subscheme $i : Y_1 \to Y$ with $i^*\mathcal{F}$ locally free. For

$$\Phi = S^\mu, \quad \Phi = \overset{\mu}{\bigotimes} \quad \text{or} \quad \Phi = \det$$

we define

$$\Phi(\mathcal{F}) = i_*\Phi(i^*\mathcal{F}) \, .$$

Let us recall two simple properties of sheaves which are ample with respect to open sets, or generically generated. A more complete list of such properties can be found in [23, §] 2. First of all the ampleness property can be expressed in a different way (see [25, 3.2], for example).

Lemma 1.2. *Let $\mathcal{H}$ be an ample invertible sheaf, and $\mathcal{F}$ a coherent torsion free sheaf on Y, whose restriction to some open dense subset $U_0 \subset Y$ is locally free. Then $\mathcal{F}$ is ample with respect to U_0 if and only if for some $\eta > 0$ there exists a morphism*

$$\bigoplus \mathcal{H} \longrightarrow S^\eta(\mathcal{F}),$$

surjective over U_0.

We will also need the following well known property of generically generated sheaves.

Lemma 1.3. *Let $\psi : Y' \to Y$ be a finite morphism and let $\mathcal{F}$ be a coherent torsion free sheaf on Y such that $\psi^* \mathcal{F}$ is generically generated. Then for some $\beta > 0$, the sheaf $S^\beta(\mathcal{F})$ is generically generated.*

Proof. We may assume that $\mathcal{F}$ is locally free, and replacing Y' by some covering, that Y' is a Galois cover of Y with Galois group G. Let

$$\pi : \mathbb{P} = \mathbb{P}(\mathcal{F}) \longrightarrow Y \quad \text{and} \quad \pi' : \mathbb{P}' = \mathbb{P}(\psi^* \mathcal{F}) \longrightarrow Y'$$

be the projective bundles. The induced covering $\psi' : \mathbb{P}' \to \mathbb{P}$ is again Galois. By assumption, for some $U_0 \subset Y$ the sheaf $\mathcal{O}_{\mathbb{P}'}(1)$ is generated by global sections over $\psi'^{-1} \pi^{-1}(U_0)$. Hence for $g = \#G$ the sheaf $\mathcal{O}_{\mathbb{P}'}(g) = \psi'^* \mathcal{O}_{\mathbb{P}}(g)$ is generated over $\psi'^{-1} \pi^{-1}(U_0)$ by G-invariant sections, hence $\mathcal{O}_{\mathbb{P}}(g)$ is globally generated by sections $s_1, \ldots, s_\ell \in H^0(\mathbb{P}, \mathcal{O}_{\mathbb{P}}(g))$ over $\pi^{-1}(U_0)$. By the Nullstellensatz, there exists some β' such that

$$S^{\beta'}\left(\bigoplus_{i=1}^\ell \mathcal{O}_{\mathbb{P}} \cdot s_i\right) \to S^{g \cdot \beta'}(\mathcal{F}) = \pi_* \mathcal{O}_{\mathbb{P}}(g \cdot \beta')$$

is surjective over U_0. $\qquad\square$

The main result of this article says, that the existence of smooth families $F : V \to U$ with $\mathrm{Var}(f) > 0$ is only possible if U carries multi-differential forms with logarithmic singularities at infinity.

Theorem 1.4. *Let Y be a projective manifold, S a reduced normal crossing divisor, and let $f : V \to U = Y \setminus S$ be a smooth family of n-dimensional projective varieties.*

i) *If $\omega_{V/U}$ is f-ample, then for some $m > 0$ the sheaf $S^m(\Omega_Y^1(\log S))$ contains an invertible sheaf $\mathcal{A}$ of Kodaira dimension $\kappa(\mathcal{A}) \geq \mathrm{Var}(f)$.*

ii) *If $\omega_{V/U}$ is f-ample and $\mathrm{Var}(f) = \dim(Y)$, then for some $0 < m \leq n$ the sheaf $S^m(\Omega_Y^1(\log S))$ contains a big coherent subsheaf $\mathcal{P}$.*

iii) *If $\omega_{V/U}$ is f-semi-ample and $\mathrm{Var}(f) = \dim(Y)$, then for some $m > 0$ the sheaf $S^m(\Omega_Y^1(\log S))$ contains a big coherent subsheaf $\mathcal{P}$.*

Base Spaces of Non-Isotrivial Families   285

iv) *Moreover under the assumptions made in iii) there exists a non-singular finite covering $\psi : Y' \to Y$ and, for some $0 < m \leq n$, a big coherent subsheaf $\mathcal{P}'$ of $\psi^* S^m(\Omega_Y^1(\log S))$.*

Before giving a guideline to the proof of 1.4, let us discuss further properties of the sheaf of one forms on U, we hope to be true.

Problem 1.5. Let Y be a projective manifold, S a reduced normal crossing divisor, and $U = Y \setminus S$. Let $\varphi : U \to M_h$ be a morphism, induced by a family $f : V \to U$. Assume that the family $f : V \to U$ induces an étale map to the moduli stack, or in down to earth terms, that the induced Kodaira Spencer map

$$T_U \longrightarrow R^1 f_* T_{V/U}$$

is injective and locally split.

a) Is $\Omega_Y^1(\log S)$ weakly positive, or perhaps even weakly positive over U?
b) Is $\det(\Omega_Y^1(\log S)) = \omega_Y(S)$ big?
c) Are there conditions on Ω_F^1, for a general fibre F of f, which imply that $\Omega_Y^1(\log S)$ is big?

As we will see in 5.1, Theorem 1.4 implies that the bigness in 1.5, b), follows from the weak positivity in a).

There is hope, that the questions a) and b), which have been raised by the first named author some time ago, will have an affirmative answer. In particular they have been verified by the second named author [26], under the additional assumption that the local Torelli theorem holds true for the general fibre F of f. The Brody hyperbolicity of moduli stacks of canonically polarized manifolds, shown in [25], the results of Kovács, and the content of this paper strengthen this hope. As S. Kovács told us, for certain divisors S in $Y = \mathbb{P}^N$, 1.5, a), holds true.

For moduli spaces of curves the sheaf $\Omega_Y^1(\log S)$ is ample with respect to U. This implies that morphisms $\pi : C_0 \to U$ are rigid (see 6.6). In the higher dimensional case the latter obviously does not hold true (see 6.3), and Problem c) asks for conditions implying rigidity.

There is no evidence for the existence of a reasonable condition in c). One could hope that "Ω_F^1 ample" or "Ω_F^1 big" will work. At least, this excludes the obvious counter examples for the ampleness of $\Omega_Y^1(\log S)$, discussed in 6.3. For a non-isotrivial smooth family $V \to U$ of varieties with Ω_F^1 ample, the restriction of Ω_V^1 to F is big, an observation which for families of curves goes back to H. Grauert [9]. Problem 1.1, c) expresses our hope that such properties of global multi-differential forms on the general fibre could be mirrored in global properties of moduli spaces.

Notations. To prove 1.4 we start by choosing any non-singular projective compactification X of V, with $\Delta = X \setminus V$ a normal crossing divisor, such that $V \to U$ extends to a morphism $f : X \to Y$. For the proof of 1.4 we are allowed to replace Y by any blowing up, if the pullback of S remains a normal crossing divisor. Moreover, as explained in the beginning of the next section, we may replace Y by the complement of a codimension two subscheme, and X by the corresponding preimage, hence to work with partial compactifications, as defined in 2.1. By abuse of notations, such a partial compactification will again be denoted by $f : X \to Y$.

In the course of the argument we will be forced to replace the morphism f by some fibred product. We will try to keep the following notations. A morphism $f' : X' \to Y'$ will denote a pullback of f under a morphism, usually dominant, $Y' \to Y$, or a desingularization of such a pullback. The smooth parts will be denoted by $V \to U$ and $V' \to U'$, respectively. $f^r : X^r \to Y$ will denote the family obtained as the r-fold fibred product over Y, and $f^{(r)} : X^{(r)} \to Y$ will be obtained as a desingularization of X^r. Usually U_0 will denote an open dense subscheme of Y, and $\tilde{U}$ will be a blowing up of U.

At several places we need in addition some auxiliary constructions. In Sect. 2 this will be a family $g : Z \to Y'$, dominating birationally $X' \to Y'$ and a specific model $g' : Z' \to Y'$. For curves C mapping to Y, the desingularization of the induced family will be $h : W \to C$, where again some $'$ is added whenever we have to consider a pullback family over some covering C' of C.

Finally, in Sect. 4 $h : W \to Y$ will be a blowing up of $X \to Y$ and $g : Z \to Y$ will be obtained as the desingularization of a finite covering of W.

Outline of the Proof of 1.4. Let us start with 1.4, iii). In Sect. 3 we will formulate and recall certain positivity properties of direct image sheaves. In particular, by [20] the assumptions in iii) imply that $\det(f_*\omega^\nu_{X/Y})$ is big, for some $\nu \gg 2$. This in turn implies that the sheaf $f_*\omega^\nu_{X/Y}$ is big. In 3.9 we will extend this result to the slightly smaller sheaf $f_*(\Omega^n_{X/Y}(\log \Delta))^\nu$. Hence for an ample invertible sheaf $\mathcal{A}$ on Y and for some $\mu \gg 1$ the sheaf $S^\mu(f_*(\Omega^n_{X/Y}(\log \Delta))^\nu) \otimes \mathcal{A}^{-1}$ will be globally generated over an open dense subset U_0. Replacing $f : X \to Y$ by a partial compactification we will assume this sheaf to be locally free. Then, for ν sufficiently large and divisible, $\Omega^n_{X/Y}(\log \Delta))^{\nu\mu} \otimes f^*\mathcal{A}^{-1}$ will be globally generated over $f^{-1}(U_0)$, for some $U_0 \neq \emptyset$.

This statement is not strong enough. We will need that

$$\Omega^n_{X/Y}(\log \Delta)^\nu \otimes f^*\mathcal{A}^{-\nu} \text{ is globally generated over } f^{-1}(U_0) , \qquad (1.5.1)$$

for some $\nu \gg 2$ and for some ample invertible sheaf $\mathcal{A}$. To this aim we replace in 3.9 the original morphism $f : X \to Y$ by the r-th fibred product $f^{(r)} : X^{(r)} \to Y$, for some $r \gg 1$.

The condition (1.5.1) will reappear in Sect. 4 in (4.3.1). There we study certain Higgs bundles $\bigoplus F^{p,q}$. (4.3.1) will allow to show that $\mathcal{A} \otimes \bigoplus F^{p,q}$ is contained in a Higgs bundles induced by a variation of Hodge structures. The latter is coming from a finite cyclic covering Z of X. The negativity theorem in [26] will finish the proof of 1.4 iii).

For each of the other cases in 1.4 we need some additional constructions, most of which are discussed in Sect. 2. In ii) and iv) (needed to prove 0.3, a) we have to bound m by the fibre dimension, hence we are not allowed to replace $f : X \to Y$ by the fibre product $f^{(r)} : X^{(r)} \to Y$. Instead we choose a suitable covering $Y' \to Y$, in such a way that the assumption (4.3.1) holds true over the covering. For i) we have to present the family $X' \to Y'$ as the pullback of a family of maximal variation.

In Sect. 2 we also recall the weak semi-stable reduction theorem due to Abramovich and Karu [1] and some of its consequences. In particular it will allow to construct a generically finite dominant morphism $Y' \to Y$ such that for a desingularization of the pullback family $f' : X' \to Y'$ the sheaves $\bigotimes^\mu f'_* \omega^\nu_{X'/Y'}$ are reflexive. This fact was used in [25] in the proof of 3.9. As mentioned, we can restrict ourselves to partial compactifications $f : X \to Y$, and repeating the arguments from [25] in this case, we would not really need the weakly semi-stable reduction. However, the proof of the finiteness theorem 6.2 is based on this method.

As in [24] it should be sufficient in 1.4, iii) and iv), to require that the fibres F of f are of general type, or in the case $0 \le \kappa(F) < \dim(F)$ that F is birational to some F' with $\omega_{F'}$ semi-ample. We do not include this, since the existence of relative base loci make the notations even more confusing than they are in the present version. However, comparing the arguments in [24, §3], with the ones used here, it should not be too difficult to work out the details.

2 Mild Morphisms

As explained at the end of the last section it will be convenient, although not really necessary, to use for the proof of 1.4 some of the results and constructions contained in [25], in particular the weak semi-stable reduction theorem due to Abramovich and Karu [1]. It will allow us to formulate the strong positivity theorem 3.9 for product families, shown in [25, 4.1.], and it will be used in the proof of the boundedness of the functor of homomorphism in 6.2. We will use it again to reduce the proof of 1.4, i), to the case of maximal variation, although this part could easily be done without the weak semi-stable reduction. We also recall Kawamata's covering construction. The content of this section will be needed in the proof of parts i), ii) and iv) of 1.4, but not for iii).

Definition 2.1.

a) Given a family $\tilde{V} \to \tilde{U}$ we will call $V \to U$ a birational model if there exist compatible birational morphisms $\tau : U \to \tilde{U}$ and $\tau' : V \to \tilde{V} \times_{\tilde{U}} U$.

If we underline that U and $\tilde{U}$ coincide, we want τ to be the identity. If $\tilde{V} \to \tilde{U}$ is smooth, we call $V \to U$ a smooth birational model, if τ' is an isomorphism.

b) If $V \to U$ is a smooth projective family of quasi-projective manifolds, we call $f : X \to Y$ a partial compactification, if
  i) X and Y are quasi-projective manifolds, and $U \subset Y$.
  ii) Y has a non-singular projective compactification $\bar{Y}$ such that S extends to a normal crossing divisor and such that $\mathrm{codim}(\bar{Y} \setminus Y) \geq 2$.
  iii) f is a projective morphism and $f^{-1}(U) \to U$ coincides with $V \to U$.
  iv) $S = Y \setminus U$, and $\Delta = f^* S$ are normal crossing divisors.

c) We say that a partial compactification $f : X \to Y$ is a good partial compactification if the condition iv) in b) is replaced by
  iv) f is flat, $S = Y \setminus U$ is a smooth divisor, and $\Delta = f^* S$ is a relative normal crossing divisor, i.e. a normal crossing divisor whose components, and all their intersections are smooth over components of S.

d) The good partial compactification $f : X \to Y$ is semi-stable, if in c), iv), the divisor $f^* S$ is reduced.

e) An arbitrary partial compactification of $V \to U$ is called semi-stable in codimension one, if it contains a semi-stable good partial compactification.

Remark 2.2. The second condition in b) or c) allows to talk about invertible sheaves $\mathcal{A}$ of positive Kodaira dimension on Y. In fact, $\mathcal{A}$ extends in a unique way to an invertible sheaf $\bar{\mathcal{A}}$ on $\bar{Y}$ and

$$H^0(Y, \mathcal{A}^\nu) = H^0\left(\bar{Y}, \bar{\mathcal{A}}^\nu\right) ,$$

for all ν. So we can write $\kappa(\mathcal{A}) := \kappa(\bar{\mathcal{A}})$, in case Y allows a compactification satisfying ii). If $\tau : \bar{Y}' \to \bar{Y}$ is a blowing up with centers in $\bar{Y} \setminus Y$, and if $\bar{\mathcal{A}}'$ is an extension of $\mathcal{A}$ to $\bar{Y}'$, then $\kappa(\mathcal{A}) \geq \kappa(\bar{\mathcal{A}}')$.

In a similar way, one finds a coherent sheaf $\mathcal{F}$ on Y to be semi-ample with respect to $U_0 \subset Y$ (or weakly positive over U_0), if and only if its extension to $\bar{Y}$ has the same property.

Kawamata's covering construction will be used frequently throughout this article. First of all, it allows the semi-stable reduction in codimension one, and secondly it allows to take roots out of effective divisors.

Lemma 2.3.

a) *Let Y be a quasi-projective manifold, S a normal crossing divisor, and let $\mathcal{A}$ be an invertible sheaf, globally generated over Y. Then for all μ there exists a non-singular finite covering $\psi : Y' \to Y$ whose discriminant $\Delta(Y'/Y)$ does not contain components of S, such that $\psi^*(S + \Delta(Y'/Y))$ is a normal crossing divisor, and such that $\psi^* \mathcal{A} = \mathcal{O}_{Y'}(\mu \cdot A')$ for some reduced non-singular divisor A' on Y'.*

b) *Let $f : X \to Y$ be a partial compactification of a smooth family $V \to U$.
Then there exists a non-singular finite covering $\psi : Y' \to Y$, and a
desingularization $\psi' : X' \to X \times_Y Y'$ such that the induced family f' :
$X' \to Y'$ is semi-stable in codimension one.*

Proof. Given positive integers ϵ_i for all components S_i of S, Kawamata constructed a finite non-singular covering $\psi : Y' \to Y$ (see [23], 2.3), with
$\psi^*(S + \Delta(Y'/Y))$ a normal crossing divisor, such that all components of
$\psi^* S_i$ are ramified of order exactly ϵ_i.

In a) we choose A to be the zero-divisor of a general section of $\mathcal{A}$, and
we apply Kawamata's construction to $S + A$, where the ϵ_i are one for the
components of S and where the prescribed ramification index for A is μ.

In b) the semi-stable reduction theorem for families over curves (see [11])
allows to choose the ϵ_i such that the family $f' : X' \to Y'$ is semi-stable in
codimension one. $\qquad\square$

Unfortunately in 2.3, b), one has little control on the structure of f' over
the singularities of S. Here the weak semi-stable reduction theorem will be
of help. The pullback of a weakly semi-stable morphism under a dominant
morphism of manifolds is no longer weakly semi-stable. However some of the
properties of a weakly semi-stable morphism survive. Those are collected in
the following definition, due again to Abramovich and Karu [1].

Definition 2.4. A projective morphism $g' : Z' \to Y'$ between quasi-projective
varieties is called mild, if

a) g' is flat, Gorenstein with reduced fibres.
b) Y' is non-singular and Z' normal with at most rational singularities.
c) Given a dominant morphism $Y_1' \to Y'$ where Y_1' has at most rational
   Gorenstein singularities, $Z' \times_{Y'} Y_1'$ is normal with at most rational singularities.
d) Let Y_0' be an open subvariety of Y', with $g'^{-1}(Y_0') \to Y_0'$ smooth. Given
   a non-singular curve C' and a morphism $\pi : C' \to Y'$ whose image meets
   Y_0', the fibred product $Z' \times_{Y'} C'$ is normal, Gorenstein with at most
   rational singularities.

The mildness of g' is, more or less by definition, compatible with pullback.
Let us rephrase three of the properties shown in [25, 2.2].

Lemma 2.5. *Let Z and Y be quasi-projective manifolds, $g : Z \to Y'$ be a
projective, birational to a projective mild morphism $g' : Z' \to Y'$. Then one
has:*

i) *For all $\nu \geq 1$ the sheaf $g_* \omega_{Z/Y'}^\nu$ is reflexive and isomorphic to $g'_* \omega_{Z'/Y'}^\nu$.*
ii) *If $\gamma : Y'' \to Y'$ is a dominant morphism between quasi-projective mani-
    folds, then the morphism $pr_2 : Z \times_{Y'} Y'' \to Y''$ is birational to a projective
    mild morphism to Y''.*

290 Eckart Viehweg and Kang Zuo

iii) *Let $Z^{(r)}$ be a desingularization of the r-fold fibre product $Z \times_{Y'} \cdots \times_{Y'} Z$. Then the induced morphism $Z^{(r)} \to Y'$ is birational to a projective mild morphism over Y'.*

One consequence of the weakly semi-stable reduction says that, changing the birational model of a morphism, one always finds a finite cover of the base such that the pullback is birational to a mild morphism (see [25], 2.3).

Lemma 2.6. *Let $V \to U$ be a smooth family of projective varieties. Then there exists a quasi-projective manifold $\tilde{U}$ and a smooth birational model $\tilde{V} \to \tilde{U}$, non-singular projective compactifications Y of $\tilde{U}$ and X of $\tilde{V}$, with $S = Y \setminus \tilde{U}$ and $\Delta = X \setminus \tilde{V}$ normal crossing divisors, and a diagram of projective morphisms*

$$
\begin{array}{ccccc}
X & \xleftarrow{\psi'} & X' & \xleftarrow{\sigma} & Z \\
\downarrow{\scriptstyle f} & & \downarrow{\scriptstyle f'} & \nearrow{\scriptstyle g} & \\
Y & \xleftarrow{\psi} & Y' & &
\end{array}
\tag{2.6.1}
$$

with:

a) *Y' and Z are non-singular, X' is the normalization of $X \times_Y Y'$, and σ is a desingularization.*
b) *g is birational to a mild morphism $g' : Z' \to Y'$.*
c) *For all $\nu > 0$ the sheaf $g_* \omega_{Z/Y'}^\nu$ is reflexive and there exists an injection*

$$
g_* \omega_{Z/Y'}^\nu \longrightarrow \psi^* f_* \omega_{X/Y}^\nu .
$$

d) *For some positive integer N_ν, and for some invertible sheaf λ_ν on Y*

$$
\det \left(g_* \omega_{Z/Y'}^\nu \right)^{N_\nu} = \psi^* \lambda_\nu .
$$

e) *Moreover, if $\bar{Y}$ is a given projective compactification of U, we can assume that there is a birational morphism $Y \to \bar{Y}$.*

This diagram has been constructed in [25, §2]. Let us just recall that the reflexivity of $g_* \omega_{Z/Y'}^\nu$ in c) is a consequence of b), using 2.5, i). $\qquad \square$

Unfortunately the way it is constructed, the mild model $g' : Z' \to Y'$ might not be smooth over $\psi^{-1}(\tilde{U})$. Moreover even in case U is non-singular one has to allow blowing ups $\tau : \tilde{U} \to U$. Hence starting from $V \to U$ we can only say that U contains some "good" open dense subset U_g for which τ is an isomorphism between $\tilde{U}_g := \tau^{-1}(U_g)$ and U_g, and for which

$$
g'^{-1} \psi^{-1} \left(\tilde{U}_g \right) \longrightarrow \psi^{-1} \left(\tilde{U}_g \right)
$$

is smooth. The next construction will be needed in the proof of 6.2.

Base Spaces of Non-Isotrivial Families   291

Corollary 2.7. *Let C be a non-singular projective curve, $C_0 \subset C$ an open dense subset, and let $\pi_0 : C_0 \to U$ be a morphism with $\pi_0(C_0) \cap U_g \neq \emptyset$. Hence there is a lifting of π_0 to $\tilde{U}$ and an extension $\pi : C \to Y$ with $\pi_0 = \tau \circ \pi|_{C_0}$. Let $h : W \to C$ be the family obtained by desingularizing the main component of the normalization of $X \times_Y C$, and let λ_ν and N_ν be as in 2.6, d). Then*

$$\deg(\pi^*\lambda_\nu) \leq N_\nu \cdot \deg\left(\det\left(h_*\omega_{W/C}^\nu\right)\right).$$

Proof. Let $\rho : C' \to C$ be a finite morphism of non-singular curves such that π lifts to $\pi' : C' \to Y'$, and let $h' : W' \to C'$ be the family obtained by desingularizing the main component of the normalization of $X' \times_{Y'} C'$. By condition d) in the definition of a mild morphism, and by the choice of U_g, the family h' has $Z' \times_{Y'} C'$ as a mild model. Applying 2.6, c), to h and h', and using 2.6, d), we find

$$\deg(\rho) \cdot \deg(\pi^*\lambda_\nu) = N_\nu \cdot \deg\left(\pi'^*g_*\omega_{Z'/Y'}^\nu\right) \quad \text{and}$$

$$\deg\left(h'_*\omega_{W'/C'}^\nu\right) \leq \deg(\rho) \cdot \deg\left(h_*\omega_{W/C}^\nu\right).$$

Moreover, by 2.6, c), and by base change, one obtains a morphism of sheaves

$$\pi'^*g_*\omega_{Z/Y'}^\nu \simeq \pi'^*g'_*\omega_{Z'/Y'}^\nu \longrightarrow pr_{2*}\omega_{Z'\times_{Y'}C'/C'}^\nu \simeq h'_*\omega_{W'/C'}^\nu, \quad (2.7.1)$$

which is an isomorphism over some open dense subset. Let r denote the rank of those sheaves.

It remains to show, that $(2.7.1)$ induces an injection from $\pi'^* \det(g'_*\omega_{Z'/Y'}^\nu)$ into $\det(pr_{2*}\omega_{Z'\times_{Y'}C'/C'}^\nu)$. To this aim, as in 2.5, iii), let "(r)" stand for "taking a desingularization of the r-th fibre product". Then $g^{(r)} : Z^{(r)} \to Y'$ is again birational to a mild morphism over Y'. As in [25, 4.1.1], flat base change and the projection formula give isomorphisms

$$h'^{(r)}_*\omega_{W'^{(r)}/C'}^\nu \simeq \bigotimes^r h'_*\omega_{W'/C'}^\nu \quad \text{and} \quad g^{(r)}_*\omega_{Z^{(r)}/Y'}^\nu \simeq \bigotimes^r g_*\omega_{Z/Y'}^\nu,$$

the second one outside of a codimension two subscheme. Since both sheaves are reflexive, the latter extends to Y'.

The injection $(2.7.1)$, applied to $g^{(r)}$ and $h^{(r)}$ induces an injective morphism

$$\pi'^* \bigotimes^r g_*\omega_{Z/Y'}^\nu \longrightarrow \bigotimes^r h'_*\omega_{W'/C'}^\nu. \quad (2.7.2)$$

The left hand side contains $\pi'^* \det(g'_*\omega_{Z'/Y'}^\nu)$ as direct factor, whereas the righthand contains $\det(pr_{2*}\omega_{Z'\times_{Y'}C'/C'}^\nu)$, and we obtain the injection for the determinant sheaves as well. $\qquad\square$

292 Eckart Viehweg and Kang Zuo

The construction of (2.6.1) in [25] will be used to construct a second diagram (2.8.1). There we do not insist on the projectivity of the base spaces, and we allow ourselves to work with good partial compactifications of an open subfamily of the given one. This quite technical construction will be needed to prove 1.4, i).

Lemma 2.8. *Let $V \to U$ be a smooth family of canonically polarized manifolds. Let $\bar{Y}$ and $\bar{X}$ be non-singular projective compactifications of U and V such that both, $\bar{Y} \setminus U$ and $\bar{X} \setminus V$, are normal crossing divisors and such that $V \to U$ extends to $\bar{f} : \bar{X} \to \bar{Y}$. Blowing up $\bar{Y}$ and $\bar{X}$, if necessary, one finds an open subscheme Y in $\bar{Y}$ with $\mathrm{codim}(\bar{Y} \setminus Y) \geq 2$ such that the restriction $f : Y \to X$ is a good partial compactification of a smooth birational model of $V \to U$, and one finds a diagram of morphisms between quasi-projective manifolds*

$$
\begin{array}{ccccccc}
X & \xleftarrow{\psi'} & X' & \xleftarrow{\sigma} & Z & \xrightarrow{\eta'} & Z^{\#} \\
\downarrow{\scriptstyle f} & & \downarrow{\scriptstyle f'} & \swarrow{\scriptstyle g} & & \swarrow{\scriptstyle g^{\#}} & \\
Y & \xleftarrow{\psi} & Y' & \xrightarrow{\eta} & Y^{\#} & &
\end{array}
\tag{2.8.1}
$$

with:

a) *$g^{\#}$ is a projective morphism, birational to a mild projective morphism $g^{\#'} : Z^{\#'} \to Y^{\#}$.*

b) *$g^{\#}$ is semi-stable in codimension one.*

c) *$Y^{\#}$ is projective, η is dominant and smooth, η' factors through a birational morphism $Z \to Z^{\#} \times_{Y^{\#}} Y'$, and ψ is finite.*

d) *X' is the normalization of $X \times_Y Y'$ and σ is a blowing up with center in $f'^{-1}\psi^{-1}(S')$. In particular f' and g' are projective.*

e) *Let $U^{\#}$ be the largest subscheme of $Y^{\#}$ with*

$$
V^{\#} = g^{\#^{-1}}\left(U^{\#}\right) \longrightarrow U^{\#}
$$

smooth. Then $\psi^{-1}(U) \subset \eta^{-1}(U^{\#})$, and $U^{\#}$ is generically finite over M_h.

f) *For all $\nu > 0$ there are isomorphisms*

$$
g_* \omega^{\nu}_{Z/Y'} \simeq \eta^* g^{\#}_* \omega^{\nu}_{Z^{\#}/Y^{\#}} \quad \text{and} \quad \det\left(g_* \omega^{\nu}_{Z/Y'}\right) \simeq \eta^* \det\left(g^{\#}_* \omega^{\nu}_{Z^{\#}/Y^{\#}}\right).
$$

g) *For all $\nu > 0$ there exists an injection*

$$
g_*\left(\omega^{\nu}_{Z/Y'}\right) \longrightarrow \psi^* f_*\left(\omega^{\nu}_{X/Y}\right).
$$

For some positive integer N_{ν}, and for some invertible sheaf λ_{ν} on Y

$$
\det\left(g_*\left(\omega^{\nu}_{Z/Y'}\right)\right)^{N_{\nu}} = \psi^* \lambda_{\nu}.
$$

Base Spaces of Non-Isotrivial Families   293

Proof. It remains to verify, that the construction given in [25, §2], for (2.6.1) can be modified to guaranty the condition e) along with the others.

Let $\varphi : U \to M_h$ be the induced morphism to the moduli scheme. Seshadri and Kollár constructed a finite Galois cover of the moduli space which is induced by a family (see [23, 9.25], for example). Hence there exists some manifold $U^\#$, generically finite over the closure of $\varphi(U)$ such that the morphism $U^\# \to M_h$ is induced by a family $V^\# \to U^\#$. By [25, 2.3], blowing up $U^\#$, if necessary, we find a projective compactification $Y^\#$ of $U^\#$ and a covering $Y^{\#'}$, such that

$$V^\# \times_{Y^\#} Y^{\#'} \longrightarrow Y^{\#'}$$

is birational to a projective mild morphism over $Y^{\#'}$. Replacing $U^\#$ by some generically finite cover, we can assume that $V^\# \to U^\#$ has such a model already over $Y^\#$.

Next let Y' be any variety, generically finite over $\bar{Y}$, for which there exists a morphism $\eta : Y' \to Y^\#$. By 2.5, ii), we are allowed to replace $Y^\#$ by any manifold, generically finite over $Y^\#$, without loosing the mild birational model. Doing so, we can assume the fibres of $Y' \to Y^\#$ to be connected. Replacing Y' by some blowing up, we may assume that for some non-singular blowing up $Y \to \bar{Y}$ the morphism $Y' \to \bar{Y}$ factors through a finite morphism $Y' \to Y$.

Next choose a blowing up $Y^{\#'} \to Y^\#$ such that the main component of $Y' \times_{Y^\#} Y^{\#'}$ is flat over $Y^\#$, and Y'' to be a desingularization. Hence changing notations again, and dropping one prime, we can assume that the image of the largest reduced divisor E in Y' with $\mathrm{codim}(\eta(E)) \geq 2$ maps to a subscheme of Y of codimension larger that or equal to 2. This remains true, if we replace $Y^\#$ and Y' by finite coverings. Applying 2.3, b), to $Y' \to Y^\#$, provides us with non-singular covering of $Y^\#$ such that a desingularization of the pullback of $Y' \to Y^\#$ is semi-stable in codimension one. Again, this remains true if we replace $Y^\#$ by a larger covering, and using 2.3, b), a second time, now for $Z^\# \to Y^\#$, we can assume that this morphism is as well semi-stable in codimension one.

Up to now, we succeeded to find the manifolds in (2.8.1) such that a) and b) hold true. In c), the projectivity of $Y^\#$ and the dominance of Y' over $Y^\#$ follow from the construction. For the divisor E in Y' considered above, we replace Y by $Y \setminus \psi(E)$ and Y' by $Y' \setminus E$, and of course X, X' and Z by the corresponding preimages. Then the non-equidimensional locus of η in Y' will be of codimension larger than or equal to two. ψ is generically finite, by construction, hence finite over the complement of a codimension two subscheme of Y. Replacing again Y by the complement of codimension two subscheme, we can assume η to be equidimensional, hence flat, and ψ to be finite. The morphism η has reduced fibres over general points of divisors in $Y^\#$, hence it is smooth outside a codimension two subset of Y', and replacing Y by the complement of its image, we achieved c).

294   Eckart Viehweg and Kang Zuo

Since $V \to U$ is smooth, the pullback of $X \to Y$ to Y' is smooth outside of $\psi^{-1}(S)$. Moreover the induced morphism to the moduli scheme M_h factors through an open subset of $Y^\#$. Since by construction $U^\#$ is proper over its image in M_h, the image of $\psi^{-1}(U)$ lies in $U^\#$ and we obtain d) and e).

For f) remark that the pullback $Z' \to Y'$ of the mild projective morphism $g^\# : Z^{\#'} \to Y^\#$ to Y' is again mild, and birational to $Z \to Y'$. By flat base change,

$$g'_* \omega^\nu_{Z'/Y'} \simeq \eta^* g^{\#'}_* \omega^\nu_{Z^{\#'}/Y^\#} \ .$$

Since Z' and $Z^{\#'}$ are normal with at most rational Gorenstein singularities, we obtain (as in [25, 2.3]) that the sheaf on the right hand side is $g_* \omega^\nu_{Z/Y'}$ whereas the one on the right hand side is $\eta^* g^\#_* \omega^\nu_{Z^\#/Y^\#}$.

g) coincides with 2.6, d), and it has been verified in [25, 2.4]. as a consequence of the existence of a mild model for g over Y'.   □

3   Positivity and Ampleness

Next we will recall positivity theorems, due to Fujita, Kawamata, Kollár and the first named author. Most of the content of this section is well known, or easily follows from known results.

As in 1.4 we will assume throughout this section that U is the complement of a normal crossing divisor $\bar{S}$ in a manifold $\bar{Y}$, and that there is a smooth family $V \to U$ with $\omega_{V/U}$ relative semi-ample. Leaving out a codimension two subset in $\bar{Y}$ we find a good partial compactification $f : X \to Y$, as defined in 2.1.

For an effective $\mathbb{Q}$-divisor $D \in \mathrm{Div}(X)$ the integral part $[D]$ is the largest divisor with $[D] \leq D$. For an effective divisor Γ on X, and for $N \in \mathbb{N} - \{0\}$ the algebraic multiplier sheaf is

$$\omega_{X/Y}\Big\{\frac{-\Gamma}{N}\Big\} = \psi_*\Big(\omega_{T/Y}\Big(-\Big[\frac{\Gamma'}{N}\Big]\Big)\Big)$$

where $\psi : T \to X$ is any blowing up with $\Gamma' = \psi^* \Gamma$ a normal crossing divisor (see for example [6, 7.4], or [23, §5.3]).

Let F be a non-singular fibre of f. Using the definition given above for F, instead of X, and for a divisor Π on F, one defines

$$e(\Pi) = \mathrm{Min}\Big\{N \in \mathbb{N} \setminus \{0\}; \ \omega_F\Big\{\frac{-\Pi}{N}\Big\} = \omega_F\Big\} \ .$$

By [6] or [23, §5.4], $e(\Gamma|_F)$ is upper semi-continuous, and there exists a neighborhood V_0 of F with $e(\Gamma|_{V_0}) \leq e(\Gamma|_F)$. If $\mathcal{L}$ is an invertible sheaf on F, with $H^0(F, \mathcal{L}) \neq 0$, one defines

$$e(\mathcal{L}) = \mathrm{Max}\big\{e(\Pi); \ \Pi \text{ an effective divisor and } \mathcal{O}_F(\Pi) = \mathcal{L}\big\} \ .$$

Base Spaces of Non-Isotrivial Families   295

Proposition 3.1 ([25, 3.3]). *Let $\mathcal{L}$ be an invertible sheaf, let Γ be a divisor on X, and let $\mathcal{F}$ be a coherent sheaf on Y. Assume that, for some $N > 0$ and for some open dense subscheme U_0 of U, the following conditions hold true:*

a) *$\mathcal{F}$ is weakly positive over U_0 (in particular $\mathcal{F}|_{U_0}$ is locally free).*
b) *There exists a morphism $f^*\mathcal{F} \to \mathcal{L}^N(-\Gamma)$, surjective over $f^{-1}(U_0)$.*
c) *None of the fibres F of $f : V_0 = f^{-1}(U_0) \to U_0$ is contained in Γ, and for all of them*
$$e(\Gamma|_F) \leq N .$$

Then $f_*(\mathcal{L} \otimes \omega_{X/Y})$ is weakly positive over U_0.

As mentioned in [25, 3.8], the arguments used in [23, 2.45], carry over to give a simple proof of the following, as a corollary of 3.1.

Corollary 3.2 ([22, 3.7]). *$f_*\omega_{X/Y}^\nu$ is weakly positive over U.*

In [21], for families of canonically polarized manifolds and in [10], in general, one finds the strong positivity theorem saying:

Theorem 3.3. *If $\omega_{V/U}$ is f-semi-ample, then for some η sufficiently large and divisible,*
$$\kappa(\det(f_*\omega_{X/Y}^\eta)) \geq \mathrm{Var}(f) .$$

In case $\mathrm{Var}(f) = \dim(Y)$, 3.3 implies that $\det(f_*\omega_{X/Y}^\eta)$ is ample with respect to some open dense subset U_0 of Y. If the general fibre of f is canonically polarized, and if the induced map $\varphi : U \to M_h$ quasi-finite over its image, one can choose $U_0 = U$, as follows from the last part of the next proposition.

Proposition 3.4. *Assume that $\mathrm{Var}(f) = \dim(Y)$, and that $\omega_{V/U}$ is f-semi-ample. Then:*

i) *The sheaf $f_*\omega_{X/Y}^\nu$ is ample with respect to some open dense subset U_0 of Y for all $\nu > 1$ with $f_*\omega_{X/Y}^\nu \neq 0$.*
ii) *If B is an effective divisor, supported in S then for all ν sufficiently large and divisible, the sheaf $\mathcal{O}_Y(-B) \otimes f_*\omega_{X/Y}^\nu$ is ample with respect to some open dense subset U_0.*
iii) *If the smooth fibres of f are canonically polarized and if the induced morphism $\varphi : U \to M_h$ is quasi-finite over its image, then one can chose $U_0 = U$ in i) and ii).*

Proof. For iii) one uses a variant of 3.3, which has been shown [22, 1.19]. It also follows from the obvious extension of the ampleness criterion in [23, 4.33], to the case "ample with respect to U":

296   Eckart Viehweg and Kang Zuo

Claim 3.5. Under the assumption made in 3.4, iii), for all η sufficiently large and divisible, there exist positive integers a, b and μ such that

$$\det\left(f_*\omega_{X/Y}^{\mu\eta}\right)^a \otimes \det\left(f_*\omega_{X/Y}^{\eta}\right)^b$$

ample with respect to U. $\qquad\square$

Since we do not want to distinguish between the two cases i) and iii) in 1.4, we choose $U_0 = U$ in iii), and we allow $a = \mu = 0$ in case i). By 3.5 and 3.3, respectively, in both cases the sheaf $\det(f_*\omega_{X/Y}^{\mu\eta})^a \otimes \det(f_*\omega_{X/Y}^{\eta})^b$ is ample with respect to U_0.

By [6, §7], or [23, §5.4], the number $e(\omega_F^{\mu\eta})$ is bounded by some constant e, for all smooth fibres of f. We will choose e to be divisible by η and larger than $\mu\eta$.

Replacing a and b by some multiple, we may assume that there exists a very ample sheaf $\mathcal{A}$ and a morphism

$$\mathcal{A} \longrightarrow \det\left(f_*\omega_{X/Y}^{\mu\eta}\right)^a \otimes \det\left(f_*\omega_{X/Y}^{\eta}\right)^b$$

which is an isomorphism over U_0, and that b is divisible by μ.

By 2.3, a), there exists a non-singular covering $\psi : Y' \to Y$ and an effective divisor H with $\psi^*\mathcal{A} = \mathcal{O}_{Y'}(e \cdot (\nu - 1) \cdot H)$, and such that the discriminant locus $\Delta(Y'/Y)$ does not contain any of the components of S. Replacing Y by a slightly smaller scheme, we can assume that $\Delta(Y'/Y) \cap S = \emptyset$, hence $X' = X \times_Y Y'$ is non-singular and by flat base change

$$pr_{2*}\omega_{X'/Y'}^{\sigma} = \psi^* f_*\omega_{X/Y}^{\sigma}$$

for all σ. The assumptions in 3.4, i), ii) or iii) remain true for $pr_2 : X' \to Y'$, and by [23, 2.16], it is sufficient to show that the conclusions in 3.4 hold true on Y' for $\psi^{-1}(U_0)$.

Dropping the primes, we will assume in the sequel that $\mathcal{A}$ has a section whose zero-divisor is $e \cdot (\nu - 1) \cdot H$ for a non-singular divisor H.

Let $r(\sigma)$ denote the rank of $f_*\omega_{X/Y}^{\sigma}$. We choose

$$r = r(\eta) \cdot \frac{b}{\mu} + r(\mu\eta) \cdot a ,$$

consider the r-fold fibre product

$$f^r : X^r = X \times_Y X \ldots \times_Y X \longrightarrow Y ,$$

and a desingularization $\delta : X^{(r)} \to X^r$. Using flat base change, and the natural maps

$$\mathcal{O}_{X^r} \longrightarrow \delta_*\mathcal{O}_{X^{(r)}} \quad \text{and} \quad \delta_*\omega_{X^{(r)}} \longrightarrow \omega_{X^r} ,$$

one finds morphisms

$$\bigotimes^{r} f_* \omega_{X/Y}^{\mu\eta} \longrightarrow f_*^{(r)} \delta^* \omega_{X^r/Y}^{\mu\eta} \qquad \text{and} \qquad (3.5.1)$$

$$f_*^{(r)} \delta^* (\omega_{X^r/Y}^{\nu-1} \otimes \omega_{X^{(r)}/Y}) \longrightarrow f_*^r \omega_{X^r/Y}^{\nu} = \bigotimes^{r} f_* \omega_{X/Y}^{\nu}, \qquad (3.5.2)$$

and both are isomorphism over U. We have natural maps

$$\det(f_* \omega_{X/Y}^{\mu\eta}) \xrightarrow{\ \subset\ }^{r(\mu\eta)} \bigotimes^{r} f_* \omega_{X/Y}^{\mu\eta} \qquad \text{and} \qquad (3.5.3)$$

$$\det(f_* \omega_{X/Y}^{\eta})^{\mu} \xrightarrow{\ \subset\ }^{r(\eta)\cdot\mu} \bigotimes^{r} f_* \omega_{X/Y}^{\eta} \xrightarrow{\ }^{r(\eta)} \bigotimes^{r} f_* \omega_{X/Y}^{\mu\eta}, \qquad (3.5.4)$$

where the last morphism is the multiplication map. Hence we obtain

$$\mathcal{A} = \mathcal{O}_Y(e \cdot (\nu - 1) \cdot H) \longrightarrow \det(f_* \omega_{X/Y}^{\mu\eta})^a \otimes \det(f_* \omega_{X/Y}^{\eta})^b$$

$$\longrightarrow \bigotimes^{r} f_* \omega_{X/Y}^{\mu\eta} \longrightarrow f_*^{(r)} \delta^* \omega_{X^r/Y}^{\mu\eta}.$$

Thereby the sheaf $f^{(r)*} \mathcal{A}$ is a subsheaf of $\delta^* \omega_{X^r/Y}^{\mu\eta}$. Let Γ be the zero divisor of the corresponding section of

$$f^{(r)*} \mathcal{A}^{-1} \otimes \delta^* \omega_{X^r/Y}^{\mu\eta} \qquad \text{hence} \qquad \mathcal{O}_{X^{(r)}}(-\Gamma) = f^{(r)*} \mathcal{A} \otimes \delta^* \omega_{X^r/Y}^{-\mu\eta}.$$

For the sheaf

$$\mathcal{M} = \delta^* \left(\omega_{X^r/Y} \otimes \mathcal{O}_{X^r} (-f^{r*} H) \right)$$

one finds

$$\mathcal{M}^{e\cdot(\nu-1)}(-\Gamma) = \delta^* \omega_{X^r/Y}^{e\cdot(\nu-1)} \otimes f^{(r)*} \mathcal{A}^{-1} \otimes \mathcal{O}_{X^{(r)}}(-\Gamma) = \delta^* \omega_{X^r/Y}^{e\cdot(\nu-1)-\mu\eta}.$$

By the assumption 3.4, i), and by the choice of e we have a morphism

$$f^* f_* \omega_{X/Y}^{e\cdot(\nu-1)-\mu\eta} \longrightarrow \omega_{X/Y}^{e\cdot(\nu-1)-\mu\eta},$$

surjective over $f^{-1}(U_0)$. The sheaf

$$\mathcal{F} = \bigotimes^{r} f_* \omega_{X/Y}^{e\cdot(\nu-1)-\mu\eta}$$

is weakly positive over U_0 and there is a morphism

$$f^{(r)*} \mathcal{F} \longrightarrow \mathcal{M}^{e\cdot(\nu-1)}(-\Gamma)$$

surjective over $f^{-1}(U_0)$. Since the morphism of sheaves in (3.5.3), as well as the first one in (3.5.4), split locally over U_0 the divisor Γ cannot contain a fibre F of

$$f^{(r)-1}(U_0) \longrightarrow U_0,$$

and by [6, §7], or [23, 5.21], for those fibres

$$e(\Gamma|_{F^r}) \leq e(\omega_{F^r}^{\mu\eta}) = e(\omega_F^{\mu\eta}) \leq e \, .$$

Applying 3.1 to $\mathcal{L} = \mathcal{M}^{\nu-1}$ one obtains the weak positivity of the sheaf

$$f_*^{(r)}(\mathcal{M}^{\nu-1} \otimes \omega_{X^{(r)}/Y}) = f_*^{(r)}(\delta^*(\omega_{X^r/Y}^{\nu-1} \otimes \omega_{X^{(r)}/Y})) \otimes \mathcal{O}_Y(-(\nu-1) \cdot H)$$

over U_0. By (3.5.2) one finds morphisms, surjective over U_0

$$f_*^{(r)}(\delta^*(\omega_{X/Y}^{\nu-1} \otimes \omega_{X^{(r)}/Y})) \otimes \mathcal{O}_Y(-(\nu-1) \cdot H)$$

$$\longrightarrow f_*^r(\omega_{X^r/Y}^{\nu}) \otimes \mathcal{O}_Y(-(\nu-1) \cdot H)$$

$$= \left(\bigotimes^r f_* \omega_{X/Y}^{\nu} \right) \otimes \mathcal{O}_Y(-(\nu-1) \cdot H)$$

$$\longrightarrow S^r(f_* \omega_{X/Y}^{\nu}) \otimes \mathcal{O}_Y(-(\nu-1) \cdot H) \, .$$

Since the quotient of a weakly positive sheaf is weakly positive, the sheaf on the right hand side is weakly positive over U_0, hence $f_* \omega_{X/Y}^{\nu}$ is ample with respect to U_0. For ν sufficiently large $\mathcal{O}_Y((\nu-1) \cdot H - S)$ is ample, and one obtains the second part of 3.4. $\qquad \square$

If $f : X \to Y$ is not semi-stable in codimension one, the sheaf of relative n-forms $\Omega_{X/Y}^n(\log \Delta)$ might be strictly smaller than the relative dualizing sheaf $\omega_{X/Y}$. In fact, comparing the first Chern classes of the entries in the tautological sequence

$$0 \longrightarrow f^* \Omega_Y^1(\log S) \longrightarrow \Omega_X^1(\log \Delta) \longrightarrow \Omega_{X/Y}^1(\log \Delta) \longrightarrow 0 \qquad (3.5.5)$$

one finds for $\Delta = f^*S$

$$\Omega_{X/Y}^n(\log \Delta) = \omega_{X/Y}(\Delta_{\mathrm{red}} - \Delta) \, . \qquad (3.5.6)$$

Corollary 3.6. *Under the assumptions made in 3.4, i), ii) or iii), for all ν sufficiently large and divisible, the sheaf $f_* \Omega_{X/Y}^n(\log \Delta)^{\nu}$ is ample with respect to U_0.*

Before proving 3.6 let us start to study the behavior of the relative q-forms under base extensions. Here we will prove a more general result than needed for 3.6, and we will not require $Y \setminus U$ to be smooth.

Assumptions 3.7. Let $f : X \to Y$ be any partial compactification of a smooth family $V \to U$, let $\psi : Y' \to Y$ be a finite covering with Y' non singular, and let $\tilde{X}$ be the normalization of $X \times_Y Y'$. Consider a desingularization $\varphi : X' \to \tilde{X}$, where we assume the center of φ to lie in the singular locus of $\tilde{X}$. The induced morphisms are denoted by

$$X' \xrightarrow{\varphi} \tilde{X} \xrightarrow{\tilde{\varphi}} X \times_Y Y' \xrightarrow{\pi_1} X$$

$$\begin{array}{ccc} & f' \searrow \quad \tilde{f} \downarrow \quad \pi_2 \nearrow \qquad \qquad \nearrow f \\ & Y' \xrightarrow{\psi} \quad Y. \end{array}$$

(3.7.1)

Let us define $\delta = \tilde{\varphi} \circ \varphi : X' \to X \times_Y Y'$ and $\psi' = \pi_1 \circ \delta : X' \to X$. Finally we write $S' = \psi^* S$ and $\Delta' = \psi'^* \Delta$. The discriminant loci of ψ and ψ' will be $\Delta(X'/X)$, and $\Delta(Y'/Y)$, respectively. We will assume that $S + \Delta(Y'/Y)$ and $\Delta + \Delta(X'/X)$, as well as their preimages in Y' and X', are normal crossing divisors.

Lemma 3.8. *Using the assumptions and notations from 3.7,*

i) *there exists for all p an injection*

$$\psi'^* \Omega^p_{X/Y}(\log \Delta) \xrightarrow{\subset} \Omega^p_{X'/Y'}(\log \Delta') \, ,$$

which is an isomorphism over $\psi'^{-1}(X \setminus \mathrm{Sing}(\Delta))$.
ii) *there exists for all $\nu > 0$ an injection*

$$f'_* \left(\Omega^n_{X'/Y'}(\log \Delta')^{(\nu-1)} \otimes \omega_{X'/Y'} \right) \longrightarrow \psi^* f_* \left(\Omega^n_{X/Y}(\log \Delta)^{(\nu-1)} \otimes \omega_{X/Y} \right) \, ,$$

which is an isomorphism over $\psi^{-1}(U)$.

Proof. If one replaces in the tautological sequence (3.5.5) the divisor S by a larger one, the sheaf on the right hand side does not change, hence

$$\Omega^1_{X/Y}(\log \Delta) = \Omega^1_{X/Y}(\log(\Delta + \Delta(X'/X))) \, .$$

Both, $\Omega^1_Y(\log(S + \Delta(Y'/Y)))$ and $\Omega^1_X(\log(\Delta + \Delta(X'/X)))$ behave well under pullback to X' (see [6, 3.20], for example). To be more precise, there exists an isomorphism

$$\psi^* \Omega^1_Y(\log(S + \Delta(Y'/Y))) \simeq \Omega^1_{Y'}(\log(S' + \psi^* \Delta(Y'/Y)))$$

and an injection

$$\psi'^* \Omega^1_X(\log(\Delta + \Delta(X'/X))) \xrightarrow{\subset} \Omega^1_{X'}(\log \psi'^*(\Delta + \Delta(X'/X)))$$

which is an isomorphism over the largest open subscheme V_1', where ψ' is an isomorphism. Since $\tilde{X}$ is non-singular outside of Δ, and since the singularities of $\tilde{X}$ can only appear over singular points of the discriminant $\Delta(X'/X)$, we find $\psi'^{-1}(X \setminus \mathrm{Sing}(\Delta)) \subset V_1'$.

For ii) we use again that $\tilde{X}$ is non-singular over $X \setminus \mathrm{Sing}(\Delta)$. So part i) induces an isomorphism

$$\varphi_* \left(\Omega^n_{X'/Y'}(\log \Delta')^{(\nu-1)} \otimes \omega_{X'/Y'} \right) \xrightarrow{\simeq} \tilde{\varphi}^* \pi_1^* \left(\Omega^n_{X/Y}(\log \Delta)^{(\nu-1)} \right) \otimes \omega_{\tilde{X}/Y'} .$$

The natural map $\tilde{\varphi}_* \omega_{\tilde{X}/Y'} \longrightarrow \omega_{X \times_Y Y'/Y'}$ and the projection formula give

$$\delta_* \left(\Omega^n_{X'/Y'}(\log \Delta')^{(\nu-1)} \otimes \omega_{X'/Y'} \right) \longrightarrow \pi_1^* \left(\Omega^n_{X/Y}(\log \Delta)^{(\nu-1)} \otimes \omega_{X/Y} \right),$$

and ii) follows by flat base change. □

Proof of 3.6. Applying 2.3, b), one finds a finite covering $\psi : Y' \to Y$ such that the family $f' : X' \to Y'$ is semi-stable in codimension one, hence (3.5.6) implies $\omega_{X'/Y'} = \Omega^n_{X'/Y'}(\log \Delta')$, whereas $\omega_{X/Y} \otimes f^* \mathcal{O}_Y(-S) \subset \Omega^n_{X/Y}(\log(\Delta))$. So 3.8, ii), gives a morphism of sheaves

$$f'_*(\omega^\nu_{X'/Y'}) \otimes \mathcal{O}_{Y'}(-\psi^* S) \longrightarrow \psi^* \left(f_*(\Omega^n_{X/Y}(\log \Delta)^{(\nu-1)} \otimes \omega_{X/Y}) \otimes \mathcal{O}_Y(-S) \right)$$

$$\longrightarrow \psi^* f_* \left(\Omega^n_{X/Y}(\log \Delta)^\nu \right)$$

By 3.4, iii), for some $\nu \gg 0$ the sheaf on the left hand side will be ample with respect to $\psi^{-1}(U_0)$, hence the sheaf on the right hand side has the same property. □

A positivity property, similar to the last one, will be expressed in terms of fibred products of the given family. It will be used in the proof of 1.4, iii). We do not need it in its full strength, just "up to codimension two in Y". Nevertheless, in order to be able to refer to [25], we formulate it in a more general setup.

Let $V \to U$ be a smooth family with $\omega_{V/U}$ f-semi-ample. By 2.6 we find a smooth birational model $\tilde{V} \to \tilde{U}$ whose compactification $f : X \to Y$ fits into the diagram (2.6.1):

$$
\begin{array}{ccccc}
X & \xleftarrow{\psi'} & X' & \xleftarrow{\sigma} & Z \\
\downarrow{\scriptstyle f} & & \downarrow{\scriptstyle f'} & \diagup{\scriptstyle g} & \\
Y & \xleftarrow{\psi} & Y' . &
\end{array}
$$

Let us choose any $\nu \geq 3$ such that

$$f^* f_* \omega^\nu_{X/Y} \longrightarrow \omega^\nu_{X/Y}$$

is surjective over $\tilde{V}$, and that the multiplication map

$$S^\eta(f_* \omega^\nu_{X/Y}) \longrightarrow f_* \omega^{\eta \cdot \nu}_{X/Y}$$

is surjective over $\tilde{U}$. By definition one has $\mathrm{Var}(f) = \mathrm{Var}(g)$. If $\mathrm{Var}(f) = \dim(Y)$, applying 3.4, i), to g one finds that the sheaf λ_ν, defined in 2.6, d), is of maximal Kodaira dimension. Hence some power of λ_ν is of the form $\mathcal{A}(D)$, for an ample invertible sheaf $\mathcal{A}$ on Y and for an effective divisor D on Y. We may assume moreover, that $D \geq S$ and, replacing the number N_ν in 2.6 by some multiple, that

$$\det(g_*\omega^\nu_{Z/Y'})^{N_\nu} = \mathcal{A}(D)^{\nu \cdot (\nu - 1) \cdot e}$$

where $e = \text{Max}\{e(\omega^\nu_F); F \text{ a fibre of } V \to U\}$.

Proposition 3.9. *For* $r = N_\nu \cdot \text{rank}(f_*\omega^\nu_{X/Y})$, *let* $X^{(r)}$ *denote a desingularization of the r-th fibre product* $X \times_Y \ldots \times_Y X$ *and let* $f^{(r)} : X^{(r)} \to Y$ *be the induced family. Then for all* β *sufficiently large and divisible the sheaf*

$$f^{(r)}_* \left(\Omega^{r \cdot n}_{X^{(r)}/Y}(\log \Delta)^{\beta \cdot \nu} \right) \otimes \mathcal{A}^{-\beta \cdot \nu \cdot (\nu - 2)}$$

is globally generated over some non-empty open subset U_0 *of* $\tilde{U}$, *and the sheaf*

$$\Omega^{r \cdot n}_{X^{(r)}/Y}(\log \Delta)^{\beta \cdot \nu} \otimes f^{(r)*}\mathcal{A}^{-\beta \cdot \nu \cdot (\nu - 2)}$$

is globally generated over $f^{(r)-1}(U_0)$.

Proof. By [25, 4.1], the sheaf

$$f^{(r)}_*(\omega^{\beta \cdot \nu}_{X^{(r)}/Y}) \otimes \mathcal{A}^{-\beta \cdot \nu \cdot (\nu - 2)} \otimes \mathcal{O}_Y(-\beta \cdot \nu \cdot (\nu - 1) \cdot D)$$

is globally generated over some open subset. However, by (3.5.6)

$$\omega^{\beta \cdot \nu}_{X^{(r)}/Y} \otimes f^*\mathcal{O}_Y(-\beta \cdot \nu \cdot S)$$

is contained in

$$\Omega^{r \cdot n}_{X^{(r)}/Y}(\log \Delta)^{\beta \cdot \nu}.$$

Since

$$\beta \cdot \nu \cdot (\nu - 1) \cdot D \geq \beta \cdot \nu \cdot S$$

one obtains 3.9, as stated. $\qquad\square$

4 Higgs Bundles and the Proof of 1.4

As in [24] and [25], in order to prove 1.4 we have to construct certain Higgs bundles, and we have to compare them to one, induced by a variation of Hodge structures. For 1.4, iii), we will just use the content of the second half of Sect. 3. For iv) we need in addition Kawamata's covering construction, as explained in 2.3. The reduction steps contained in the second half of Sect. 2 will be needed for 1.4, i).

302   Eckart Viehweg and Kang Zuo

So let U be a manifold and let Y be a smooth projective compactification with $Y \setminus U$ a normal crossing divisor. Starting with a smooth family $V \to U$ with $\omega_{V/U}$ relative semi-ample over U, we first choose a smooth projective compactification X of V, such that $V \to U$ extends to $f : X \to Y$.

In the first half of the section, we will work with good partial compactifications as defined in 2.1. Hence leaving out a codimension two subscheme of Y, we will assume that the divisor $S = Y \setminus U$ is smooth, that f is flat and that $\Delta = X \setminus V$ is a relative normal crossing divisor. The exact sequence (3.5.5) induces a filtration on the wedge product $\Omega^p_{X/Y}(\log \Delta)$, and thereby the tautological sequences

$$0 \to f^* \Omega^1_Y(\log S) \otimes \Omega^{p-1}_{X/Y}(\log \Delta) \to \mathfrak{gr}(\Omega^p_X(\log \Delta)) \to \Omega^p_{X/Y}(\log \Delta) \to 0 \,,$$

$$(4.0.1)$$

where

$$\mathfrak{gr}(\Omega^p_X(\log \Delta)) = \Omega^p_X(\log \Delta)/f^* \Omega^2_Y(\log S) \otimes \Omega^{p-2}_{X/Y}(\log \Delta) \,.$$

Given an invertible sheaf $\mathcal{L}$ on X we will study in this section various sheaves of the form

$$F_0^{p,q} := R^q f_*(\Omega^p_{X/Y}(\log \Delta) \otimes \mathcal{L}^{-1})/_{\text{torsion}}$$

together with the edge morphisms

$$\tau^0_{p,q} : F_0^{p,q} \longrightarrow F_0^{p-1,q+1} \otimes \Omega^1_Y(\log S) \,,$$

induced by the exact sequence (4.0.1), tensored with $\mathcal{L}^{-1}$.

First we have to extend the base change properties for direct images, studied in 3.8, ii), to higher direct images.

Lemma 4.1. *Keeping the notations and assumptions from 3.7, let Y_1' be the largest open subset in Y' with $X \times_Y Y_1'$ normal. We write*

$$\iota : \psi^* \Omega^1_Y(\log S) \longrightarrow \Omega^1_{Y'}(\log S')$$

for natural inclusion, and we consider an invertible sheaf $\mathcal{L}$ on X, and its pullback $\mathcal{L}' = \psi'^* \mathcal{L}$ to X'.

Then for all p and q, there are morphisms

$$\psi^* F_0^{p,q} \xrightarrow{\zeta_{p,q}} F_0'^{p,q} := R^q f_*'(\Omega^p_{X'/Y'}(\log \Delta') \otimes \mathcal{L}'^{-1})/_{\text{torsion}} \,,$$

whose restriction to Y_1' are isomorphisms, and for which the diagram

$$
\begin{array}{ccc}
\psi^* F_0^{p,q} & \xrightarrow{\psi^*(\tau^0_{p,q})} & \psi^* F_0^{p-1,q+1} \otimes \Omega^1_Y(\log S)) \\
\zeta_{p,q} \downarrow & & \zeta_{p-1,q+1} \otimes \iota \downarrow \\
F_0'^{p,q} & \xrightarrow{\tau'^0_{p,q}} & F_0'^{p-1,q+1} \otimes \Omega^1_{Y'}(\log S')
\end{array}
\qquad (4.1.1)
$$

commutes. Here $\tau'^0_{p,q}$ is again the edge morphism induced by the exact sequence on X', corresponding to (4.0.1) and tensored with $\mathcal{L}'^{-1}$.

Proof. We use the notations from (3.7.1), i.e.

$$X' \xrightarrow{\ \varphi\ } \tilde{X} \xrightarrow{\ \tilde{\varphi}\ } X \times_Y Y' \xrightarrow{\ \pi_1\ } X$$

with f', $\tilde{f}$, π_2, f, $Y' \xrightarrow{\ \psi\ } Y$

and $\psi' = \varphi \circ \tilde{\varphi} \circ \pi_1$. As in the proof of 3.8, in order to show the existence of the morphisms $\zeta_{p,q}$ and the commutativity of the diagram (4.1.1) we may enlarge S and S' to include the discriminant loci, hence assume that

$$\psi^* \Omega_Y^1(\log S) = \Omega_{Y'}^1(\log S') \ .$$

By the generalized Hurwitz formula [6, 3.21],

$$\psi'^* \Omega_X^p(\log \Delta) \subset \Omega_{X'}^p(\log \Delta') \ ,$$

and by [5, Lemme 1.2],

$$R^q \varphi_* \Omega_{X'}^p(\log \Delta') = \begin{cases} \tilde{\varphi}^* \pi_1^* \Omega_X^p(\log \Delta) & \text{for } q = 0 \\ 0 & \text{for } q > 0 \ . \end{cases}$$

The tautological sequence

$$0 \to f'^* \Omega_{Y'}^1(\log(S')) \to \Omega_{X'}^1(\log \Delta') \to \Omega_{X'/Y'}^1(\log \Delta') \to 0$$

defines a filtration on $\Omega_{X'}^p(\log \Delta')$, with subsequent quotients isomorphic to

$$f'^* \Omega_{Y'}^\ell(\log S') \otimes \Omega_{X'/Y'}^{p-\ell}(\log \Delta') \ .$$

Induction on p allows to deduce that

$$R^q \varphi_* \Omega_{X'/Y'}^p(\log \Delta') = \begin{cases} \tilde{\varphi}^* \pi_1^* \Omega_{X/Y}^p(\log \Delta) & \text{for } q = 0 \\ 0 & \text{for } q > 0 \ . \end{cases} \qquad (4.1.2)$$

On the other hand, the inclusion $\mathcal{O}_{Z \times_Y Y'} \to \tilde{\varphi}_* \mathcal{O}_{\tilde{Z}}$ and flat base change gives

$$\psi^* F_0^{p,q} = \psi^* R^q f_* \left(\Omega_{X/Y}^p(\log \Delta) \otimes \mathcal{L}^{-1} \right)$$
$$\xrightarrow{\ \cong\ } R^q \pi_{2*} \left(\pi_1^* \left(\Omega_{X/Y}^p(\log \Delta) \otimes \mathcal{L}^{-1} \right) \right)$$
$$\to R^q \tilde{f}_* \left(\tilde{\varphi}^* \left(\pi_1^* \left(\Omega_{X/Y}^p(\log \Delta) \otimes \mathcal{L}^{-1} \right) \right) \right) = F_0'^{p,q} \ , \qquad (4.1.3)$$

hence $\zeta_{p,q}$. The second morphism in (4.1.3) is an isomorphism on the largest open subset where $\tilde{\varphi}$ is an isomorphism, in particular on $\tilde{f}^{-1}(Y_1')$.

The way we obtained (4.1.2) the morphisms are obviously compatible with the different tautological sequences. Since we assumed S to contain the discriminant locus, the pull back of (4.0.1) to $\tilde{X}$ is isomorphic to

$$0 \to \varphi_* \left(f'^* \Omega^1_{Y'} (\log S') \otimes \Omega^{p-1}_{X'/Y'} (\log \Delta') \right) \to \varphi_* \left(\mathfrak{gr} \left(\Omega^p_{X'} (\log \Delta') \right) \right)$$
$$\to \varphi_* \left(\Omega^p_{X'/Y'} (\log \Delta') \right) \to 0 ,$$

and the diagram (4.1.1) commutes. $\qquad\square$

Remark 4.2. If $\psi : Y' \to Y$ is any smooth morphism, then again $X \times_Y Y'$ non-singular. The compatibility of the $F^{p,q}_0$ with pullback, i.e. the existence of an isomorphism $\zeta_{p,q} : \psi^* F^{p,q}_0 \to F'^{p,q}_0$, and the commutativity of (4.1.1) is also guaranteed, in this case. In fact, both φ and $\tilde{\varphi}$ are isomorphisms, as well as the two morphisms in (4.1.3).

Corollary 4.3. *Keeping the assumptions from 4.1, assume that $X \times_Y Y'$ is normal. Then the image of*

$$F'^{p,q}_0 \xrightarrow{\tau'^0_{p,q}} F'^{p-1,q+1}_0 \otimes \Omega^1_{Y'} (\log S')$$

lies in $F'^{p-1,q+1}_0 \otimes \psi^* (\Omega^1_Y (\log S))$.

In the sequel we will choose $\mathcal{L} = \Omega^n_{X/Y} (\log \Delta)$. Let us consider first the case that for some $\nu \gg 1$ and for some invertible sheaf $\mathcal{A}$ on Y the sheaf

$$\mathcal{L}^\nu \otimes f^* \mathcal{A}^{-\nu} \quad \text{is globally generated over} \quad V_0 = f^{-1}(U_0) , \qquad (4.3.1)$$

for some open dense subset U_0 of Y.

We will recall some of the constructions performed in [25, §6]. Let H denote the zero divisor of a section of $\mathcal{L}^\nu \otimes f^* \mathcal{A}^{-\nu}$, whose restriction to a general fibre of f is non-singular. Let T denote the closure of the discriminant of $H \cap V \to U$. Leaving out some more codimension two subschemes, we may assume that $S + T$ is a smooth divisor. We will write $\Sigma = f^* T$ and we keep the notation $\Delta = f^*(S)$.

Let $\delta : W \to X$ be a blowing up of X with centers in $\Delta + \Sigma$ such that $\delta^* (H + \Delta + \Sigma)$ is a normal crossing divisor. We write

$$\mathcal{M} = \delta^* \left(\Omega^n_{X/Y} (\log \Delta) \otimes f^* \mathcal{A}^{-1} \right) .$$

Then for $B = \delta^* H$ one has $\mathcal{M}^\nu = \mathcal{O}_W(B)$. As in [6, §3], one obtains a cyclic covering of W, by taking the ν-th root out of B. We choose Z to be a desingularization of this covering and we denote the induced morphisms by $g : Z \to Y$, and $h : W \to Y$. Writing $\Pi = g^{-1}(S \cup T)$, the restriction of g to $Z_0 = Z \setminus \Pi$ will be smooth.

For the normal crossing divisor B we define

$$\mathcal{M}^{(-1)} = \mathcal{M}^{-1} \otimes \mathcal{O}_W \left(\left[\frac{B}{\nu} \right] \right), \quad \text{and} \quad \mathcal{L}^{(-1)} = \delta^* (\mathcal{L}^{-1}) \otimes \mathcal{O}_W \left(\left[\frac{B}{\nu} \right] \right) .$$

In particular the cokernel of the inclusion $\delta^* \mathcal{L}^{-1} \subset \mathcal{L}^{(-1)}$ lies in $h^{-1}(S+T)$. The sheaf

$$
\begin{aligned}
F^{p,q} &= R^q h_* (\delta^* (\Omega^p_{X/Y}(\log \Delta)) \otimes \mathcal{M}^{(-1)}) \otimes \mathcal{A}^{-1}/\text{torsion} \\
&= R^q h_* (\delta^* (\Omega^p_{X/Y}(\log \Delta)) \otimes \mathcal{L}^{(-1)})/\text{torsion}
\end{aligned}
$$

contains the sheaf $F^{p,q}_0$ and both are isomorphic outside of $S+T$. The edge morphism

$$
\tau_{p,q} : F^{p,q} \longrightarrow F^{p-1,q+1} \otimes \Omega^1_Y(\log S)
$$

given by the tautological exact sequence

$$
\begin{aligned}
0 &\to h^* \Omega^1_Y(\log S) \otimes \delta^* (\Omega^{p-1}_{X/Y}(\log \Delta)) \otimes \mathcal{L}^{(-1)} \\
&\to \delta^* (\mathfrak{gr}(\Omega^p_X(\log \Delta))) \otimes \mathcal{L}^{(-1)} \to \delta^* (\Omega^p_{X/Y}(\log \Delta)) \otimes \mathcal{L}^{(-1)} \to 0
\end{aligned}
$$

is compatible with $\tau^0_{p,q}$. Let us remark, that the sheaves $F^{p,q}$ depend on the choice of the divisor H and they can only be defined assuming (4.3.1).

Up to now, we constructed two Higgs bundles

$$
F_0 = \bigoplus F^{p,q}_0 \xrightarrow{\subset} F = \bigoplus F^{p,q} .
$$

We will see below, that $\mathcal{A} \otimes F$ can be compared with a Higgs bundle E, given by a variation of Hodge structures. This will allow to use the negativity of the kernel of Kodaira-Spencer maps (see [26]), to show that $\mathrm{Ker}(\tau_{p,q})^\vee$ is big.

By [4], for all $k \geq 0$, the local constant system $R^k g_* \mathbb{C}_{Z_0}$ gives rise to a local free sheaf $\mathcal{V}_k$ on Y with the Gauß-Manin connection

$$
\nabla : \mathcal{V}_k \longrightarrow \mathcal{V}_k \otimes \Omega^1_Y(\log(S+T)) .
$$

We assume that $\mathcal{V}_k$ is the quasi-canonical extension of

$$
(R^k g_* \mathbb{C}_{Z_0}) \otimes_{\mathbb{C}} \mathcal{O}_{Y \backslash (S \cup T)} ,
$$

i.e. that the real part of the eigenvalues of the residues around the components of $S+T$ lie in $[0,1)$.

Since we assumed $S+T$ to be non-singular, $\mathcal{V}_k$ carries a filtration $\mathcal{F}^p$ by subbundles (see [18]). So the induced graded sheaves $E^{p,k-p}$ are locally free, and they carry a Higgs structure with logarithmic poles along $S+T$. Let us denote it by

$$
(\mathfrak{gr}_{\mathcal{F}}(\mathcal{V}_k), \mathfrak{gr}_{\mathcal{F}}(\nabla)) = (E, \theta) = \left(\bigoplus_{q=0}^{k} E^{k-q,q} , \bigoplus_{q=0}^{k} \theta_{k-q,q} \right) .
$$

As well-known (see for example [7], page 130) the bundles $E^{p,q}$ are given by

$$
E^{p,q} = R^q g_* \Omega^p_{Z/Y}(\log \Pi).
$$

Writing again $\mathfrak{gr}(\_)$ for "modulo the pullback of 2-forms on Y", the Gauß-Manin connection is the edge morphism of

$$0 \to g^*\Omega^1_Y(\log(S+T)) \otimes \Omega^{\bullet-1}_{Z/Y}(\log \Pi) \to \mathfrak{gr}(\Omega^\bullet_Z(\log \Pi))$$
$$\to \Omega^\bullet_{Z/Y}(\log \Pi) \to 0 .$$

Hence the Higgs maps

$$\theta_{p,q} : E^{p,q} \longrightarrow E^{p-1,q+1} \otimes \Omega^1_Y(\log(S+T))$$

are the edge morphisms of the tautological exact sequences

$$0 \to g^*\Omega^1_Y(\log(S+T)) \otimes \Omega^{p-1}_{Z/Y}(\log \Pi)$$
$$\to \mathfrak{gr}(\Omega^p_Z(\log \Pi)) \to \Omega^p_{Z/Y}(\log \Pi) \to 0 .$$

In the sequel we will write $T_*(-\log **)$ for the dual of $\Omega^1_*(\log **)$.

Lemma 4.4. *Under the assumption (4.3.1) and using the notations introduced above, let*

$$\iota : \Omega^1_Y(\log S) \longrightarrow \Omega^1_Y(\log(S+T))$$

be the natural inclusion. Then there exist morphisms $\rho_{p,q} : \mathcal{A} \otimes F^{p,q} \to E^{p,q}$ such that:

i) *The diagram*

$$
\begin{array}{ccc}
E^{p,q} & \xrightarrow{\theta_{p,q}} & E^{p-1,q+1} \otimes \Omega^1_Y(\log(S+T)) \\
\rho_{p,q} \uparrow & & \uparrow \rho_{p-1,q+1} \otimes \iota \\
\mathcal{A} \otimes F^{p,q} & \xrightarrow{\mathrm{id}_\mathcal{A} \otimes \tau_{p,q}} & \mathcal{A} \otimes F^{p-1,q+1} \otimes \Omega^1_Y(\log S) .
\end{array}
$$

commutes.

ii) $F^{n,0}$ *has a section $\mathcal{O}_Y \to F^{n,0}$, which is an isomorphism on $Y \setminus (S \cup T)$.*

iii) $\tau_{n,0}$ *induces a morphism*

$$\tau^\vee : T_Y(-\log S) = (\Omega^1_Y(\log S))^\vee \longrightarrow F^{n,0^\vee} \otimes F^{n-1,1} ,$$

which coincides over $Y \setminus (S \cup T)$ with the Kodaira-Spencer map

$$T_Y(-\log S) \longrightarrow R^1 f_* T_{X/Y}(-\log \Delta) .$$

iv) $\rho_{n,0}$ *is injective. If the general fibre of f is canonically polarized, then the morphisms $\rho_{n-m,m}$ are injective, for all m.*

v) *Let $\mathcal{K}^{p,q} = \mathrm{Ker}(E^{p,q} \xrightarrow{\theta_{p,q}} E^{p-1,q+1} \otimes \Omega^1_Y(\log(S+T)))$. Then the dual $(\mathcal{K}^{p,q})^\vee$ is weakly positive with respect to some open dense subset of Y.*

Base Spaces of Non-Isotrivial Families 307

vi) *The composite*

$$\theta_{n-q+1,q-1} \circ \cdots \circ \theta_{n,0} : E^{n,0} \longrightarrow E^{n-q,q} \otimes \bigotimes^{q} \Omega_Y^1(\log(S+T))$$

factors like

$$E^{n,0} \xrightarrow{\theta^q} E^{n-q,q} \otimes S^q \Omega_Y^1(\log(S+T)) \xrightarrow{\subset} E^{n-q,q} \otimes \bigotimes^{q} \Omega_Y^1(\log(S+T)) \ .$$

Proof. The properties i)–iv) have been verified in [25, 6.3] in case the general fibre is canonically polarized. So let us just sketch the arguments.

By [6] (see also [25, 6.2]) the sheaf

$$R^q h_* \left(\Omega_{W/Y}^p(\log(B + \delta^*\Delta + \delta^*\Sigma)) \otimes \mathcal{M}^{(-1)} \right)$$

is a direct factor of $E^{p,q}$. The morphism $\rho_{p,q}$ is induced by the natural inclusions

$$\delta^* \Omega_{X/Y}^p(\log \Delta) \to \delta^* \Omega_{X/Y}^p(\log(\Delta + \Sigma))$$
$$\to \Omega_{W/Y}^p(\log(\delta^*\Delta + \delta^*\Sigma)) \to \Omega_{W/Y}^p(\log(B + \delta^*\Delta + \delta^*\Sigma)) \ ,$$

tensored with $\mathcal{M}^{(-1)} = \mathcal{L}^{(-1)} \otimes h^*\mathcal{A}$.

Such an injection also exist for Y replaced by $\mathrm{Spec}(\mathbb{C})$. Since the different tautological sequences are compatible with those inclusions one obtains i). Over $Y \setminus (S \cup T)$ the kernel of $\rho_{n-m,m}$ is a quotient of the sheaf

$$R^{m-1}(h|_B)_* \left(\Omega_{B/Y}^{n-m-1} \otimes \mathcal{M}^{-1}|_B \right) \ .$$

In particular $\rho_{n,0}$ is injective. The same holds true for all the $\rho_{n-m,m}$ in case $\mathcal{M}$ is fibre wise ample, by the Akizuki-Kodaira-Nakano vanishing theorem.

By definition

$$F^{n,0} = h_* \left(\delta^*(\Omega_{X/Y}^n(\log \Delta)) \otimes \mathcal{L}^{(-1)} \right) = h_* \mathcal{O}_W \left(\left[\frac{B}{\nu} \right] \right) \ ,$$

and ii) holds true.

For iii), recall that over $Y \setminus (S \cup T)$ the morphism

$$\delta^*(\mathcal{L} \otimes f^*\mathcal{A}^{-1}) = \mathcal{M}^{-1} \to \mathcal{M}^{(-1)}$$

is an isomorphism. By the projection formula the morphism $\tau_{n,0}|_{Y \setminus (S \cup T)}$ is the restriction of the edge morphism of the short exact sequence

$$0 \to f^*\Omega_U^1 \otimes \Omega_{V/U}^{n-1} \otimes \mathcal{L}^{-1} \to \mathfrak{gr}(\Omega_V^n) \otimes \mathcal{L}^{-1} \to \Omega_{V/U}^n \otimes \mathcal{L}^{-1} \to 0 \ .$$

The sheaf on the right hand side is $\mathcal{O}_V$ and the one on the left hand side is $f^*\Omega_U^1 \otimes T_{V/U}$. For $r = \dim(U)$, tensoring the exact sequence with

$$f^*T_U = f^*(\Omega_U^{r-1} \otimes \omega_U^{-1})$$

308   Eckart Viehweg and Kang Zuo

and dividing by the kernel of the wedge product

$$f^*\Omega_U^1 \otimes f^*(\Omega_U^{r-1} \otimes \omega_U^{-1}) = f^*\Omega_U^1 \otimes f^*T_U \longrightarrow \mathcal{O}_V$$

on the left hand side, one obtains an exact sequence

$$0 \longrightarrow T_{V/U} \longrightarrow \mathcal{G} \longrightarrow f^*T_U \longrightarrow 0 , \qquad (4.4.1)$$

where $\mathcal{G}$ is a quotient of $\mathfrak{gr}(\Omega_V^n)\otimes\omega_V^{-1}\otimes f^*\Omega_U^{r-1}$. By definition, the restriction to $Y \setminus (S \cup T)$ of the morphism considered in iii) is the first edge morphism in the long exact sequence, obtained by applying $R^\bullet f_*$ to (4.4.1).

The wedge product induces a morphism

$$\Omega_V^n \otimes \omega_V^{-1} \otimes f^*\Omega_U^{r-1} \longrightarrow \Omega_V^{n+r-1} \otimes \omega_V^{-1} = T_V .$$

This morphism factors through $\mathcal{G}$. Hence the exact sequence (4.4.1) is isomorphic to the tautological sequence

$$0 \longrightarrow T_{V/U} \longrightarrow T_V \longrightarrow f^*T_U \longrightarrow 0 . \qquad (4.4.2)$$

The edge morphism $T_U \to R^1 f_* T_{V/U}$ of (4.4.2) is the Kodaira-Spencer map.

In order to prove v), we use as in the proof of 2.3, b), Kawamata's covering construction to find a non-singular finite covering $\rho : Y' \to Y$ such that for some desingularization Z' of $Z \times_Y Y'$ the induced variation of Hodge structures has uni-potent monodromy, and such that $g' : Z' \to Y'$ is semi-stable.

From 4.1, applied to Z, $\mathcal{O}_Z$ and Π instead of X, $\mathcal{L}$ and Δ one obtains a commutative diagram

$$
\begin{array}{ccc}
\rho^* E^{p,q} & \xrightarrow{\rho^*\theta_{p,q}} & \rho^* E^{p-1,q+1} \otimes \Omega_Y^1(\log S + T) \\
\cap \downarrow & & \cap \downarrow \\
E'^{p,q} & \xrightarrow{\theta'_{p,q}} & E'^{p-1,q+1} \otimes \Omega_{Y'}^1(\log S') ,
\end{array}
$$

where $S' = \psi^*(S+T)$, where $E'^{p,q} = R^q g'_* \Omega_{Z'/Y'}^p(\log \Pi')$, and where $\theta'_{p,q}$ is the edge-morphism.

In particular the pullback of the kernel of $\theta_{p,q}$, the sheaf $\rho^*\mathcal{K}^{p,q}$, lies in the kernel $\mathcal{K}'^{p,q}$ of $\theta'_{p,q}$. Leaving out some codimension two subschemes of Y and Y', we may assume that $\mathcal{K}'^{p,q}$ is a subbundle of $E'^{p,q}$. Choose a smooth extension $\bar{Y}'$ of Y' such that the closure of $S' \cup (\bar{Y}' - Y')$ is a normal crossing divisor, and let $\bar{E}'^{p,q}$ be the Higgs bundle, corresponding to the canonical extension of the variation of Hodge structures. For some choice of the compactification $\mathcal{K}'^{p,q}$ will extend to a subbundle $\bar{\mathcal{K}}'^{p,q}$ of $\bar{E}'^{p,q}$. By [26, 1.2], the dual $(\bar{\mathcal{K}}'^{p,q})^\vee$ is numerically effective, hence weakly positive. Thereby $\rho^*(\mathcal{K}^{p,q})^\vee$ is weakly positive over some open subset, and the compatibility of weak positivity with pullback shows v).

For vi) one just has to remark that on page 12 of [19] it is shown that $\theta \wedge \theta = 0$ for

$$\theta = \bigoplus_{q=0}^{n} \theta_{n-q,q} \ .$$

$\square$

Corollary 4.5. *Assume (4.3.1) holds true for some ample invertible sheaf $\mathcal{A}$ and for some $\nu \gg 1$. Assume moreover that there exists a locally free subsheaf Ω of $\Omega_Y^1(\log S)$ such that $\mathrm{id}_{\mathcal{A}} \otimes \tau_{p,q}$ factors through*

$$\mathcal{A} \otimes F^{n-q,q} \longrightarrow \mathcal{A} \otimes F^{n-q-1,q+1} \otimes \Omega \ ,$$

for all q. Then for some $0 < m \leq n$ there exists a big coherent subsheaf $\mathcal{P}$ of $S^m(\Omega)$.

Proof. Using the notations from 4.4, write $\mathcal{A} \otimes \tilde{F}^{n-q,q} = \rho_{n-q,q}(\mathcal{A} \otimes F^{n-q,q})$. By 4.4, i), and by the choice of Ω

$$\theta_{n-q,q}(\mathcal{A} \otimes \tilde{F}^{n-q,q}) \subset \mathcal{A} \otimes \tilde{F}^{n-q-1,q+1} \otimes \Omega \ .$$

By 4.4, ii) and iv), there is a section $\mathcal{O}_Y \to F^{n,0} \simeq \tilde{F}^{n,0}$, generating $\tilde{F}^{n,0}$ over $Y \setminus (S \cup T)$, and by 4.4, v), $\mathcal{A} \otimes \tilde{F}^{n,0}$ cannot lie in the kernel of $\theta_{n,0}$. Hence the largest number m with $\theta^m(\mathcal{A} \otimes \tilde{F}^{n,0}) \neq 0$ satisfies $1 \leq m \leq n$. By the choice of m

$$\theta^{m+1}(\mathcal{A} \otimes \tilde{F}^{n,0}) = 0 \ ,$$

and 4.4, vi) implies that $\theta^m(\mathcal{A} \otimes \tilde{F}^{n,0})$ lies in

$$\left(\mathcal{K}^{n-m,m} \cap \mathcal{A} \otimes \tilde{F}^{n-m,m} \right) \otimes S^m(\Omega) \subset \mathcal{K}^{n-m,m} \otimes S^m(\Omega) \ .$$

We obtain morphisms of sheaves

$$\mathcal{A} \otimes \left(\mathcal{K}^{n-m,m} \right)^{\vee} \xrightarrow{\subset} \mathcal{A} \otimes \tilde{F}^{n,0} \otimes \left(\mathcal{K}^{n-m,m} \right)^{\vee} \xrightarrow{\neq 0} S^m(\Omega) \ .$$

By 4.4, v), the sheaf on the left hand side is big, hence its image $\mathcal{P} \subset S^m(\Omega)$ is big as well. $\square$

Proof of 1.4, iii). Let Y be the given smooth projective compactification of U with $Y \setminus U$ a normal crossing divisor. In order to prove iii) we may blow up Y. Hence given a morphism $V \to U$ with $\omega_{V/U}$ semi-ample, by abuse of notations we will assume that $V \to U$ itself fits into the diagram (2.6.1). So we may apply 3.9 and replace X by $X^{(r)}$ for r sufficiently large. In this way we loose control on the dimension of the fibres, but we enforce the existence of a family for which (4.3.1) holds true. We obtain the big coherent subsheaf $\mathcal{P}$, asked for in 1.4, ii), by 4.5, applied to $\Omega = \Omega_Y^1(\log S)$. $\square$

310   Eckart Viehweg and Kang Zuo

In order to prove 1.4, iv), we have to argue in a slightly different way, since we are not allowed to perform any construction, changing the dimension of the general fibre.

Proof of 1.4, iv). We start again with a smooth projective compactifications X of V, such that $V \to U$ extends to $f : X \to Y$. Recall that for $\mathcal{L} = \Omega^n_{X/Y}(\log \Delta)$, we found in 3.6 some $\nu \gg 1$ and an open dense subset U_0 of Y such that

$$f_* \mathcal{L}^\nu = f_* \Omega^n_{X/Y}(\log \Delta)^\nu \quad \text{is ample with respect to} \quad U_0 \quad (4.5.1)$$

$$\text{and} \quad f^* f_* \mathcal{L}^\nu \longrightarrow \mathcal{L}^\nu \quad \text{is surjective over} \quad V_0 = f^{-1}(U_0) . \quad (4.5.2)$$

Given a very ample sheaf $\mathcal{A}$ on Y, lemma 1.2 implies that for some μ' the sheaf $\mathcal{A}^{-1} \otimes S^{\mu'}(f_* \mathcal{L}^\nu)$ is globally generated over U_0. Lemma 2.3, a), allows to find some smooth covering $\psi : Y' \to Y$ such that $\psi^* \mathcal{A} = \mathcal{A}'^\mu$ for an invertible ample sheaf $\mathcal{A}'$ on Y' and for $\mu = \mu' \cdot \nu$. We will show, that for this covering 1.4, iv), holds true. To this aim, we are allowed to replace Y by the complement of a codimension two subscheme, hence assume that $f : X \to Y$ is a good partial compactification, as defined in 2.1. In particular, we can assume $f_* \mathcal{L}^\nu$ to be locally free. Then the sheaf $\mathcal{L}^\mu \otimes f^* \mathcal{A}^{-1}$ is globally generated over $f^{-1}(U_0)$. Let H be the zero divisor of a general section of this sheaf, and let T denote the non-smooth locus of $H \to Y$. Leaving out some additional codimension two subset, we may assume that the discriminant $\Delta(Y'/Y)$ does not meet T and the boundary divisor S, hence in particular that the fibred product $X' = X \times_Y Y'$ is smooth. If $\psi' : X' \to X$ and $f' : X' \to Y'$ denote the projections, we write $S' = \psi^* S$, $T' = \psi^* T$, $\Delta' = \psi'^*(\Delta) = h^*(S')$,

$$\mathcal{L}' = \Omega^n_{X'/Y'}(\log \Delta') = \psi'^* \mathcal{L} ,$$

and so on. The sheaf

$$\mathcal{L}'^\mu \otimes f'^* \mathcal{A}'^{-\mu} = \psi'^* \left(\mathcal{L}^\mu \otimes f^* \mathcal{A}^{-1} \right)$$

is globally generated over $\psi^{-1}(V_0)$ and (4.3.1) holds true on Y'. So we can repeat the construction made above, this time over Y' and for the divisor $H' = \psi'^* H$, to obtain the sheaf

$$F'^{p,q} = R^q h'_* \delta'^* \left(\Omega^p_{X'/Y'}(\log \Delta') \otimes \mathcal{L}'^{(-1)} \right) / \text{torsion} ,$$

together with the edge morphism

$$\tau'_{p,q} : F'^{p,q} \longrightarrow F'^{p-1,q+1} \otimes \Omega^1_{Y'}(\log S') ,$$

induced by the exact sequence

$$0 \to h'^* \Omega^1_{Y'}(\log S') \otimes \delta'^* \left(\Omega^{p-1}_{X'/Y'}(\log \Delta') \right) \otimes \mathcal{L}'^{(-1)}$$

$$\to \delta'^* \left(\mathfrak{gr} \left(\Omega^p_{X'}(\log \Delta') \right) \right) \otimes \mathcal{L}'^{(-1)} \to \delta'^* \left(\Omega^p_{X'/Y'}(\log \Delta') \right) \otimes \mathcal{L}'^{(-1)} \to 0 .$$

Returning to the notations from 4.1, the sheaf $F_0'^{p,q}$ defined there is a subsheaf of $F'^{p,q}$, both are isomorphic outside of $S' + T'$ and $\tau_{p,q}'$ commutes with $\tau'^0_{p,q}$. By 4.3 the image of $\tau_{p,q}'$ lies in $\psi^*(\Omega_Y^1) \otimes \mathcal{O}_{Y'}(*(S' + T'))$, hence in

$$\left(\psi^*\left(\Omega_Y^1\right) \otimes \mathcal{O}_{Y'}(*(S'+T'))\right) \cap \Omega_{Y'}(\log S') = \psi^*\left(\Omega_Y^1(\log S)\right) .$$

By 4.5, for some $1 \leq m \leq n$ the m-th symmetric product of the sheaf

$$\Omega = \psi^*\left(\Omega_Y^1(\log S)\right)$$

contains a big coherent subsheaf $\mathcal{P}$, as claimed. $\qquad\square$

Assume from now on, that the fibres of the smooth family $V \to U$ are canonically polarized, and let $f : X \to Y$ be a partial compactification. The injectivity of $\rho_{n-m,m}$ in 4.4, iv), gives another method to bound the number m in 1.4, iii) and to prove 1.4, ii). For i) we will use in addition, the diagram (2.8.1).

Lemma 4.6. *Using the notations from 4.1, the composite* $\tau^0_{n-q+1,q-1} \circ \cdots \circ \tau^0_{n,0}$ *factors like*

$$F_0^{n,0} = \mathcal{O}_Y \xrightarrow{\tau_0^q} F_0^{n-q,q} \otimes S^q\left(\Omega_Y^1(\log(S))\right) \overset{\subset}{\longrightarrow} F_0^{n-q,q} \otimes \bigotimes^q \Omega_Y^1(\log(S)) .$$

Proof. The equality $F_0^{n,0} = \mathcal{O}_Y$ is obvious by definition. Moreover all the sheaves in 4.6 are torsion free, hence it is sufficient to verify the existence of τ_0^q on some open dense subset. So we may replace Y by an affine subscheme, and (4.3.1) holds true for $\mathcal{A} = \mathcal{O}_Y$. By 4.4, iv), the sheaves $F_0^{p,q}$ embed in the sheaves $E^{p,q}$, in such a way that $\theta_{p,q}$ restricts to $\tau_{p,q}^0$. One obtains 4.6 from 4.4, vi). $\qquad\square$

Proof of 1.4, i) and ii). Replacing Y by the complement of a codimension two subscheme, we may choose a good partial compactification $f : X \to Y$ of $V \to U$. Define

$$\mathcal{N}_0^{p,q} = \mathrm{Ker}\left(\tau_{p,q}^0 : F_0^{p,q} \to F_0^{p-1,q+1} \otimes \Omega_Y^1(\log S)\right) .$$

Claim 4.7. Assume (4.3.1) to hold true for some invertible sheaf $\mathcal{A}$, and let $(\mathcal{N}_0^{p,q})^\vee$ be the dual of the sheaf $\mathcal{N}_0^{p,q}$. Then $\mathcal{A}^{-1} \otimes (\mathcal{N}_0^{p,q})^\vee$ is weakly positive.

Proof. Recall that under the assumption (4.3.1) we have considered above the slightly different sheaf

$$F^{p,q} = R^q h_*\left(\delta^*(\Omega_{X/Y}^p(\log \Delta)) \otimes \mathcal{L}^{(-1)}\right)/\text{torsion} ,$$

for $\delta^*\mathcal{L}^{-1} \subset \mathcal{L}^{(-1)}$. So $F_0^{p,q}$ is a subsheaf of $F^{p,q}$ of full rank. The compatibility of $\tau_{p,q}^0$ and $\tau_{p,q}$ implies that $\mathcal{N}_0^{p,q}$ is a subsheaf of

$$\mathcal{N}^{p,q} = \mathrm{Ker}\left(\tau_{p,q} : F^{p,q} \longrightarrow F^{p-1,q+1} \otimes \Omega_Y^1(\log S)\right) .$$

312   Eckart Viehweg and Kang Zuo

of maximal rank. Hence the induced morphism

$$(\mathcal{N}^{p,q})^\vee \to (\mathcal{N}_0^{p,q})^\vee$$

is an isomorphism over some dense open subset. By 4.4, iv), the sheaf $\mathcal{A} \otimes F^{p,q}$ is a subsheaf of $E^{p,q}$ and by 4.4, i), the restriction $\theta_{p,q}|_{F^{p,q}}$ coincides with $\mathrm{id}_\mathcal{A} \otimes \tau_{p,q}$. Using the notations from 4.4, v), one obtains

$$\mathcal{A} \otimes \mathcal{N}^{p,q} = \mathcal{A} \otimes F^{p,q} \cap \mathcal{K}^{p,q} \overset{\subseteq}{\longrightarrow} \mathcal{K}^{p,q} . \tag{4.6.1}$$

By 4.4, v), the dual sheaf $(\mathcal{K}^{p,q})^\vee$ is weakly positive. (4.6.1) induces morphisms

$$(\mathcal{K}^{p,q})^\vee \longrightarrow \mathcal{A}^{-1} \otimes (\mathcal{N}^{p,q})^\vee \longrightarrow \mathcal{A}^{-1} \otimes (\mathcal{N}_0^{p,q})^\vee,$$

surjective over some dense open subset, and we obtain 4.7.   □

Claim 4.8.

  i) If $\mathrm{Var}(f) = \dim(Y)$, then $(\mathcal{N}_0^{p,q})^\vee$ is big.
 ii) In general, for some $\alpha > 0$ and for some invertible sheaf λ of Kodaira dimension $\kappa(\lambda) \geq \mathrm{Var}(f)$ the sheaf

$$S^\alpha \left((\mathcal{N}_0^{p,q})^\vee\right) \otimes \lambda^{-1}$$

  is generically generated.

Proof. Let us consider as in 3.7 and 4.1 some finite morphism $\psi : Y' \to Y$, a desingularization X' of $X \times_Y Y'$, the induced morphisms $\psi' : X' \to X$ and $f' : X' \to Y'$, $S' = \psi^{-1}(S)$, and $\Delta' = \psi'^{-1}\Delta$. In 4.1 we constructed an injection

$$\psi^*(F_0^{p,q}) \overset{\zeta_{p,q}}{\longrightarrow} F_0'^{p,q} = R^q f'_* \left(\Omega_{X'/Y'}^p(\log \Delta') \otimes \mathcal{L}'^{-1}\right)/\text{torsion} ,$$

compatible with the edge morphisms $\tau_{p,q}^0$ and $\tau'^0_{p,q}$. Thereby we obtain an injection

$$\psi^*\mathcal{N}_0^{p,q} \overset{\zeta'_{p,q}}{\longrightarrow} \mathcal{N}_0'^{p,q} := \mathrm{Ker}\left(\tau'^0_{p,q}\right) .$$

In case $X \times_Y Y'$ is non-singular, $\zeta_{p,q}$ and hence $\zeta'_{p,q}$ are isomorphisms.

If $\mathrm{Var}(f) = \dim(Y)$, the conditions (4.5.1) and (4.5.2) hold true. As in the proof of 1.4, iv), there exists a finite covering $\psi : Y' \to Y$ with $X \times_Y Y'$ non-singular, such that (4.3.1) holds true for the pullback family $X' \to Y'$. Since a sheaf is ample with respect to some open set, if and only if it has the property on some finite covering, we obtain the bigness of $(\mathcal{N}_0^{p,q})^\vee$ by applying 4.7 to $\psi^*(\mathcal{N}_0^{p,q})^\vee = (\mathcal{N}_0'^{p,q})^\vee$.

In general 2.2 allows to assume that $X \to Y$ fits into the diagram (2.8.1) constructed in 2.8. Let us write $F_0^{\#p,q}$, and $\mathcal{N}_0^{\#p,q}$ for the sheaves corresponding to $F_0^{p,q}$ and $\mathcal{N}_0^{\#p,q}$ on $Y^\#$ instead of Y.

Base Spaces of Non-Isotrivial Families   313

As we have seen in 4.2 the smoothness of η implies that $\eta^* F^{\#p,q} = F'^{p,q}$, and

$$\eta^* \mathcal{N}_0^{\#p,q} \simeq \mathcal{N}_0'^{p,q} \supset \psi^* \mathcal{N}_0^{p,q} . \tag{4.6.2}$$

On $Y^\#$ we are in the situation where the variation is maximal, hence i) holds true and the dual of the kernel $\mathcal{N}_0^{\#p,q}$ is big. So for any ample invertible sheaf $\mathcal{H}$ we find some $\alpha > 0$ and a morphism

$$\overset{r}{\bigoplus} \mathcal{H} \longrightarrow S^\alpha \left((\mathcal{N}_0^{\#p,q})^\vee \right)$$

which is surjective over some open set. Obviously the same holds true for any invertible sheaf $\mathcal{H}$, independent of the ampleness. In particular we may choose for any $\nu > 1$ with $f_* \omega_{X/Y}^\nu \neq 0$ and for the number N_ν given by 2.8, g) the sheaf

$$\mathcal{H} = \det \left(g_*^\# \omega_{Z^\#/Y^\#}^\nu \right)^{N_\nu} .$$

By 2.8, f) and by (4.6.2), applied to $Y' \to Y^\#$, the sheaf

$$\eta^* \left(S^\alpha \left(\left(\mathcal{N}_0^{\#p,q} \right)^\vee \right) \otimes \det \left(g_*^\# \omega_{Z^\#/Y^\#}^\nu \right)^{-N_\nu} \right)$$

$$= 7 S^\alpha \left((\mathcal{N}_0'^{p,q})^\vee \right) \otimes \det \left(g_* \omega_{Z/Y'}^\nu \right)^{-N_\nu}$$

$$\subset \psi^* \left(S^\alpha \left((\mathcal{N}_0^{p,q})^\vee \right) \otimes \lambda_\nu^{-1} \right)$$

is generically generated. By 1.3 the same holds true for some power of

$$S^\alpha \left((\mathcal{N}_0^{p,q})^\vee \right) \otimes \lambda_\nu^{-1} .$$

By 3.3 and by the choice of λ_ν in 2.8, g), one finds $\kappa(\lambda_\nu) \geq \mathrm{Var}(f)$. $\square$

To finish the proof of 1.4, i) and ii) we just have to repeat the arguments used to prove 4.5, using 4.6. By 4.8 $\mathcal{O}_Y = F_0^{n,0}$ cannot lie in the kernel of $\tau_{n,0}$. We choose $1 \leq m \leq n$ to be the largest number with $\tau^m(F_0^{n,0}) \neq 0$. Then $\tau^m(F_0^{n,0})$ is contained in $\mathcal{N}_0^{n-m,m} \otimes S^m(\Omega_Y^1(\log S))$, and we obtain morphisms of sheaves

$$\left(\mathcal{N}_0^{n-m,m} \right)^\vee \longrightarrow F_0^{n,0} \otimes \left(\mathcal{N}_0^{n-m,m} \right)^\vee \overset{\neq 0}{\longrightarrow} S^m \left(\Omega_Y^1(\log S) \right) . \tag{4.6.3}$$

Under the assumptions made in 1.4, ii) we take $\mathcal{P}$ to be the image of this morphism. By 4.8, i), this is the image of a big sheaf, hence big.

If $\mathrm{Var}(f) < \dim(Y)$ 4.8, ii) implies that $S^\alpha((\mathcal{N}_0^{n-m,m})^\vee) \otimes \lambda^{-1}$ is globally generated, for some $\alpha > 0$, and by (4.6.3) one obtains a non-trivial morphism

$$\overset{r}{\bigoplus} \lambda \longrightarrow S^{\alpha \cdot m} \left(\Omega_Y^1(\log S) \right) .$$ $\square$

5 Base Spaces of Families of Smooth Minimal Models

As promised in section one, we will show that in problem 1.5, the bigness in b) follows from the weak positivity in a). The corresponding result holds true for base spaces of morphisms of maximal variation whose fibres are smooth minimal models.

Throughout this section Y denotes a projective manifold, and S a reduced normal crossing divisor in Y.

Corollary 5.1. *Let* $f : V \to U = Y \setminus S$ *be a smooth family of n-dimensional projective manifolds with* $\mathrm{Var}(f) = \dim(Y)$ *and with* $\omega_{V/U}$ *f-semi-ample. If* $\Omega^1_Y(\log S)$ *is weakly positive, then* $\omega_Y(S)$ *is big.*

Proof. By 1.4, iii), there exists some $m > 0$, a big coherent subsheaf $\mathcal{P}$, and an injective map

$$\mathcal{P} \xhookrightarrow{\subset} S^m\left(\Omega^1_Y(\log S)\right) .$$

Its cokernel $\mathcal{C}$, as the quotient of a weakly positive sheaf, is weakly positive, hence $\det(S^m(\Omega^1_Y(\log S)))$ is the tensor product of the big sheaf $\det(\mathcal{P})$ with the weakly positive sheaf $\det(\mathcal{C})$. $\qquad\square$

Corollary 5.2 (Kovács, [14], for $S = \emptyset$).
If $T_Y(-\log S)$ *is weakly positive, then there exists*

a) *no non-isotrivial smooth projective family* $f : V \to U$ *of canonically polarized manifolds.*

b) *no smooth projective family* $f : V \to U$ *with* $\mathrm{Var}(f) = \dim(U)$ *and with* $\omega_{V/U}$ *f-semi-ample.*

Proof. In both cases 1.4 would imply for some $m > 0$ that $S^m(\Omega^1_Y(\log S))$ has a subsheaf $\mathcal{A}$ of positive Kodaira dimension. But $\mathcal{A}^\vee$, as a quotient of a weakly positive sheaf, must be weakly positive, contradicting $\kappa(\mathcal{A}) > 0$. $\qquad\square$

There are other examples of varieties U for which $S^m(\Omega^1_Y(\log S))$ cannot contain a subsheaf of strictly positive Kodaira dimension or more general, for which

$$H^0\left(Y, S^m\left(\Omega^1_Y(\log S)\right)\right) = 0 \quad \text{for all} \quad m > 0 . \tag{5.2.1}$$

The argument used in 5.2 carries over and excludes the existence of families, as in 5.2, a) or b). For example, (5.2.1) has been verified by Brückmann for $U = H$ a complete intersection in $\mathbb{P}^N$ of codimension $\ell < N/2$ (see [3] for example). As a second application of this result, one can exclude certain discriminant loci for families of canonically polarized manifolds in $\mathbb{P}^N$. If $H = H_1 + \cdots + H_\ell$ is a normal crossing divisor in $\mathbb{P}^N$, and $\ell < N/2$, then for $U = \mathbb{P}^N \setminus H$ the conclusions in 5.2 hold true. In order allow a proof by induction, we formulate both results in a slightly more general setup. $\qquad\square$

Base Spaces of Non-Isotrivial Families   315

Corollary 5.3. *For $\ell < N/2$ let $H = H_1 + \cdots + H_\ell$ be a normal crossing divisor in $\mathbb{P}^N$. For $0 \le r \le l$ define*

$$H = \bigcap_{j=r+1}^{\ell} H_j, \quad S_i = H_i|_H, \quad S = \sum_{i=1}^{r} S_i, \quad and \quad U = H \setminus S$$

(where for $l = r$ the intersection with empty index set is $H = \mathbb{P}^N$). Then there exists

a) *no non-isotrivial smooth projective family $f : V \to U$ of canonically polarized manifolds.*
b) *no smooth projective family $f : V \to U$ with $\mathrm{Var}(f) = \dim(U)$ and with $\omega_{V/U}$ f-semi-ample.*

Proof. Let $\mathcal{A}$ be an invertible sheaf of Kodaira dimension $\kappa(\mathcal{A}) > 0$. Replacing $\mathcal{A}$ by some power, we may assume that $\dim(H^0(H,\mathcal{A})) > r+1$. We have to verify that there is no injection $\mathcal{A} \to S^m(\Omega^1_H(\log S))$.

For $r = 0$ such an injection would contradict the vanishing (5.2.1) shown in [3]. Hence starting with $r = 0$ we will show the non-existence of the subsheaf $\mathcal{A}$ by induction on $\dim(H) = N - \ell + r$ and on r.

The exact sequence

$$0 \to \Omega^1_{S_r}(\log(S_1 + \cdots + S_{r-1})) \to \Omega^1_H(\log S)|_{S_r} \to \mathcal{O}_{S_r} \to 0$$

induces a filtration on $S^m(\Omega^1_H(\log S))|_{S_r}$ with subsequent quotients

$$S^\mu\left(\Omega^1_{S_r}(\log(S_1 + \cdots + S_{r-1}))\right)$$

for $\mu = 0, \cdots, m$. By induction none of those quotients can contain an invertible subsheaf of positive Kodaira dimension. Hence either the restriction of $\mathcal{A}$ to S_r is a sheaf with $\kappa(\mathcal{A}|_{S_r}) \le 0$, hence $\dim(H, \mathcal{A}(-S_r)) > r$, or the image of $\mathcal{A}$ in $S^m(\Omega^1_H(\log S))|_{S_r}$ is zero. In both cases

$$S^m\left(\Omega^1_H(\log(S)) \otimes \mathcal{O}_H(-S_r)\right)$$

contains an invertible subsheaf $\mathcal{A}_1$ with at least two linearly independent sections, hence of positive Kodaira dimension. Now we repeat the same argument a second time:

$$\left(S^m\left(\Omega^1_H(\log S)\right) \otimes \mathcal{O}_H(-S_r)\right)|_{S_r}$$

has a filtration with subsequent quotients

$$\left(S^\mu\left(\Omega^1_{S_r}(\log(S_1 + \cdots + S_{r-1}))\right) \otimes \mathcal{O}_H(-S_r))|_{S_r}\right).$$

$\mathcal{O}_H(S_r)$ is ample, hence by induction none of those quotients can have a nontrivial section. Repeating this argument m times, we find an invertible sheaf which is contained in

$$S^m\left(\Omega^1_H(\log S)\right) \otimes \mathcal{O}_H(-m \cdot S_r) = S^m\left(\Omega^1_H(\log S) \otimes \mathcal{O}_H(-S_r)\right)$$
$$\subset S^m\left(\Omega^1_H(\log(S_1 + \cdots + S_{r-1}))\right),$$

316   Eckart Viehweg and Kang Zuo

and which is of positive Kodaira dimension, contradicting the induction hypothesis.   □

6 Subschemes of Moduli Stacks of Canonically Polarized Manifolds

Let M_h denote the moduli scheme of canonically polarized n-dimensional manifolds with Hilbert polynomial h. In this section we want to apply 1.4 to obtain properties of submanifolds of the moduli stack. Most of those remain true for base spaces of smooth families with a relatively semi-ample dualizing sheaf, and of maximal variation.

Assumptions 6.1. Let Y be a projective manifold, S a normal crossing divisor and $U = Y - S$. Consider the following three setups:

a) There exists a quasi-finite morphism $\varphi : U \to M_h$ which is induced by a smooth family $f : V \to U$ of canonically polarized manifolds.
b) There exists a smooth family $f : V \to U$ with $\omega_{V/U}$ f-semi-ample and with $\mathrm{Var}(f) = \dim(U)$.
c) There exists a smooth family $f : V \to U$ with $\omega_{V/U}$ f-semi-ample and some $\nu \geq 2$ for which the following holds true. Given a non-singular projective manifold Y', a normal crossing divisors S' in Y', and a quasi-finite morphism $\psi' : U' = Y' \setminus S' \longrightarrow U$, let X' be a non-singular projective compactification of $V \times_U U'$ such that the second projection induces a morphism $f' : X' \to Y'$. Then the sheaf $\det(f'_* \omega_{X'/Y'}^\nu)$ is ample with respect to U'.

Although we are mainly interested in the cases 6.1, a) and b), we included the quite technical condition c), since this is what we really need in the proofs.

The assumption made in a) implies the one in c). In fact, if φ is quasi-finite, the same holds true for $\varphi \circ \psi' : U' \to M_h$, and by 3.4, iii), $\det(f'_* \omega_{X'/Y'}^\nu)$ is ample with respect to U'.

Under the assumption b), it might happen that we have to replace U in c) by some smaller open subset $\tilde{U}$. To this aim start with the open set U_g considered in 2.7. Applying 3.4, i), to the family $g : Z \to Y'$ in (2.6.1), one finds an open subset $\tilde{U}'$ of Y' with $g_* \omega_{Z/Y'}^\nu$ ample with respect to $\tilde{U}'$. We may assume, of course, that $\tilde{U}'$ is the preimage of $\tilde{U} \subset U_g$. Since $g : Z \to Y'$ is birational to a mild morphism over Y', the same holds true for all larger coverings, and the condition c) follows by flat base change, for $\tilde{U}$ instead of U.

Let us start with a finiteness result for morphisms from curves to M_h, close in spirit to the one obtained in [2, 4.3], in case that M_h is the moduli space of surfaces of general type. Let C be a projective non-singular curve and let C_0 be a dense open subset of C. By [8] the morphisms

$$\pi : C \to Y \quad \text{with} \quad \pi(C_0) \subset U$$

Base Spaces of Non-Isotrivial Families 317

are parameterized by a scheme $\mathbf{H} := \mathbf{Hom}((C, C_0), (Y, U))$, locally of finite type.

Theorem 6.2.

i) *Under the assumptions made in 6.1, a) or c), the scheme $\mathbf{H}$ is of finite type.*
ii) *Under the assumption 6.1, b), there exists an open subscheme U_g in U such that there are only finitely many irreducible components of $\mathbf{H}$ which parameterize morphisms $\pi : C \to Y$ with $\pi(C_0) \subset U$ and $\pi(C_0) \cap U_g \neq \emptyset$.*

Proof. Let us return to the notations introduced in 2.6 and 2.7. There we considered an open dense non-singular subvariety U_g of U, depending on the construction of the diagram (2.6.1). In particular U_g embeds to Y.

Let $\mathcal{H}$ be an ample invertible sheaf on Y. In order to prove i) we have to find some constant c which is an upper bound for $\deg(\pi^*\mathcal{H})$, for all morphisms $\pi : C \to Y$ with $\pi(C_0) \subset U$. For part ii) we have to show the same, under the additional assumption that $\pi(C) \cup U_g \neq \emptyset$. Let us start with the latter.

By 3.4, iii) in case 6.1, a), or by assumption in case 6.1, c), one finds the sheaf λ_ν, defined in 2.6, d), to be ample with respect to U_g. For part ii) of 6.2, i.e. if one just assumes that $\mathrm{Var}(f) = \dim(U)$, we may use 3.3, and choose U_g a bit smaller to guarantee the ampleness of $g_*\omega_{Z/Y'}^\nu$ over $\psi^{-1}(U_g)$.

Replacing N_ν by some multiple and λ_ν by some tensor power, we may assume that $\lambda_\nu \otimes \mathcal{H}^{-1}$ is generated by global sections over U_g.

Assume first that $\pi(C_0) \cap U_g \neq \emptyset$. Let $h : W \to C$ be a morphism between projective manifolds, obtained as a compactification of $X \times_Y C_0 \to C_0$. By definition h is smooth over C_0. In 2.7 we have shown, that

$$\deg(\pi^*\lambda_\nu) \leq N_\nu \cdot \deg\left(\det\left(h_*\omega_{W/C}^\nu\right)\right) .$$

On the other hand, upper bounds for the right hand side have been obtained for case a) in [2], [13] and in general in [24]. Using the notations from [24],

$$\deg\left(\det(h_*\omega_{W/C}^\nu)\right) \leq (n \cdot (2g(C) - 2 + s) + s) \cdot \nu \cdot \mathrm{rank}\left(h_*\omega_{W/C}^\nu\right) \cdot e ,$$

where $g(C)$ is the genus of C, where $s = \#(C - C_0)$, and where e is a positive constant, depending on the general fibre of h. In fact, if F is a general fibre of h, the constant e can be chosen to be $e(\omega_F^\nu)$. Since the latter is upper semicontinous in smooth families (see [6] or [23, 5.17]) there exists some e which works for all possible curves. Altogether, we found an upper bound for $\deg(\pi^*\mathcal{H})$, whenever the image $\pi(C)$ meets the dense open subset U_g of U.

In i), the assumptions made in 6.1, a) and c) are compatible with restriction to subvarieties of U, and we may assume by induction, that we already obtained similar bounds for all curves C with $\pi(C_0) \subset (U \setminus U_g)$. $\square$

318   Eckart Viehweg and Kang Zuo

From now on we fix again a projective non-singular compactification with $S = Y \setminus U$ a normal crossing divisor. Even if $\varphi : U \to M_h$ is quasi finite, one cannot expect $\Omega^1_Y(\log S)$ to be ample with respect to U, except for $n = 1$, i.e. for moduli of curves. For $n > 1$ there are obvious counter examples.

Example 6.3. Let $g_1 : Z_1 \to C_1$ and $g_2 : Z_2 \to C_2$ be two non-isotrivial families of curves over curves C_1 and C_2, with degeneration loci S_1 and S_2, respectively. We assume both families to be semi-stable, of different genus, and we consider the product

$$f : X = Z_1 \times Z_2 \longrightarrow C_1 \times C_2 = Y \ ,$$

the projections $p_i : Y \to C_i$, and the discriminant locus $S = p_1^{-1}(S_1) \cup p_2^{-1}(S_2)$. For two invertible sheaves $\mathcal{L}_i$ on C_i we write

$$\mathcal{L}_1 \boxplus \mathcal{L}_2 = p_1^*\mathcal{L}_1 \oplus p_2^*\mathcal{L}_2 \quad \text{and} \quad \mathcal{L}_1 \boxtimes \mathcal{L}_2 = p_1^*\mathcal{L}_1 \otimes p_2^*\mathcal{L}_2 \ .$$

For example

$$S^2(\mathcal{L}_1 \boxplus \mathcal{L}_2) = p_1^*\mathcal{L}_1^2 \oplus p_2^*\mathcal{L}_2^2 \oplus \mathcal{L}_1 \boxtimes \mathcal{L}_2 \ .$$

The family f is non-isotrivial, and it induces a generically finite morphism to the moduli space of surfaces of general type M_h, for some h. Obviously,

$$
\begin{aligned}
\Omega^1_Y(\log S) \ &= \Omega^1_{C_1}(\log S_1) \boxplus \Omega^1_{C_2}(\log S_2) \\
&:= p_1^* \left(\Omega^1_{C_1}(\log S_1) \right) \oplus p_2^* \left(\Omega^1_{C_2}(\log S_2) \right)
\end{aligned}
$$

cannot be ample with respect to any open dense subset.

Let us look, how the edge morphisms $\tau_{p,q}$ defined in Sect. 4 look like in this special case. To avoid conflicting notations, we write $G_i^{p,q}$ instead of $F^{p,q}$, for the two families of curves, and

$$\sigma_i : G_i^{1,0} = g_{i*}\mathcal{O}_{Z_i} = \mathcal{O}_{C_i} \longrightarrow G_i^{0,1} \otimes \Omega^1_{C_i}(\log S_i)$$

for the edge morphisms. The morphism

$$\tau^2 = \tau_{1,1} \circ \tau_{2,0} : F^{2,0} = \mathcal{O}_Y \longrightarrow F^{0,2} \otimes S^2 \left(\Omega^1_Y(\log S) \right) \ ,$$

considered in the proof of 1.4, i) and ii), thereby induces three maps,

$$t_i : F^{0,2\vee} \to S^2 \left(p_i^* \Omega^1_{C_i}(\log S_i) \right) \ ,$$

for $i = 1,\ 2$, and

$$t : F^{0,2\vee} \longrightarrow \Omega^1_{C_1}(\log S_1) \boxtimes \Omega^1_{C_2}(\log S_2) \ .$$

Since $F^{0,2\vee} = g_{1*}\omega^2_{Z_1/C_1} \boxtimes g_{2*}\omega^2_{Z_2/C_2}$ is ample the first two morphisms t_1 and t_2 must be zero.

$$F^{1,1} = R^1 f_* \left(\left(\omega_{Z_1/C_1} \boxplus \omega_{Z_2/C_2} \right) \otimes \omega_{X/Y}^{-1} \right) \simeq R^1 f_* \left(\omega_{Z_1/C_1}^{-1} \boxplus \omega_{Z_2/C_2}^{-1} \right) \ ,$$

Base Spaces of Non-Isotrivial Families   319

where the isomorphism interchanges the two factors. In particular

$$F^{1,1} = G_1^{0,1} \boxplus G_2^{0,1} \; ,$$

and one has $\tau_{2,0} = \sigma_1 \boxplus \sigma_2$. Its image lies in the direct factor

$$G' := G_1^{0,1} \otimes \Omega^1_{C_1}(\log S_1) \boxplus G_2^{0,1} \otimes \Omega^1_{C_2}(\log S_2)$$

of $F^{1,1} \otimes \Omega^1_Y(\log S)$. The picture should be the following one:

$$F^{0,2} \otimes \Omega^1_Y(\log S) = \left(G_1^{0,1} \boxtimes G_2^{0,1} \right) \otimes \left(\Omega^1_{C_1}(\log S_1) \boxplus \Omega^1_{C_2}(\log S_2) \right) \; ,$$

and $\tau_{1,1}|_{G_1^{0,1}} = \mathrm{id}_{G_1^{0,1}} \otimes \sigma_2$ with image in

$$\left(G_1^{0,1} \boxtimes G_2^{0,1} \right) \otimes \Omega^1_{C_2}(\log S_2) \; .$$

Hence $\tau_{1,1} \circ \tau_{2,0}$ is the sum of the two maps $\tau_{1,1}|_{G_i^{0,1}} \circ \sigma_i$, both with image in

$$\left(G_1^{0,1} \boxtimes G_2^{0,1} \right) \boxtimes \Omega^1_{C_1}(\log S_1) \otimes \Omega^1_{C_2}(\log S_2) \; .$$

In general, when there exists a generically finite morphism $\varphi : C_1 \times C_2 \to M_h$ induced by $f : V \to U$, the picture should be quite similar, however we were unable to translate this back to properties of the general fibre of f. However, for moduli of surfaces there cannot exist a generically finite morphism from the product of three curves. More generally one obtains from 1.4:

Corollary 6.4. *Let* $U = C_1^0 \times \cdots C_\ell^0$ *be the product of* ℓ *quasi-projective curves, and assume there exists a smooth family* $f : V \to U$ *with* $\omega_{V/U}$ f-*semi-ample and with* $\mathrm{Var}(f) = \dim(U)$. *Then* $\ell \leq n = \dim(V) - \dim(U)$.

Proof. For C_i, the non-singular compactification of C_i^0, and for $S_i = C_i \setminus C_i^0$, a compactification of U is given by $Y = C_1 \times \cdots \times C_\ell$ with boundary divisor $S = \sum_{i=1}^{\ell} pr_i^* S_i$. Then

$$S^m \left(\Omega^1_Y(\log S) \right) = \bigoplus S^{j_1} \left(pr_1^* \Omega^1_{C_1}(\log S_1) \right) \otimes \cdots \otimes S^{j_\ell} \left(pr_\ell^* \Omega^1_{C_\ell}(\log S_\ell) \right)$$

where the sum is taken over all tuples $j_1, \ldots, j_\ell$ with $j_1 + \cdots + j_\ell = m$. If $\ell > m$, each of the factors is the pullback of some sheaf on a strictly lower dimensional product of curves, hence for $\ell > m$ any morphism from a big sheaf $\mathcal{P}$ to $S^m(\Omega^1_Y(\log S))$ must be trivial. If $\psi : Y' \to Y$ is a finite covering, the same holds true for $\psi^* S^m(\Omega^1_Y(\log S))$. By 1.4, iv), there exists such a covering, some $m \leq n$ and a big subsheaf of $\psi^* S^m(\Omega^1_Y(\log S))$, hence $\ell \leq m \leq n$. $\qquad\square$

320   Eckart Viehweg and Kang Zuo

The next application of 1.4 is the rigidity of generic curves in moduli stacks. If $\varphi : U \to M_h$ is induced by a family, 1.4, ii) provides us with a big subsheaf $\mathcal{P}$ of $S^m(\Omega_Y^1(\log S))$, and if we do not insist that $m \leq n$, the same holds true whenever there exists a family $V \to U$, as in 6.1, b). In both cases, replacing m by some multiple, we find an ample invertible sheaf $\mathcal{H}$ on Y and an injection

$$\iota : \bigoplus \mathcal{H} \longrightarrow S^m \left(\Omega_Y^1(\log S) \right) .$$

Let U_1 be an open dense subset in U, on which ι defines a subbundle.

Corollary 6.5. *Under the assumption a) or b) in 6.1 there exists an open dense subset U_1 and for each point $y \in U_1$ a curve $C_0 \subset U$, passing through y, which is rigid, i.e.: If for a reduced curve T_0 and for $t \in T_0$ there exists a morphism $\rho : T_0 \times C_0 \to U$ with $\rho(\{t_0\} \times C_0) = C_0$, then ρ factors through $pr_2 : T_0 \times C_0 \to C_0$.*

Proof. Let $\pi : \mathbb{P} = \mathbb{P}(\Omega_Y^1(\log S)) \to Y$ be the projective bundle.

$$\iota : \bigoplus \mathcal{H} \longrightarrow \pi_* \mathcal{O}_{\mathbb{P}}(m)$$

defines sections of $\mathcal{O}_Y(m) \otimes \pi^* \mathcal{H}^{-1}$, which are not all identically zero on $\pi^{-1}(y)$ for $y \in U_1$. Hence there exists a non-singular curve $C_0 \subset U$ passing through y, such that the composite

$$\bigoplus \mathcal{H}|_U \longrightarrow S^m \left(\Omega_U^1 \right) \longrightarrow S^m \left(\Omega_{C_0}^1 \right) \tag{6.5.1}$$

is surjective over a neighborhood of y. For a nonsingular curve T_0 and $t_0 \in T_0$ consider a morphism $\phi_0 : T_0 \times C_0 \to U$, with $\phi_0(\{t_0\} \times C_0) = C_0$. Let T and C be projective non-singular curves, containing T_0 and C_0 as the complement of divisors Θ and Γ, respectively. On the complement W of a codimension two subset of $T \times C$ the morphism ϕ_0 extends to $\phi : W \to Y$. Then ι induces a morphism

$$\phi^* \bigoplus \mathcal{H} \longrightarrow \phi^* S^m \left(\Omega_Y^1(\log S) \right) \longrightarrow S^m \left(\Omega_T^1(\log \Theta) \boxplus \Omega_C^1(\log \Gamma) \right)|_W$$

whose composite with

$$S^m \left(\Omega_T^1(\log \Theta) \boxplus \Omega_C^1(\log \Gamma) \right)|_W \longrightarrow S^m \left(pr_2^* \Omega_C^1(\log \Gamma) \right)|_W$$
$$\longrightarrow S^m \left(\Omega_{\{t\} \times C}(\log \Gamma) \right)|_{W \cap \{t\} \times C}$$

is non-zero for all t in an open neighborhood of t_0, hence

$$\phi^* \bigoplus \mathcal{H} \longrightarrow pr_2^* S^m \left(\Omega_C^1(\log \Gamma) \right)|_W$$

is surjective over some open dense subset. Since $\mathcal{H}$ is ample, this is only possible if $\phi : W \to Y$ factors through the second projection $W \to T \times C \to C$. $\qquad \square$

Base Spaces of Non-Isotrivial Families   321

Assume we know in 6.5 that $\Omega^1_Y(\log S)$ is ample over some dense open sub-scheme U_2. Then the morphism (6.5.1) is non-trivial for all curves C_0 meeting U_2, hence the argument used in the proof of 6.5 implies, that a morphisms $\pi : C_0 \to U$ with $\pi(C_0) \cap U_2 \neq \emptyset$ has to be rigid. If $U_2 = U$, this, together with 6.2 proves the next corollary.

Corollary 6.6. *Assume in 6.2, i), that $\Omega^1_Y(\log S)$ is ample with respect to U. Then* **H** *is a finite set of points.*

The generic rigidity in 6.5, together with the finiteness result in 6.2, implies that subvarieties of the moduli stacks have a finite group of automorphism. Again, a similar statement holds true under the assumption 6.1, b).

Theorem 6.7. *Under the assumption 6.1, a) or b) the automorphism group* $\mathrm{Aut}(U)$ *of U is finite.*

Proof. Assume $\mathrm{Aut}(U)$ is infinite, and choose an infinite countable subgroup

$$G \subset \mathrm{Aut}(U).$$

Let U_1 be the open subset of U considered in 6.5, and let U_g be the open subset from 6.2, b). We may assume that $U_1 \subset U_g$ and write $\Gamma = U \setminus U_1$. Since

$$\bigcup_{g \in G} g(\Gamma) \neq U$$

we can find a point $y \in U_1$ whose G-orbit is an infinite set contained in U_1. By 6.5 there are rigid smooth curves $C_0 \subset U$ passing through y. Obviously, for all $g \in G$ the curve $g(C_0) \subset U$ is again rigid, it meets U_1, hence U_g and the set of those curves is infinite, contradicting 6.2, b). $\square$

7   A Vanishing Theorem for Sections of Symmetric Powers of Logarithmic One Forms

Proposition 7.1. *Let Y be a projective manifold and let $D = D_1 + \ldots + D_r$ and $S = S_1 + \ldots + S_\ell$ be two reduced divisors without common component. Assume that*

i) $S + D$ *is a normal crossing divisor.*
ii) *For no subset $J \subseteq \{1, \ldots, \ell\}$ the intersection*

$$S_J = \bigcap_{j \in J} S_i$$

  is zero dimensional.
iii) $T_Y(-\log D) = (\Omega^1_Y(\log D))^\vee$ *is weakly positive over $U_1 = Y - D$.*

322   Eckart Viehweg and Kang Zuo

Then for all ample invertible sheaves $\mathcal{A}$ and for all $m \geq 1$

$$H^0 \left(Y, S^m \left(\Omega_Y^1(\log(D+S))\right) \otimes \mathcal{A}^{-1}\right) = 0 .$$

Corollary 7.2. *Under the assumption i), ii) and iii) in 7.1 there exists no smooth family $f : V \to U = Y \setminus (S+D)$ with $\mathrm{Var}(f) = \dim Y$. In particular there is no generically finite morphism $U \to M_h$, induced by a family.*

Proof. By 1.4, iii) the existence of such a family implies that for some $m > 0$ the sheaf $S^m(\Omega_Y^1(\log(D+S)))$ contains a big coherent subsheaf $\mathcal{P}$. Replacing m by some multiple, one can assume that $\mathcal{P}$ is ample and invertible, contradicting 7.1. $\square$

Proof of 0.2 and of the second part of 0.4. Since for an abelian variety Y the sheaf Ω_Y^1 is trivial, and since the condition ii) in 7.1 is obvious for $\ell < \dim Y$, part a) of 0.2 is a special case of 7.2.

For b) again i) and ii) hold true by assumption. For iii) we remark, that

$$\Omega_{\mathbb{P}^{\nu_i}}^1 \left(\log D^{(\nu_i)}\right) = \oplus^{\nu_i} \mathcal{O}_{\mathbb{P}^{\nu_i}} ,$$

hence $\Omega_Y^1(\log D)$ is again a direct sum of copies of $\mathcal{O}_Y$. Assume that $\mathcal{A}$ is an invertible subsheaf of $S^m(\Omega_Y^1(\log(D+S)))$, for some $m > 0$. If $\kappa(\mathcal{A}) > 0$, then for some $\mu_i \in \mathbb{N}$,

$$\mathcal{A} = \mathcal{O}_Y(\mu_1, \ldots, \mu_k) = \bigotimes_{i=1}^k pr_i^* \mathcal{O}_{\mathbb{P}^{\nu_i}}(\mu_i) .$$

By 7.1 not all the μ_i can be strictly larger than zero, hence

$$\kappa(\mathcal{A}) \leq \mathrm{Max}\{\dim(Y) - \nu_i; \; i = 1, \ldots, k\} = M .$$

By 1.4, i), for any morphism $\varphi : U \to M_h$, induced by a family $f : V \to U$, one has

$$\mathrm{Var}(f) = \dim(\varphi(U)) \leq M .$$

1.4, iii), implies that there exists no smooth family $f : V \to U$ of maximal variation, and with $\omega_{V/U}$ f-semi-ample. $\square$

Remark 7.3. In 0.2, a), one can also show, that for an abelian variety Y and for a morphism $\varphi : Y \to M_h$, induced by a family,

$$\dim(\varphi(Y)) \leq \dim(Y) - \nu ,$$

where ν is the dimension of the smallest simple abelian subvariety of Y. In fact, by the Poincaré decomposition theorem, Y is isogenous to the product of simple abelian varieties, hence replacing Y by an étale covering, we may assume that

$$Y = Y_1 \times \ldots \times Y_k$$

Base Spaces of Non-Isotrivial Families   323

with Y_i simple abelian, and with

$$\nu = \dim(Y_1) \leq \dim(Y_2) \leq \ldots \leq \dim Y_k \,.$$

Since an invertible sheaf of positive Kodaira dimension on a simple abelian variety must be ample, one finds that for an non-ample invertible sheaf $\mathcal{A}$ on Y

$$\kappa(\mathcal{A}) \leq \dim(Y) - \nu \,.$$

Before proving 7.1 let us show that for $Y = \mathbb{P}^2$ we can not allow S to have two irreducible components of high degree, even if $D = 0$.

Example 7.4. Given a surface Y and a normal crossing divisor $S + D$, with $S = \sum_{i=1}^{\ell} S_i$, consider the two exact sequences

$$0 \to \mathcal{O}_Y(-S_i) \to \mathcal{O}_Y \to \mathcal{O}_{S_i} \to 0$$

and

$$0 \to \Omega_Y^1(\log D) \to \Omega_Y^1(\log(D+S)) \to \bigoplus_{i=1}^{\ell} \mathcal{O}_{S_i} \to 0 \,.$$

Writing $c(\mathcal{E}) = 1 + c_1(\mathcal{E}) + c_2(\mathcal{E})$ for a sheaf $\mathcal{E}$ on Y, one finds

$$c(\mathcal{O}_{S_i}) = 1 + S_i + S_i^2$$

and

$$c\left(\Omega_Y^1(\log(D+S))\right) = c\left(\Omega_Y^1(\log D)\right) \cdot \prod_{i=1}^{\ell}\left(1 + S_i + S_i^2\right) \,.$$

Hence

$$c_2\left(\Omega_Y^1(\log(D+S))\right) = c_2\left(\Omega_Y^1(\log D)\right)$$
$$+ \sum_{i=1}^{\ell} c_1\left(\Omega_Y^1(\log D)\right).S_i + \sum_{i<j} S_i.S_j + \sum_{i=1}^{\ell} S_i^2$$

and $c_1(\Omega_Y^1(\log(D+S))) = c_1(\Omega_Y^1) + D + S$. The Riemann-Roch theorem for vector bundles on surfaces and the isomorphism

$$S^m\left(\Omega_Y^1(\log(D+S))^\vee\right) \otimes \omega_Y = S^m\left(\Omega_Y^1(\log(D+S))\right) \otimes \omega_Y \otimes \omega_Y(D+S)^{-m}$$

imply that for an invertible sheaf $\mathcal{A}$

$$h^0\left(Y, S^m\left(\Omega_Y^1(\log(D+S))\right) \otimes \mathcal{A}^{-1}\right)$$
$$+ h^0\left(Y, S^m\left(\Omega_Y^1(\log(D+S))\right) \otimes \omega_Y \otimes \omega_Y(D+S)^{-m} \otimes \mathcal{A}\right)$$
$$\geq \frac{m^3}{6}\left(c_1\left(\Omega_Y^1(\log(D+S))\right)^2 - c_2\left(\Omega_Y^1(\log(D+S))\right)\right) + O(m^2) \,,$$

where $O(m^2)$ is a sum of terms of order ≤ 2 in m. If $\omega_Y(D+S)$ is big and if

$$c_1\left(\Omega_Y^1(\log(D+S))\right)^2 > c_2\left(\Omega_Y^1(\log(D+S))\right) \tag{7.4.1}$$

then for $m \gg 0$ the sheaf $\omega_Y \otimes \omega_Y(D+S)^{-m} \otimes \mathcal{A}$ is a subsheaf of $\mathcal{A}^{-1}$ and

$$h^0\left(Y, S^m\left(\Omega_Y^1(\log(D+S))\right) \otimes \mathcal{A}^{-1}\right) \neq 0.$$

For $Y = \mathbb{P}^2$ and for a coordinate system $D = D_0 + D_1 + D_2$,

$$c_1\left(\Omega_Y^1(\log(D+S))\right)^2 - c_2\left(\Omega_Y^1(\log D+S))\right)$$
$$= \left(\sum_{i=1}^\ell S_i\right)^2 - \sum_{i=1}^\ell S_i^2 - \sum_{i<j} S_i.S_j = \sum_{i<j} S_i.S_j,$$

and as soon as S has more than one component, (7.4.1) holds true. So in 7.1, for $Y = \mathbb{P}^2$ and $\ell > 1$, the arguments used to prove 0.2 fail. We do not know however, whether $U = \mathbb{P}^2 \setminus (S_1 + S_2)$ can be the base of a non-isotrivial family of canonically polarized manifolds.

For $D = 0$ and $Y = \mathbb{P}^2$, one finds

$$c_1\left(\Omega_Y^1(\log S)\right)^2 - c_2(\Omega_Y^1(\log S)) = 6 - 3 \cdot \deg(S) + \sum_{i<j} S_i.S_j$$
$$= 3 \cdot (2 - \deg(S)) + \sum_{i<j} \deg(S_i) \cdot \deg(S_j).$$

Assume that $2 \leq \deg(S_1) \leq \deg(S_2) \ldots \leq \deg(S_\ell)$. Then the only cases where (7.4.1) does not hold true are $\ell = 2$ and $\deg(S_1) = 2$, or $\ell = 3$ and $\deg(S_i) = 2$ for $i = 1, 2, 3$. Again we do not know any example of a non-isotrivial family over $U = \mathbb{P}^2 \setminus S$.

As a first step in the proof of 7.1 we need

Lemma 7.5. *Let $\mathcal{E}$ and $\mathcal{F}$ be locally free sheaves on Y. Assume that, for a non-singular divisor B, for some ample invertible sheaf $\mathcal{A}$, and for all $m \geq 0$*

$$H^0(Y, S^m(\mathcal{F}) \otimes \mathcal{A}^{-1}) = H^0(B, S^m(\mathcal{F}) \otimes \mathcal{A}^{-1}|_B) = 0.$$

Assume moreover that there exists an exact sequence

$$0 \to \mathcal{F} \to \mathcal{E} \to \mathcal{O}_B \to 0.$$

Then for all $m > 0$

$$H^0(Y, S^m(\mathcal{E}) \otimes \mathcal{A}^{-1}) = 0.$$

Proof. Write $\pi : \mathbb{P} = \mathbb{P}(\mathcal{E}) \to Y$. The surjection $\mathcal{E} \to \mathcal{O}_B$ defines a morphism $s : B \to \mathbb{P}$. For the ideal I of $s(B)$ the induced morphism $\pi^* \mathcal{F} \to I \otimes \mathcal{O}_{\mathbb{P}}(1)$ is surjective, as well as the composite

$$\tilde{\pi}^* \mathcal{F} \to \delta^* (I \otimes \mathcal{O}_{\mathbb{P}}(1)) \to \mathcal{O}_{\tilde{\mathbb{P}}}(-E) \otimes \delta^* \mathcal{O}_{\mathbb{P}}(1) \,,$$

where $\delta : \tilde{\mathbb{P}} \to \mathbb{P}$ is the blowing up of I with exceptional divisor E, and where $\tilde{\pi} = \pi \circ \delta$.

Let us write $M + 1$ for the rank of $\mathcal{E}$. For $y \in B$ and $p = s(y)$ let

$$\delta_y : \tilde{\mathbb{P}}_y \to \mathbb{P}^M = \pi^{-1}(y)$$

be the blowing up of p, with exceptional divisor F. Then

$$\tilde{\pi}^{-1}(y) = \tilde{\mathbb{P}}_y \cup \mathbb{P}^M \quad \text{with} \quad F = \tilde{\mathbb{P}}_y \cap \mathbb{P}^M \,.$$

In particular $\tilde{\pi}$ is equidimensional, hence flat. For $0 \le \mu \le m$ and for $i > 0$

$$H^i(\tilde{\mathbb{P}}_y, \mathcal{O}_{\tilde{\mathbb{P}}_y}(-(\mu + 1) \cdot F) \otimes \delta_y^* \mathcal{O}_{\mathbb{P}^M}(m)) = 0$$

and

$$H^0 \left(\tilde{\mathbb{P}}_y, \mathcal{O}_{\tilde{\mathbb{P}}_y}(-(\mu + 1) \cdot F) \otimes \delta_y^* \mathcal{O}_{\mathbb{P}^M}(\mu) \right) = 0 \,.$$

One has an exact sequence

$$0 \to \mathcal{O}_{\tilde{\mathbb{P}}_y}(-(\mu + 1) \cdot F) \otimes \delta_y^* \mathcal{O}_{\mathbb{P}^M}(m)$$
$$\to \mathcal{O}_{\tilde{\mathbb{P}}}(-\mu \cdot E) \otimes \delta^* \mathcal{O}_{\mathbb{P}}(m)|_{\tilde{\pi}^{-1}(y)} \to \mathcal{O}_{\mathbb{P}^M}(\mu) \to 0$$

and $H^1(\tilde{\pi}^{-1}(y), \mathcal{O}_{\tilde{\mathbb{P}}}(-\mu \cdot E) \otimes \delta^* \mathcal{O}_{\mathbb{P}}(m)|_{\tilde{\pi}^{-1}(y)}) = 0$. By flat base change one finds

$$R^1 \tilde{\pi}_* \left(\mathcal{O}_{\tilde{\mathbb{P}}}(-\mu \cdot E) \otimes \delta^* \mathcal{O}_{\mathbb{P}}(m) \right) = 0 \,.$$

Moreover

$$\tilde{\pi}_* \left(\mathcal{O}_{\tilde{\mathbb{P}}}(-\mu \cdot E) \otimes \delta^* \mathcal{O}_{\mathbb{P}}(m) \right) |_y \to \tilde{\pi}_* \mathcal{O}_E(-\mu \cdot E)|_y \cong H^0 \left(\mathbb{P}^M, \mathcal{O}_{\mathbb{P}^M}(\mu) \right)$$

is an isomorphism. The inclusion

$$S^\mu(\mathcal{F}) \longrightarrow \tilde{\pi}_* \delta^* \mathcal{O}_{\mathbb{P}}(\mu)) \cong S^\mu(\mathcal{E})$$

factors through

$$S^\mu(\mathcal{F}) \overset{\subset}{\longrightarrow} \tilde{\pi}_* \left(\mathcal{O}_{\tilde{\mathbb{P}}}(-\mu \cdot E) \otimes \delta^* \mathcal{O}_{\mathbb{P}}(\mu) \right) \,.$$

This map is an isomorphism. We know the surjectivity of

$$\mathcal{F}|_y \longrightarrow \tilde{\pi}_* \left(\mathcal{O}_{\tilde{\mathbb{P}}}(-E) \otimes \delta^* \mathcal{O}_{\mathbb{P}}(1) \right) |_y \cong H^0 \left(\mathbb{P}^M, \mathcal{O}_{\mathbb{P}^M}(1) \right) \,,$$

so for $\mu > 1$ the morphism from $S^\mu(\mathcal{F})|_y$ to

$$\tilde{\pi}_* \left(\mathcal{O}_{\tilde{\mathbb{P}}}(-\mu \cdot E) \otimes \delta^* \mathcal{O}_{\mathbb{P}}(\mu) \right)|_y \cong H^0 \left(\mathbb{P}^M, \mathcal{O}_{\mathbb{P}^M}(\mu) \right) = S^\mu \left(H^0 \left(\mathbb{P}^M, \mathcal{O}_{\mathbb{P}^M}(1) \right) \right)$$

is surjective as well. By the choice of $s(B)$ one has

$$\mathcal{O}_{\mathbb{P}}(1)|_{s(B)} = \mathcal{O}_{s(B)} \quad \text{and} \quad \delta^* \mathcal{O}_{\mathbb{P}}(1)|_E = \mathcal{O}_E \,.$$

Starting with $\mu = m$, assume by descending induction that

$$H^0 \left(Y, \tilde{\pi}_* \left(\mathcal{O}_{\tilde{\mathbb{P}}}(-\mu \cdot E) \otimes \delta^* \mathcal{O}_{\mathbb{P}}(m) \right) \otimes \mathcal{A}^{-1} \right) = 0 \,.$$

Since

$$H^0 \left(E, \tilde{\pi}_* \left(\mathcal{O}_{\tilde{\mathbb{P}}}(-(\mu - 1) \cdot E) \otimes \delta^* \mathcal{O}_{\mathbb{P}}(m) \right) \otimes \mathcal{A}^{-1}|_E \right)$$
$$= H^0 \left(E, \tilde{\pi}_* \left(\mathcal{O}_{\tilde{\mathbb{P}}}(-(\mu - 1) \cdot E) \otimes \delta^* \mathcal{O}_{\mathbb{P}}(\mu - 1) \right) \otimes \mathcal{A}^{-1}|_E \right)$$
$$= H^0 \left(B, S^{\mu-1}(\mathcal{F}) \otimes \mathcal{A}^{-1}|_B \right) = 0$$

one finds

$$H^0 \left(Y, \tilde{\pi}_* \left(\mathcal{O}_{\tilde{\mathbb{P}}}(-(\mu - 1) \cdot E) \otimes \delta^* \mathcal{O}_{\mathbb{P}}(m) \right) \otimes \mathcal{A}^{-1} \right) = 0 \,,$$

as well. $\qquad\square$

Proof of 7.1. Let us fix some $J \subseteq \{1, \ldots, n\}$ with $S_J \neq \emptyset$. We will write $S_\emptyset = Y$. The sheaf $S^m(T_Y(-\log D)) \otimes \mathcal{A}$ is ample with respect to $U_1 = Y - D$. Since $S + D$ is a normal crossing divisor, $S_J \cap U_1 \neq \emptyset$ and since $\dim(S_J) \geq 1$,

$$H^0 \left(S_J, S^m \left(\Omega_Y^1(\log D) \right) \otimes \mathcal{A}^{-1}|_{S_J} \right) = 0 \,.$$

Assume, by induction on ρ, that

$$H^0 \left(S_{J'}, S^m \left(\Omega_Y^1(\log(D + S_1 + \ldots + S_{\rho-1})) \right) \otimes \mathcal{A}^{-1}|_{S_{J'}} \right) = 0 \,,$$

for all $m \geq 0$, and all $J' \subseteq \{\rho, \ldots, \ell\}$ with $S_{J'} \neq \emptyset$. For $J \subseteq \{\rho + 1, \ldots, \ell\}$ assume $T = S_J \neq \emptyset$. If $T_\rho = S_{J \cup \{\rho\}} = \emptyset$, i.e. if $S_\rho \cap T = \emptyset$, then

$$\Omega_Y^1(\log(D + S_1 + \ldots + S_\rho))|_{S_J} = \Omega_Y^1(\log(D + S_1 + \ldots + S_{\rho-1}))|_{S_J}$$

and there is nothing to prove. Otherwise $T_\rho = S_\rho|_T$ is a divisor and the restriction of

$$0 \to \Omega_Y^1(\log(D + S_1 + \ldots + S_{\rho-1})) \to \Omega_Y^1(\log(D + S_1 + \ldots + S_\rho)) \to \mathcal{O}_{S_\rho} \to 0$$

to T remains exact. Hence for

$$\mathcal{F} = \Omega_Y^1(\log(D + S_1 + \ldots + S_{\rho-1}))|_T \quad \text{and} \quad \mathcal{E} = \Omega_Y^1(\log(D + S_1 + \ldots + S_\rho))|_T$$

$$0 \to \mathcal{F} \to \mathcal{E} \to \mathcal{O}_{T_s} \to 0$$

is exact, $H^0(T, S^m(\mathcal{F}) \otimes \mathcal{A}^{-1}|_T) = 0$ and

$$H^0 \left(T_\rho, S^m(\mathcal{F}) \otimes \mathcal{A}^{-1}|_{T_s} \right)$$
$$= H^0 \left(S_{J \cup \{\rho\}}, S^m \left(\Omega_Y^1 \left(\log \left(D + S_1 + \ldots + S_{\rho-1} \right) \right) \otimes \mathcal{A}^{-1}|_{S_{J \cup \{\rho\}}} \right) \right) = 0 \,.$$

Using 7.5 we obtain

$$H^0\left(S_J, S^m\left(\Omega_Y^1\left(\log\left(D + S_1 + \ldots + S_\rho\right)\right)\right) \otimes \mathcal{A}^{-1}|_{S_J}\right) = 0.$$ $\qquad\square$

Remark 7.6. The assumption "$\mathcal{A}$ ample" was not really needed in the proof of 7.1. It is sufficient to assume that

$$\kappa(\mathcal{A}|_{S_J}) \geq 1, \text{ for all } J \text{ with } S_J \neq \emptyset.$$

References

[1] Abramovich, D., Karu, K., Weak semi-stable reduction in characteristic 0, *Invent. Math.* **139** (2000), 241–273.

[2] Bedulev, E., Viehweg, E., On the Shafarevich conjecture for surfaces of general type over function fields, *Invent. Math.* **139** (2000), 603–615.

[3] Brückmann, P., Rackwitz, H.-G., T-symmetrical tensor forms on complete intersections, *Math. Ann.* **288** (1990), 627–635.

[4] Deligne, P., Équations différentielles à points singuliers réguliers, *Lecture Notes in Math.* **163** (1970), Springer, Berlin Heidelberg New York.

[5] Esnault, H., Viehweg, E., Revêtement cycliques, in *Algebraic Threefolds, Proc. Varenna 1981, Springer Lect. Notes in Math.* **947** (1982), 241–250.

[6] Esnault, H., Viehweg, E., Lectures on vanishing theorems, *DMV Seminar* **20** (1992), Birkhäuser, Basel Boston.

[7] Griffiths, P. (Editor), Topics in transcendental algebraic geometry, *Ann. of Math. Stud.* **106** (1984), Princeton Univ. Press, Princeton, NJ.

[8] Grothendieck, A., Techniques de construction et théorèmes d'existence en géométrie algébrique, IV: Les schémas de Hilbert, *Sém. Bourbaki* **221** (1960/61), in: *Fondements de la Géométrie Algébrique* Sém. Bourbaki, Secrétariat, Paris 1962.

[9] Grauert, H., Mordells Vermutung über rationale Punkte auf algebraischen Kurven und Funktionenkörper, *Publ. Math. IHES* **25** (1965), 131–149.

[10] Kawamata, Y., Minimal models and the Kodaira dimension of algebraic fibre spaces, *Journ. Reine Angew. Math.* **363** (1985), 1–46.

[11] Kempf, G., Knudsen, F., Mumford, D. and Saint-Donat, B., Toroidal embeddings I, *Lecture Notes in Math.* **339** (1973), Springer, Berlin Heidelberg New York.

[12] Kovács, S., Algebraic hyperbolicity of fine moduli spaces, *J. Alg. Geom.* **9** (2000), 165–174.

[13] Kovács, S., Logarithmic vanishing theorems and Arakelov-Parshin boundedness for singular varieties, preprint (AG/0003019), to appear in *Comp. Math.*

[14] Kovács, S., Families over a base with a birationally nef tangent bundle, *Math. Ann.* **308** (1997), 347–359.

[15] Lu, S.S-Y., On meromorphic maps into varieties of log-general type, *Proc. Symp. Amer. Math. Soc.* **52** (1991), 305–333.

[16] Migliorini, L., A smooth family of minimal surfaces of general type over a curve of genus at most one is trivial, *J. Alg. Geom.* **4** (1995), 353-361.

[17] Oguiso, K., Viehweg, E., On the isotriviality of families of elliptic surfaces, *J. Alg. Geom.* **10** (2001), 569–598.

328   Eckart Viehweg and Kang Zuo

[18] Schmid, W., Variation of Hodge structure: The singularities of the period mapping, *Invent. Math.* **22** (1973), 211–319.

[19] Simpson, C., Higgs bundles and local systems, *Publ. Math. I.H.E.S* **75** (1992), 5–95.

[20] Viehweg, E., Weak positivity and the additivity of the Kodaira dimension for certain fibre spaces, in: *Algebraic Varieties and Analytic Varieties, Advanced Studies in Pure Math.* **1** (1983), 329–353.

[21] Viehweg, E., Weak positivity and the additivity of the Kodaira dimension for certain fibre spaces II. The local Torelli map, in: *Classification of Algebraic and Analytic Manifolds, Progress in Math.* **39** (1983), 567–589.

[22] Viehweg, E., Weak positivity and the stability of certain Hilbert points, *Invent. Math.* **96** (1989), 639–667.

[23] Viehweg, E., Quasi-projective Moduli for Polarized Manifolds, *Ergebnisse der Mathematik, 3, Folge* **30** (1995), Springer Verlag, Berlin-Heidelberg-New York.

[24] Viehweg, E., Zuo K., On the isotriviality of families of projective manifolds over curves, *J. Alg. Geom.* **10** (2001), 781 – 799.

[25] Viehweg, E., Zuo K, On the Brody hyperbolicity of moduli spaces for canonically polarized manifolds, preprint (AG/0101004).

[26] Zuo, K., On the negativity of kernels of Kodaira-Spencer maps on Hodge bundles and applications, *Asian J. of Math.* **4** (2000), 279–302.

Uniform Vector Bundles on Fano Manifolds and an Algebraic Proof of Hwang-Mok Characterization of Grassmannians

Jarosław A. Wiśniewski

Institute of Mathematics, Warsaw University, Banacha 2, 02-097 Warszawa, Poland
E-mail address: jarekw@mimuw.edu.pl

Table of Contents

Introduction 329
1 $\mathcal{M}$-Uniform Manifolds 331
2 Atiyah Extension and Twisted Trivial Bundles 333
3 Characterization of Grassmann Manifolds 336
4 Characterization of Isotropic Grassmann Manifolds 337
References 339

0 Introduction

A projective manifold X is called Fano if its anticanonical divisor $-K_X$ is ample. Fano manifolds form a very distinguished class: in each dimension there is only a finite number of deformation classes of them and they are classified in dimension ≤ 3, the case $\dim X = 3$ due to Fano, Roth, Iskovskih and Shokurov. In dimension ≥ 4 not much is known about Fano manifolds in general. However, due to results of Mori, Kawamata and Shokurov, Fano manifolds with Picard number $\rho(X)$ bigger than 1 admit special morphisms, called Fano-Mori contractions, which can be used to study the structure of such Fano's. The case $\rho(X) = 1$ seems to be harder to approach, see [IP] for an overview on Fano varieties.

If G is a semisimple algebraic group and $P \subset G$ its parabolic subgroup then G/P is a homogeneous rational variety and a Fano manifold. If G is simple and P maximal parabolic then $\rho(G/P) = 1$. A classification of such manifolds is very well known and they can be enumerated in terms of vertices of Dynkin diagrams associated to the simple linear groups, see e.g. [FH, p. 394].

It is a natural question to ask about intrinsic properties of Fano's which distinguish homogeneous manifolds among them. Here, "intrinsic" means defined in terms of numerical properties of TX rather than in terms of its sections. A prototype for the expected characterization of homogeneous manifolds is the following characterization of the projective space, conjectured

2000 *Mathematics Subject Classification*: 14J45, 14J60, 14M15.

330   Jarosław A. Wiśniewski

by Frankel and Hartshorne, and proved by Mori [Mo]: if TX is ample then X is the projective space. Subsequently, in [CP] Campana and Peternell conjectured that if X is Fano and TX is a nef vector bundle then X is a homogeneous manifold.

Mori's proof of Frankel-Hartshorne conjecture provides existence of rational curves on Fano manifolds; the characterization of $\mathbf{P}^n$ is eventually obtained by analyzing properties of a family of rational curves on X. Since Mori's paper [Mo] his technique has been further extended and it proved to be a powerful method of dealing with Fano manifolds and other manifolds whose canonical divisor is not nef.

Thus, in the present paper we propose a characterization of homogeneous Fano manifolds in terms of rational curves contained in X. We conjecture that X is homogeneous if TX is uniform with respect to some dominating unsplit family of rational curves (these notions are explained in the subsequent section), which means that TX is the same on any curve from this family.

This view on homogeneous Fano manifolds is supported by recent works of Hwang and Mok as well as Landsberg and Manivel. Hwang and Mok, see [HM1], [HM2], [Hw], analyse the tangents of rational curves from minimal dominating families of rational curves in order to get information on the global features of homogeneous Fano's. On the other hand the works of Landsberg and Manivel [LM1], [LM2] give some intriguing insight on relations of of homogeneous manifolds and their line tangents. Thus one may hope that the minimal family of rational curves on a Fano manifold X carry enough information to recover the most essential features of X — so that uniformity of TX on such a family would imply homogenity of X.

In the present paper we test the uniformity assumption by dealing with some special cases. All of the treated cases are already known — without assuming uniformity — for complex varieties. Case of the projective space is just a part of Mori's argument in [Mo]. Characterisation of hyperquadrics follows by Cho-Sato [CS], Ye [Ye], Hwang–Mok [HM1], or Wierzba [Wi]. Cases of Grassmann manifolds and isotropic Grassmann manifolds are covered by Hwang and Mok in [HM1]. Actually, the characterisation of Grassmanians in [HM1] does not require anything like uniformity and it is obtained as a corollary to a general theorem on flatness of G-structures. Here "flatness" is understood in the sense of a local (analytic) biholomorphism which "flattens" the G-structure.

The approach we take is of an algebraic nature (we work over an algebraically closed field of characteristic zero) and it is based on two technical tools: (1) a characterization of twisted trivial vector bundles, from a recent joint paper with Andreatta [AW], and (2) the Atiyah extension, $\Omega_X \to \mathcal{L} \to \mathcal{O}_X$, which has been recently used in a joint paper with Kebekus, Peternell and Sommese [KPSW]. Both tools are explained in the sequel. The

Uniform Vector Bundles on Fano Manifolds   331

use of Atiyah extension was inspired by work of Krzysztof Jaczewski, who applied it to get a characterization of toric varieties [Ja].

The idea is as follows: assume that we have a characteristic embedding of a homogeneous manifold, $X \to \mathbf{P}(V)$, provided by a complete linear system $V \otimes \mathcal{O}_X \to L$ which can be "recovered" or characterized in terms of Atiyah extension on X. The framework for "recovering procedure" of $V \otimes \mathcal{O}_X \to L$ is hoped to be provided in general by Beilinson and Bernstein-Gelfand-Gelfand theory but here we do not have to deal with this problem since in our case it is given explicitly. Now we can characterize X in terms of a dominating family of rational curves. The characterization is the following: take a Fano manifold X with a dominating and unsplit family of rational curves; suppose that TX is uniform with respect to this family and the "recovering procedure" on a rational curve C from the family produces $V_C \to L_C$, where V_C is trivial on C. Then the same procedure produces $V \to L$ globally and the bundle V is trivial hence, possibly, we can get the original $X \to \mathbf{P}(V)$. The point is that the "recovering procedure" is based on the Atiyah extension class which is "remembered" on each single line C from the family in question; since TX is uniform the resulting $V_C \to L_C$ is the same on each C thus V is trivial by the characterization of twisted trivial bundles, [AW, (1.2)].

The Grassmannians considered in the present paper are particularly convenient, since they are characterized by their universal bundles. Thus, the "recovering procedure" is actually applied to get the universal vector bundle sequence.

Dedication and thanks. The paper is dedicated to Professor Hans Grauert. It was my personal pleasure to spend a year, as an Alexander von Humboldt fellow, in Göttingen, in the environment developed under the Professor Grauert's mathematical and personal influence. Some ideas contained in the present paper appeared to me during this stay. I would like to thank Alexander von Humboldt Foundation for supporting my stay in Göttingen as well as Polish KBN for financial support in my home country (project 2P03A 02216).

1 $\mathcal{M}$-Uniform Manifolds

Let X be a smooth projective variety over an algebraically closed field of characteristic zero. Throughout the present paper we shall assume that X is Fano with Picard number $\rho(X) = 1$, that is $-K_X$ is ample and $\operatorname{Pic} X = \mathbf{Z}$. By n we will denote the dimension of X and we will assume $n \geq 3$. By H we will denote the ample line bundle (or divisor) which generates $\operatorname{Pic} X$.

Let $\mathcal{M} \subset \operatorname{Hom}(\mathbf{P}^1, X)$ be an irreducible component of the scheme $\operatorname{Hom}(\mathbf{P}^1, X)$ parameterizing morphisms $f : \mathbf{P}^1 \to X$. By $F_{\mathcal{M}} : \mathcal{M} \times \mathbf{P}^1 \to X$ we denote the evaluation morphisms $F_{\mathcal{M}}(f, p) = f(p) \in X$. We say that $\mathcal{M}$ dominates X if $F_{\mathcal{M}}$ dominates X. We shall be interested in rational curves

332   Jarosław A. Wiśniewski

parameterized by morphisms from $\mathcal{M}$ and therefore, for brevity, we will call such $\mathcal{M}$ a family of rational curves on X.

Given $f \in \mathcal{M}$ we can associate its cycle image $[f(\mathbf{P}^1)] \in \mathrm{Chow}(X)$. We say that the family of rational curves $\mathcal{M}$ is unsplit if all morphisms in $\mathcal{M}$ are birational onto their image and the image of $\mathcal{M}$ in $\mathrm{Chow}(X)$ is proper. If $\mathcal{M}$ is unsplit and dominates X then $F_{\mathcal{M}}$ is surjective and X is rationally connected with respect to $\mathcal{M}$, that is any two points of X can be joint by a connected chain of curves parameterized by morphisms from $\mathcal{M}$, see [KMM]. In the present paper we shall assume that such a family $\mathcal{M}$ exists and we will refer to such $\mathcal{M}$ as an unsplit and dominating family of rational curves.

Let us recall that any vector bundle E over $\mathbf{P}^1$ splits into a direct sum of line bundles $E \simeq \mathcal{O}(a_1) \oplus \cdots \oplus \mathcal{O}(a_r)$ where the sequence of integers $a_1 \geq \cdots \geq a_r$ is uniquely defined and is called the splitting type of E. If integers a_i appear k_i times in the splitting type of E then will write $(a_1^{k_1}, \ldots, a_s^{k_s})$. We say that the splitting type is non-negative, or the bundle is semipositive, if all a_i are non-negative.

If $\mathcal{M}$ is an unsplit and dominating family of rational curves on X then for a generic $f \in \mathcal{M}$ the splitting type of f^*TX is $(2, 1^a, 0^b)$, with $a + b + 1 = n$ [Ko, IV.2.9]; we call this splitting type of f^*TX minimal.

Let X be a Fano manifold such that TX is nef, then for any $f : \mathbf{P}^1 \to X$ the pull-back f^*TX has a non-negative splitting type, hence (see e.g. [Ko, IV.1.9]) any irreducible component $\mathcal{M}$ of $\mathrm{Hom}(\mathbf{P}^1, X)$ dominates X. If we choose the component $\mathcal{M}$ such that $\deg(f^*(-K_X))$ is minimal then it is unsplit and dominates X. Thus, whenever X is assumed to have TX nef we shall choose such a family.

The motivation for the subsequent definition is taken from the projective space case, see [OSS].

Definition 1.1. Suppose that $\mathcal{M}$ is an unsplit and dominating family of rational curves, as above. By the degree of a vector bundle E with respect to $\mathcal{M}$ we understand the number $\deg(\det(f^*E))$ for $f \in \mathcal{M}$. A vector bundle E over X is called $\mathcal{M}$-uniform if the splitting type of f^*E is the same for any $f \in \mathcal{M}$; then we will call it the splitting type of E over $\mathcal{M}$. The manifold X is called $\mathcal{M}$-uniform if the tangent bundle TX is $\mathcal{M}$-uniform.

Let X be a homogeneous manifold with $\rho(X) = 1$, that is $X \simeq G/P$ where G is a simple algebraic group and P its maximal parabolic subgroup. It is clear that then TX is nef. Campana and Peternell [CP] conjectured that this can be conversed.

Conjecture A (Campana, Peternell). If X is a Fano manifold with TX nef then it is homogeneous.

If $X = G/P$ is homogeneous then the ample generator H of $\mathrm{Pic} X$ is very ample and if we embed $X \hookrightarrow \mathbf{P}(H^0(X, \mathcal{O}(H))^*)$ then the image contains lines (that is curves whose intersection with H is 1), see e.g. [LM1]. Of course TX

is then nef and lines give rise to an unsplit and dominating family of minimal rational curves in the sense of the previous definition. The normal bundle of the line in X is a sub-bundle of its normal in the projective space, which is equal to $\mathcal{O}(1)^{\oplus N}$, and therefore the splitting type of TX on any line is minimal. Henceforth if X is homogeneous then it is uniform with respect to lines (though, it is worthwhile to note that the variety parameterizing lines on X may not be homogeneous, c.f. [LM1]). We conjecture that the converse is true.

Conjecture B. If X is Fano with $\rho = 1$ and uniform with respect to some unsplit and dominating family of rational curves then X is homogeneous.

Let us point out that the assumption in the above conjecture refers to the behavior of TX on a fixed family of rational curves while Campana-Peternell conjecture assumes a property of TX on X as whole. Thus, it may seem that Conjecture B is stronger than Conjecture A. However — as it will be seen in this paper — the uniformity assumption in Conjecture B is still rather well manageable.

2 Atiyah Extension and Twisted Trivial Bundles

In the present paper we want to discuss some features of manifolds with nef or uniform tangent bundle. Our main technical tools are the following two observations.

2.1 Atiyah Extension

([KPSW]) Given $c_1(L) \in H^1(X, \Omega_X) = Ext^1(TX, \mathcal{O}_X)$ we can produce an extension

$$0 \longrightarrow \mathcal{O}_X \longrightarrow E \longrightarrow TX \longrightarrow 0$$

with E a vector bundle of rank $n + 1$; the bundle $E^* \otimes L$ can be identified as the 1-st jets of sections of L. If $C \subset X$ is a curve such that $C \cdot L \neq 0$ then the restriction of the above extension is non-trivial. In particular, for $f : \mathbf{P}^1 \to X$ such that $\deg(f^*L) \neq 0$ the pull-back extension

$$0 \longrightarrow \mathcal{O} \longrightarrow f^*E \longrightarrow f^*TX \longrightarrow 0$$

is nontrivial and it pulls back, via the tangent map $Tf : T\mathbf{P}^1 \to f^*TX$, to the extension

$$0 \longrightarrow \mathcal{O} \longrightarrow \mathcal{O}(1) \oplus \mathcal{O}(1) \longrightarrow T\mathbf{P}^1 \simeq \mathcal{O}(2) \longrightarrow 0$$

334   Jarosław A. Wiśniewski

2.2 Twisted Trivial Bundles

([AW, (1.2)]) Let $\mathcal{M} \subset \operatorname{Hom}(\mathbf{P}^1, X)$ be an irreducible component which is unsplit and dominates X. Suppose that E is a rank r vector bundle which is $\mathcal{M}$ uniform with the splitting type $(a^r) = (a, \ldots, a)$. Then E is a twisted trivial bundle, that is there exists a line bundle L, such that $\deg(f^*L) = a$ for any $f \in \mathcal{M}$ and $E \simeq L^{\oplus r}$.

Let us explain our line of argument by discussing an alternative for a part of Mori's proof of the characterization of projective spaces.

Theorem 2.1. ([Mo]) *Let X be a Fano manifold, $\rho(X) = 1$ and $\dim X \geq 2$, with $\mathcal{M}$ unsplit and dominating family of rational curves. If TX is $\mathcal{M}$-uniform of degree $n + 1$ then $X \simeq \mathbf{P}^n$.*

Proof. First let us note that by [Ko, IV.2.9] the splitting type of TX on $\mathcal{M}$ is minimal so equal to $(2, 1, \ldots, 1) = (2, 1^{n-1})$. Let us take the Atiyah extension

$$0 \longrightarrow \mathcal{O}_X \longrightarrow E \longrightarrow TX \longrightarrow 0$$

then, by what we have said, the pull back of this extension via any $f \in \mathcal{M}$ is nontrivial. Therefore E is $\mathcal{M}$-uniform with splitting type (1^{n+1}). By the characterization of the twisted trivial bundles $E \simeq L^{\oplus(n+1)} \simeq W \otimes L$, where L is of degree 1 with respect to $\mathcal{M}$ and W is a vector space of dimension $n + 1$. Dualising and twisting the Atiyah extension we get

$$0 \longrightarrow \Omega_X \otimes L \longrightarrow W^* \otimes \mathcal{O}_X \longrightarrow L \longrightarrow 0$$

and therefore L is spanned by $n + 1$ sections and we have a morphism $\varphi : X \longrightarrow \mathbf{P}(W^*)$. Since L is ample the morphism φ is finite. On the other hand from the above sequence we find out that $K_X = \det(\Omega_X) = L^{\otimes(-n-1)} = \varphi^*(\mathcal{O}(-n-1)) = \varphi^*(K_{\mathbf{P}^n})$ so φ is unramified. Since $\mathbf{P}^n$ is (algebraically) simply connected then φ is an isomorphism.

Arguments, similar to these used above, lead to a characterization of quadrics. Let $\mathbf{Q}^n \subset \mathbf{P}^{n+1} = \mathbf{P}(\mathcal{W})$ be a smooth n-dimensional quadric defined by a quadratic form $\alpha \in H^0(\mathbf{P}^{n+1}, \mathcal{O}(2)) \simeq S^2\mathcal{W}$. We assume that $n \geq 3$. The associated bilinear form α on $\mathcal{W}^*$ descends to a nondegenerate bilinear pairing $T\mathbf{Q}^n \times T\mathbf{Q}^n \to \mathcal{O}(2)$, hence it defines an isomorphism $T\mathbf{Q}^n \simeq \Omega_{\mathbf{Q}^n}(2)$. This can be observed by chasing the Euler and normal-tangent sequences or, directly, by the following geometric argument. At each point $x \in \mathbf{Q}^n$ the intersection of the projective tangent hyperplane and the quadric is a cone over a non-singular quadric. Hence we have a nondegenerate symmetric form on $T_x\mathbf{Q}^n$. Allowing to change x we get a symmetric twisted form at $H^0(\mathbf{Q}^n, S^2(\Omega)(a))$ whose determinant is in $H^0(\mathbf{Q}^n, 2K_{\mathbf{Q}^n} \otimes \mathcal{O}(na))$. The form is nowhere degenerate, hence $a = 2$.

Uniform Vector Bundles on Fano Manifolds   335

Theorem 2.2. ([Ye]) *Let X be Fano manifold, $\rho(X) = 1$ and $\dim X \geq 3$, with $\mathcal{M}$ an unsplit and dominating family of rational curves. Suppose that TX is uniform with respect to $\mathcal{M}$. If $TX \simeq \Omega_X \otimes L$ for an ample line bundle L then $X \simeq \mathbf{Q}^n$.*

Proof. The splitting type of TX on $\mathcal{M}$ is minimal, so of the form $(2, 1^d, 0^k)$, hence the splitting type of Ω_X is $(0^k, (-1)^d, (-2))$. Thus the degree of L over $\mathcal{M}$ is 2 and the splitting type of TX over $\mathcal{M}$ is $(2, 1^{n-2}, 0)$.

As before, let us consider the extension associated to $c_1(H) \in \mathrm{Ext}^1(\mathcal{O}_X, \Omega_X) = H^1(X, \Omega_X)$

$$0 \longrightarrow \Omega_X \longrightarrow V_1 \longrightarrow \mathcal{O}_X \longrightarrow 0 \,. \tag{2.2.1}$$

Dualising and twisting 2.2.1, and using the isomorphism $TX \otimes L^{-1} \simeq \Omega_X$ we get

$$0 \longrightarrow L^{-1} \longrightarrow V_1^* \otimes L^{-1} \longrightarrow \Omega_X \longrightarrow 0$$

since by Kodaira vanishing $H^i(X, L^{-1}) = 0$ for $i = 1$, 2 thus $\mathrm{Ext}^1(L, V_1^*) = H^1(X, V_1^* \otimes L^{-1}) \simeq H^1(X, \Omega_X)$ so we may use the class of $c_1(H)$ to produce yet another extension

$$0 \longrightarrow V_1^* \longrightarrow V \longrightarrow L \longrightarrow 0 \,. \tag{2.2.2}$$

Claim V is $\mathcal{M}$-uniform of splitting type (1^{n+2}).

Proof of Claim. Consider $f \in \mathcal{M}$. We trace the construction of V pulling it back via f. By the argument from the previous section $c_1(H)$ pulls back to to the unique nontrivial element in $H^1(\mathbf{P}^1, f^* \Omega_X)$. Thus the first exact sequence pulls back to

$$0 \to \mathcal{O} \oplus \mathcal{O}(-1)^{\oplus (n-2)} \oplus \mathcal{O}(-2) \to f^* V_1 = \mathcal{O} \oplus \mathcal{O}(-1)^{\oplus n} \to \mathcal{O} \to 0. \tag{2.2.3}$$

Then dualising and twisting the above sequence we see that the nontrivial element in the group $H^1(\mathbf{P}^1, f^* TX \otimes L^{-1}) = H^1(\mathbf{P}^1, f^* \Omega_X)$ lifts-up to the unique non-trivial element in $H^1(\mathbf{P}^1, f^* V_1^*(-2))$. Thus we get that the restriction of sequence 2.2.2 is as follows

$$0 \longrightarrow \mathcal{O}(1)^{\oplus n} \oplus \mathcal{O} \longrightarrow f^* V = \mathcal{O}(1)^{\oplus (n+2)} \longrightarrow \mathcal{O}(2) \longrightarrow 0 \tag{2.2.4}$$

which concludes the proof of claim.

As before, by the characterization of twisted trivial bundles we get $V \simeq H^{\oplus (n+2)}$ where the line bundle H has degree 1 on $\mathcal{M}$ and $L = H^{\otimes 2}$. So we get $-K_X = nH$. Twisting the sequence (2) by H^{-1} we get that H is spanned. We conclude in a standard way: by adjunction a general intersection of $n-1$ divisors from $|H|$ is a smooth curve with canonical divisor $-H$, hence the curve is rational the self-intersection of H is 2. Now the morphism $X \to \mathbf{P}^n$, given by n dimensional sub-system of $|H|$ is of degree 2 and (again, by restricting to curves) branched along a quadric.

336   Jarosław A. Wiśniewski

3   Characterization of Grassmann Manifolds

Theorem 3.1. ([HM1]) *Let X be a Fano manifold, $\rho(X) = 1$, of dimension $n = u \cdot v$, with u, $v > 1$. Suppose that there exist vector bundles U and V of rank u, v, respectively, such that $TX \simeq U \otimes V$. Assume moreover that either TX is nef or it is uniform with respect to some dominating and unsplit family of rational curves. Then $X \simeq \text{Grass}(u, u+v) \simeq \text{Grass}(v, u+v)$, and for some $a \in \mathbf{Z}$ the twisted bundles $U \otimes H^{\otimes a}$ and $V \otimes H^{\otimes -a}$ are isomorphic to the respective universal quotients.*

Proof. The proof was hinted by an exercise in a book of Debarre, [Db], the task of which is to show that no Fano manifold has the tangent bundle which is a non-trivial product of more than two bundles. Let $\mathcal{M}$ denote the unsplit and dominating family of rational curves: in the case TX is nef this is just any minimal family of rational curves, in the other case this is the family with respect to which TX is uniform. As usually, H is the generator of $\text{Pic}X$; let d denote the degree of H on $\mathcal{M}$. We can twist both U and V by a multiple of H, calling the results U and V again, so that still $TX = U \otimes V$ and for a general $f \in \mathcal{M}$ the pull-back f^*U is semi-positive but $f^*(U \otimes H^*)$ is not.

Claim. There exists a non-negative integer $a < d$ such that bundles U and V are $\mathcal{M}$ uniform with the splitting type $(a+1, a^{u-1})$ and $(1-a, (-a)^{v-1})$, respectively.

Proof of Claim. We can write $-K_X = rH$ and then for a general $f \in \mathcal{M}$ the bundle f^*TX is of splitting type $(2, 1^{rd-2}, 0^{uv-rd+1})$ (if TX is $\mathcal{M}$-uniform then this is the case for any $f \in \mathcal{M}$). Let $(a_1, \ldots a_u)$ and $(b_1, \ldots b_v)$ be splitting types of f^*U and f^*V respectively. Then, $a_1 + b_1 = 2$, $a_2 < a_1$, $b_2 < b_1$ and thus $a_i + b_j = 0$ for $i = 2, \ldots, u$ and $j = 2, \ldots, v$. Thus the splitting types for f^*U and f^*V are $(a+1, a^{u-1})$ and $(-a+1, (-a)^{v-1})$, respectively. This concludes them claim if TX is $\mathcal{M}$ uniform.

In case TX is nef, but possibly not $\mathcal{M}$ uniform, to prove that the bundles are uniform we let $f \in \mathcal{M}$ be arbitrary, with $(a_1, \ldots a_u)$ and $(b_1, \ldots b_v)$ be splitting types of f^*U and f^*V respectively. By the deformation theory $a_u \leq a$ and $b_v \leq -a$. Since, by nefness, $a_u + b_v \geq 0$ it follows that $a_u = a = -b_v$. Then, since $\sum a_i = ua + 1$ and $\sum b_i = -va + 1$, the rest of the splitting type can be recovered.

Proof of Theorem Continued. Let us consider an extension

$$0 \longrightarrow V^* \longrightarrow W \longrightarrow U \longrightarrow 0 \tag{3.1.1}$$

associated to

$$c_1(H) \in H^1(X, \Omega_X) = H^1(X, U^* \otimes V^*) = \text{Ext}^1(U, V^*) \,.$$

For any $f \in \mathcal{M}$ this extension pulls back to

$$0 \to \mathcal{O}(a-1) \oplus \mathcal{O}(a)^{\oplus(v-1)} \to f^*W \to \mathcal{O}(a+1) \oplus \mathcal{O}(a)^{\oplus(u-1)} \to 0 \,. \tag{3.1.2}$$

Now let us look at the composition

$$H^1(X, \Omega_X) \xrightarrow{\quad f^* \quad} H^1(\mathbf{P}^1, f^*(\Omega_X)) \xrightarrow{df=(Tf)^*} H^1(\mathbf{P}^1, \Omega_{\mathbf{P}^1})$$

which sends $c_1(H)$ to $c_1(f^*H)$. Since the latter is nontrivial it follows that $f^*(c_1(H))$ is non-trivial in $H^1(\mathbf{P}^1, f^*(\Omega_X)) \simeq \mathrm{Ext}^1(f^*U, f^*V^*)$.

Thus the extension (3.1.2) is non-trivial and the only such non-trivial extension produces $f^*W \simeq \mathcal{O}(a)^{\oplus(v+w)}$. Subsequently W is uniform of splitting type (a^{u+v}). Therefore, by the characterization of the twisted trivial bundles, the $W \simeq L^{\oplus(u+v)}$ for some line bundle of degree a on $\mathcal{M}$. By our assumption on a and H it follows that $a = 0$ hence W is actually trivial, that is $W \simeq X \times V$, where V is a vector space of dimension $u + v$.

Now we can look at X as a space parametrizing linear subspaces of dimension v inside V thus, by the universal property of Grassmann manifolds, we have a morphism $\varphi : X \to \mathrm{Grass}(v, V)$ and the pull-backs of the universal sub-bundle and the universal quotient from $\mathrm{Grass}(v, V)$ to X are V^* and U, respectively. The morphism has finite fibers (because $\rho(X) = 1$ and φ is nontrivial) and it is actually finite because $\dim X = \dim \mathrm{Grass}(v, V)$. By comparing $\det(T(\mathrm{Grass}))$ with $\det(TX)$ we see that $\varphi^*(K_{\mathrm{Grass}}) = K_X$, hence φ is unramified, hence an isomorphism.

4 Characterization of Isotropic Grassmann Manifolds

Let $\mathcal{W}$ be a linear space of dimension $2v$ with a non-degenerate 2-form ω which is either symmetric or skew-symmetric. The form ω defines an isomorphism $\omega : \mathcal{W} \to \mathcal{W}^*$. A v-dimensional subspace $i : W \hookrightarrow \mathcal{W}$ is called isotropic with respect to ω if the following ω-autodual sequence is exact

$$0 \longrightarrow W \xrightarrow{\quad i \quad} \mathcal{W} = \mathcal{W}^* \xrightarrow{\quad i^* \quad} W^* \longrightarrow 0 .$$

The (connected) set of ω-isotropic n-spaces forms a sub-manifold in $\mathrm{Grass}(n, \mathcal{W})$, which is called the isotropic Grassmann manifold $\mathrm{Grass}(\omega, \mathcal{W})$. If ω is symmetric then the tangent space to $\mathrm{Grass}(\omega, \mathcal{W})$ at $[W]$ can be identified naturally with $\Lambda^2 W^*$. If ω is skew-symmetric then the tangent at $[W]$ is identified with $S^2 W^*$. If ω is symmetric then the respective isotropic Grassmannian is also called spinor variety and the isotropic linear spaces are projectivised to projective spaces $\mathbf{P}^{v-1}$ lying in a quadric $\mathbf{Q}^{2n-2} \subset \mathbf{P}^{2n-1}$ defined by the associated quadratic form. If ω is skew-symmetric then the linear spaces parametrized by $\mathrm{Grass}(\omega, \mathcal{W})$ are called Lagrangian (with respect to ω).

Theorem 4.1. ([HM1]). *Let X be a Fano manifold, $\rho(X) = 1$, of dimension $n = v(v+1)/2$ or $n = v(v-1)/2$, where $v \geq 2$ or $v \geq 4$, respectively. Suppose that there exists a vector bundle V of rank v such that $TX \simeq S^2 V$ or $TX \simeq \Lambda^2 V$, respectively. If moreover X is uniform with respect to some unsplit and dominating family of rational curves then X is isomorphic to the*

338   Jarosław A. Wiśniewski

respective isotropic Grassmann manifold, with V isomorphic to the respective universal bundle. In case $TX = S^2V$ the same conclusion is true if we assume that TX is nef instead of uniform.

The proof is similar to the previous one. The only substantial difference is the following general observation which is probably well known but I know no reference for it.

Lemma 4.2. *Suppose that V is a rank v vector bundle over a manifold Y. If the extension $0 \longrightarrow V^* \longrightarrow W \longrightarrow V \longrightarrow 0$ is defined by a symmetric or skew-symmetric cocycle in $\mathrm{Ext}^1(V, V^*) = H^1(Y, V^* \otimes V^*)$ then W admits a skew-symmetric, respectively, symmetric 2-form which makes the above exact sequence auto-dual, i.e. V is an isotropic sub-bundle of W.*

Proof. We use Čech cohomology. Suppose that the bundle V is trivialised in a covering $\{\mathcal{U}_i\}$ with transition matrices $g_{ji} : V_i \to V_j$ where by V_i we denote the restriction of the bundle V over the open set $\mathcal{U}_i$ over which the bundle is trivial. The dual bundle of V has transition matrices g_{ji}^* equal to $(g_{ji}^{-1})^T = g_{ij}^T$. Let us consider a Čech cocycle $(\alpha_{ji}) \in Z^1(\{\mathcal{U}_i\}, \mathrm{Hom}(V, V^*))$; it consists of matrices α_{ji} defined over $\mathcal{U}_i \cap \mathcal{U}_j$, that is $\alpha_{ji} : (V_i)_{\mathcal{U}_i \cap \mathcal{U}_j} \to (V_j^*)_{\mathcal{U}_i \cap \mathcal{U}_j}$, which satisfy the cocycle condition

$$\alpha_{ki} = \alpha_{kj} \cdot g_{ji} + g_{kj}^* \cdot \alpha_{ji} .$$

The cocycle is symmetric or skew-symmetric if $\alpha_{ji} \cdot g_{ij} = \pm(\alpha_{ji} \cdot g_{ij})^T$, or equivalently, $\alpha_{ji} = \pm g_{ji}^* \cdot \alpha_{ji}^T \cdot g_{ji}$, where $+$ sign is in the symmetric case and $-$ sign in skew-symmetric case, respectively.

The vector bundle W associated to the extension given by α has transition matrices

$$A_{ji} = \begin{pmatrix} g_{ji}^* & \alpha_{ji} \\ 0 & g_{ji} \end{pmatrix}$$

where $V^* \to W$ is an inclusion into first v coordinates while $W \to V$ is the projection onto last v coordinates. The dual bundle W^* is then given by

$$A_{ji}^* = (A_{ji}^{-1})^T = \begin{pmatrix} g_{ji} & 0 \\ -g_{ji}^* \cdot \alpha_{ji}^T \cdot g_{ji} & g_{ji}^* \end{pmatrix} .$$

One verifies that the above condition for α being symmetric or skew-symmetric is equivalent to

$$A_{ji}^* \cdot E = E \cdot A_{ji} \quad \text{with} \quad E = \begin{pmatrix} 0 & I \\ \pm I & 0 \end{pmatrix}$$

and I denoting the $v \times v$ identity matrix; the $+$ sign in E appears in the skew-symmetric case, while the $-$ sign comes in the symmetric case. Thus E defines a nowhere degenerate symmetric or, respectively, skew-symmetric 2-form on W and V^* is clearly isotropic sub-bundle with respect to E.

Proof of Theorem. Let $\mathcal{M} \subset \mathrm{Hom}(\mathbf{P}^1, X)$ be unsplit and dominating. The proof of the first claim is the same as in the previous case and we omit it.

Claim. V is $\mathcal{M}$-uniform with splitting type $(1, 0^{v-1})$, if $TX \simeq S^2V$, or $(1, 1, 0^{v-2})$ if $TX \simeq \Lambda^2 V$.

As in the previous case we consider the extension

$$0 \longrightarrow V^* \longrightarrow W \longrightarrow V \longrightarrow 0 \tag{4.2.1}$$

associated to $c_1(H)$.

Claim. The bundle W is trivial $W \simeq X \times \mathcal{W}$ and the vector space $\mathcal{W}$ is equipped with a symmetric or, respectively, skew-symmetric form ω which makes the above sequence auto-dual.

Proof of Claim. As in the case of (absolute) Grassmanianns we see that f^* pull-back of the extension 4.2.1 is nontrivial. This is enough to conclude that $f^*W \simeq \mathcal{O}^{\oplus 2v}$ in the case of the splitting type $(1, 0^{v-1})$, because the dimension of $\mathrm{Ext}^1(f^*V, f^*V^*)$ is 1. In the case of the splitting type $(1, 1, 0^{v-2})$ we have $\dim \mathrm{Ext}^1(f^*V, f^*V^*) = 4$. However, $\dim H^1(\Lambda^2(f^*V^*)) = 1$. So among all non-trivial extensions $\mathcal{O}(1)^2 \oplus \mathcal{O}^{\oplus(v-2)}$ by $\mathcal{O}^{\oplus(v-2)} \oplus \mathcal{O}(-1)^2$ only one comes from a skew-symmetric form — and this is the one which produces $f^*W \simeq \mathcal{O}^{2v}$.

Thus W is $\mathcal{M}$-uniform with splitting type (0^{2v}) hence by the characterization of twisted trivial bundles it is actually trivial. The existence of the form ω follows from the lemma which we have proved before.

Conclusion of the Proof of Theorem. The conclusion is the same as in the (absolute) Grassmannian case: by the universal property of Grassmannians we have a morphism $\varphi : X \to \mathrm{Grass}(v, \mathcal{W})$, which maps X into isotropic Grassmannian of ω, with V becoming the pull-back of the universal quotient bundle. By arguments similar to these before the map is finite and unramified hence an isomorphism.

References

[AW]   Andreatta, M., Wiśniewski, J. A., On manifolds whose tangent bundle contains an ample locally free subsheaf, *Invent. Math.* **146** (2001), 209–217.

[CP]   Campana, F., Peternell, Th., Projective manifolds whose bundles are numerically effective, *Math. Ann.* **289** (1991), 169–187.

[Db]   Debarre, O., *Higher dimensional algebraic geometry*, Springer Universitext Series, Springer Verlag 2001.

[FH]   Fulton, W., Harris, J., *Representation Theory. A First Course, Graduate Text in Math* **129** (1999), Springer Verlag.

[Hw]   Hwang, J.-M., Geometry of minimal rational curves on Fano manifolds, Notes of a lecture given at the School on Vanishing Theorem etc., ICTP Trieste, 2000.

[HM1]   Hwang, J.-M., Mok, N., Uniruled projective manifolds with irreducible reductive G-structure, *J. reine angew. Math.* **490** (1997), 55–64.

340   Jarosław A. Wiśniewski

[HM2] Hwang, J.-M., Mok, N., Holomorphic maps from rational homogeneous spaces of Picard number 1 onto projective manifolds, *Invent. Math.* **136** (1999), 208–236.

[IP]   Iskovskikh, V. A., Prokhorov, Yu. G., Fano varieties, *Encyclopeadia of Math. Sciences, Algebraic Geometry V*, ed. Parshin, Shafarevich, Springer Verlag, 1999.

[Ja]   Jaczewski, K., Generalized Euler sequence and toric varieties, *Classification of Algebraic Varieties, L'Aquila, 1992*, 227–247, *Contemp. Math.*, **162** (1994), AMS, Providence, RI.

[KPSW] Kebekus, S., Peternell, Th., Sommese, A.J., Wiśniewski J. A., Projective contact manifolds, *Invent. Math.* **142** (2000), 1–15.

[Ko]   Kollár, J., *Rational Curves on Algebraic Varieties, Ergebnisse der Math.* **32** (1995), Springer Verlag.

[KMM]   Kollár, J., Miyaoka, Y., Mori, Sh., Rational curves on Fano varieties, in: *Clasification of irregular varities, minimal models and abelian varieties. Proceedings Trento 1990, Lect. Notes in Math.* **1515** (1990), 100-105.

[LM1]   Landsberg, J. M., Manivel, L., On the projective geometry of rational homogeneous varieties, E-print math.AG/9810140.

[LM2]   Landsberg, J. M., Manivel, L., Classification of complex Lie algebras via projective geometry, E-print math.AG/9902102.

[Mo]   Mori, Sh., Projective manifolds with ample tangent bundles, *Ann. of Math.* **110** (1979), 593–606.

[OSS]   Okonek, Ch., Schneider, M., Spindler H., Vector bundles on complex projective spaces, *Prog. in Math.* **3** (1980), Birkhäuser.

[Wi]   Wierzba, J., On 4-dimensional isolated symplectic singularities and a characterization of quadrics, preprint, 2000.

[Ye]   Ye, Y., Extremal rays and null geodesics on a complex conformal manifold, *Inter. J. Math.* **5** (1994), 141–168.